# Carbon Balance Management: A Critical Environmental Approach

# Carbon Balance Management: A Critical Environmental Approach

Editor: Eugene Richards

Callisto Reference

www.callistoreference.com

**Callisto Reference,**
118-35 Queens Blvd., Suite 400,
Forest Hills, NY 11375, USA

Visit us on the World Wide Web at:
www.callistoreference.com



ISBN: 978-1-64116-134-3 (Hardback)



Printed in China.

**Cataloging-in-Publication Data**

Carbon balance management : a critical environmental approach / edited by Eugene Richards.
p. cm.
Includes bibliographical references and index.
ISBN 978-1-64116-134-3
1. Global environmental change. 2. Atmospheric carbon dioxide--Environmental aspects.
3. Carbon cycle (Biogeochemistry)--Environmental aspects. 4. Environmental chemistry.
I. Richards, Eugene.

GE149 .C37 2019
363.7--dc23

# Table of Contents

# Preface

It is often said that books are a boon to mankind. They document every progress and pass on the knowledge from one generation to the other. They play a crucial role in our lives. Thus I was both excited and nervous while editing this book. I was pleased by the thought of being able to make a mark but I was also nervous to do it right because the future of students depends upon it. Hence, I took a few months to research further into the discipline, revise my knowledge and also explore some more aspects. Post this process, I begun with the editing of this book.

Carbon cycle is the process by which carbon is exchanged between the biosphere, earth, air and water. Carbon balance refers to the balance in the percentage of carbon dioxide in the air. The processes of exchange of carbon are respiration, transpiration, combustion and decomposition. The widespread combustion of fossil fuels has caused carbon to accumulate in the atmosphere. Carbon cycle rebalancing can be initiated by a number of progressive measures like rehabilitation of forests, adoption of sustainable energy, carbon capture, etc. This book contains some path-breaking studies in the field of carbon management. The topics included herein are of utmost significance and bound to provide incredible insights to readers. It will provide a comprehensive knowledge to all readers involved in this field at various levels.

I thank my publisher with all my heart for considering me worthy of this unparalleled opportunity and for showing unwavering faith in my skills. I would also like to thank the editorial team who worked closely with me at every step and contributed immensely towards the successful completion of this book. Last but not the least, I wish to thank my friends and colleagues for their support.

**Editor**

# 1

# Comparisons of allometric and climate-derived estimates of tree coarse root carbon stocks in forests of the United States

Matthew B. Russell[1*], Grant M. Domke[2], Christopher W. Woodall[2] and Anthony W. D'Amato[3]

**Abstract**

**Background:** Refined estimation of carbon (C) stocks within forest ecosystems is a critical component of efforts to reduce greenhouse gas emissions and mitigate the effects of projected climate change through forest C management. Specifically, belowground C stocks are currently estimated in the United States' national greenhouse gas inventory (US NGHGI) using nationally consistent species- and diameter-specific equations applied to individual trees. Recent scientific evidence has pointed to the importance of climate as a driver of belowground C stocks. This study estimates belowground C using current methods applied in the US NGHGI and describes a new approach for merging both allometric models with climate-derived predictions of belowground C stocks.

**Results:** Climate-adjusted predictions were variable depending on the region and forest type of interest, but represented an increase of 368.87 Tg of belowground C across the US, or a 6.4 % increase when compared to currently-implemented NGHGI estimates. Random forests regressions indicated that aboveground biomass, stand age, and stand origin (i.e., planted versus artificial regeneration) were useful predictors of belowground C stocks. Decreases in belowground C stocks were modeled after projecting mean annual temperatures at various locations throughout the US up to year 2090.

**Conclusions:** By combining allometric equations with trends in temperature, we conclude that climate variables can be used to adjust the US NGHGI estimates of belowground C stocks. Such strategies can be used to determine the effects of future global change scenarios within a C accounting framework.

**Keywords:** Belowground biomass, Carbon accounting, Carbon-climate, Root:shoot ratio, Forest Inventory and Analysis

## Background

The management of forest ecosystems and their associated carbon (C) stocks has become an important global strategy for reducing greenhouse gas (GHG) emissions and possibly mitigating future effects of climate change [1–3]. Societal demands and trends in land use, in combination with future global change scenarios, may reduce the amount of C stored in forests and associated wood products [4]. As a result, there are substantial knowledge gaps regarding the C implications of various forest management activities, which may arise from the complex pathways of C emissions and sequestration in forest ecosystems [2, 3]. In addition, the logistical and methodological constraints associated with estimating C in certain forest ecosystem components across large areas, namely belowground pools, has hampered the development of accurate estimates, creating a need for refined modeling approaches to quantify belowground C stocks.

From an ecological perspective, the use of plant characteristics such as root to shoot ratio and root mass fraction have aided investigators in understanding belowground C stocks associated with coarse roots. For example, Poorter et al. [5] presented a mean root mass fraction of 0.21 for temperate and boreal forests and Smyth et al. [6] calculated a root to shoot ratio of 0.256 across Canada.

*Correspondence: russellm@umn.edu
[1] Department of Forest Resources, University of Minnesota, St. Paul, MN 55108, USA
Full list of author information is available at the end of the article

These proportions lend insight into partitioning effects and belowground C stores. Various studies have examined the degree to which these ratios are altered under various stand and environmental conditions. Litton et al. [7] found that partitioning to belowground components increased with stand density in lodgepole pine forests (*Pinus contorta* Dougl. ex Loud. var. *latifolia* Engelm.), which may in turn be related to tree size. Root mass fractions have been found to decrease with total plant biomass consistently among angiosperm and gymnosperm forests [5, 8]. In tropical systems, 15–20 year-old plantations allocated more C belowground when compared to mature broad-leaved forests [9], highlighting the importance of accounting for management scenarios in assessments of belowground C stores. However, application of these findings to forest C accounting activities has been limited as few studies measure all components of the C budget (e.g., biomass, flux, and partitioning; [7]).

The monitoring of belowground C has incorporated a number of these ecological insights using a variety of approaches at different scales. Allometric equations designed at the individual-tree level are common for determining belowground C [e.g., 10, 11]. Throughout the United States (US), belowground biomass of coarse roots is commonly estimated using the equations of Jenkins et al. [12] as a ratio of total aboveground biomass and tree diameter at breast height (DBH; [13]). Although allometric equations tend to account for a large portion of the apparent variability associated with belowground biomass (e.g., $R^2$ values range from 0.77 to 0.96; Litton et al. [11]), there are a few drawbacks to this approach. First, these equations tend to rely on DBH and are not explicitly constructed to estimate belowground C [14]. Secondly, allometric equations have not historically incorporated climate information that integrates differences in ecosystem productivity and allows for evaluations of future climate change scenarios on global C cycles. Highlighting this concern, Reich et al. [8] recently compiled a global dataset and concluded that forest biomass found in roots was inversely related to mean annual temperature, suggesting that climate may act as a driver of belowground C allocation.

Globally, there has been an increased interest in recent years for refining forest carbon estimation to understand greenhouse gas emissions in support of the United Nations Framework Convention on Climate Change [15, 16]. Forest C stocks in the US are estimated using data collected by the US Forest Service, Forest Inventory and Analysis (FIA) program. In the current national greenhouse gas inventory (NGHGI; [17]), belowground stocks are estimated in two stages by first quantifying total aboveground biomass using allometric equations then estimating a ratio of coarse root to total aboveground biomass [12, 13]. As observations of belowground tree biomass and C are often limited [14], relying on allometric equations has been necessary to obtain estimates from strategic-scale forest inventories such as FIA's. At the same time, the lack of empirical information across a diverse array of tree species in temperate forests such as those found throughout North America encourages researchers to test alternative approaches for quantifying belowground biomass and C. Exploring belowground C modeling approaches that incorporate climatic attributes may both adjust our estimates of coarse root C stocks at national scales (i.e., application in the US NGHGI) while enhancing evaluations of future climate change scenarios on forest C cycles.

The overall objective of this research is to adjust belowground C estimation procedures for reporting in the US NGHGI. Specific objectives are to (1) estimate belowground C stocks by employing individual tree- and stand-level methodologies, (2) adjust estimates of belowground C stocks by combining allometric and climate-derived approaches using current and projected climate attributes, and (3) compare alternative estimation approaches for belowground C stocks for future application in the US NGHGI.

## Results

Estimates of belowground carbon (BGC) from approaches currently employed in the US NGHGI suggest that C stocks are dependent on geographic region and forest type. Mean values of belowground carbon in the US greenhouse gas inventory ($BGC_{NGHGI}$) were small in short-statured, open forests such as pinyon-juniper and woodland hardwood types (typically less than 2 Mg ha$^{-1}$). Mean $BGC_{NGHGI}$ was largest in hemlock-Sitka spruce forests in the Pacific Northwest [40.76 ± 0.96 Mg ha$^{-1}$ (mean ± SE)] and redwood forests in the Pacific Southwest (59.27 ± 7.06 Mg ha$^{-1}$). For climate-derived estimates of belowground C, belowground carbon from climate-derived models ($BGC_{Clim}$) stock estimates were slightly smaller in magnitude compared to $BGC_{NGHGI}$ estimates [e.g., hemlock-Sitka spruce (33.82 ± 0.80 Mg ha$^{-1}$) and redwood forests (45.64 ± 5.44 Mg ha$^{-1}$)] and generally showed decreasing C at lower latitudes (Fig. 1). On average, $BGC_{Clim}$ estimates were 0.60 Mg ha$^{-1}$ greater than current $BGC_{NGHGI}$ models when considering all forest types (Additional file 1: Table S1).

The adjustment factors used to align climate-derived predictions of belowground C with the US NGHGI approach ranged from 0.77 to 1.60 with little variability within a region of interest (Fig. 2; Additional file 1: Table S1). Equivalence tests used to contrast the two approaches for estimating belowground C with a null

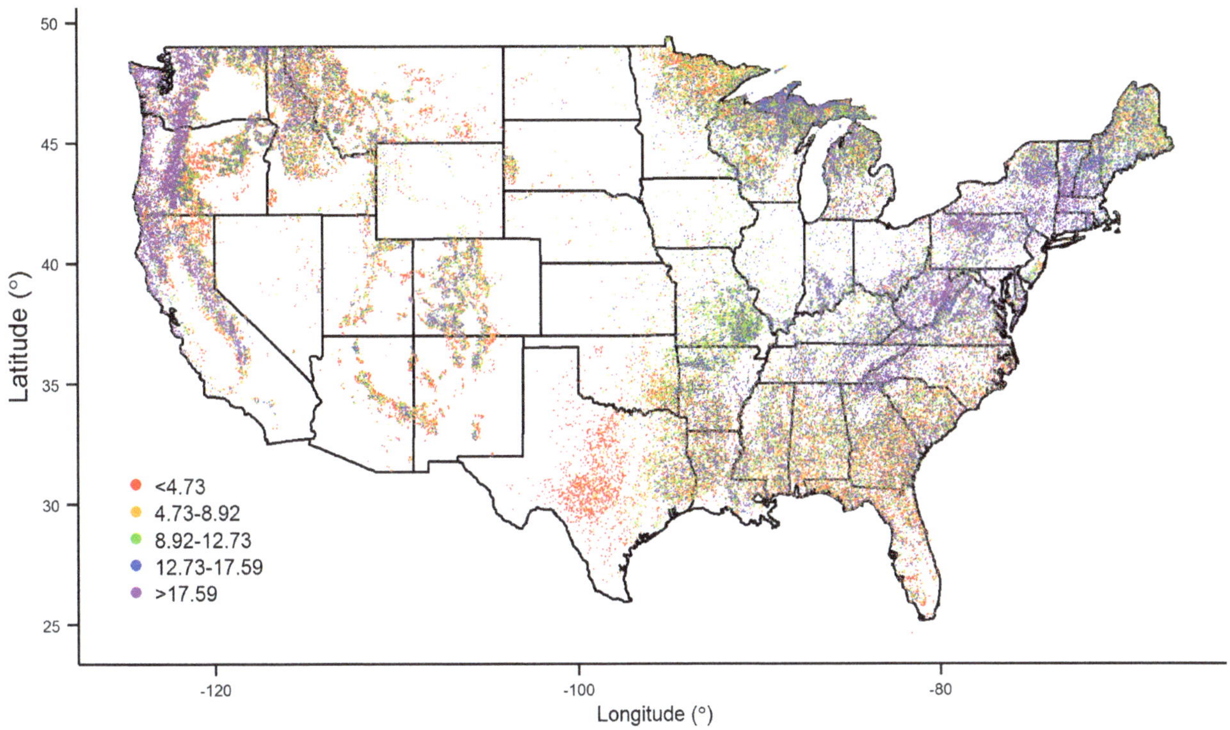


**Fig. 1** Distribution of live-tree belowground C estimates from the model of Reich et al. [8] ($BGC_{Clim}$; Mg $ha^{-1}$)

hypothesis of dissimilarity and a threshold of ±25 % were rejected for 20 out of the 78 forest types examined. Equivalence tests were generally rejected for forest types that displayed relatively low and high BGC stocks, e.g., pinyon-juniper and woodland hardwoods (low C stocks) and redwood forest types (high C stocks). Mean differences were generally largest across the Pacific Northwest (Westside), indicating climate-adjusted predictions estimated less belowground C compared to allometric-derived estimates in this region. Generally, negative mean differences were observed across most forest types, indicating that climate-adjusted models predict greater belowground C stocks (e.g., mean percent difference was −5.2 % greater across all forest types; Table 1). Compared to current NGHGI models, model differences showed greater belowground C stocks occurring in the Appalachian Mountain region and areas where northern hardwood forests are common, e.g., in the upper Midwest and northeastern US states. Conversely, areas of smaller belowground C stocks were identified across the Pacific Northwest and Southeast US (Fig. 3). This was further reflected when population estimates were scaled to the state level. The states of Oregon and Washington were predicted to display the largest negative mean difference in belowground C stocks (−10.6 and −10.7 %, respectively). Conversely, the largest mean positive difference in belowground C stocks was in the states of Kentucky, Tennessee, and Oklahoma (28.0, 26.7, and 22.6 %, respectively). This represents a total estimated increase of 368.87 Tg of belowground C across the US, or a 6.4 % increase when compared to currently implemented NGHGI models (Table 2).

Results from the random forests (RF) regressions indicated that aboveground biomass, stand age, and stand origin (i.e., planted versus natural regeneration) were useful predictors of climate-adjusted models of belowground carbon ($BGC_{ClimAdj}$), as measured by their importance scores (Table 3). These variables accounted for 87 % of the total variation in belowground C stocks. Without employing aboveground biomass, 47 % of the total variability was accounted for, indicating that surrogates of climate (e.g., latitude and longitude) and knowledge of stand structure and management history (e.g., stand age and origin) may aid in understanding belowground C stocks.

The largest differences in projected live-tree belowground C stocks under future mean annual temperature (MAT) changes were positive, indicating decreases in belowground C stocks up to year 2090 (Fig. 4). Differences in projected belowground C stocks were similar across regions, with belowground stocks displaying approximately 0.1 Mg $ha^{-1}$ less C in 2030 than assuming current normal climates. In 2090, differences comparing current versus future climates were highest in the

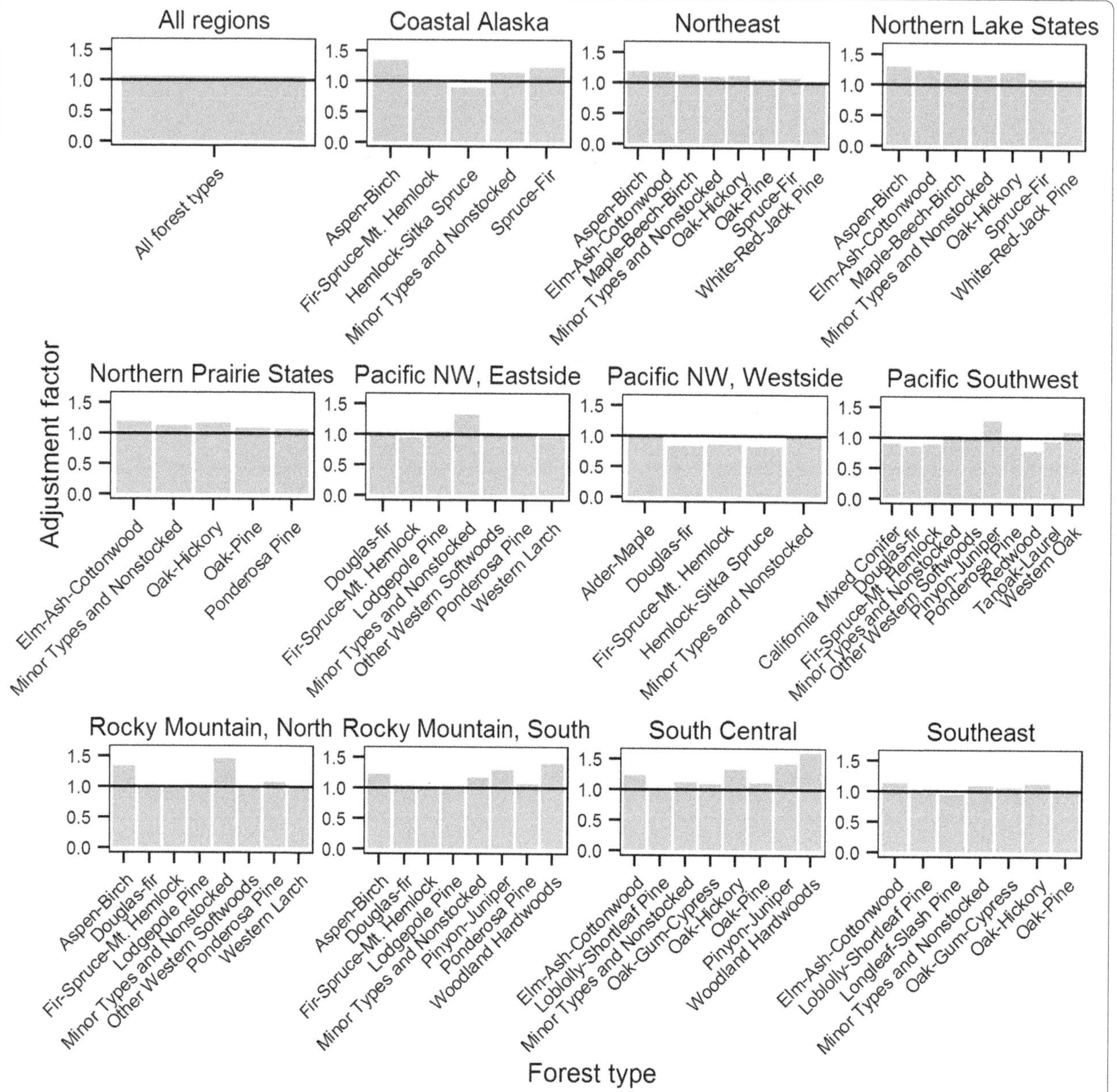


**Fig. 2** Adjustment factors used to amend live-tree belowground C estimates from the current US national greenhouse gas inventory ($BGC_{NGHGI}$; Smith et al. [13]) with climate-derived predictions ($BGC_{Clim}$; Reich et al. [8]) by US region and forest type. Adjustment factors less than and greater than one indicate less and more belowground C when climate-derived predictions are used. See Additional file 1: Table S1 for mean belowground C values and adjustment factors

US Northeast, representing $0.50 \pm 0.13$ Mg $ha^{-1}$ less C (assuming an RCP8.5 scenario), or a reduction of approximately 3.4 % in belowground C stocks in the region.

## Discussion

Live tree belowground C estimated using allometric equations developed for individual trees and climate-specific predictions made at the forest stand level show markedly different patterns across various geographic regions and forest types in the US. Adjustments to the US' approach to estimating live tree belowground C resulted in a C density estimate of 12 Mg $ha^{-1}$ across the US and a 6.4 % increase in this forest C national stock when compared to current estimation strategies employed in the US NGHGI.

The largest differences between current and adjusted estimates of live tree belowground C were observed in the states of Oregon, Washington, and California, indicating

**Table 1 Equivalence test results (critical threshold of ±25 %) comparing live-tree belowground C estimates from the current US national greenhouse gas inventory ($BGC_{NGHGI}$) [13] to adjusted estimates ($BGC_{ClimAdj}$)**

| Region | Forest type | *n* | $BGC_{NGHGI} - BGC_{ClimAdj}$ | | | |
|---|---|---|---|---|---|---|
| | | | Mean difference | Mean % difference | SE difference | Result[a] |
| All regions | All forest types | 70,126 | −0.58 | −5.2 | 0.01 | E |
| Coastal Alaska | Aspen-Birch | 78 | −1.71 | −34.5 | 0.16 | NE |
| Coastal Alaska | Fir-Spruce-Mt. Hemlock | 342 | 0.41 | 3.4 | 0.02 | E |
| Coastal Alaska | Hemlock-Sitka Spruce | 571 | 2.75 | 10.6 | 0.07 | E |
| Coastal Alaska | Minor types and nonstocked | 85 | −0.77 | −13.8 | 0.13 | NE |
| Coastal Alaska | Spruce-Fir | 105 | −0.60 | −23.0 | 0.05 | NE |
| Northeast | Aspen-Birch | 367 | −1.58 | −18.2 | 0.05 | E |
| Northeast | Elm-Ash-Cottonwood | 205 | −1.76 | −16.9 | 0.08 | NE |
| Northeast | Maple-Beech-Birch | 3845 | −1.78 | −12.8 | 0.01 | E |
| Northeast | Minor types and nonstocked | 346 | −0.97 | −8.5 | 0.03 | E |
| Northeast | Oak-Hickory | 2509 | −1.67 | −11.1 | 0.01 | E |
| Northeast | Oak-Pine | 247 | −0.68 | −5.0 | 0.02 | E |
| Northeast | Spruce-Fir | 900 | −0.59 | −6.6 | 0.01 | E |
| Northeast | White-Red-Jack Pine | 380 | 0.00 | −0.1 | 0.00 | E |
| Northern Lake States | Aspen-Birch | 2477 | −1.82 | −28.8 | 0.02 | NE |
| Northern Lake States | Elm-Ash-Cottonwood | 971 | −1.91 | −22.7 | 0.04 | NE |
| Northern Lake States | Maple-Beech-Birch | 2956 | −2.08 | −18.1 | 0.02 | E |
| Northern Lake States | Minor types and nonstocked | 460 | −0.94 | −15.2 | 0.04 | E |
| Northern Lake States | Oak-Hickory | 1845 | −2.05 | −20.0 | 0.03 | E |
| Northern Lake States | Spruce-Fir | 1854 | −0.51 | −8.4 | 0.01 | E |
| Northern Lake States | White-Red-Jack Pine | 935 | −0.52 | −5.5 | 0.01 | E |
| Northern Prairie States | Elm-Ash-Cottonwood | 342 | −1.88 | −19.0 | 0.06 | NE |
| Northern Prairie States | Minor types and nonstocked | 465 | −1.26 | −13.0 | 0.04 | E |
| Northern Prairie States | Oak-Hickory | 3265 | −1.80 | −17.2 | 0.01 | E |
| Northern Prairie States | Oak-Pine | 207 | −0.76 | −9.0 | 0.03 | E |
| Northern Prairie States | Ponderosa Pine | 164 | −0.54 | −7.6 | 0.03 | E |
| Pacific Northwest, Eastside | Douglas-fir | 992 | 0.27 | 2.3 | 0.01 | E |
| Pacific Northwest, Eastside | Fir-Spruce-Mt. Hemlock | 948 | 0.82 | 4.5 | 0.02 | E |
| Pacific Northwest, Eastside | Lodgepole Pine | 571 | −0.37 | −4.6 | 0.01 | E |
| Pacific Northwest, Eastside | Minor types and nonstocked | 213 | −0.91 | −33.1 | 0.09 | NE |
| Pacific Northwest, Eastside | Other Western Softwoods | 505 | −0.06 | −0.5 | 0.00 | E |
| Pacific Northwest, Eastside | Ponderosa Pine | 1446 | −0.36 | −4.2 | 0.01 | E |
| Pacific Northwest, Eastside | Western Larch | 136 | 0.65 | 4.3 | 0.03 | E |
| Pacific Northwest, Westside | Alder-Maple | 226 | 0.18 | 0.5 | 0.01 | E |
| Pacific Northwest, Westside | Douglas-fir | 2130 | 5.79 | 16.3 | 0.09 | E |
| Pacific Northwest, Westside | Fir-Spruce-Mt. Hemlock | 553 | 4.27 | 13.6 | 0.13 | E |
| Pacific Northwest, Westside | Hemlock-Sitka Spruce | 552 | 6.93 | 16.7 | 0.16 | E |
| Pacific Northwest, Westside | Minor types and nonstocked | 355 | 0.28 | 1.8 | 0.01 | E |
| Pacific Southwest | California Mixed Conifer | 946 | 2.30 | 9.9 | 0.05 | E |
| Pacific Southwest | Douglas-fir | 139 | 4.45 | 14.1 | 0.27 | E |
| Pacific Southwest | Fir-Spruce-Mt. Hemlock | 243 | 2.81 | 10.7 | 0.12 | E |
| Pacific Southwest | Minor types and nonstocked | 116 | −0.25 | −2.8 | 0.04 | NE |
| Pacific Southwest | Other Western Softwoods | 382 | −0.16 | −2.3 | 0.01 | E |
| Pacific Southwest | Pinyon-Juniper | 31 | −0.57 | −26.7 | 0.12 | NE |
| Pacific Southwest | Ponderosa Pine | 242 | 0.00 | 0.0 | 0.00 | E |
| Pacific Southwest | Redwood | 64 | 13.63 | 22.6 | 1.62 | NE |
| Pacific Southwest | Tanoak-Laurel | 207 | 1.51 | 5.6 | 0.07 | E |

**Table 1 continued**

| Region | Forest type | *n* | $BGC_{NGHGI} - BGC_{ClimAdj}$ Mean difference | Mean % difference | SE difference | Result[a] |
|---|---|---|---|---|---|---|
| Pacific Southwest | Western Oak | 778 | −0.98 | −9.9 | 0.03 | E |
| Rocky Mountain, North | Aspen-Birch | 98 | −1.46 | −32.7 | 0.15 | NE |
| Rocky Mountain, North | Douglas-fir | 1659 | −0.11 | −1.4 | 0.00 | E |
| Rocky Mountain, North | Fir-Spruce-Mt. Hemlock | 1467 | 0.00 | 0.3 | 0.00 | E |
| Rocky Mountain, North | Lodgepole Pine | 828 | −0.30 | −2.6 | 0.01 | E |
| Rocky Mountain, North | Minor types and nonstocked | 331 | −0.70 | −45.8 | 0.06 | NE |
| Rocky Mountain, North | Other Western Softwoods | 329 | 0.23 | 2.1 | 0.01 | E |
| Rocky Mountain, North | Ponderosa Pine | 499 | −0.47 | −6.6 | 0.02 | E |
| Rocky Mountain, North | Western Larch | 128 | 0.26 | 1.7 | 0.02 | E |
| Rocky Mountain, South | Aspen-Birch | 647 | −1.77 | −22.4 | 0.05 | NE |
| Rocky Mountain, South | Douglas-fir | 420 | −0.41 | −3.5 | 0.01 | E |
| Rocky Mountain, South | Fir-Spruce-Mt. Hemlock | 1061 | −0.34 | −3.4 | 0.01 | E |
| Rocky Mountain, South | Lodgepole Pine | 359 | −0.41 | −4.0 | 0.01 | E |
| Rocky Mountain, South | Minor types and nonstocked | 166 | −0.76 | −16.7 | 0.07 | NE |
| Rocky Mountain, South | Pinyon-Juniper | 647 | −0.41 | −28.9 | 0.02 | NE |
| Rocky Mountain, South | Ponderosa Pine | 924 | −0.45 | −6.1 | 0.01 | E |
| Rocky Mountain, South | Woodland Hardwoods | 345 | −0.78 | −40.0 | 0.05 | NE |
| South Central | Elm-Ash-Cottonwood | 737 | −1.68 | −22.7 | 0.05 | NE |
| South Central | Loblolly-Shortleaf Pine | 3291 | 0.30 | 3.3 | 0.00 | E |
| South Central | Minor types and nonstocked | 842 | −0.94 | −11.6 | 0.03 | E |
| South Central | Oak-Gum-Cypress | 1264 | −1.02 | −8.1 | 0.02 | E |
| South Central | Oak-Hickory | 5806 | −2.61 | −32.3 | 0.03 | NE |
| South Central | Oak-Pine | 1093 | −0.81 | −10.2 | 0.02 | E |
| South Central | Pinyon-Juniper | 298 | −0.30 | −41.7 | 0.02 | NE |
| South Central | Woodland Hardwoods | 268 | −0.28 | −59.6 | 0.02 | NE |
| Southeast | Elm-Ash-Cottonwood | 173 | −1.22 | −12.7 | 0.07 | E |
| Southeast | Loblolly-Shortleaf Pine | 2667 | 0.39 | 4.1 | 0.01 | E |
| Southeast | Longleaf-Slash Pine | 1314 | 0.42 | 5.2 | 0.01 | E |
| Southeast | Minor types and nonstocked | 286 | −0.68 | −8.4 | 0.04 | E |
| Southeast | Oak-Gum-Cypress | 1311 | −0.68 | −5.0 | 0.01 | E |
| Southeast | Oak-Hickory | 3077 | −1.34 | −10.9 | 0.01 | E |
| Southeast | Oak-Pine | 982 | −0.27 | −2.5 | 0.01 | E |

[a] Equivalent (E) or not equivalent (NE)

climate-adjusted predictions estimated less belowground C compared to allometric-derived estimates in this region. Comparatively, this region contains the largest belowground C stocks in the US, quantified using model imputation strategies [18]. Similar differences but of a lesser magnitude were observed across the Coastal Plain region in the southeastern US. Such differences in these regions (i.e., the Pacific Northwest and Southeast) could point to the reliance on the allometric equations driven by aboveground metrics (i.e., tree diameter at breast height) in these carbon-dense stands. This may especially be the case for common tree species in the Pacific Northwest where the development of such allometric relationships is complicated by their relatively large above and belowground dimensions. In addition, intensive forest management regimes relying on artificial regeneration are common throughout these regions such that stand origin emerged as a useful predictor of live tree belowground C. Such a finding is analogous to recent work indicating greater partitioning to belowground components in managed plantations compared to natural-origin mature forests [9]. These patterns of partitioning may be reflective of the increased nutrient demands required to sustain the elevated levels of aboveground net primary production found in plantation systems [19], particularly on sites with lower nutrient capital such as in those found across the southeastern US. The importance of variables describing forest management regime and other forest

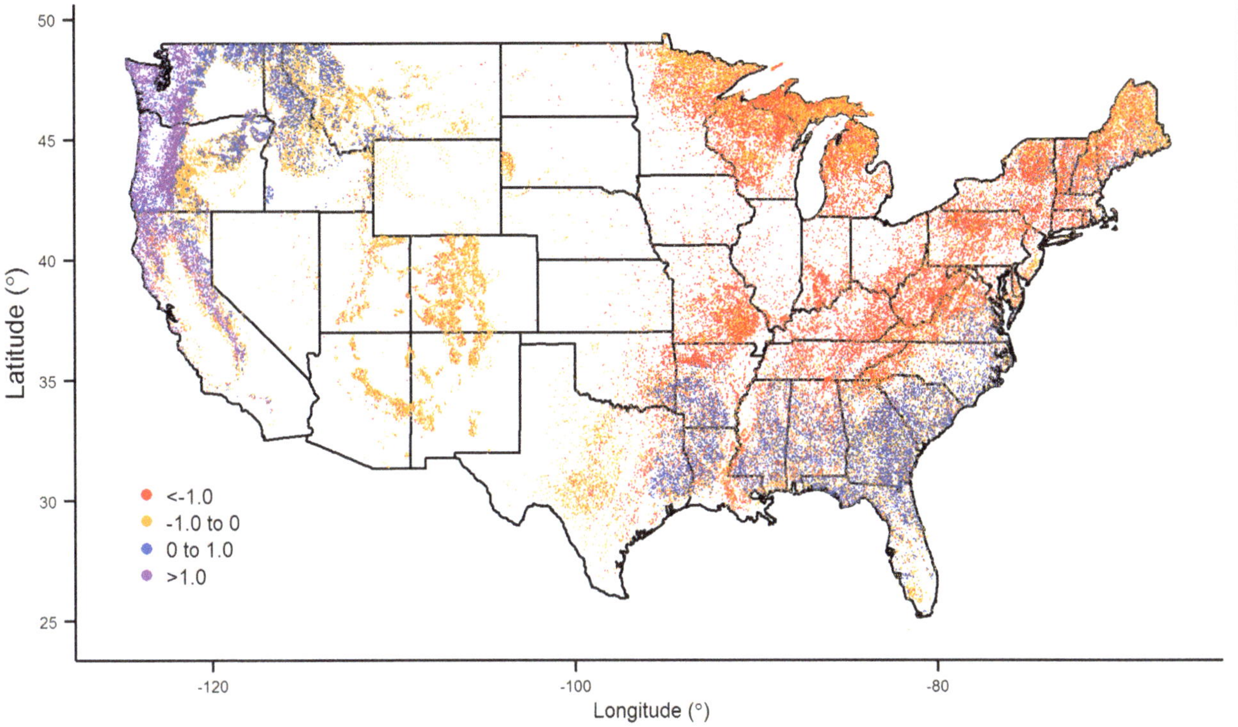


**Fig. 3** Distribution of differences between live-tree belowground C estimates and adjusted estimates ($BGC_{NGHGI}$ − $BGC_{ClimAdj}$; Mg $ha^{-1}$), with *red colors* indicating higher estimated C and *purple colors* less C

conditions such as stand age in our random forests model output suggests that accounting for these aspects of forest origin and development is useful for understanding broad scale patterns in live tree belowground C.

The utilization of nationally consistent allometric equations [e.g., 12, 20] provides a general assessment of current C stocks. However, the appeal of employing $BGC_{ClimAdj}$ estimates in the US NGHGI centers on utilizing standard forest inventory data collected in national forest inventories along with climate information. It is important to note that the model of Reich et al. [8] used here may be considered an empirical model. Such an empirical model may work well for data-rich countries with established national forest inventories. Other strategies, including process models, could potentially be employed to examine the carbon–climate relationship in forests. Incorporation of climate data enables evaluation of future climate change scenarios [e.g., 21] and their impacts on C monitoring efforts (e.g., United Nations Framework Convention on Climate Change submissions). As there are few studies that allow empirical observation of "true" coarse root C given the time and effort involved in destructively sampling belowground components, model validation procedures are limited for assessments at national scales. As an alternative, model-based approaches can be specified to be dynamic by incorporating future forest conditions and global change scenarios to determine their implications on C stocks and sequestration patterns. Such an approach using current climate conditions could provide general estimates of C stocks with associated uncertainty bounds for the temperate forests which occupy the US.

The majority of forest types displayed negative mean differences between current NGHGI and climate-adjusted models, indicating greater live tree belowground C stocks when using the adjusted models. The larger stocks in climate-adjusted models is partially a reflection of the ability of this framework to account for temperature-related shifts in patterns of belowground allocation within a species; a relationship held constant in current NGHGI models. In particular, a key component of the climate-adjusted model is an increased level of belowground allocation with decreasing temperatures given increasing nutrient limitation with colder temperatures [8]. These adjustments are reflected in the greater climate-adjusted estimates for northern latitude and high elevation forests (Fig. 2). Given the importance of climate variables in driving other belowground processes such as coarse root decomposition [22], the integration of these variables in models for describing other belowground stocks (e.g., dead roots) may not only aid in understanding C stock differences across large geographic scales

**Table 2 Estimates of belowground carbon stocks (Tg) and associated sampling errors (SE; %) for current US national greenhouse gas inventory ($BGC_{NGHGI}$) [13] and adjusted estimates ($BGC_{ClimAdj}$) by state**

| State | $BGC_{NGHGI}$ (SE) | $BGC_{ClimAdj}$ (SE) | Mean % difference |
|---|---|---|---|
| Alabama | 176.23 (1.38) | 198.13 (1.4) | 12.4 |
| Alaska[a] | 234.3 (2.63) | 252.84 (2.69) | 7.9 |
| Arizona | 31.93 (4.27) | 34.94 (4.2) | 9.4 |
| Arkansas | 142.92 (1.61) | 164.02 (1.59) | 14.8 |
| California | 456.22 (1.37) | 428.69 (1.27) | −6.0 |
| Colorado | 128.66 (1.92) | 139.28 (1.9) | 8.3 |
| Connecticut | 25.9 (3.51) | 28.67 (3.5) | 10.7 |
| Delaware | 4.66 (6.09) | 5.13 (6.11) | 10.1 |
| Florida | 105.88 (2.14) | 111.18 (2.15) | 5.0 |
| Georgia | 191.36 (1.48) | 198.67 (1.48) | 3.8 |
| Idaho | 209.16 (1.77) | 212.81 (1.75) | 1.7 |
| Illinois | 38.95 (2.86) | 45.65 (2.86) | 17.2 |
| Indiana | 45.67 (1.88) | 53.27 (1.88) | 16.6 |
| Iowa | 18.21 (4.21) | 21.34 (4.22) | 17.2 |
| Kansas | 14.7 (4.2) | 17.26 (4.23) | 17.4 |
| Kentucky | 109.67 (1.83) | 140.42 (1.84) | 28.0 |
| Louisiana | 133.93 (1.65) | 145.01 (1.64) | 8.3 |
| Maine | 127.35 (1.42) | 140.69 (1.41) | 10.5 |
| Maryland | 35.24 (3.31) | 38.9 (3.31) | 10.4 |
| Massachusetts | 43.72 (2.55) | 47.78 (2.55) | 9.3 |
| Michigan | 150.1 (1.38) | 176.7 (1.39) | 17.7 |
| Minnesota | 82.15 (1.35) | 98.24 (1.34) | 19.6 |
| Mississippi | 179.82 (1.3) | 199.05 (1.32) | 10.7 |
| Missouri | 103.99 (1.46) | 121.17 (1.46) | 16.5 |
| Montana | 197.84 (1.43) | 202.28 (1.42) | 2.2 |
| Nebraska | 6.64 (7.21) | 7.65 (7.27) | 15.2 |
| Nevada | 4.21 (13.01) | 4.71 (12.85) | 11.9 |
| New Hampshire | 62.93 (1.82) | 69.28 (1.82) | 10.1 |
| New Jersey | 23.15 (3.71) | 25.57 (3.73) | 10.5 |
| New Mexico | 45.86 (3.43) | 50.12 (3.35) | 9.3 |
| New York | 235.57 (1.05) | 262.74 (1.05) | 11.5 |
| North Carolina | 203.82 (1.23) | 214.85 (1.24) | 5.4 |
| North Dakota | 2.55 (11.19) | 2.98 (11.26) | 16.9 |
| Ohio | 93.07 (1.75) | 108 (1.75) | 16.0 |
| Oklahoma | 47.62 (2.54) | 58.36 (2.54) | 22.6 |
| Oregon | 478.52 (1.03) | 427.63 (0.99) | −10.6 |
| Pennsylvania | 214 (1.1) | 238.49 (1.1) | 11.4 |
| Rhode Island | 4.96 (5.77) | 5.44 (5.76) | 9.7 |
| South Carolina | 109.52 (1.92) | 112.13 (1.93) | 2.4 |
| South Dakota | 7.86 (6.02) | 8.68 (6.02) | 10.4 |
| Tennessee | 160.47 (1.27) | 203.39 (1.29) | 26.7 |
| Texas | 129.47 (1.68) | 149.23 (1.62) | 15.3 |
| Utah | 34.85 (3.9) | 38.62 (3.84) | 10.8 |
| Virginia | 191.11 (1.18) | 205.22 (1.19) | 7.4 |
| Washington | 410.46 (1.29) | 366.61 (1.26) | −10.7 |
| West Virginia | 169.05 (1.18) | 188.03 (1.18) | 11.2 |
| Wisconsin | 108.69 (1.14) | 128.72 (1.13) | 18.4 |
| Wyoming | 65.9 (5.54) | 69.14 (5.49) | 4.9 |
| Total | 5798.84 | 6167.71 | 6.4 |

[a] Coastal Alaska, only

**Table 3 Summary of random forests model output and their importance scores (% IncMSE) for predicting refined estimates belowground C using basic stand structure, management, and physiographic variables**

| Variable | % IncMSE | Variable | % IncMSE |
|---|---|---|---|
| *With aboveground biomass* | | *Without aboveground biomass* | |
| Aboveground biomass | 34.5 | Stand age | 33.8 |
| Stand age | 20.7 | Stand origin | 29.1 |
| Stand origin | 19.0 | Longitude | 27.5 |
| Latitude | 16.8 | Elevation | 19.8 |
| Longitude | 15.9 | Latitude | 16.4 |
| Elevation | 13.6 | Hopkins index | 14.8 |
| Hopkins index | 12.8 | | |
| $R^2$ | 0.87 | $R^2$ | 0.47 |
| RMSE (Mg C ha-1) | 3.72 | RMSE (Mg C ha-1) | 7.47 |

but may also inform our understanding of other belowground processes.

The forecast of decreases in belowground C stocks up to year 2090 across all US regions and forest types is a function of projected increases in MAT and its role in the distribution of coarse root C stocks. National patterns in C stocks observed with US data indicated similar trends to the models presented in Reich et al. [8] showing that the proportion of total biomass to roots is greater in increasingly cold climates. Such assessments conducted in this analysis include a dynamic climate but assume constant aboveground biomass stocks, and a stand origin and phylogeny (e.g., conifer- or hardwood-dominated forest type). For example, the economic incentives of increasing C stocks to meet increased demand for wood for bioenergy [e.g., 23] has the potential to alter the proportion of planted compared to natural-origin stands at a national level. Although aridity was not found to influence the global distribution of coarse root biomass [8], what role might precipitation and/or its interaction with temperature and other climate variables play in determining future C sequestration patterns and stocks? Through designing national-scale models that incorporate climate parameters, the ability to quantify C stocks using relationships observed between climate and C dynamics is possible but is otherwise impracticable using allometric equations alone.

A particular concern from a physiological perspective relies in assessing carbon allocation tradeoffs in concert with changing climates. Changing allocation to belowground components can alter biomass accumulation and nutrient uptake [9], but without an assessment of additional carbon stocks from other pools (e.g., foliage and soil components), the role that future climate may play in determining overall stocks may be somewhat limited. Our approach in using current belowground C stocks from temperate forests across the US, with a range of current and projected climates, management histories, and stand structures, serves as a preliminary investigation of the role that climate may play on coarse root systems (e.g., Fig. 4). Carbon allocation to soil microbial biomass may be limited at northern sites with cooler climates [24] which would seemingly influence nutrient uptake and allocation within tree components. Process models [e.g., 25] may be well suited to examine such carbon allocation-climate tradeoffs, presenting a different approach compared to the empirical models examined here. In the interim, additional biomass data collected on all components from a range of species across sites with different climate regimes and management histories will aid in improving our understanding of carbon allocation patterns related to climate [26].

The finding that surrogates of climate (e.g., latitude and longitude) and knowledge of forest structure and management history (e.g., stand age and origin) were useful in predicting belowground C is encouraging when considering approaches to constructing a NGHGI. For example, any new strategy for estimating forest C pools in the US NGHGI requires "back casting" estimates to the 1990 reporting year [17]. Obtaining climate data from past years at large geographic scales may present less of a barrier to refinement of NGHGIs than other efforts (e.g., in situ field inventories in the 1990's) to reduce the uncertainty associated with estimates of forest C and harvested wood products. Other nations could similarly adjust their

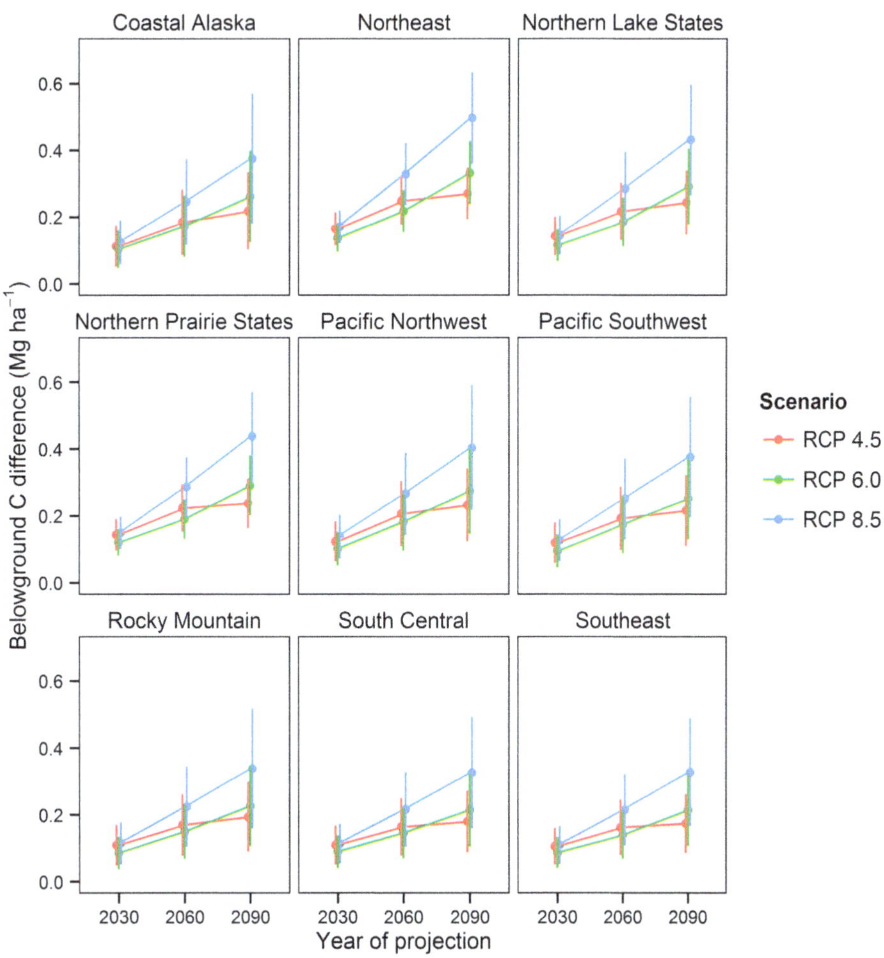


**Fig. 4** Regional differences in projected live-tree belowground C stocks compared to climate-sensitive estimates of belowground C derived from 30-year normal conditions (1960–1990). *Error bars* denote one standard deviation

C accounting practices using insight from this analysis to determine belowground stocks. Regardless of whether or not tree-level information from a national forest inventory is available, incorporating indicators of forest management history (e.g., proportion of planted versus natural origin stands) and variability in climate within a country (e.g., general trends in temperature), estimates of belowground C stocks could be developed that theoretically represent site and regional differences. In an era where destructive samples of belowground tree components is incomplete across the world's forests, estimation strategies that merge the attributes of both allometric equations and C–climate relationships may refine forest C stocks estimations especially given emerging science that supports the relationships between climate, forest biomass, and C pools [1, 3, 8, 27].

## Conclusions

Numerous findings emerged from our investigation by incorporating climate variables into the estimation of belowground C stocks. First, climate variables can be used to adjust the US NGHGI estimates of belowground C stocks. Specifically, adjustment factors were specified to amend current coarse root C stocks estimated from allometric equations by incorporating mean annual temperature at various locations across the US. Second, for the US NGHGI, incorporating mean annual temperature increased national belowground C stocks by 6.4 %. In contrast, coarse root C stocks were projected to decrease through 2090, primarily due to lower partitioning to belowground components under warmer conditions. Third, whether or not a forest was planted or from natural origin, and its stand age were influential variables in determining belowground C stocks. Future work that integrates both climate and stand origin will increase our ability to predict belowground C stocks across regions containing a mixture of management and climate regimes. Finally, as a means of refining NGHGIs, climate-adjusted models depicting belowground C stocks should be adopted to incorporate the impacts of future global change and management scenarios on C sequestration patterns and stocks.

## Methods

### Study area

Forests across the US range are characterized by four major ecoclimatic zones, including polar, temperate humid, arid, and tropical humid types [28, 29]. The study area investigated here included forestlands across the contiguous US, spanning approximately 24° latitude (LAT) and 58° longitude (LONG) in addition to coastal Alaska (mean coordinates 57.87°N, 138.60°W). Mean annual temperature (MAT) ranged from −3.0 to 24.9 °C and precipitation (MAP) from 18 to 420 cm [30, 31]. Nine broad geographic regions were identified across the study area, ultimately containing 78 unique forest types [13] (Additional file 1: Table S1).

### Forest Inventory and Analysis data

The FIA program within the US Forest Service monitors forests by establishing permanent sample plots across the US in three phases [32]. During the inventory's first phase, sample plot locations are established at an intensity of approximately 1 plot per 2400 ha. If the plot lies partially or wholly within a forested area, field personnel visit the site and establish a phase two (P2) inventory plot. Standard P2 inventory plots consist of four 7.32-m fixed radius subplots for a total plot area of approximately 0.07 ha where standing tree and site attributes are measured. Live trees with a DBH of at least 12.7 cm are measured on these subplots. Within each subplot a 2.07-m microplot is established where saplings with a DBH between 2.5 and 12.7 cm are measured.

All data were obtained from the publically-available FIA database (FIADB; [33]; http://apps.fs.fed.us/fiadb-downloads/datamart.html; download date 14 May 2014). If an FIA plot was remeasured at any point, only the most recent measurement was used in the analysis. Using the individual tree measurements, aboveground live-tree biomass ($BIO_{AG;}$ Mg $ha^{-1}$) was estimated by summing the bole, stump, top (excluding foliage), saplings, and woodland tree species (primarily those from dryland forests) components of each plot. Additional condition-level information including stand age (STDAGE; years) and a binary variable depicting stand origin (NAT; 1 = natural, 0 = artificial regeneration) were subsequently analyzed for each plot. As a bioclimatic measure, The Hopkins index ([34]; HI) standardizes the onset of spring for a given region and was computed for each FIA plot sampled relative to the mean LAT (40.35), LONG (−95.84), and elevation (ELEV; 2283 feet):

$$HI = \left(\frac{ELEV - 2283}{100}\right) + 4\,(LAT - 40.35) + 1.25\,(-95.84 - LONG) \qquad (1)$$

The HI variable assumes that spring is delayed by one day for each 100-foot rise in ELEV, four days for each 1° increase in LAT, and 1.25 days for each 1° increase in LONG (e.g., more westward; [34]). In total, 70,126 FIA plots were analyzed for their belowground C.

### Belowground C in the US greenhouse gas inventory ($BGC_{NGHGI}$)

The Intergovernmental Panel on Climate Change's (IPCC) Good Practice Guidance considers forest C stocks associated with live aboveground, live belowground, dead wood, litter, and soil organic pools [35]. Hence, the focus of this analysis is on adjusting estimates of live belowground C, defined as all coarse living roots greater than 2 mm diameter [13]. Estimates of aboveground live C in the NGHGI are calculated using the component ratio method [36], but do not ultimately influence the analysis for belowground C here.

Belowground C for live trees is estimated in two stages using allometric equations. First, total aboveground biomass is estimated as a function of tree DBH [12]:

$$\text{Biomass} = \exp\left(\alpha_{1,i} + \alpha_{2,i} \ln \text{DBH}\right) \tag{2}$$

where $\alpha_{1,i}$ and $\alpha_{2,i}$ are parameters for one of ten species groups [four hardwood groups (aspen/alder/cottonwood/willow, soft maple/birch, mixed hardwood, and hard maple/oak/hickory/beech), five conifer groups (cedar/larch, Douglas-fir, true fir/hemlock, pine, and spruce), and one woodland species group (juniper, oak, mesquite)]. Second, belowground root biomass is estimated as a ratio ($BG_{RATIO}$) of root to total aboveground biomass [12]:

$$BG_{RATIO} = \begin{cases} \exp\left(-1.6911 + \dfrac{0.8160}{\text{DBH}}\right), & \text{for hardwood species} \\ \exp\left(-1.5619 + \dfrac{0.6614}{\text{DBH}}\right), & \text{for conifer species} \end{cases} \tag{3}$$

Hence, parameters indicate $BG_{RATIO}$ will decrease for larger DBH trees and that for a fixed DBH, $BG_{RATIO}$ will be larger for conifer compared to hardwood species. Belowground biomass was estimated by multiplying the values obtained from Eqs. 2 and 3, then converted to C by multiplying by 0.5, assuming 50 % of biomass is C [35]. Estimates of belowground C were scaled to the plot level and are hereby abbreviated as $BGC_{NGHGI}$.

#### Belowground C from climate-derived models ($BGC_{Clim}$)

Recent investigations of the global distribution of biomass within forests have provided insight for comparing size- and species-specific predictions (i.e., allometric equations and proportional ratios) with climate-sensitive estimations of belowground C [e.g., 8]. Using the relationship observed between mean annual temperature and root mass fraction, the model used to determine a climate-derived estimate of belowground C was parameterized with global data compiled from various sources including Usoltsev [37], Luo et al. [38], Cannell [39], and from over 1000 additional forest stands including the US [8]. A total of 3043 of these stands contained measurements of belowground biomass. The motivation for the development of this model was to assess the distribution of biomass in roots along a temperature and precipitation spectrum [8]. We estimated a climate-sensitive prediction of belowground biomass ($BGB_{Clim}$) using the model of Reich et al. [8] (Table 4). In addition to MAT and NAT, a dummy variable indicating whether or not the FIA plot was primarily dominated by hardwoods or conifers (HDWD) and stem biomass ($BIO_{STEM}$; Mg $ha^{-1}$) of live trees were used to estimate $BGB_{Clim}$. Thirty-year (1961–1990) climate data (i.e., MAT) were obtained by specifying LAT, LONG, and ELEV of each FIA plot location to a spline surface model developed from climate station data across forests of North America [30, 31]. We assigned the HDWD variable using the FIA forest type code [40] by separating conifer-dominated forest type codes (i.e., FORTYPCD $\leq$ 409) with hardwood-dominated codes (FORTYPCD $\geq$ 500). By incorporating measures of aboveground biomass, these allometric relationships allow one to capture the variability observed across a range of stand structures and ages. Continuous independent variables were centered prior to applying the Reich et al. [8] model to the FIA plots. When standardized for a given MAT and aboveground stem biomass, the model of Reich et al. [8] indicates conifer forests tended to have a smaller root mass fraction than hardwood forests. Values for $BGB_{Clim}$ were converted to $BGC_{Clim}$ by multiplying by 0.5 [35].

#### Climate-adjusted models of belowground C ($BGC_{ClimAdj}$)

Current models of belowground C in the NGHGI (i.e., $BGC_{NGHGI}$) could likely be adjusted by incorporating climate-derived estimates of belowground C (i.e., $BGC_{Clim}$). From a NGHGI reporting perspective, estimates of BGC would need to be made on individual plots, then "back cast" to contemporary estimates using the 1990 baseline reporting year [17]. Adjustment factors were estimated to align allometric- and climate-derived estimates:

$$\text{AdjFactor} = \frac{BGC_{Clim}}{BGC_{NGHGI}} \tag{4}$$

where AdjFactor is the ratio of climate- to allometric-derived belowground C for a specific forest type found in a given geographic region. New climate-adjusted estimates of belowground C ($BGC_{ClimAdj}$) are then:

$$BGC_{ClimAdj} = BGC_{NGHGI} \times \text{AdjFactor} \tag{5}$$

where $BGC_{ClimAdj}$ was computed for each FIA plot record.

#### Analyzing belowground C model differences

We conducted equivalence tests comparing $BGC_{NGHGI}$ and $BGC_{ClimAdj}$ model predictions using two one-sided tests [41]. Equivalence tests are commonly applied in the forest science literature and are advantageous in that they can be used in model validation by assuming a null hypothesis of dissimilarity [42]. Equivalence tests

**Table 4 Model parameters and model form Reich et al. [8] used in this analysis for estimating climate-adjusted belowground biomass ($BGB_{Clim}$)**

| Term | Parameter | Value | SE |
|---|---|---|---|
| $b_0$ | Intercept | −0.18088 | 0.021062 |
| $b_1$ | HDWD | 0.0172682 | 0.005315 |
| $b_2$ | NAT | 0.0018117 | 0.005256 |
| $b_3$ | MAT | −0.003032 | 0.000564 |
| $b_4$ | $\log(BIO_{STEM})$ | 0.7940911 | 0.011187 |
| $b_5$ | HDWD × NAT | −0.000591 | 0.005291 |
| $b_6$ | HDWD × (MAT − 9.1374) | −0.001423 | 0.000566 |
| $b_7$ | HDWD × ($\log(BIO_{STEM}$ − 1.88807)) | −0.031736 | 0.008798 |
| $b_8$ | NAT × (MAT − 9.1374) | −0.000555 | 0.000574 |
| $b_9$ | NAT × ($\log(BIO_{STEM}$ − 1.88807)) | 0.021458 | 0.010028 |
| $b_{10}$ | (MAT − 9.1374) × ($\log(BIO_{STEM}$ − 1.88807)) | 0.0020443 | 0.001155 |
| $b_{11}$ | HDWD × NAT × (MAT − 9.1374) | 0.0020594 | 0.000566 |
| $b_{12}$ | HDWD × NAT × ($\log(BIO_{STEM}$ − 1.88807)) | 0.0269889 | 0.010957 |
| $b_{13}$ | HDWD × (MAT − 9.1374) × ($\log(BIO_{STEM}$ − 1.88807)) | 0.0050601 | 0.001197 |
| $b_{14}$ | NAT × (MAT − 9.1374) × ($\log(BIO_{STEM}$ − 1.88807)) | 0.0016555 | 0.001206 |

Parameters are: a binary variable depicting general forest type (HDWD; 1 = hardwood-dominated, 0 = conifer-dominated), stem biomass of live trees ($BIO_{STEM}$; Mg $ha^{-1}$), a binary variable depicting stand origin (NAT; 1 = natural, 0 = artificial regeneration), mean annual temperature (MAT; °C), and log is to the base 10

In

$$\begin{aligned}\log(BGBClim) = {} & b0 + b1(HDWD) + b2(NAT) + b3(MAT) + b4(\log(BIOSTEM)) + b5((HDWD \times NAT))+\\ & b6((HDWD \times (MAT - 9.1374))) + b7((HDWD \times (\log(BIOSTEM) - 1.88807)))+\\ & b8((NAT \times (MAT - 9.1374))) + b9((NAT \times (\log(BIOSTEM) - 1.88807)))+\\ & b10(((MAT - 9.1374) \times (\log(BIOSTEM) - 1.88807)))+\\ & b11((HDWD) \times (NAT) \times (MAT - 8.8508))+\\ & b12((HDWD) \times (NAT) \times (\log(BIOSTEM) - 1.88807))+\\ & b13((HDWD) \times (MAT - 8.8508) \times (\log(BIOSTEM) - 1.88807))+\\ & b14((NAT) \times (MAT - 8.8508) \times (\log(BIOSTEM) - 1.88807))\end{aligned}$$

are unlike statistical goodness-of-fit approaches and instead examine dissimilarity. Dissimilarity in the equivalence test was specified using a threshold of ±25 %. This threshold allows for a moderate amount of disagreement between the various model predictions, with non-equivalence suggesting biological disparities in C stocks. Differences between $BGC_{NGHGI}$ and $BGC_{ClimAdj}$ models were mapped across the US to examine geographic trends in estimates of belowground C when using each approach. We computed US state-level population estimates of belowground C (Tg) using $BGC_{NGHGI}$ and $BGC_{ClimAdj}$ models and compared mean percent differences for the two estimation strategies.

Nonparametric random forests (RF; [43]) were implemented in R [44] to identify variables that were effective in describing $BGC_{ClimAdj}$. Recognizing that not all users may implement climate data in determining forest C stocks, seven variables from the FIADB ($BIO_{AG}$, STDAGE, NAT, LAT, LONG, ELEV, and HI) were chosen for incorporation into the RF. This method involved building a set of regression trees based on bootstrapped samples of the belowground C data. We similarly fit a RF model without $BIO_{AG}$ to examine how belowground C can be predicted without knowledge of aboveground biomass stocks.

### Belowground C in future scenarios

Current CMIP5 models [45] as described in the fifth assessment report (AR5) of the IPCC [46] were obtained using three scenarios (RCP 4.5, RCP 6.0, RCP 8.5; [30]). An ensemble of 17 AR5 model predictions was used for each RCP scenario. Provided that differences in belowground C could exist in future global change scenarios (i.e., changes in MAT at various locations in the US), climate data were obtained for the 30-year normal (1961–1990) and years 2030, 2060, and 2090. To gain insight into temperature-related patterns and their influence on belowground C, climate-sensitive estimates were made using these new MAT values in future years (while holding fixed the variables HDWD, NAT, and $BIO_{STEM}$) to assess the variation in belowground C assuming future scenarios. While $BIO_{STEM}$ will fluctuate in response to trends in forest growth and mortality, this simulation was specifically designed to understand the influence of MAT in future belowground C stocks.

## Additional file

**Additional file 1: Table S1.** Summary of live-tree belowground C estimates (Mg ha$^{-1}$) from the current US national greenhouse gas inventory ($BGC_{NGHGI}$; Smith et al. 2013), climate-derived predictions ($BGC_{Clim}$; Reich et al. 2014), and adjustment factors (AdjFactor) used to amend $BGC_{NGHGI}$ estimates by taking into account climate-derived predictions.

### Abbreviations

C: carbon; US NGHGI: United States' National Greenhouse Gas Inventory; GHG: greenhouse gas; FIA: Forest Inventory and Analysis; LAT: latitude; LONG: longitude; MAT: mean annual temperature; MAP: mean annual precipitation; $BIO_{AG}$: aboveground live-tree biomass; STDAGE: stand age; NAT: natural regeneration; RF: random forests; HI: Hopkins index; ELEV: elevation; $BGC_{NGHGI}$: belowground carbon in the US greenhouse gas inventory; IPCC: Intergovernmental Panel on Climate Change; DBH: tree diameter at breast height; $BG_{RATIO}$: belowground root biomass ratio; $BGC_{Clim}$: belowground carbon from climate-derived models; $BGB_{Clim}$: belowground biomass from climate-derived models; HDWD: hardwood forest type; $BIO_{STEM}$: stem biomass; FORTYPCD: forest type code; $BGC_{ClimAdj}$: climate-adjusted models of belowground carbon; AdjFactor: the ratio of climate- to allometric-derived belowground carbon.

### Authors' contributions

MBR, GMD, and CWW designed the study and performed the research; MBR and GMD analyzed data and contributed new methods; all authors discussed the results of the study and contributed to writing the manuscript. All authors read and approved the final manuscript.

### Author details

[1] Department of Forest Resources, University of Minnesota, St. Paul, MN 55108, USA. [2] USDA Forest Service, Northern Research Station, St. Paul, MN 55108, USA. [3] Rubenstein School of Environment and Natural Resources, University of Vermont, Burlington, VT 05405, USA.

### Acknowledgements

We thank Brian Clough, Christopher Oswalt, and John Stanovick for comments that improved this work and Georgii Alexandrov and two anonymous reviewers for their evaluation of this paper.

### Compliance with ethical guidelines

### Competing interests

The authors declare that they have no competing interests.

### References

1. Ryan MG, Harmon ME, Birdsey RA, Giardina CP, Heath LS, Houghton RA, et al. A synthesis of the science on forests and carbon for U.S. forests. Ecol Soc Am. 2010;13:1–16.
2. Malmsheimer RW, Bowyer JL, Fried JS, Gee E, Izlar RL, Miner RA, et al. Managing forests because carbon matters: integrating energy, products, and land management policy. J For. 2011;109:S7–50.
3. McKinley DC, Ryan MG, Birdsey RA, Giardina CP, Harmon ME, Heath LS, et al. A synthesis of current knowledge on forests and carbon storage in the United States. Ecol Appl. 2011;21:1902–24.
4. Joyce LA, Running SW, Breshears DD, Dale VH, Malmsheimer RW, Sampson RN, et al. Ch. 7: Forests. Climate change impacts in the United States: the third national climate assessment. In: Melillo JM, Richmond TC, Yohe GW, editors. Washington, D.C: U.S. Global Change Research Program; 2014. p. 175–94. doi:10.7930/J0Z60KZC.
5. Poorter H, Niklas KJ, Reich PB, Oleksyn J, Poot P, Mommer L. Biomass allocation to leaves, stems and roots: meta-analyses of interspecific variation and environmental control. New Phytol. 2012;193(1):30–50. doi:10.1111/j.1469-8137.2011.03952.x.
6. Smyth CE, Kurz WA, Neilson ET, Stinson G. National-scale estimates of forest root biomass carbon stocks and associated carbon fluxes in Canada. Glob Biogeochem Cycles. 2013;27(4):2012GB004536. doi:10.1002/2012gb004536.
7. Litton CM, Raich JW, Ryan MG. Carbon allocation in forest ecosystems. Glob Chang Biol. 2007;13(10):2089–109. doi:10.1111/j.1365-2486.2007.01420.x.
8. Reich PB, Luo Y, Bradford JB, Poorter H, Perry CH, Oleksyn J. Temperature drives global patterns in forest biomass distribution in leaves, stems, and roots. Proc Natl Acad Sci USA. 2014;111(38):13721–6.
9. Raich JW, Clark DA, Schwendenmann L, Wood TE. Aboveground tree growth varies with belowground carbon allocation in a tropical rainforest environment. PLoS One. 2014;9(6):e100275. doi:10.1371/journal.pone.0100275.
10. Bond-Lamberty B, Wang C, Gower ST. Aboveground and belowground biomass and sapwood area allometric equations for six boreal tree species of northern Manitoba. Can J For Res. 2002;32(8):1441–50. doi:10.1139/x02-063.
11. Litton CM, Ryan MG, Tinker DB, Knight DH. Belowground and aboveground biomass in young postfire lodgepole pine forests of contrasting tree density. Can J For Res. 2003;33(2):351–63. doi:10.1139/X02-181.
12. Jenkins JC, Chojnacky DC, Heath LS, Birdsey RA. National-scale biomass estimators for United States tree species. For Sci. 2003;49(1):12–35.
13. Smith JE, Heath LS, Hoover CM. Carbon factors and models for forest carbon estimates for the 2005–2011 National Greenhouse Gas Inventories of the United States. For Ecol Manag. 2013;307:7–19.
14. Zianis D, Muukkonen P, Mäkipää R, Mencuccini M. Biomass and stem volume equations for tree species in Europe. Silva Fenn Monog. 2005;4:63.
15. Beets PN, Brandon AM, Goulding CJ, Kimberley MO, Paul TSH, Searles N. The inventory of carbon stock in New Zealand's post-1989 planted forest for reporting under the Kyoto protocol. For Ecol Manag. 2011;262(6):1119–30. doi:10.1016/j.foreco.2011.06.012.
16. Green C, Tobin B, O'Shea M, Farrell EP, Byrne KA. Above- and belowground biomass measurements in an unthinned stand of Sitka spruce (Picea sitchensis (Bong) Carr.). Eur J For Res. 2006;126:179–88.
17. US EPA. Inventory of U.S. Greenhouse Gas Emissions and Sinks: 1990–2012. Chapter 7. Land Use, Land-Use Change, and Forestry. Annex 3.12. Methodology for Estimating Net Carbon Stock Changes in Forest Land Remaining Forest Lands. #430-R-14-003. Washington, DC: U.S. Environmental Protection Agency; 2014. http://www.epa.gov/climatechange/Downloads/ghgemissions/US-GHG-Inventory-2012-Annex-3-Additional-Source-or-Sink-Categories.pdf. Accessed 3 July 2014.
18. Wilson BT, Woodall CW, Griffith D. Imputing forest carbon stock estimates from inventory plots to a nationally continuous coverage. Carbon Balance Manag. 2013;8:1.
19. McMurtrie RE, Dewar RC. New insights into carbon allocation by trees from the hypothesis that annual wood production is maximized. New Phytol. 2013;199:981–90.
20. Ter-Mikaelian MT, Korzukhin MD. Biomass models for sixty-five North American tree species. For Ecol Manag. 1997;97(1):1–24.
21. Melillo JM, Richmond TC, Yohe GW. Climate change impacts in the United States: the third national climate assessment. U.S. Global Change Research Program. 2014. p. 841. http://ncadac.globalchange.gov/.
22. Garrett LG, Kimberley MO, Oliver GR, Pearce SH, Beets PN. Decomposition of coarse woody roots and branches in managed Pinus radiata plantations in New Zealand—a time series approach. For Ecol Manag. 2012;269:116–23. doi:10.1016/j.foreco.2011.12.030.
23. Sedjo R, Tian X. Does wood bioenergy increase carbon stocks in forests? J For. 2012;110(6):304–11.
24. Gorissen A, Tietema A, Joosten NN, Estiarte M, Peñuelas J, Sowerby A, et al. Climate change affects carbon allocation to the soil in shrublands. Ecosystems. 2004;7(6):650–61. doi:10.1007/s10021-004-0218-4.
25. Running SW, Gower ST. FOREST-BGC, A general model of forest ecosystem processes for regional applications. II. Dynamic carbon allocation and nitrogen budgets. Tree Physiol. 1991;9(1-2):147–60. doi:10.1093/treephys/9.1-2.147.

26. Weiskittel AR, MacFarlane DW, Radtke PJ, Affleck DLR, Temesgen H, Woodall CW, et al. A call to improve methods for estimating tree biomass for regional and national assessments. J For. 2015;113(4):414–24. doi:10.5849/jof.14-091.
27. Nemani RR, Keeling CD, Hashimoto H, Jolly WM, Piper SC, Tucker CJ, et al. Climate-driven increases in global terrestrial net primary production from 1982 to 1999. Science. 2003;300(5625):1560–3. doi:10.1126/science.1082750.
28. Smith WB, Miles PD, Perry CH, Pugh SA. Forest resources of the United States, 2007. USDA For. Serv. Gen. Tech. Rep. WO-78. 2009. p. 336.
29. Bailey RG. Ecosystem geography. New York: Springer; 1996. p. 216.
30. US Forest Service. Research on forest climate change: potential effects of global warming on forests and plant climate relationships in western North America and Mexico. Rocky Mountain Research Station, Moscow Laboratory. 2014. http://forest.moscowfsl.wsu.edu/climate/. Accessed 28 Mar 2014.
31. Rehfeldt GE. A spline model of climate for the western United States. USDA For. Serv. Gen. Tech. Rep. RMRS-165. 2006.
32. Bechtold WA, Patterson PL, editors. Forest Inventory and Analysis national sample design and estimation procedures. USDA For. Serv. Gen. Tech. Rep. SRS-GTR-80. 2005.
33. US Forest Service. The Forest Inventory and Analysis database: database description and user guide version 6.0 for phase 2. Washington DC; 2014. p. 624.
34. Hopkins AD. Bioclimatics, a science of life and climate relations. USDA Misc. Publ. 280. USDA, Washington, DC; 1938. p. 188.
35. Intergovernmental Panel on Climate Change (IPCC). Good practice guidance for land use, land use change and forestry. Institute for Global Environmental Studies (IGES). 2003. http://www.ipcc-nggip.iges.or.jp/public/gpglulucf/gpglulucf.html. Accessed 21 May 2014.
36. Woodall CW, Heath LS, Domke GM, Nichols MC. Methods and equations for estimating aboveground volume, biomass, and carbon for trees in the U.S. forest inventory, 2010. Gen. Tech. Rep. NRS-88. Newtown Square, PA: U.S. Department of Agriculture, Forest Service, Northern Research Station (1 CD-ROM). 2011. p. 30.
37. Usoltsev VA. Forest biomass of northern Eurasia. Russian Academy of Sciences, Ural Branch, Botanical Garden and Ministry of Education of Russian Federation, Ural State Forest Engineering University, Yekaterienburg) (in Russian). 2001.
38. Luo Y, Wang X, Zhang X, Booth T, Lu F. Root:shoot ratios across China's forests: forest type and climatic effects. For Ecol Manag. 2012;269:19–25.
39. Cannell MGR. World forest biomass and primary production data. London: Academic Press; 1982. p. 391.
40. Woudenberg SW, Conkling BL, O'Connell BM, LaPoint EB, Turner JA, Waddell KL. The Forest Inventory and Analysis Database: database description and users manual version 4.0 for phase 2. US Department of Agriculture, Forest Service Gen. Tech. Rep. RMRS-245. 2010. p. 339.
41. Wellek S. Testing statistical hypotheses of equivalence. London: Chapman and Hall; 2003.
42. Robinson AP, Froese RE. Model validation using equivalence tests. Ecol Model. 2004;176(3–4):349–58. doi:10.1016/j.ecolmodel.2004.01.013.
43. Breimen L. Random forests. Mach Learn. 2001;45:5–32.
44. Liaw A, Wiener M. Classification and regression by randomForest. R News. 2002;2(3):18–22.
45. Taylor KE, Stouffer RJ, Meehl GA. An overview of CMIP5 and the experiment design B. Am Meteorol Soc. 2012;93:485–98.
46. Intergovernmental Panel on Climate Change (IPCC). Climate change 2013: the physical science basis. Contribution of Working Group I to the Fifth Assessment Report of the Intergovernmental Panel on Climate Change. In: Stocker TF, Qin D, Plattner G-K, Tignor M, Allen SK, Boschung J, Nauels A, Xia Y, Bex V, Midgley PM, editors. Cambridge: Cambridge University Press; 2013. p. 1535.

2

# Time-series maps of aboveground carbon stocks in the forests of central Sumatra

Rajesh Bahadur Thapa*, Takeshi Motohka, Manabu Watanabe and Masanobu Shimada

**Abstract**

**Background:** Efforts to reduce emissions from deforestation and forest degradation in tropical Asia require accurate high-resolution mapping of forest carbon stocks and predictions of their likely future variation. Here we combine radar and LiDAR with field measurements to create a high-resolution aboveground forest carbon stock (AFCS) map and use spatial modeling to present probable future AFCS changes for the Riau province of central Sumatra.

**Results:** Our map provides spatially explicit estimates of the AFCS with an accuracy of ±23.5 Mg C ha$^{-1}$. According to this map, the natural forests in the province currently store 265 million Mg C, with a density of 72 Mg C ha$^{-1}$, as aboveground biomass. Using a spatially explicit modeling technique we derived time-series AFCS maps up to the year 2030 under three forest policy scenarios: business as usual, conservation, and concession. The spatial patterns of AFCS and their trends under different scenarios vary on a local scale, and some areas are highlighted that are at eminent risk of carbon emission. Based on the business as usual scenario, the current AFCS could decrease by 75 %, which may lead to the release of 747 million Mg $CO_2$. The other two scenarios, conservation and concession, suggest the risk reductions by 11 and 59 %, respectively.

**Conclusion:** The time-series AFCS maps provide spatially explicit scenarios of changes in AFCS. These data may aid in planning Reducing Emissions from Deforestation and forest Degradation in developing countries projects in the study area, and stimulate the development of AFCS maps for other regions of tropical Asia.

**Keywords:** Forest carbon, Deforestation, PALSAR, AFCS, Aboveground biomass, REDD+, Riau

## Background

Across the world, existing tropical forest landscapes are undergoing rapid deforestation due to natural disasters, as well as the increasing demand for agricultural land, wood products, energy, and developmental projects. Currently, global forest areas account for 3.85 billion ha, or 26 % of the Earth's land surface [1], but this area is decreasing at around 13 million ha per year [2]. As deforestation continues, the Earth becomes more susceptible to potentially negative impacts on ecosystems and the overall climate system due to the associated effects on carbon balance, biodiversity, soil, water regulation, and weather patterns. Currently, emissions caused by deforestation worldwide are considered to be very high, and are likely to continue in this way for the coming decades. Tropical forest regions in particular are major potential sources of carbon emissions [3–7]. Reducing Emissions from Deforestation and forest Degradation in developing countries (REDD+) [8] is one of the key global initiatives that aims to conserve forests and reduce carbon emissions. The goal of REDD+ is to connect investors to forest users and offer an economic portfolio for the retention of forest carbon and the avoidance of deforestation, while also slowing the drivers of land use change. As a result, the initiative contributes indirectly to biodiversity conservation by helping to reduce habitat loss and ensure the continuation of normal ecosystem services; hence, it is considered a sustainable option for the maintenance of forests. Meaningful implementation of REDD+ requires accurate, high-resolution, spatially explicit maps of forested areas and forest carbon stocks, as well as predictions of their change in the future. Therefore, efforts to improve the methods for mapping forest extents and

*Correspondence: rajesh.thapa@jaxa.jp; thaparb@gmail.com
Earth Observation Research Center, Japan Aerospace Exploration Agency (JAXA), 2-1-1 Sengen, Tsukuba, Ibaraki 305-8505, Japan

forest-related carbon stocks, as well as identifying their changes, have been advancing in many parts of the world, including tropical Asia [5, 9–13]. Remote sensing and spatial modeling techniques offer a practical means to monitor and examine changes in forest cover, analyze the implications of forest policies, predict spatial patterns of forest cover in the future, and relate these patterns to carbon stock densities [9, 14–16].

Accurate mapping of aboveground forest carbon stocks (AFCS) using spaceborne satellite data is still very challenging due to the requirement of a large amount of in situ data for forest carbon estimation model calibration and validation. Although traditional plot-based field measurements of AFCS have proven most accurate, they are costly and difficult to implement for large areas with dense tropical forests. Studies [13, 14, 17] have shown that light detection and ranging (LiDAR) techniques allow the accurate measurements of geographically referenced vertical forest structures, including canopy height, volume, and biomass. Using LiDAR data, an allometric model for AFCS can be developed with a relatively small number of field measurements [13, 17]. Modeling results can be used to extend the field data, providing spatially extensive and detailed forest attribute data, that can be used to calibrate AFCS predictive models build around a wide variety of spaceborne data, including synthetic aperture radar (SAR) and optical imageries covering larger areas [14, 18].

The integration of both airborne and spaceborne remote sensing techniques has offered the opportunity to more precisely map forest cover and related carbon stocks over wider areas at suitable spatiotemporal scales [6, 13, 14, 19]. However, the potential application of optical spaceborne remote sensing data in Asian tropical forest regions is limited, due to the frequent appearance of clouds and haze, as well as the insensitivity of sensing systems to the variability of biomass with a multi-layer canopy in highly dense forests. In contrast, spaceborne SAR is not limited by these factors as it penetrates clouds to image the Earth's surface regardless of weather conditions or solar illumination. Among the available spaceborne SAR systems, the Advanced Land Observation Satellite (ALOS) Phase Arrayed L-band SAR (PALSAR) operating at a wavelength of 23.6 cm is very sensitive to forest structure, yielding valuable information for the mapping of forest cover [20–22] and AFCS measurements [23, 24]. However, studies have shown that saturation remains a dominant issue when directly estimating AFCS using SAR data in high biomass areas [25–29]. A consideration of multi-temporal SAR data with multiple polarizations and the use of rule-based algorithms can help to mitigate the saturation problem and improve AFCS estimations [14, 24].

In addition to remote sensing techniques, spatial modeling is required in order to visualize and quantify the future variations of AFCS [3, 9, 15]. Future trends are reliant on the past processes of deforestation, and represent a consolidation of the relationships between time, space, and driving factors. A logically developed spatial model incorporates these relationships and extrapolates the likelihoods of various forest spatial patterns into the future [15, 16]. Such models offer a means of examining the implications of different forest policies on AFCS, allowing appropriate measures to control deforestation and retain AFCS to be formulated. In this study, our aim is to create a baseline AFCS map of a tropical forest in Asia and to estimate its future AFCS patterns under different forest policy frameworks. The Riau Province in Indonesia (Fig. 1) was chosen as the study site, due to its high carbon emissions as a result of deforestation compared with other provinces in the country [30]. Currently, 5.54 million people live in the province, with an annual growth rate of 3.6 % [16, 31]. Pulp and paper, oil-palm, rubber, and petroleum products are the main sources of income, while the forest landscape has also become a major provider of land in recent years. An ever increasing population and demanding economic activities have increased deforestation and forest degradation, which is ultimately threatening the forest carbon stocks, peat drainage, and biodiversity in the province.

Field measurements, LiDAR data, time-series PALSAR data, and a rule-based algorithm were used together to create a baseline AFCS map with high spatial resolution. The spatial model developed by Thapa et al. [16] was applied to visualize and assess the implications of different forest policies on future AFCS.

## Results and discussion

Table 1 summarizes the field measurement data collected over 87 plots, divided according to the various forest types in the study area. The forests are diverse and exhibit high variability in AFCS of different regions. Around 47 % of the measurement plots are within natural forests, including peat swamps (21.8), dry moist forest (10.3), regrowth (5.7), and mangrove areas (9.2). The remaining plots are within plantation forests including rubber (11.5), acacia (10.3), oil palm (23.0), and coconut (8.0) forests. Such a range of forest types is host to AFCS of 1.18–334.10 Mg C $ha^{-1}$. Among the plots, oil palm was found to have the lowest AFCS, while the natural dry moist forest areas had the highest. The regrowth forest areas had AFCS of 84.58–164.49 Mg C $ha^{-1}$, ranging between that of dry moist forests and peat swamps. Among the plantation forests, rubber plantations had both the highest mean and the maximum AFCS. Across the full range of values, the oil palm plantations had the

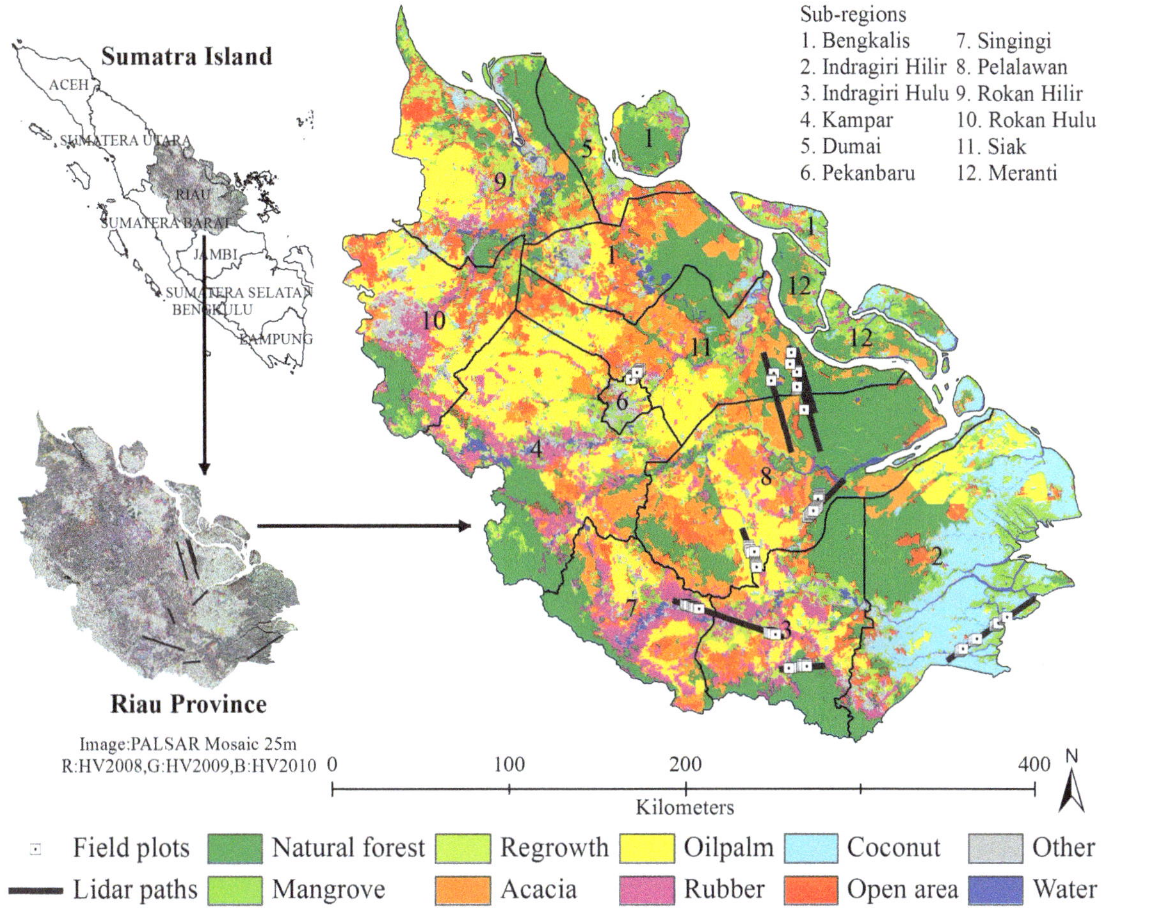


**Fig. 1** Study area, showing the location, composite PALSAR image, land use and land cover map derived from PALSAR data analysis [35], locations of field measurement plots, and LiDAR data acquisition paths

**Table 1 Summary of field measurement plots and AFCS estimates by forest types**

| Forest types | Field plots in % | AFCS (Mg C ha$^{-1}$) | |
|---|---|---|---|
| | | Mean | Range |
| *Natural forests* | | | |
| Peat swamp | 21.8 | 114.06 | 69.70–173.62 |
| Dry moist | 10.3 | 198.10 | 106.69–334.10 |
| Regrowth | 5.7 | 125.15 | 84.58–164.49 |
| Mangrove | 9.2 | 26.55 | 9.02–42.99 |
| *Plantation forests* | | | |
| Rubber | 11.5 | 38.82 | 19.10–58.88 |
| Acacia | 10.3 | 32.39 | 21.04–46.38 |
| Oil palm | 23.0 | 8.27 | 1.18–20.94 |
| Coconut | 8.0 | 16.00 | 6.87–27.60 |

AFCS is 47 % of the field measured aboveground biomass

lowest carbon stock of all the forest types in the study region.

The maximum likelihood algorithm (MLA) mapping procedure was used to create four maps using the PALSAR gamma-naught image from 2010, three AFCS maps to which the Lee, Frost, and median filters had been applied, respectively, with a 3 × 3 window, and an AFCS map without any filters for comparison. Validation was performed with three individual statistical measures to assess the filtering effects. The median filter provided the best AFCS map, with a RMSE and bias of 27.59 and −0.83 Mg C ha$^{-1}$ and an index of agreement (D) of 0.74. The other two filters, Lee and Frost, gave RMSEs with biases of 28.03 and −1.28, and 27.47 and −3.67 Mg C ha$^{-1}$, and D values of 0.71 and 0.74, respectively. The map created without the application of any filters had an

RMSE and bias of 30.09 and −2.71 Mg C $ha^{-1}$, respectively, and a D of 0.59. The D values are similar in both the Frost and the median filters, although the RMSE indicates a slightly better performance with the Frost filtered map, the bias is comparatively high. Overall, these results suggest that the median filter provides a good AFCS mapping product.

Four additional AFCS maps were created using the 2010 dataset and four increasing sizes of the filtering window with the median filter. The mapping results gradually improved as the window size increased from 3 × 3 to 9 × 9. Compared with the 3 × 3 window size, the map at the 9 × 9 window size had a reduced RMSE of 25.54 Mg C $ha^{-1}$ with a minimal bias of 0.06 Mg C $ha^{-1}$ and showed high similarity between the predicted and observed AFCS, as indicated by the index of agreement (D = 0.814). However, a decrease in performance was observed with the application of the 11 × 11 window size, with a RMSE and bias of 25.83 and 0.31 Mg C $ha^{-1}$ and a D of 0.812, respectively. This indicated that investigation into larger window sizes is unnecessary; the observed degradation is likely due to overgeneralization of the image data at larger window sizes. As such, a 9 × 9 median filter was applied to PALSAR data from 2009, which was used together with the 2010 data while running the MLA. Inclusion of the 2009 data set improved the AFCS mapping by reducing the mapping uncertainty, and the RMSE and bias dropped to 23.49 and 1.13 Mg C $ha^{-1}$. The value of D also improved to 0.843, indicating a very high similarity between the predicted and observed AFCS values (Fig. 2). Consideration of multi-temporal mosaic data sets can provide better estimates, indicating a normalization of climatic conditions and increasing the potential of replicating the modeling results in other similar tropical regions. It should be noted that after removing the bias in the RMSE, the error was reduced to 23.47 Mg C $ha^{-1}$.

The level of error in this map may have resulted from several unquantifiable factors, including differences in field measurement processes, inaccuracies in the field and LiDAR allometric equations, the slope correction method for the PALSAR mosaic data, and the time difference between field and remote sensing measurements [24]. However, the mapping error produced herein is low, and to date remains unmatched in similar studies of tropical forest regions, such as Kalimantan [27], the Amazon [17], and other areas [19]. Owing to the low mapping error and the high level of similarity between predicted and observed AFCS in such a diverse tropical forest, this map was used as a baseline for a spatial model that

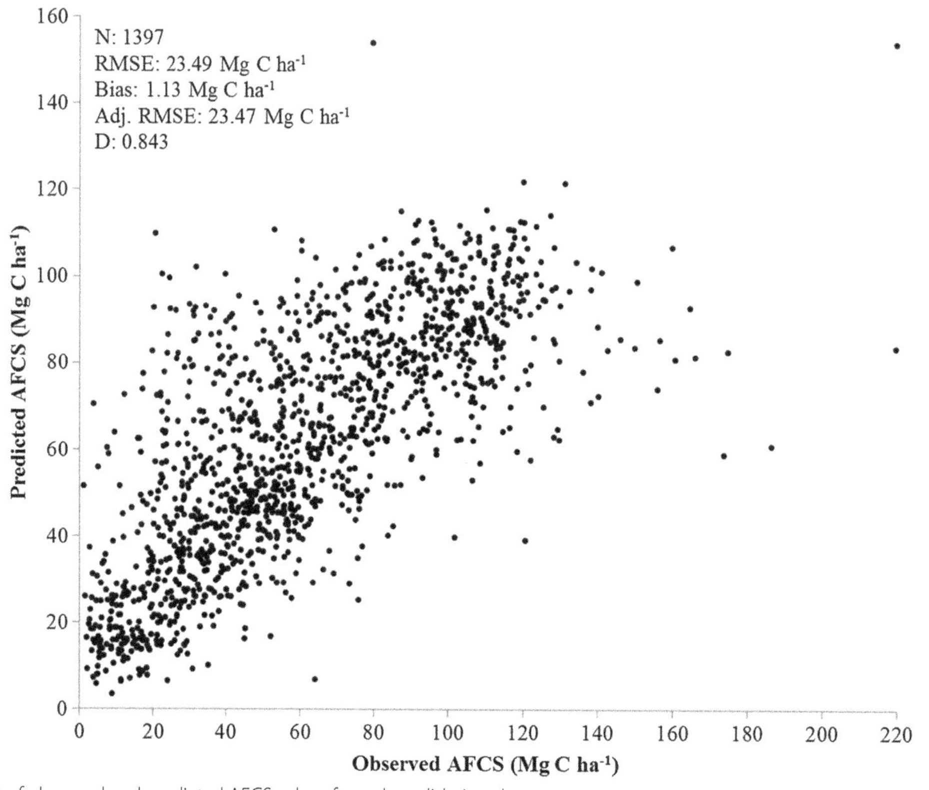


**Fig. 2** Scatterplot of observed and predicted AFCS values from the validation data set

calculates future changes in the forest carbon footprint of the study area.

Figure 3 illustrates the baseline AFCS map for the Riau province, in which interesting spatial patterns, with AFCS ranging from <1 to 334 Mg C ha$^{-1}$, can be observed. The spatial variation of AFCS across this map indicates that the majority of areas have carbon densities between 100 and 200 Mg C ha$^{-1}$. High carbon density areas are mostly found within the northern and south-eastern parts of the province, as well as along its western margin. The central area has a generally low carbon density from north to south. These patterns result from the relative distributions of natural forests and plantation forests. The low carbon density areas are mostly covered by plantation forests, agricultural land, and urban forests, while the higher density areas are correlated with existing natural forests, including peat swamps, dry moist forests, and regrowth. The slightly higher AFCS values on the islands in the central-eastern region and on the southeastern margin of the province represent mangrove forests.

The results of this study suggest that the AFCS of the natural forest areas in Riau province, which cover around 3.68 million ha, was 265.57 million Mg at the time of measurement (Table 2). The AFCS density of the natural forest is on average 71.99 Mg C ha$^{-1}$ across the province. Among the different districts, Indragiri Hilir in the southeastern part of the province has the highest AFCS density, at 77.99 Mg C ha$^{-1}$, whereas Pekanbaru in the central region has the lowest density, at 43.63 Mg C ha$^{-1}$. The Pekanbaru district is dominated by urban areas and contains fewer forested areas than the other districts, and therefore has a very little AFCS of 0.33 million Mg. In contrast, the Pelalawan district has the highest AFCS in the province, at 48.79 million Mg, which amounts to 18.37 % of the total AFCS stored in natural forests across the study area. However, the AFCS density in this district is slightly less than the provincial average. In terms of overall AFCS density, the Indragiri Hulu and Meranti districts have a similar quality of forest, as they both have a density of 75 Mg C ha$^{-1}$. Interestingly, the Bengkalis

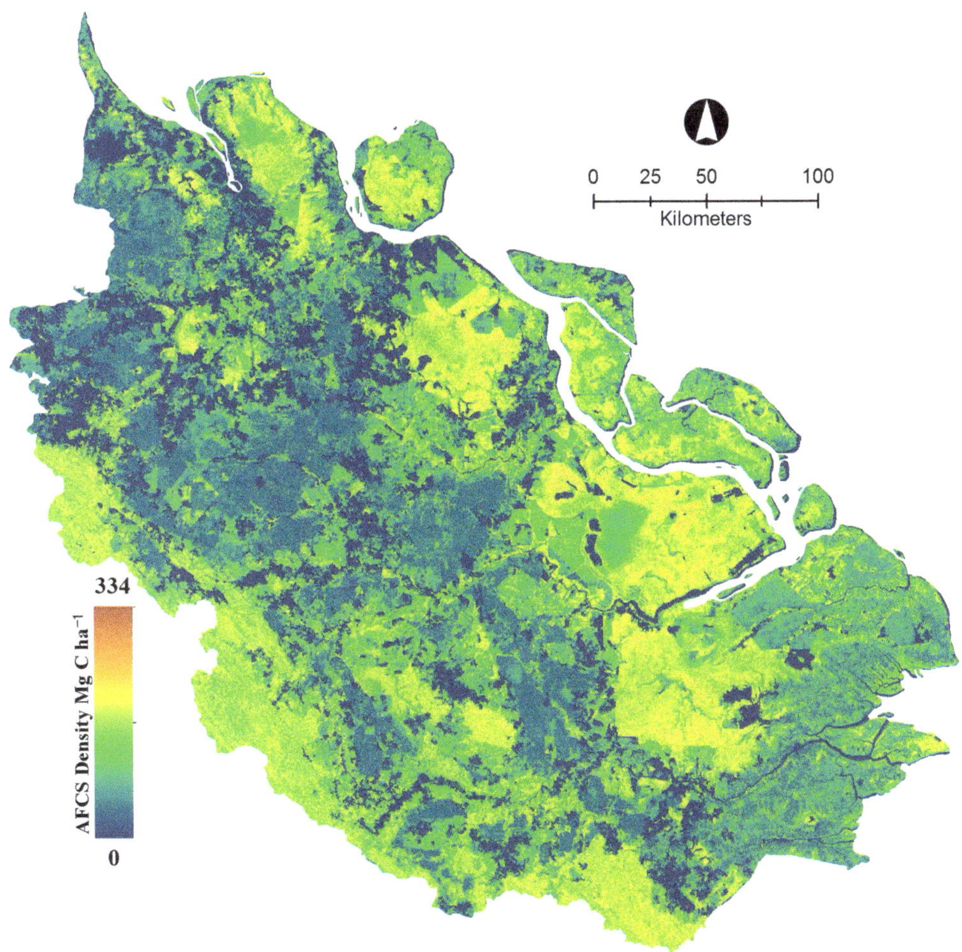


**Fig. 3** AFCS map, calculated using PALSAR mosaic data

**Table 2 Quantity of AFCSs extracted for natural forest areas and their distribution by district**

| Districts | Forest area in ha | AFCSs | | |
|---|---|---|---|---|
| | | In million Mg C | % distribution | Density in Mg C ha$^{-1}$ |
| Bengkalis | 370,636 | 26.74 | 10.07 | 72.15 |
| Indragiri Hilir | 288,914 | 22.53 | 8.49 | 77.99 |
| Indragiri Hulu | 443,704 | 33.31 | 12.54 | 75.06 |
| Kampar | 433,434 | 30.50 | 11.48 | 70.37 |
| Dumai | 101,173 | 7.22 | 2.72 | 71.36 |
| Pekanbaru | 7526 | 0.33 | 0.12 | 43.63 |
| Singingi | 307,621 | 20.93 | 7.88 | 68.04 |
| Pelalawan | 679,426 | 48.79 | 18.37 | 71.81 |
| Rokan Hilir | 276,571 | 18.57 | 6.99 | 67.13 |
| Rokan Hulu | 217,973 | 15.64 | 5.89 | 71.74 |
| Siak | 370,296 | 26.63 | 10.03 | 71.92 |
| Meranti | 191,670 | 14.39 | 5.42 | 75.07 |
| Total | 3,688,944 | 265.57 | 100.00 | 71.99 |

Natural forest area is calculated using the Thapa et al. [16] map for 2010

and Siak districts contain a similar amount of AFCS, each with around 10 % of the total carbon stocks, but the AFCS in the Bengkalis districts is slightly higher. Additionally, although the total AFCS of the Dumai district is lower, its density is higher than that in the Singingi district. The Indragiri Hulu district has the second largest AFCS overall, at 33.31 million Mg, and has a density similar to that of the Meranti district. Finally, the Kampar, Rokan Hulu, and Rokan Hilir districts within the northwestern part of the province contain considerable forest carbon stocks, but their densities are slightly lower than the provincial average.

We overlaid the AFCS baseline map derived in this study with the observed and simulated scenario-wide forest maps of Thapa et al. [16]. Figure 4 contains information on the distributions of conservation areas, concession areas, as well as illustrating the AFCS map for natural forest cover in 2010, and scenario-wide simulated AFCS maps for the years 2015, 2020, 2025, and 2030. The changes in spatial patterns of AFCS are clear when comparing the two policy scenarios, the Government-Forest Conservation (G-FC) and Government-Concession for Plantation and Logging (G-CPL) policies, with the Business as Usual (BAU) approach. If the past deforestation processes continue without any policy implementation, as shown by the BAU policy scenario, then the AFCS will be persistently released from most of the forested areas. The AFCS removal will likely occur in the most ecologically delicate areas, including the peat swamps and conservation areas in the northeast, and the dry-forested areas in the southwest of the study area. This indicates that the current land use change-inducing activities pose an extremely serious threat to AFCS, and immediate measures are required to ensure sustainability and forest protection. The spatial trends in AFCS observed under the G-CF policy scenario are somewhat comparable to those of the BAU scenario, with the exception of the forest conservation zones. The AFCS hosting areas remain fairly large under the G-CF scenario, due to the impact of the policy on forest protection. For instance, the forests in the designated conservation areas of 2010 remain intact for the forthcoming decades, retaining their AFCS. However, deforestation pressures will still affect the regions outside the conservation areas, rapidly releasing AFCS from the districts in the northern part of the study area through 2015 and 2020. Under the G-CPL scenario, the geographic distribution of AFCS across the province was better retained compared with that of the other scenarios. AFCS removal will likely occur only in the concession lands for plantations and selective logging. If the G-CPL policy is implemented without modification, then entire districts will retain a considerable amounts of their AFCS, even by the end of 2030.

Figure 5 illustrates quantifications of the predicted AFCS changes under the three policy scenarios by district for every 5 years until 2030. These charts represent important information by showing the spatial variations in future forest carbon reserves. The AFCS curves for each of the different scenarios show the dynamic variability of carbon stocks over time. In all districts and at all times, the BAU line resides at the bottom of the chart, indicating the greatest loss to AFCS. Thus, if the previous tendencies continue, the AFCS in the Pekanbaru

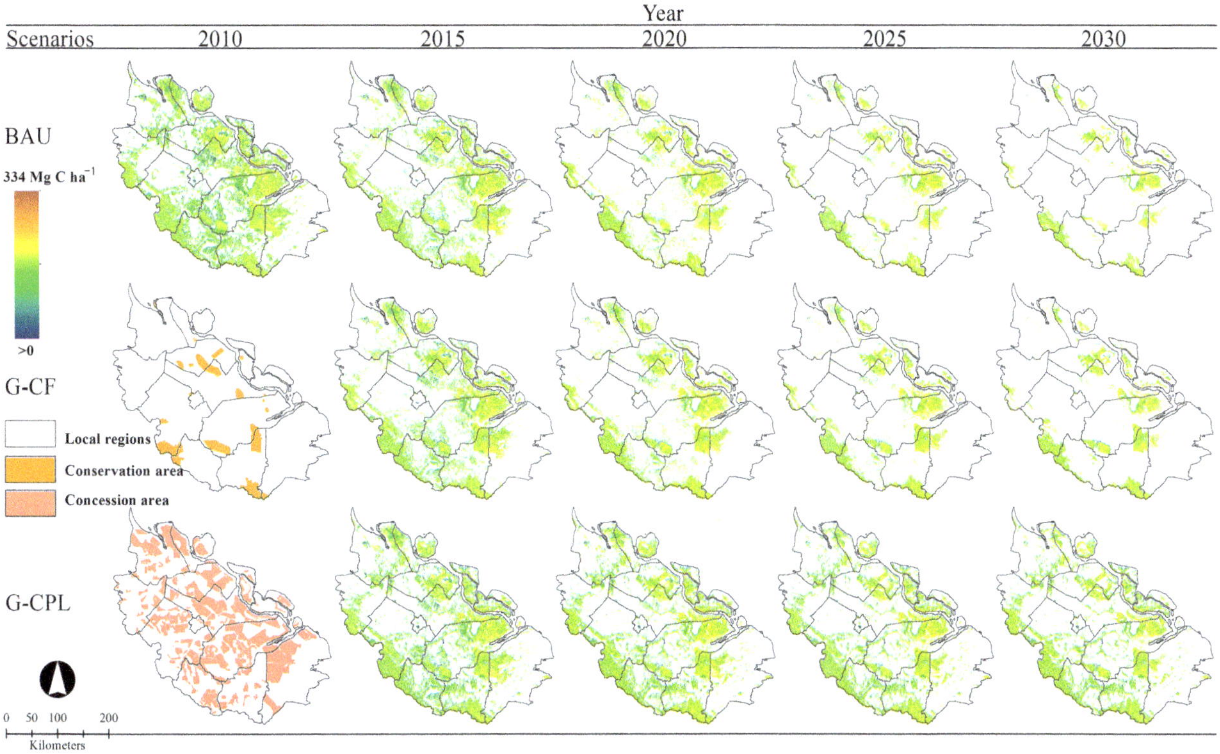


**Fig. 4** AFCS patterns from 2010 to 2030 under the three forest policy scenarios

district will be almost entirely lost by 2030. In addition, the Dumai, Meranti, Rokan Hulu, and Rokan Hilir districts are likely to face severe damage to their AFCS balances over the next two decades. These districts will also face similar problems under the conservation scenario. The AFCS emissions in the Pelalawan district under the BAU expected to be high, due to the large number of reserves in this region compared with the other districts. The most rapid declines in AFCS up to 2015 under BAU are expected in the Siak, Indragiri Hilir, and Rokan Hilir districts. Over the study period, the changes in projected AFCS between the G-CF and BAU scenarios gradually widen in the Kampar, Pelalawan, Siak, and Indragiri Hulu districts. This indicates that the G-CF scenario somehow retains the stability of forest carbon reserves to a greater degree than the BAU scenario. In comparison, the G-CPL policy scenario estimates the retention of relatively high AFCS in the Meranti, Singini, Indragiri Hulu, and Kampar districts, even by 2030. Remarkably, the estimated AFCS emissions in Indragiri Hulu, Kampar, Rokan Hulu, and Singingi districts under the G-CPL policy scenario are very low over the study period. In contrast, rapidly declining levels of AFCS are detected in the Siak, Dumai, and Indragiri Hilir districts under the G-CPL policy scenario. The remaining districts, including Indragiri Hulu, Rokan Hulu, Meranti, and Singingi, will contain higher AFCS by 2030 in the G-CPL scenario than under the other scenarios.

Figure 6 presents the expected forest carbon emissions from the province as a whole under each scenario in five-year intervals up to 2030. The effects of the different scenarios on the projected carbon emission differ with the passing time. If the existing trend continues, as evidenced by the BAU, around 747.61 million Mg $CO_2$, representing 75 % of the current AFCS, will likely be in the atmosphere by 2030. The trend indicates that the emissions will be greater in earlier years, meaning that two thirds of the forest carbon will be emitted into the air over the next 10 years, potentially resulting in globally adverse environmental consequences. In comparison, the trend of the G-CF scenario suggests some measure of success in the form of reduction in emissions of 20 million Mg $CO_2$ by 2015, although these double by 2020, and reach approximately 84.3 million Mg $CO_2$, a reduction of 11.27 % as compared to the BAU in 2030. In contrast, the emissions under the G-CPL scenario appear to differ remarkably from those under the G-CF and the BAU scenarios. The implementation of this policy will gradually slow the emission of forest carbon stocks by controlling the deforestation to a greater degree than the other scenarios. Thus, under the G-CPL scenario, the estimated carbon emissions will be around 305 million Mg $CO_2$ in

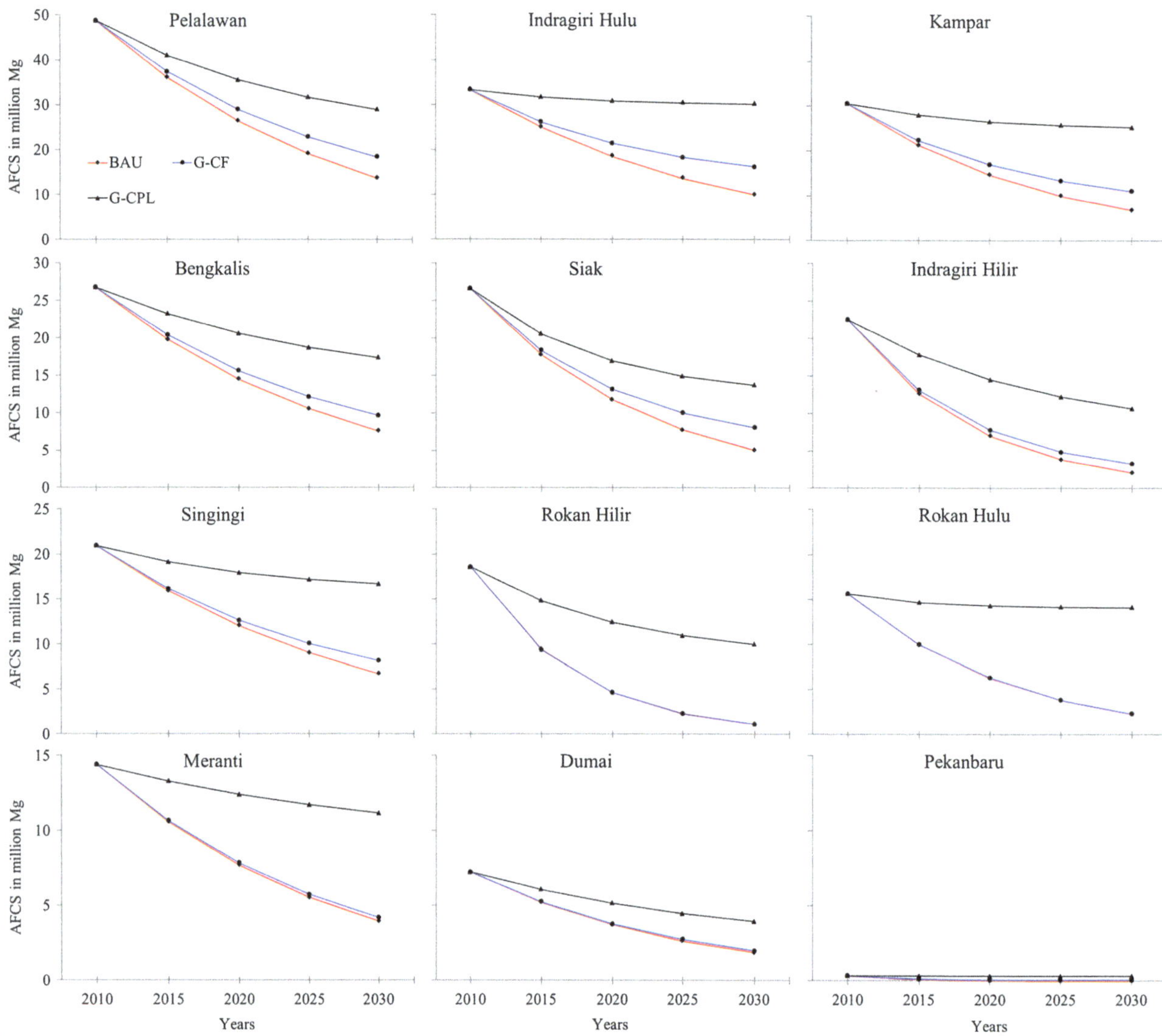


**Fig. 5** Scenario-specific expected AFCS in each district at 5-year intervals from 2010 to 2030

2030, which is only 31 % of the current AFCS; this represents a reduction of 2.5 times compared with the BAU scenario. It is worth noting that the G-CPL scenario is likely to delay the carbon emissions by a further 15 years, whereas a similar amount is expected to be released by 2015 under the present conditions. If BAU is considered as a reference scenario, then the concession policy (G-CPL) scenario is likely to reduce the $CO_2$ emissions by 59.20 % (442.6 million Mg $CO_2$) through its expected deforestation by the end of 2030.

Despite the high spatial resolution of our AFCS map, which reveals the AFCS dynamics of natural forests on a local scale, information about the dynamics of belowground carbon stocks (BGCS) in natural forest areas is lacking in this study. Estimation of BGCS in natural forest regions using remote sensing is extremely difficult. However, the general model [32, 33] for estimating the BGCS in tropical regions can be employed using the baseline map derived in this study if necessary. In addition to affecting the BGCS, the deforestation process also contributes to the release of carbon from other sources such as soils and peat lands. These additional sources may significantly increase the net carbon emissions, and the inclusion of forest fire parameters in the modeling process may improve the accuracy of future estimation. Furthermore, the majority of the natural forest land in this area has already undergone transformations to produce economically valuable industrial plantations, such as oil palm, acacia, coconut, and rubber trees. These plantation forests also store a significant amount of

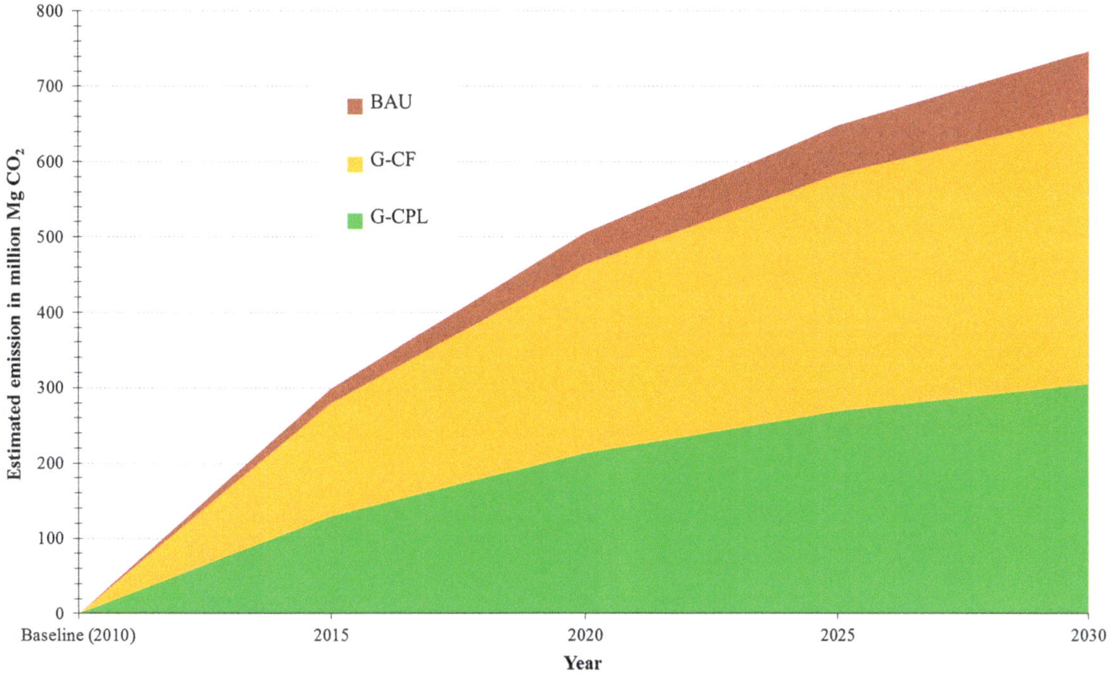


**Fig. 6** Scenario-specific estimated $CO_2$ emissions as a result of deforestation across the province from 2010 to 2030

aboveground forest carbon, as reflected in baseline map (Fig. 3) and in Table 1. From a carbon stock perspective, a consideration of the AFCS dynamics in these forests may represent a trade off in overall carbon balance to some extent. However, there is an immediate risk of carbon emission when the natural forests are cleared, even when they are replaced by plantation forests. Additionally, these plantation forests store carbon only in the short term, as they are harvested within a certain period of time. For example, an acacia plantation in the study area is harvested every 4 years. The other plantations, such as oil palm, coconut, and rubber are often harvested every 20–30 years. The plantation forests maintain greenness in the province but still have a significant impact on carbon recycling and prevent the restoration of ecosystem services.

## Conclusion

Through the integration of multiple remote sensing techniques, from airborne LiDAR to spaceborne SAR with field measurement data and a rule-based algorithm, an accurate baseline AFCS map with high spatial resolution was developed for one of the major tropical forests in Asia. This baseline map provides highly accurate, spatially explicit distributions and quantitative estimations of forest carbon stocks in the study area. The AFCS distribution varies geographically, indicating spatial variations in forest quality and vulnerability. The spatial modeling technique provides an opportunity to extrapolate the spatial trends in AFCS and examine the implications of different forest management policies on carbon stocks and emissions over the next two decades. The inherent capability of the model to distinguish local variations in future AFCS trends under different scenarios is key to identifying the areas most vulnerable to high carbon emissions, which would require immediate mitigation measures to ensure forest conservation. The model was used to predict the spatiotemporal variations and associated quantities of remaining AFCS under different scenarios up until the year 2030. These predicted spatial patterns of AFCS indicate that the forest carbon emission rate is likely to be high in the coming decades across the province. Ongoing deforestation is expected to release around 747 million Mg $CO_2$ into the atmosphere by 2030. A forest conservation policy will slow the AFCS emissions, but the reduction will be insufficient. Among the scenarios tested, the concession scenario is the most promising, halving the expected emissions if it is implemented as planned. In addition to the high-resolution AFCS map, the modeling outcomes may provide opportunities for the identification of especially vulnerable localities and focuses for the implementation of REDD+ projects to obtain the greatest benefits based on the environmental settings. For the proper execution of the REDD+ project, it is important to understand how the expected trends in AFCS distribution are likely be affected in the long term by the implementation various plans, policies, and strategies. The AFCS under the BAU scenario may provide a reference emission scenario for REDD+, while the other scenarios can be used as examples in the initial exploration of the range of potential spatiotemporal issues and outcomes. These can provide important insights for preparedness activities that mitigate the problem of forest carbon emissions. The spatially explicit AFCS map and the modeled scenario results will therefore contribute to the sustainable management of forests in the study area and to the formulation of REDD+ projects, as well as representing a methodological reference for wider audiences in tropical regions and beyond.

## Methods

### L-band SAR data collection and processing

The study area spans more than 9 million ha commonly experiences cloudy and hazy skies throughout the year [20]. This degradation in atmospheric conditions over the study area precludes use of optical remote sensing techniques in assessing forest quality and AFCS. As a result, PALSAR mosaic data used for tropical forest monitoring as they are unaffected by atmospheric condition and are available for wall-to-wall mapping [1, 20]. The mosaic data are slope-corrected and orthorectified using the widely available SRTM 90 m digital elevation model without any alteration the image quality [21]. Currently, 25 m global mosaic data in two polarizations (HH and HV) are available as one set per year, from 2007 to 2010. The mosaic data are available at a downloadable size of 1-degree tiles, equivalent to approximately 111 × 111 km [34]. Those mosaic products covering the whole province of Riau for the years 2009 and 2010 were used. The mosaic data were converted into radar backscatter coefficients using gamma-naught ($\gamma^{\circ}$) [21] due to high sensitivity to forest structure and its usefulness in forest cover analysis [20, 35].

To improve the confidence of the AFCS mapping, we also examined whether a particular filter or its size would affect the mapping results. Three filters were examined: the Lee, Frost, and median filters. These filters possess different formulations and assumptions for smooth speckled data in radar imagery. The Lee filter is a standard deviation-based filter, and filters data on the basis of statistics calculated within individual filtering windows. It conserves image sharpness and details while reducing speckle noise. The value of the pixel being filtered is replaced by a value computed using the neighboring pixels. In comparison, the Frost filter is an exponentially damped circularly symmetric filter, which utilizes local statistics. The value of the pixel being filtered is replaced by a value computed with a consideration of the damping factor, the local variance, and the distance from the filter center. This filter is able to preserve the edges in the images. The median filter reduces the speckle noise in an image by conserving edges greater than the kernel dimensions. It replaces the value of each center pixel with the median value of the neighborhood specified by the filter size. In order to investigate the impact of speckle filtering sizes on the mapping, we also evaluated five different filtering window sizes (3 × 3, 5 × 5, 7 × 7, 9 × 9, and 11 × 11).

### Field data collection and processing

The combination of LiDAR data and plot-based field measurements has emerged as a promising technique for accurately estimating AFCS [13, 17]. We conducted field measurements and airborne LiDAR surveys within the province during 2012 and 2013. Owing to the differences in forest structure and associated biomass in different land use and land cover (LULC) types, we adopted a stratified sampling approach based on the major forest types to determine the locations (Fig. 1) for field and LiDAR measurements. The major forest types in the study site were defined as natural forests, including peat swamps, dry moist, mangrove, and regrowth, and plantations including acacia, oil palm, rubber, and coconut. Based on these forest types, eight strata were created.

Across the field measurement campaigns, we made 87 biomass measurements within 1 ha-size plots that coincided with the LiDAR acquisition sites. Forest stands of all ages were inventoried, from mature to recent regrowth. Owing to the time and cost involved in conducting a census-based measurement of all trees in a 1-ha-sized plot, a sub-sampling approach was adopted using representative subplots. The sub-sampling methods differed between the natural and plantation forests.

To determine woody biomass, all living and standing deadwood trees with a diameter at breast height (at 1.3 m; DBH) ≥5 cm were measured in each subplot. We used allometric equations previously developed for the specific forest types: peat swamp forest [36], dry moist forest [37], mangrove [38], acacia [39], rubber [40], coconut and non-trees [41], oil palm [42], standing deadwood [43], lying deadwood [44], and bamboo [45]. The biomass of understory vegetation and litter was calculated by multiplying the mass of a fresh sample measured in the field by the ratio of sub-sample dry mass to sub-sample fresh mass. The plots within regrowth forests were all located in either peat swamps or dry moist forested areas. Therefore, the biomass for this class was calculated using the corresponding allometric equation, based on its location. Detailed descriptions of the field measurement method, plot specification, and allometric equations used to convert the field measured data to plot level aboveground biomass (AGB) are presented in Thapa et al. [13, 24]. In this study, the AFCS for each plot was considered to be 47 % of the field measured AGB [46].

Airborne LiDAR surveys were conducted at eight sites. LiDAR data were acquired on fine weather days in February 2012 and during November–December 2012. In February, we used an LM-5600 laser system operated at approximately 1000 m above the ground. This system captured first and last returns at a scan angle of ±20°. The average point density was 1.2 $m^{-2}$. During this measurement period, three of the eight sites were surveyed, covering 3600 ha of forested land. The remaining LiDAR data were collected using an Optech ALTM 3100EA laser system in the second survey period. This system was operated approximately 600 m above the ground, and captured full waveform data at a scan angle of ±32°. Discrete return data were recorded at average point density of 3.6 $m^{-2}$. At this time, 4472 ha of forested lands were surveyed, contributing to a total 8072 ha of LiDAR data acquired during the two periods combined. Further details of the LiDAR systems, data acquisition and processing, and the LiDAR-to-AFCS model calibration and validation are discussed in Thapa et al. [13]. In the present study, the LiDAR allometric model (Eq. 1) was used to create additional AFCS plots as the AFCS model was calibrated and validated over the same area [13].

$$\begin{aligned}\text{AFCS Mg C ha}^{-1} = {} & 259.488 - (146.373 \times \text{MCH}) \\ & + \left(4.738 \times \text{MCH}^2\right) \\ & - (4.881 \times \text{Cover}) \\ & + (3.513 \times \text{MCH\_cover}) \\ & - \left(0.0954 \times \text{MCH}^2\text{\_cover}\right) \\ & - (1.583 \times \text{QMCH\_cover}) \\ & + (22.568 \times \text{P50}) + (26.118 \times P90)\end{aligned} \tag{1}$$

where, MCH = mean canopy height, Cover = forest cover as a percentage of all returns above the MCH, MCH_cover = MCH × Cover, $\text{MCH}^2$_cover = $\text{MCH}^2$ × Cover, QMCH_cover = quadratic MCH × Cover, and P50 and P90 are the 50th and 90th percentiles of canopy height, respectively.

A total of 2716 1-ha LiDAR-based AFCS plots were created, avoiding an excess of path boundaries, field measurement plots, agricultural fields, clear cut areas, water areas, and built structures including buildings and roads. These LiDAR AFCS plots were combined with the 87 field measurement plots, resulting in a total of 2803 plots available for calibration and validation of the SAR-based AFCS baseline map.

**Mapping and validating the AFCS**

A supervised rule-based approach was adopted for the AFCS mapping, as the direct relationship between the PALSAR backscattered coefficients and field-measured AFCS in the study region is affected by saturation in higher biomass areas [24]. This approach analyzes the spatial patterns in the PALSAR mosaic data through the use of training samples and the rules formulated in a maximum likelihood algorithm (MLA) [47], as in Eq. 2, to determine the AFCS value for each PALSAR pixel. Half of the total AFCS plots were used to train the algorithm. The MLA quantitatively examines both the variance and the covariance of the patterns of corresponding backscatter in the training samples and provides the most probable AFCS values for the selected pixels. At first, the MLA was applied to the 2010 gamma naught image, which includes HH and HV polarizations. The different impacts of the three filters (Lee, Frost, and median) and five increasing window sizes on the mapping results were analyzed. Then, we investigated whether the addition of the gamma naught image of the previous year, in the form of mosaic data from 2009, would improve the mapping result. Layer stacking was performed while adding the 2009 image layers (HH and HV) to the 2010 image data set.

$$g_i(x) = \ln p(\omega_i) - \frac{1}{2}\ln\left|\sum_i\right| - \frac{1}{2}(x - m_i)^t \sum_i^{-1} (x - m_i) \tag{2}$$

where $g_i$ is the AFCS Mg C $ha^{-1}$ corresponding to the training sample, in which $i = 1 \ldots C$ (the number of AFCS training plots available); $x$ is the position, in n-dimensional data where n is the number of image data layers (for example, 2 per year); $p(\omega_i)$ is the probability that the AFCS value $\omega_i$ occurs in the image; $|\Sigma_i|$ is the covariance matrix of the image data intersected with the spatial size of AFCS plot $\omega_i$; $t$ is the transposition of the matrix; $\Sigma_i^{-1}$ is the inverse matrix; and $m_i$ is the mean backscatter vector corresponding to AFCS plot $\omega_i$.

A validation map was prepared using the remaining half of the sample plots. The AFCS map was compared with the validation map through the calculation of three statistical mapping uncertainty measures: root mean square error (RMSE, Eq. 3), bias (Eq. 4), and index of agreement (D, Eq. 5).

$$\text{RMSE (Mg C ha}^{-1}) = \sqrt{\frac{\sum_{i=1}^{n}(P_i - O_i)^2}{n}} \tag{3}$$

$$\text{Bias (Mg C ha}^{-1}) = \frac{\sum_{i=1}^{n}(P_i - O_i)}{n} \tag{4}$$

$$D = 1 - \frac{\sum_{i=1}^{n}(P_i - O_i)^2}{\sum_{i=1}^{n}(|P_i - \bar{O}| + |O_i - \bar{O}|)^2} \quad (5)$$

where $P_i$ represents predicted values, $O_i$ represents observed values, $\bar{O}$ is the observed mean, and n is the number of observations. D is the index of agreement proposed by Willmott and Wicks [48]. D ranges from 0 to 1, which corresponds to disagreement or perfect agreement between the predicted and observed values, respectively.

Bias-adjusted RMSE (Eq. 6) was calculated for the final map to ensure the high accuracy of the AFCS estimation in the map.

$$\text{Adj. RMSE (Mg C ha}^{-1}) = \sqrt{\frac{\sum_{i=1}^{n}(P_i - O_i - Bias)^2}{n}} \quad (6)$$

### Mapping of future expected AFCS footprints

In this study, we used the forest cover map derived from PALSAR mosaic data from 2010, in combination with the scenario maps for 2015, 2020, 2025, and 2030 from Thapa et al. [16] to generate the expected AFCS footprints for the various years in the future. Three policy scenarios were analyzed: BAU, corresponding to the 'business as usual policy', G-FC indicating the 'government-forest conservation policy', and G-CPL, representing the 'government-concession for plantations and logging policy'. The BAU policy scenario assumes that the deforestation process will continue with the same past trend everywhere in the province, and therefore, AFCS removal will occur in the corresponding deforested areas. The G-FC policy scenario assumes that the deforestation process does not follow the past trend and, in the future, will likely occur outside the designated forest conservation areas. In this case, the forest carbon stocks remain untouched within the conservation areas. For the G-CPL policy scenario, we assume that the concession areas are allotted for selective logging and industrial plantations, and imminent deforestation likely occurs only in the concession areas, therefore the AFCS will be untouched outside these areas. Scenario-wide AFCS maps were created at five-year intervals from 2015 to 2030. Using these maps, the AFCS was quantified at the district level to identify local variation in the carbon stocks. Furthermore, the expected $CO_2$ emissions for each scenario were computed at province level for each time interval using the approach of IPCC [46].

#### Authors' contributions

RBT, TM, MW, and MS conceived the research and analyzed the PALSAR data. RBT, TM and MW collected, processed, and analyzed the field data. RBT created the forest carbon map, performed spatial modeling, and wrote the manuscript. MS arranged necessary funds and coordinated the local collaborator for the research. All authors read and approved the final manuscript.

#### Acknowledgements

The authors would like to thank Prof. I. N. S. Jaya, E. S. Purnama and other colleagues and their students from IPB, Indonesia, Mr. K. Shono and his colleagues from Hatfield Consultants, Indonesia, Mr. Tomohiro Shiraishi from JAXA, and Mr. Takuya Itoh from RESTEC for their assistance with field data collection and for sharing their thoughts in this area of research.

#### Compliance with ethical guidelines

#### Competing interests

The authors declare that they have no competing interests.

#### References

1. Shimada M, Itoh T, Motohka T, Watanabe M, Shiraishi T, Thapa R, Lucas R. New global forest/non-forest maps from ALOS PALSAR data (2007–2010). Remote Sens Environ. 2014;155:13–31.
2. FAO. Forest resource assessment 2010. Rome: Food and Agricultural Organization of the United Nations; 2012.
3. Aguilar-Amuchastegui N, Riveros JC, Forrest JL. Identifying areas of deforestation risk for REDD+ using a species modeling tool. Carbon Bal Manag. 2014;9:10.
4. Canadell JG, Schulze ED. Global potential of biospheric carbon management for climate mitigation. Nat Commun. 2014;5:5282.
5. Grace J, Mitchard E, Gloor E. Perturbations in the carbon budget of the tropics. Glob Change Biol. 2014;20:3238–55.
6. Harris NL, Brown S, Hagen SC, Saatchi SS, Petrova S, Salas W, Hansen MC, Potapov PV, Lotsch A. Baseline map of carbon emissions from deforestation in tropical regions. Science. 2012;336:1573–6.
7. Strassburg B, Turner KR, Fisher B, Schaeffer R, Lovtt A. Reducing emissions from deforestation—The combined incentives mechanism and empirical simulations. Glob Environ Chang. 2009;19:265–78.
8. http://www.un-redd.org. Accessed 29 Sept 2014.
9. Carlson KM, Curran LM, Ratnasari D, Pittman AM, et al. Committed carbon emissions, deforestation, and community land conversion from oil palm plantation expansion in West Kalimantan, Indonesia. Proc Natl Acad Sci USA. 2012;109:7559–64.
10. Cutler M, Boyd D, Foody G, Vetrivel A. Estimating tropical forest biomass with a combination of SAR image texture and Landsat TM data: an assessment of predictions between regions. ISPRS J Photogram. 2012;70:66–77.
11. Leach M, Scoones I. Carbon forestry in West Africa: the politics of models, measures and verification processes. Glob Environ Chang. 2013;23:957–67.
12. Morel AC, Fisher JB, Malhi Y. Evaluating the potential to monitor aboveground biomass in forest and oil palm in Sabah, Malaysia, for 2000–2008 with Landsat ETM+ and ALOS-PALSAR. Int J Remote Sens. 2012;33:3614–39.
13. Thapa RB, Watanabe M, Motohka T, Shiraishi T, Shimada M. Calibration of aboveground forest carbon stock models for major tropical forests in central Sumatra using airborne LiDAR and field measurement data. IEEE J Sel Top Appl. 2015;8:661–73.
14. Asner GP, Knapp DE, Martin RE, Tupayachi R, Anderson CB, Mascaro J, Sinca F, Chadwick KD, Higgins M, Farfan W, Llactayo W, Silman MR. Targeted carbon conservation at national scales with high-resolution monitoring. Proc Natl Acad Sci USA. 2014;111:E5016–22.
15. Lin L, Sills E, Cheshire H. Targeting areas for reducing emissions from deforestation and forest degradation (REDD+) projects in Tanzania. Glob Environ Chang. 2014;24:277–86.
16. Thapa RB, Shimada M, Watanabe M, Motohka T, Shiraishi T. The tropical forest in South East Asia: monitoring and scenario modeling using Synthetic Aperture Radar data. Appl Geogr. 2013;41:168–78.
17. Asner GP, Powell GVN, Mascaro J, Knapp DE, Clark JK, Jacobson J, Kennedy-Bowdoin T, Balaji A, Paez-Acosta G, Victoria E, Secada L, Valqui M, Hughes RF. High-resolution forest carbon stocks and emissions in the Amazon. Proc Natl Acad Sci USA. 2010;107:16738–42.

18. Tsui OW, Coops NC, Wulder MA, Marshall PL. Integrating airborne LiDAR and space-borne radar via multivariate kriging to estimate above-ground biomass. Remote Sens Environ. 2013;139:340–52.
19. Mitchard ETA, Saatchi SS, Baccini A, Asner GP, Goetz SJ, Harris NL, Brown S. Uncertainty in the spatial distribution of tropical forest biomass: a comparison of pan-tropical maps. Carbon Bal Manag. 2013;8:10.
20. Thapa RB, Itoh T, Shimada M, Watanabe M, Motohka T, Shiraishi T. Evaluation of ALOS PALSAR sensitivity for characterizing natural forest cover in wider tropical areas. Remote Sens Environ. 2014;155:32–41.
21. Shimada M, Ohtaki T. Generating large-scale high-quality SAR mosaic datasets: application to PALSAR data for global monitoring. IEEE J Sel Top Appl. 2010;3:637–56.
22. Motohka T, Shimada M, Uryu Y, Setiabudi B. Using time series PALSAR gamma nought mosaics for automatic detection of tropical deforestation: a test study in Riau, Indonesia. Remote Sens Environ. 2014;155:79–88.
23. Saatchi S, Marlier M, Chazdon RL, Clark DB, Russell AE. Impact of spatial variability of tropical forest structure on radar estimation of aboveground biomass. Remote Sens Environ. 2011;115:2836–49.
24. Thapa RB, Watanabe M, Motohka T, Shimada M. Potential of high-resolution ALOS-PALSAR mosaic texture for aboveground forest carbon tracking in tropical region. Remote Sens Environ. 2015;160:122–33.
25. Dobson MC, Ulaby FT, Le Toan T, Beaudoin A, Kasischke ES, Christensen NC. Dependence of radar backscatter on conifer forest biomass. IEEE Trans Geosci Remote. 1992;30:412–5.
26. Lucas R, Armston J, Fairfax R, Fensham R, Accad A, Carreiras J, Kelley J, Bunting P, Clewley D, Bray S, Metcalfe D, Dwyer J, Bowen M, Eyre T, Laidlaw M, Shimada M. An evaluation of the ALOS PALSAR L-band backscatter—above ground biomass relationship Queensland, Australia: impact of surface moisture condition and vegetation structure. IEEE J Sel Top Appl. 2010;3:576–93.
27. Englhart S, Keuck V, Siegert F. Aboveground biomass retrieval in tropical forests—the potential of combined X- and L-band SAR data use. Remote Sens Environ. 2011;115:1260–71.
28. Luckman AJ, Baker JR, Honzák MH, Lucas RM. Tropical forest biomass density estimation using JERS-1 SAR: seasonal variation, confidence limits and application to image mosaics. Remote Sens Environ. 1998;62:126–39.
29. Watanabe M, Motohka T, Shiraishi T, Thapa RB, Kawano N, Shimada M. Dependency of forest biomass on full polarimetric parameters obtained from L-band SAR data for a natural forest in Indonesia. In: Proc Intl Geosci Remote Sens Symp 2013, 3919–3922.
30. MofF. IFCA consolidation report: reducing emissions from deforestation and forest degradation in Indonesia. Forestry Research and Development Agency, Jakarta, Indonesia; 2008.
31. BPS. Hasil Sensus Pnduduk 2010: data Agregat per Provinsi (in Indonesian language); 2010. Web: http://dds.bps.go.id/eng/download_file/SP2010_agregat_data_perProvinsi.pdf. Accessed 2012.3.7.
32. Cairns MA, Brown S, Helmer EH, Baumgardner GA. Root biomass allocation in the world's upland forests. Oecologia. 1997;111:1–11.
33. Pearson T, Walker S, Brown S. Sourcebook for land use, land-use change and forestry projects. Little Rock: Biocarbon Fund, Winrock International; 2005.
34. http://www.eorc.jaxa.jp/ALOS/en/dataset/dataset_index.htm. Accessed in 2015.6.30.
35. Shiraishi T, Motohka T, Thapa RB, Watanabe M, Shimada M. Comparative assessment of supervised classifiers for land use land cover classification in a tropical region using time-series PALSAR mosaic data. IEEE J Sel Top Appl. 2014;7:1186–99.
36. Murdiyarso D, Rosalina U, Hairiah K, Muslihat L, Suryadiputra INN, Jaya A. Petunjuk Lapangan: Pendugaan Cadangan Karbon pada Lahan Gambut. Proyek Climate Change, Forests and Peatlands in Indonesia. Wetlands International, Indonesia Programme and Wildlife Habitat Canada. Bogor, Indonesia; 2004.
37. Brown S. Estimating biomass and biomass change of tropical forests: a primer. UN FAO Forestry Paper, no. 134; 1997. p 55.
38. Komiyama A, Poungparn S, Kato S. Common allometric equations for estimating the tree weight of mangroves. J Trop Ecol. 2005;21:471–7.
39. Adiriono T. Measurement of carbon stock with carbonization method in forest plantation of acacia crassicarpa: a case study in PT Sebangun Bumi Andalas Wood Based Industries. MSc Thesis, Faculty of Forestry, Gadjah Mada University; 2009.
40. Schroth G, Angelo SAD, Teixeira WG, Haag D, Lieberei R. Conversion of secondary forest to agroforestry and monoculture plantations in Amazonia: consequences for biomass, litter and soil carbon stock after 7 years. For Ecol Manag. 2002;163:131–50.
41. Frangi JL, Lugo AE. Ecosystem dynamics of a subtropical floodplain forest. Ecol Monogr. 1985;55:351–69.
42. Yulianti N. Carbon stock of peatland in oil palm agroecosystem of PTPN IV Ajamu, Labuhan Batu, North Sumatra. MSc Thesis, Faculty of Agriculture, Bogor Agricultural University (IPB), Bogor; 2009.
43. Walker SM, Pearson TRH, Casarim FM, Harris N, Petrova S, Grais A, Swails E, Netzer M, Goslee KM, Brown S. Standard operating procedures for terrestrial carbon measurement. Little Rock: Winrock International; 2012.
44. Pearson T, Walker S, Brown S. Sourcebook for land use, land-use change and forestry (LULUCF) projects. Little Rock: BioCarbon Fund and Winrock International; 2005.
45. van Noordwijk M, Rahayu S, Hairiah K, Wulan YC, Farida A, Verbist B. Carbon stock assessment for a forest-to-coffee conversion landscape in Sumber-Jaya (Lampung, Indonesia): from allometric equations to land use change analysis. Sci China (Ser C). 2002;10:75–86.
46. IPCC. Agriculture, forestry and other land use. Guidelines for National Greenhouse Gas Inventories, 4; 2006. http://www.ipcc-nggip.iges.or.jp/public/2006gl/vol4.html. Accessed 2 Feb 2012.
47. Richards JA, Jia X. Remote sensing digital image analysis. Berlin: Springer; 2006.
48. Willmott CJ, Wicks DE. An empirical method for the spatial interpolation of monthly precipitation within California. Phys Geogr. 1980;1:59–73.

# 3

# Climate change mitigation effect of harvested wood products in regions of Japan

Chihiro Kayo[1*], Yuko Tsunetsugu[2] and Mario Tonosaki[3]

**Abstract**

**Background:** Harvested wood products (HWPs) mitigate climate change through carbon storage, material substitution, and energy substitution. We construct a model to assess the overall climate change mitigation effect (comprising the carbon storage, material substitution, and energy substitution effects) resulting from HWPs in regions of Japan. The model allows for projections to 2050 based on future scenarios relating to the domestic forestry industry, HWP use, and energy use.

**Results:** Using the production approach, a nationwide maximum figure of 2.9 MtC year$^{-1}$ for the HWP carbon storage effect is determined for 2030. The maximum nationwide material substitution effect is 2.9 MtC year$^{-1}$ in 2050. For the energy substitution effect, a nationwide maximum projection of 4.3 MtC year$^{-1}$ in 2050 is established, with at least 50 % of this figure derived from east and west Japan, where a large volume of logging residue is generated. For the overall climate change mitigation effect, a nationwide maximum projection of 8.4 MtC year$^{-1}$ in 2050 is established, equivalent to 2.4 % of Japan's current carbon dioxide emissions.

**Conclusions:** When domestic roundwood production and HWP usage is promoted, an overall climate change mitigation effect is consistently expected to be attributable to HWPs until 2050. A significant factor in obtaining the material substitution effect will be substituting non-wooden buildings with wooden ones. The policy of promoting the use of logging residue will have a significant impact on the energy substitution effect. An important future study is an integrated investigation of the climate change mitigation effect for both HWPs and forests.

**Keywords:** Harvested wood products (HWPs), Carbon storage effect, Material substitution effect, Energy substitution effect, Inter-regional flow, Production approach

## Background

The Fifth Assessment Report of the Intergovernmental Panel on Climate Change (IPCC) has indicated that forests and harvested wood products (HWPs) contribute heavily to global carbon cycles [1]. HWPs mitigate climate change via a carbon storage effect, a material substitution effect (a reduction in the consumption of fossil fuels in material production, transportation, etc., as a result of the substitution of other materials with HWPs), and an energy substitution effect (the substitution of fossil fuels as a result of the energy use of HWPs) [2].

*Correspondence: kayoc@cc.tuat.ac.jp
[1] Department of Environment Conservation, Graduate School of Agriculture, Tokyo University of Agriculture and Technology, 3-5-8 Saiwai-cho, Fuchu, Tokyo 183-8509, Japan
Full list of author information is available at the end of the article

IPCC and the United Nations Framework Convention on Climate Change (UNFCCC) have been discussing various methods of calculating the carbon balance of HWPs, including the IPCC default approach, the stock change approach, the atmospheric flow approach, the stock change approach domestic use, and the production approach [3–5]. As a result of these discussions, since the second commitment period of the Kyoto Protocol (i.e., since 2013), the change in the carbon storage volume of HWPs has been included in the calculation of each country's greenhouse gas (GHG) emissions and sinks. A modified production approach has been adopted to calculate GHG emissions and sinks, which solely includes the change in the carbon storage volume of domestically produced wood from a country's forests (including wood exported to other countries), as stipulated in Article 3, Paragraphs 3 and 4, of the Kyoto Protocol [6, 7]. As a result of such international involvement, research into carbon balancing pertaining to HWPs and consideration of relevant climate change mitigation strategies are increasingly growing in significance.

Prior research has estimated and evaluated the carbon balancing of HWPs globally [8], in EU countries [9], and in specific countries such as the United States [10], Canada [11, 12], Portugal [13], and Slovekia [14]. Other studies have conducted an integrated carbon balance assessment of forests and HWPs worldwide [15, 16] and in the United States [17, 18], Canada [19–21], Germany [22], France [23], Finland [24], Switzerland [25], and China [26].

Looking specifically at Japan, Tsunetsugu and Tonosaki [27], and Hashimoto and Moriguchi [28] have estimated the climate change mitigation effect, and carbon balance relating to HWPs for the entire country from a macro viewpoint. Meanwhile, the authors of this study have conducted research [29] (hereinafter "previous report") focusing on carbon flow as HWPs circulate between the regions of Japan and on how carbon storage is distributed among the regions. In the previous report, we divided the country into east, central, and west Japan, as shown in Fig. 1, and constructed a carbon balance estimation model (hereinafter "previous model") that takes into account the flow of HWPs between these regions. In addition, we estimated the carbon balance in each region until 2050 based on future scenarios. Clarification of the carbon balance in a region and its potential future changes allow policymakers to engage in a more practical discussion of measures for climate change mitigation using HWPs that take into account the characteristics of each region and related problems (such as carbon emissions arising from HWP transportation between regions and distribution of carbon storage volume in each region) rather than solely pursuing a macro view of the issue in Japan.

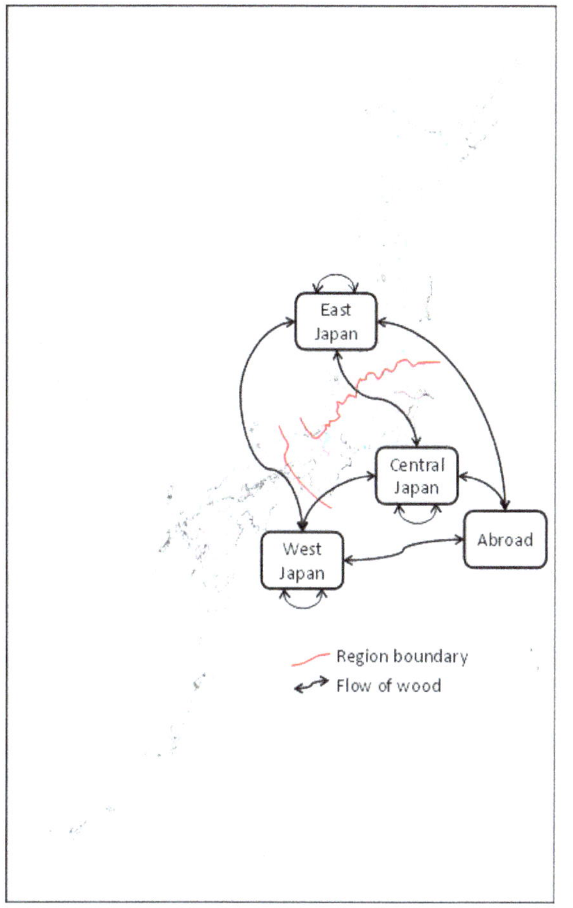


**Fig. 1** Boundaries of regions and flows of wood

### Objectives

There were two significant problems in our previous report [29]. First, despite the importance of the carbon storage volume of domestically produced HWPs from a country's forests (namely the production approach) to the UNFCCC, our previous model could not indicate the detail of HWPs produced from domestic forests, and therefore an adequate assessment of the carbon storage volume and future changes relating to HWPs derived from domestic production in each region was not possible. Second, for the climate change mitigation effect, only the carbon storage effect and energy substitution effect were targeted, and it was not feasible to address the overall mitigation effect that takes into account the material substitution effect.

In this study, therefore, the first area of concern has been addressed through a number of future scenarios relating to HWPs derived from domestic forests, which consider the future of the domestic forestry industry and of HWP use from various viewpoints. Furthermore, carbon storage volume and its future changes are estimated using the production approach with country-specific methods [7]. The second point of concern has been addressed through the evaluation of the overall climate change mitigation effect, which takes into account the material substitution effect. The initial projection year was also changed from 2005 in the previous model to 2014, to include the most recent figures available. As a result, it was possible to reflect in the projections the impact on HWP supply and demand of such things as the Lehman Brothers collapse in 2008 and the Great East Japan Earthquake of 2011. Further considerations included the civil engineering field as an application for HWPs and the use of logging residue for energy generation. As a result, the problems with the previous model were resolved, giving rise to an improved comprehensive model (hereinafter "new model"). Based on the new model, the HWP carbon balance in each region of Japan was evaluated, and the overall climate change mitigation effect to 2050 was estimated, in line with multiple future scenarios. For future studies, this new model plays the part of an integrated carbon balance model for the Japanese forest sector combining forests and HWPs [30].

### Structure of the model

The new model for evaluating the HWP carbon balance used in this study is outlined in Fig. 2. Its basic structure follows that of the previous model. The structure formulates (1) HWP flow, from roundwood production to sawnwood (including glued laminated lumber and laminated veneer lumber), plywood (including wood-based panels such as particle board and fiberboard) and chip production and HWP consumption (building construction, civil engineering, furniture, paper) and to energy use (logging residue, processing residue, waste wood); and (2) HWP stock (building construction, civil engineering, furniture, paper). Moreover, the model evaluates the climate change mitigation effect based on the carbon balance implied by this HWP flow and stock.

Because the volume of Japanese HWP imports is very large (71 % of the volume of the overall HWP supply in 2013) [31], the model targets not only domestically produced HWPs but also imported HWPs. However, when assessing carbon balance, the carbon storage in HWP derived from only domestic forests (excluding imports) and its future changes, are estimated using the production approach. The volume of Japanese HWP exports has remained below 3 % of the volume of overall HWP demand for more than 50 years [31], and the export volume for each application has been unclear. According to the 2013 Revised Supplementary Methods and Good Practice Guidance Arising from the Kyoto Protocol of the IPCC [7], the annual change in carbon storage in HWPs is assumed to be zero (instantaneous oxidation) when transparent and verifiable activity data are not available. Therefore, HWP exports were not taken into consideration.

The "Methods" section details the new model, highlighting model improvements.

### Future scenarios

In the previous model, there were two future scenarios focusing only on HWP consumption volume. However, in the new model, several scenarios were investigated relating to domestic roundwood production, HWP consumption volume, and the volume of energy use. These combinations are shown in Table 1.

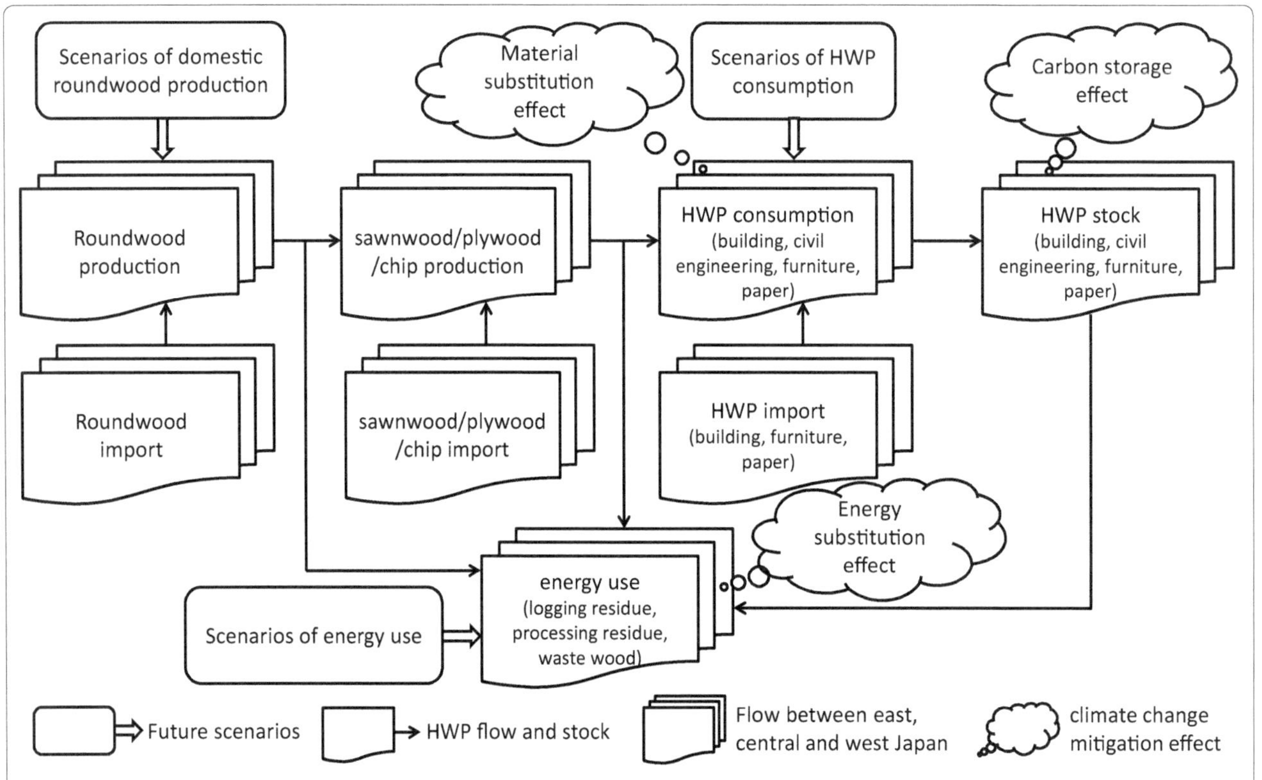


**Fig. 2** Outline of model for evaluating the HWP carbon balance in Japan (new model)

**Table 1 Combinations of future scenarios**

| Scenario combination | Scenario of domestic roundwood production | Scenario of HWP consumption | Scenario of energy use |
|---|---|---|---|
| Con & BAU | Resource conservation (Con) | Business as usual (BAU) | |
| BAU & BAU | Business as usual (BAU) | | |
| Mod & Mod | Moderate increase in harvesting (Mod) | Moderate promotion of usage (Mod) | |
| Plan & Agg | Government plan (Plan) | Aggressive promotion of usage (Agg) | |
| Int & Agg | Forest owners' intentions (Int) | | |

Increasing volume ↓

As regards the domestic roundwood production volume, we explored five future scenarios [in ascending order of volume: resource conservation (Con), business as usual (BAU), moderate increase in harvesting (Mod), in accordance with the Basic Plan for Forest and Forestry of Japanese government (Plan), in accordance with the forest owners' intentions (Int)] created by Forestry and Forest Products Research Institute [30]. In the resource conservation scenario, the volume of roundwood production declines in the future, and forest resources are conserved, with only limited harvesting of wood. In the BAU scenario, even if the future composition of forests increases, the harvesting area and the afforestation area remain unchanged at their actual recent-year level, and the forests continue to age. In the moderate harvesting increase scenario, older trees are harvested, and roundwood production volume is at least double its current level by 2050. In the government plan scenario the future production of roundwood is aggressively pursued in line with the Japanese government's Basic Plan for Forest and Forestry. In the forest owners' intentions scenario, which relies upon the results of a survey on the intentions of the forest owners affiliated with the forestry cooperatives, 15–30 % of planted trees are harvested in the 20-year period 2011–2030, and 25–30 % during 2031–2050. Roundwood production volume under these scenarios was estimated for each prefecture and the three regions; the national production is presented in Fig. 3.

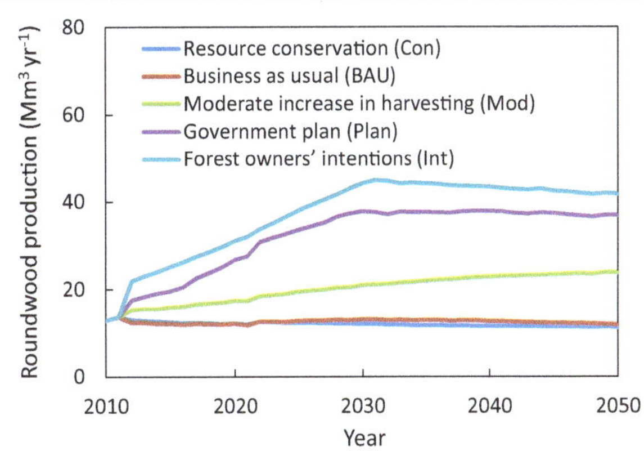


**Fig. 3** Japanese nationwide domestic roundwood production

Three scenarios were considered for HWP consumption volume, namely business as usual (BAU), moderate promotion of usage (Mod), and aggressive promotion of usage (Agg). Under the BAU scenario, the proportion of wood used in building construction, the proportion of wood used in furniture, and the volume of HWPs consumed in civil engineering projects all remain unchanged at the current level until 2050. Under the moderate usage promotion scenario, the proportion of wood used in building construction and in furniture making increases from 35 to 50 % by 2050 and the volume of HWPs consumed in civil engineering projects increases to 3 Mm$^3$ by 2050 from the current 1 Mm$^3$. Under the aggressive usage promotion scenario, the proportion of wood used in building construction and in furniture making increases to 70 % by 2050, and the volume of HWPs consumed in civil engineering projects increases to 6 $Mm^3$ by 2050. No distinction was made between scenarios for the volume of paper consumption, with no change in per-person paper consumption anticipated through 2050. Details on HWP consumption volume are explained in the "Methods" section.

Three similar scenarios were also assumed for the volume of energy usage, namely business as usual (BAU), moderate promotion of usage (Mod), and aggressive promotion of usage (Agg). The BAU scenario entails no change in energy usage until 2050 with the current proportion of logging residue, processing residue, and waste wood. In the moderate usage promotion scenario, in 2050, in addition to the current energy-usage proportion, half of the unutilized proportion is used. In the aggressive usage promotion scenario, in 2050, in addition to the current energy-usage proportion, the whole of the unutilized proportion is used. Details of the volume of energy usage are explained in the "Methods" section.

## Results and discussion

### Carbon storage effect

Figure 4 shows carbon storage volume attributable to HWPs derived from domestic forests by region. Under the BAU and BAU scenario, it is assumed that the future carbon storage volume for the whole of Japan declines compared to the actual value of 48 MtC in 2013 to the equivalent of 98 % of this level in 2030, and the equivalent of 96 %, or 46 MtC, in 2050. This decline is influenced by the downward trend in demand for wood, and is in line with declines in population and building construction up until 2013. However, under the scenarios whereby domestic roundwood production and HWP usage is promoted (the Mod & Mod, the Plan & Agg, and the Int & Agg scenarios), we assume an increasing trend until 2050. In 2050, the highest carbon storage volume occurs under the Int & Agg scenario, at a value of 125 MtC and equivalent to 262 % of the 2013 level. Focusing on regional distribution, although results vary according to the scenario and timescale, 16–20 % of the nationwide carbon storage volume is assumed to be in east Japan, 56–62 % in central Japan, and 21–24 % in west Japan, with the carbon storage volume high in central Japan, which includes the Tokyo metropolitan area. In addition, our breakdown of the carbon storage volume by application is as follows: building construction 68–88 %, civil engineering 0–12 %, furniture production 4–12 %, and paper making 7–13 %.

Figure 5 shows annual changes in the nationwide carbon storage volume attributable to HWPs derived from domestic forests. Under the Con & BAU and the BAU & BAU scenarios, as a result of a decline in carbon storage volume, the annual change has a negative value up until 2050, and a carbon storage effect cannot be expected. However, under the Mod & Mod, the Plan & Agg, and the Int & Agg scenarios, the annual change has a positive value until 2050, and a carbon storage effect can be anticipated. In particular, under the Int & Agg scenario, 2.9 MtC $year^{-1}$ is estimated for 2030, the biggest effect among the scenarios, which is in line with the trend in the volume of domestic roundwood production (Fig. 3). There is significant scope for growth in the HWP stock derived from domestic forests. However, because the volume of domestic roundwood production reduces thereafter, until 2050, the carbon storage effect decreases.

### Material substitution effect

The volume of annual carbon emissions reduction due to material substitution in each region compared to the BAU scenario is illustrated in Fig. 6 and the nationwide annual reduction volume is shown in Fig. 7. It was evident that, compared to the BAU scenario, a nationwide material substitution effect of 0.6 MtC $year^{-1}$ in 2030 and

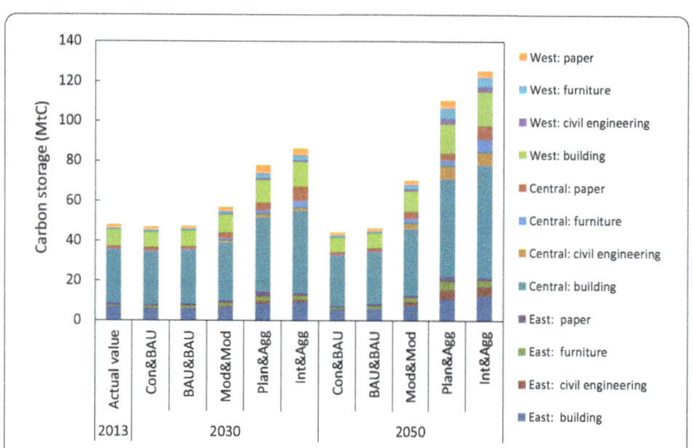


**Fig. 4** Carbon storage for HWPs derived from domestic forests, by region

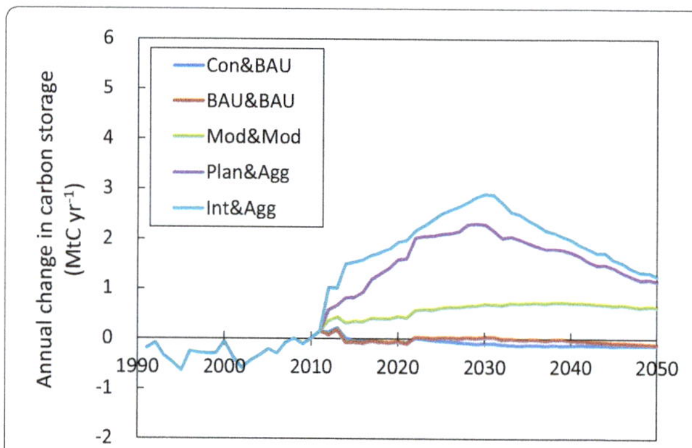


**Fig. 5** Annual change in nationwide carbon storage for HWPs derived from domestic forests

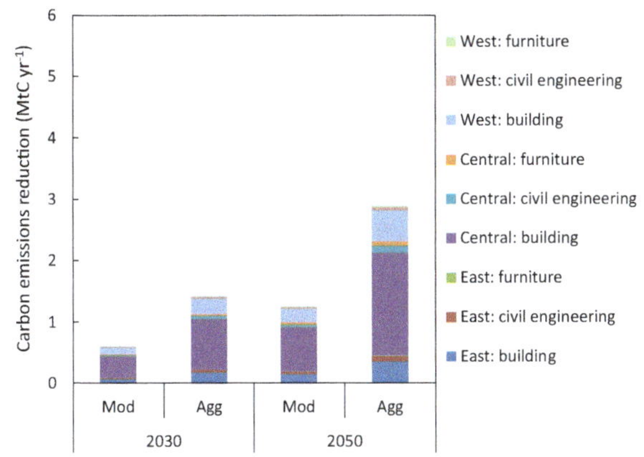


**Fig. 6** Annual carbon emissions reduction attributable to material substitution, by region (difference from the BAU scenario)

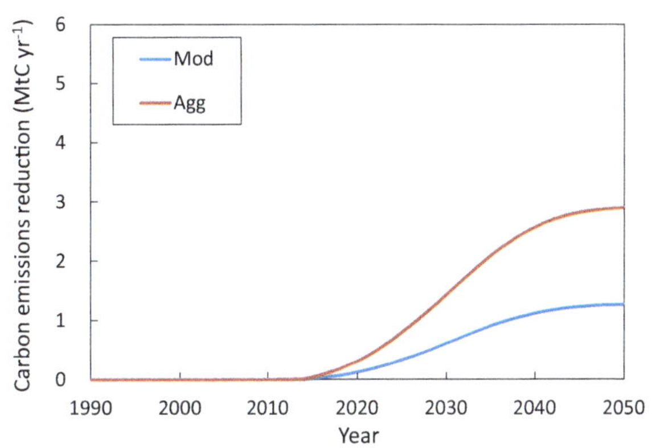


**Fig. 7** Nationwide annual carbon emissions reduction attributable to material substitution (difference from the BAU scenario)

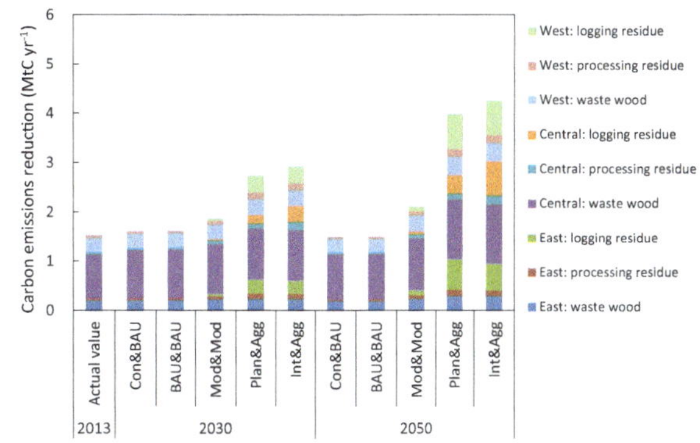


**Fig. 8** Annual carbon emissions reduction attributable to energy substitution, by region

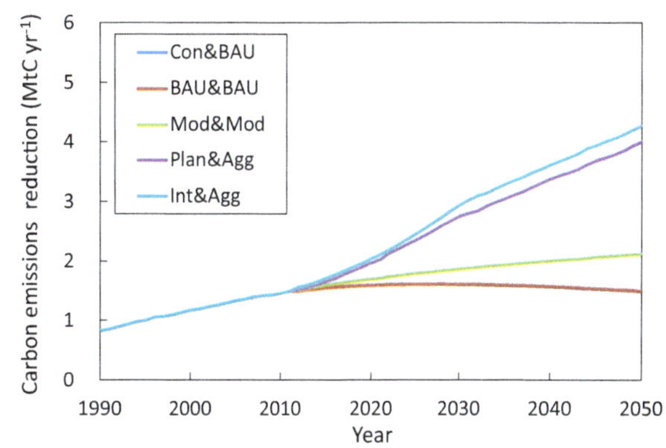


**Fig. 9** Nationwide annual carbon emissions reduction attributable to energy substitution

of 1.3 MtC year$^{-1}$ in 2050 could be expected under the Mod scenario, and an effect of 1.4 MtC year$^{-1}$ in 2030 and 2.9 MtC year$^{-1}$ in 2050 under the Agg scenario. Under all scenarios, the regional breakdown of the volume of annual emissions reduction was 17 % derived from east Japan, 63 % derived from central Japan, and 20 % derived from west Japan. In addition, the breakdown by application was 88 % derived from building construction, 8 % derived from civil engineering, and 4 % derived from furniture making, emphasizing the significance of the reduction effect attributable to the substitution of wooden building construction for non-wooden building construction.

### Energy substitution effect

The volume of annual carbon emissions reduction attributable to energy substitution for each region is shown in Fig. 8 and the nationwide annual reduction is shown in Fig. 9. Under the Con & BAU and the BAU & BAU scenarios, the annual carbon emissions reduction remained more or less flat through 2050 with little visible change. However, under the three scenarios of the Mod & Mod, the Plan & Agg and the Int & Agg, an increase in the annual carbon emissions reduction was seen, mainly as a result of a large increase in the volume of energy generation from logging residue. For 2050, a nationwide annual emissions reduction volume of 4.0 MtC year$^{-1}$ was obtained under the Plan & Agg scenario, and of 4.3 MtC year$^{-1}$ under the Int & Agg scenario, the latter figure being equivalent to 279 % of that for 2013. The regional breakdown of the actual figure for nationwide annual carbon emissions reduction in 2013 was 17, 61, and 22 % for east, central, and west Japan, respectively. Under the Plan & Agg and the Int & Agg scenarios, the proportion of reduction volume derived from central Japan declined to 43 and 49 %, while the proportions from east and west Japan increased. Under these scenarios, as a result of the aforementioned increase in energy-use volume of logging residue, the annual emissions reduction volume rose in east and west Japan, where there is a high volume of roundwood production, which generates logging residue. Within the carbon emissions reduction volume, the proportion derived from logging residue was high, at 43 % in 2050 under the Plan & Agg scenario, and 46 % under the Int & Agg scenario.

In the present study, the term "logging residue" refers to items such as branches and leaves generated after tree harvesting; it does not refer to items such as unused thinning wood. If we also take into account the residue resulting from thinning standing trees, the volume of residue available for energy use is likely to be greater. The degree to which such logging residue, seldom used in Japan and often left on the forest floor, is used in the future will have a significant impact on the energy substitution effect. At the same time, removing logging residue from the forest floor decreases the carbon storage in dead organic matter and soil in forests [32]. A future study needs to investigate a trade-off relationship between the energy use of logging residue and carbon storage in forests.

### Climate change mitigation effect

The nationwide total of the carbon storage effect, the material substitution effect, and the energy substitution effect attributable to HWPs derived from domestic forests (the annual change in carbon storage relates to HWPs derived from domestic forests, and the annual carbon emissions reduction attributable to material and energy substitution) is shown in Fig. 10. The mitigation effect under the Con & BAU scenario was the smallest at 1.3 MtC year$^{-1}$ in 2050. The mitigation effect under the Int & Agg scenario was the largest at 8.4 MtC year$^{-1}$ in 2050, equivalent to 2.4 % of the nationwide carbon dioxide emissions volume in 2013 [33]. Looking at the breakdown of three effects, under the same scenario, 15 % is seen as derived from carbon storage, 34 % from material substitution, and 51 % from energy substitution.

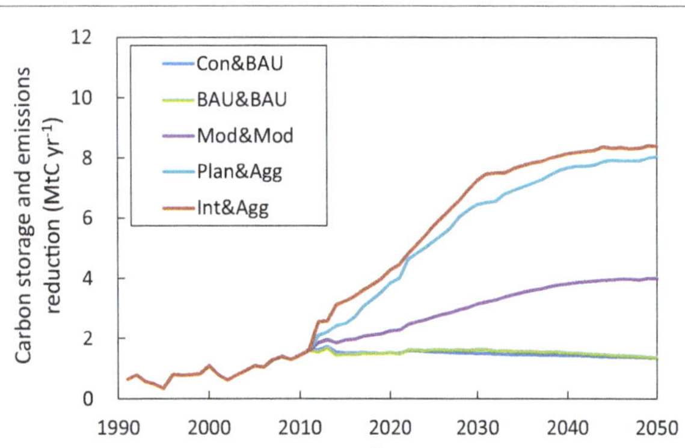


**Fig. 10** Nationwide HWP climate change mitigation effect (annual change in carbon storage relates to HWPs derived from domestic forests and annual carbon emissions reduction attributable to material and energy substitution)

Because the decline of the forestry and wood industries in Japan is currently an issue [34], it is imperative to promote the use of HWPs derived from domestic forests to contribute toward revitalizing these industries. Therefore, the aggressive usage scenarios have the potential to produce the combined effect of both revitalizing the domestic forestry and wood industries and achieving carbon emissions reduction related to HWPs. On the other hand, the volume of the usage of HWP derived from domestic forests has a trade-off relationship with the volume of carbon storage in domestic forests [35, 36]. According to the results estimated from a carbon balance model for Japanese forests (including both vegetation and soil) [30], carbon removals (annual changes in carbon storage) by domestic forests declined from around 20 MtC year$^{-1}$ under the BAU scenario to around 10 MtC year$^{-1}$ (approximately 50 % decrease) under the aggressive wood harvesting scenario in 2050. Our estimated results showed that the volume of the overall climate change mitigation effect of HWPs could be expected to be over 8 MtC year$^{-1}$ in 2050 under these aggressive usage scenarios, and mostly compensate for the decrease in domestic forests. An important future consideration is an integrated investigation of the carbon balance for both HWPs and forests [21, 37].

## Conclusions

Considering the carbon storage effect of HWPs derived from domestic forests according to the production approach, under the future scenarios of conservation of domestic forest resources and of business as usual as regards and HWP usage, the carbon storage effect will not be obtained over the span of time until 2050. However, under the future scenarios whereby domestic roundwood production and HWP usage is promoted, the carbon storage effect is consistently expected up until 2050, with an anticipated nationwide maximum of 2.9 MtC year$^{-1}$ in 2030.

As regards the material substitution effect, in the future scenarios in which HWP usage is promoted, the maximum nationwide carbon emissions reduction volume is estimated at 2.9 MtC year$^{-1}$ in 2050. In addition, the substitution of wooden building construction for non-wooden building construction could be expected to prompt a large reduction and could prove to be an effective mitigation strategy.

As regards the energy substitution effect, under future scenarios in which domestic roundwood production, HWP usage, and energy use are all promoted, a maximum nationwide carbon emissions reduction volume of 4.3 MtC year$^{-1}$ is obtained by 2050, with the greatest contribution coming from increased use of logging residue. In addition, upwards of 50 % of the reduction volume is derived from east and west Japan, influenced by the fact that these regions have a large volume of roundwood production, which is a source of logging residue.

Taken together, the carbon storage effect, the material substitution effect, and the energy substitution effect for HWPs derived from domestic forests under future scenarios that promote domestic roundwood production, HWP usage, and energy use, a maximum nationwide climate change mitigation effect of 8.4 MtC year$^{-1}$ can be achieved in 2050, which is equivalent to 2.4 % of Japan's total carbon dioxide emissions volume in 2013. In this case, the volume of the climate change mitigation effect of HWPs is comparable to the volume of the carbon removal effect of Japanese forests.

In the present study, the overall climate change mitigation effect attributable to HWPs in each region of Japan was clarified, and future projected changes until 2050 were explored. A major issue for the future entails the integrated investigation of the carbon balance of HWPs and forests and relevant changes, together with a consideration of their trade-off relationships [21, 35–37]. Furthermore, it is also important to consider the leakage of the international HWP trade on each country's forests [38, 39], where declining carbon storage in forests in countries, which export HWP to Japan, is balanced with limited domestic wood harvesting and conservation of Japanese forests.

## Methods

### HWP consumption volume and stock

#### *Building construction*

The volume of HWP (sawnwood, plywood) consumption in building construction was estimated based on our previous report, using Eq. (1).

$$UWB_{t,i,l,m} = CFA_{t,i,m} \cdot UIW_{l,m} \quad (1)$$

where $UWB$ (m$^3$ year$^{-1}$) represents the volume of HWP (sawnwood, plywood) consumption in building construction, $CFA$ (m$^2$ year$^{-1}$) is the floor area of new construction, and $UIW$ (m$^3$ m$^{-2}$) is the wood usage volume per unit of floor area [40]. Further, $t$ represents year, $i$ represents the region of final HWP consumption (east, central, or west Japan), $l$ represents produced from domestic forests, imported, and $m$ represents wooden construction, non-wooden construction.

Actual values were used through 2013 [41], and future scenarios for 2014–2050 were assigned according to the proportion of wooden construction within the floor area of new construction ($CFA_{t,i,l,m}$). Under the BAU scenario, it was assumed that there was no change in the 35 % wooden construction ratio between 2014 and 2050, and under the moderate usage promotion (Mod) scenario, an increase from 35 % in 2014 to 50 % by 2050 was assumed, following an S-shaped curve [27], while under the aggressive usage promotion (Agg) scenario an increase to 70 % by 2050 was assumed.

The volume of HWP stock (sawnwood, plywood) in building construction was estimated using Eqs. (2) and (3), based on the previous report.

$$SWB_{t,i,l,m} = \sum_{n} \{UWB_{t-n,i,l,m} \cdot BL_{t,m}(n)\} \quad (2)$$

$$BL_{t,m}(n) = \mathrm{Exp}\{-r(n-a)\}/[1+\mathrm{Exp}\{-r(n-a)\}] \quad (3)$$

where $SWB$ (m$^3$) represents the volume of HWP stock (sawnwood, plywood) in buildings, $BL$ (−) is the building lifetime function, $n$ (year) is the number of years elapsed, $r$ (−) is the rate of decrease (0.2), and $a$ (year) is the building lifetime (half-life) (wooden construction: 35 years; non-wooden construction: 30 years).

#### *Civil engineering*

In the model, no HWPs (roundwood, cylindrical poles) were assumed to be consumed in civil engineering ($UWC_{t,i,l,p}$) until 2009 because of the lack of reliable figures from statistical data, and 1 Mm$^3$ was used for 2010 [42]. From 2011, future scenarios were assigned and, under the BAU scenario, it was assumed that there was no increase from the 1 Mm$^3$ in 2010 until 2050, while, under the Mod scenario, an increase following an S-shaped curve was assumed from 2011, reaching 3 Mm$^3$ in 2050. Under the Agg scenario, an increase to 6 Mm$^3$ by 2050 was assumed. Since the current and future breakdown of consumption volume between each civil engineering application is unclear, 50 % was assumed for soil liquefaction countermeasure piles and 50 % for cylindrical poles used in wooden road safety guardrails. However, the cylindrical pole yield from roundwood was set at 0.8 [43], while the actual use in wooden guardrails is set at 40 %; the remaining 10 % assumed to be processing residue.

HWP (roundwood, cylindrical poles) stock volume in civil engineering was estimated using Eqs. (4) and (5).

$$SWC_{t,i,l,p} = \sum_{n} \{UWC_{t-n,i,l,p} \cdot CL_{t,p}(n)\} \quad (4)$$

$$CL_{t,p}(n) = \mathrm{Exp}\{-r(n-b)\}/[1+\mathrm{Exp}\{-r(n-b)\}] \quad (5)$$

where $SWC$ (m$^3$) represents the HWP (roundwood, cylindrical poles) stock volume in civil engineering, $UWC$ (m$^3$ year$^{-1}$) is the HWP (roundwood, cylindrical poles) consumption volume in civil engineering, and $CL$ (−) is the civil engineering lifetime function. Furthermore, $p$ represents piles and wooden guardrails and $b$ (year) represents civil engineering lifetime (half-life) (piles: perpetual; wooden guardrails: 10 years).

For lifetime (half-life) ($b$), it was assumed that the piles will remain permanently anchored in the ground [44], while, for the wooden guardrails, 10 years was assumed based on observations in Japan [43].

### Furniture

The volume of HWP (sawnwood, plywood) consumption in furniture making was estimated using Eq. (6), with reference to the previous report.

$$UWF_{t,i,l} = UWF_{t-1,i,l} \cdot WFP_t \cdot POP_{t,i} / POP_{t-1,i} \quad (6)$$

where $UWF$ (m$^3$ year$^{-1}$) represents the volume of HWP (sawnwood, plywood) consumption in furniture making, $WFP$ (−) represents the proportion of wood used in furniture, and $POP$ (persons) represents the population [45].

Actual values [46–48] were used until 2013, while future scenarios from 2014 were assigned according to the proportion of wood used in furniture ($WFP_t$). In the BAU scenario, it was assumed that there was no change in the current 35 % proportion of wood used in furniture during 2014–2050; in the Mod scenario, a 50 % increase in the proportion following an S-shaped curve was assumed from 2014 to 2050; and in the Agg scenario, an increase to 70 % by 2050 was assumed.

We estimated the volume of HWP stock (sawnwood, plywood) in furniture making using Eqs. (7) and (8).

$$SWF_{t,i,l} = \sum_n \{UWF_{t-n,i,l} \cdot FL_t(n)\} \quad (7)$$

$$FL_t(n) = \mathrm{Exp}\{-r(n-c)\}/[1 + \mathrm{Exp}\{-r(n-c)\}] \quad (8)$$

where $SWF$ (m$^3$) represents the volume of HWP stock (sawnwood, plywood) in furniture, $FL$ (−) represents the furniture lifetime function, and $c$ (year) represents the furniture lifetime (half-life) (20 years).

### Paper

Paper product consumption volume (UPP) (t year$^{-1}$) was estimated using Eq. (9), based on the previous report.

$$UPP_{t,i,l} = UPP_{t-1,i,l} \cdot POP_{t,i} / POP_{t-1,i} \quad (9)$$

Real values [49] were used until 2013; for future scenarios from 2014, we explored only the BAU scenario. It was assumed that paper product consumption per person would remain unchanged at its 2013 level in 2014 and beyond and that consumption would change in line with future changes in population [45].

Paper product stock volume was estimated using Eqs. (10) and (11).

$$SPP_{t,i,l} = \sum_n \{UPP_{t-n,i,l} \cdot PL_t(n)\} \quad (10)$$

$$PL_t(n) = \mathrm{Exp}\{-\mathrm{Ln}(2)/d\}^n \quad (11)$$

where $SPP$ (t) represents paper product stock volume, $PL$ (−) represents the paper lifetime function, and $d$ (year) represents paper lifetime (half-life) (2 years).

### Energy use volume

Energy use volume was estimated using Eqs. (12)–(19).

$$UWE_{t,i,j,k,q,e} = (OLR_{t,k} + OWR_{t,j,q} + ORB_{t,i} + ORF_{t,i} + WWB_{t,i} + WWC_{t,i} + WWF_{t,i}) \cdot E_e \quad (12)$$

$$OLR_{t,k} = PR_{t,k}/Y \cdot BEF - PR_{t,k} \quad (13)$$

$$OWR_{t,j,q} = PSP_{t,j,q}/Z_q - PSP_{t,j,q} \quad (14)$$

$$ORB_{t,i} = UWB_{t,i}/X - UWB_{t,i} \quad (15)$$

$$ORF_{t,i} = UWF_{t,i}/V - UWF_{t,i} \quad (16)$$

$$WWB_{t,i} = SWB_{t-1,i} - SWB_{t,i} + UWB_{t,i} \quad (17)$$

$$WWC_{t,i} = SWC_{t-1,i} - SWC_{t,i} + UWC_{t,i} \quad (18)$$

$$WWF_{t,i} = SWF_{t-1,i} - SWF_{t,i} + UWF_{t,i} \quad (19)$$

where $UWE$ (m$^3$ year$^{-1}$) represents energy use volume, $OLR$ (m$^3$ year$^{-1}$) is the volume of logging residue generated, $OWR$ (m$^3$ year$^{-1}$) is the volume of wood processing residue generated, $ORB$ (m$^3$ year$^{-1}$) is the volume of residue generated during building construction, $ORF$ (m$^3$ year$^{-1}$) is the volume of residue generated during furniture production, $WWB$ (m$^3$ year$^{-1}$) is the volume of waste wood generated from buildings after use, $WWC$ (m$^3$ year$^{-1}$) is the volume of waste wood generated from civil engineering after use, $WWF$ (m$^3$ year$^{-1}$) is the volume of waste wood generated from furniture after use, and $E$ (−) is the rate of energy use. Furthermore, $j$ represents the region of sawnwood or plywood production (east, central, or west Japan), $k$ is the region of roundwood production (east, central, or west Japan), $q$ is sawnwood or plywood, and $e$ is logging residue, processing residue, and waste wood. $PR$ (m$^3$ year$^{-1}$) represents the volume of domestic roundwood production, $Y$ (−) is the volume of roundwood yielded by a given tree trunk volume (0.856) [44], and $BEF$ (−) is the coefficient (1.23) [50] that expands a given tree trunk volume to the whole tree including branches and leaves. $PSP$ (m$^3$ year$^{-1}$) represents the volume of domestic sawnwood and plywood production and $Z$ (−) represents the volume of sawnwood and plywood yielded by a given volume of roundwood (sawnwood: 0.637; plywood: 0.618) [31]. $X$ (−) represents the volume of building construction use yielded by a given volume of sawnwood and plywood (0.9), while $V$ (−) represents the volume of furniture use yielded by a given volume of sawnwood and plywood (0.717) [48].

Future scenarios were assigned according to the rate of energy use ($E_e$). Under the BAU scenario, it was assumed that there would be no change in the current rate of energy use (0 % for logging residue, 21 % for processing residue, and 83 % for waste wood) [51, 52] between 2014 and 2050. Under the Mod scenario, an increase in usage from 2014 was assumed, reaching the current rate of use plus half the unused proportion in 2050 (49 % for logging residue, 24 % for processing residue, and 87 % for waste wood) [51, 52]. Under the Agg scenario, an increase in usage was assumed, reaching the current rate of energy use plus the whole unused proportion in 2050 (99 % for logging residue, 27 % for processing residue, and 90 % for waste wood) [51, 52]. The rate of use for each year from 2014 to 2050 was set using linear interpolation.

### Domestic production volume within HWP consumption volume and stock volume

Figure 11 shows the estimation procedure for volume of production from domestic forests within HWP consumption and stock volume as relates to the production approach. Steps (1)–(4) were followed to arrive at estimates for each region and each future scenario. Both domestically produced and imported wood is used in HWP consumption for building construction, furniture, and paper. However, wooden piles and guardrails are generally made from wood produced from domestic forests, so it was assumed that only domestically produced wood was used in civil engineering. Estimates were uniformly converted into roundwood terms, and the roundwood conversion coefficients for each HWP were as follows: for sawnwood, 1.570 $m^3$-roundwood $m^{-3}$-sawnwood; for plywood, 1.618 $m^3$-roundwood $m^{-3}$-plywood; and for paper, 3.300 $m^3$-roundwood $t^{-1}$-paper [31].

The steps are as follows:

1. The volume of roundwood and cylindrical poles production for civil engineering in each region was taken from the volume of domestically produced roundwood indicated in the Future scenarios section.
2. The remaining domestic roundwood production volume was distributed as production volume for sawnwood, plywood and chips used in building construction, furniture, and paper. The relevant proportions were set at sawnwood-use 60 %, plywood-use 15 %, and chip-use 25 %, with reference to past actual values [53].
3. Production volume of sawnwood and plywood derived from domestic roundwood was distributed among production for building construction and for furniture. Relevant proportions of HWP consumption attributable to building construction and to furniture production mentioned in the HWP consumption volume and stock section were used. Moreover, the volume of chip production derived for domestic roundwood was allocated exclusively to paper.
4. HWP import volume or roundwood import volume was assumed to be what remains when the volume of sawnwood, plywood, and chip production derived from domestic roundwood is deducted from the HWP consumption volume. However, since this study used the production approach to account for the carbon storage in HWPs, imports were not taken into consideration.

### Inter-regional flow

The flow of HWPs derived from domestic forests between regions of Japan (east, central, and west) was

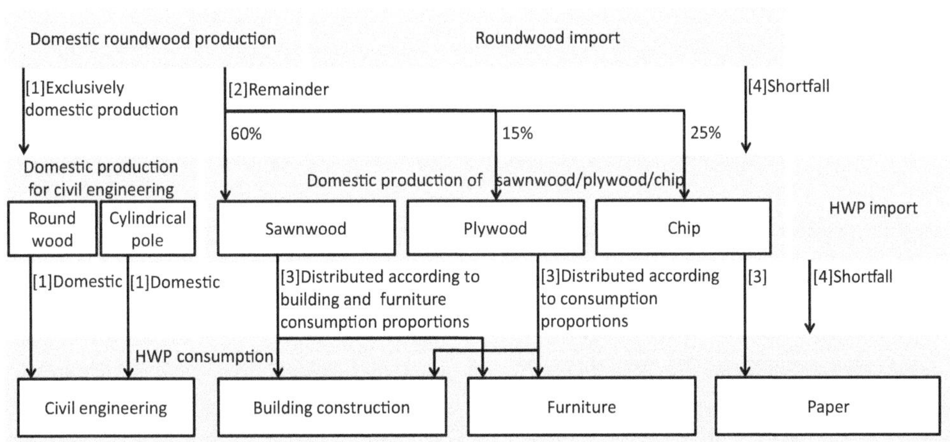


**Fig. 11** Procedure for estimating the volume of production from domestic forests within HWP consumption volume and stock volume

estimated, with reference to past actual flow data [53, 54]. The flow of roundwood from its location to the production location of resulting sawnwood, plywood, or chips was taken to be entirely intra-regional, rather than inter-regional. Looking at the flow of sawnwood from the location of its production to the location of HWP consumption, 15 % of consumption volume in central Japan was taken to be from east Japan, 70 % from central Japan, and 15 % from west Japan. Looking at the flow of plywood from the location of production to the location of HWP consumption, 15 % of consumption volume in central Japan was taken to be from east Japan, 75 % from central Japan, and 10 % from west Japan. The flow of chips from the location of their production to the location of paper product consumption was taken to be intra-regional, rather than inter-regional. Looking at the flow of paper products from the location of their production to the location of paper product consumption, 25 % of consumption volume in central Japan was taken to be from east Japan, 55 % from central Japan, and 20 % from west Japan. In addition, it was assumed that logging residue, processing residue, and waste wood were used for energy in the region where they were generated, and they were not included in consideration of inter-regional flow.

### Climate change mitigation effect attributable to HWPs

#### *Carbon storage effect*

The carbon storage volume was evaluated for only HWPs derived from domestic forests (i.e., the production approach). HWP carbon storage volume and its annual change were estimated using Eqs. (20) and (21).

$$CSW_{t,i,l} = \left(SWB_{t,i,l} + SWC_{t,i,l} + SWF_{t,i,l} + SPP_{t,i,l}\right) \cdot SG_s \cdot CC \tag{20}$$

$$ACS_{t,i,l} = CSW_{t,i,l} - CSW_{t-1,i,l} \tag{21}$$

where $CSW$ (tC) represents HWP carbon storage volume, $SG$ (t m$^{-3}$) is bulk density [28], $CC$ (tC t$^{-1}$) is carbon content (0.5), $s$ is roundwood, sawnwood, plywood, and paper, and $ACS$ (tC year$^{-1}$) is the annual change in HWP carbon storage volume.

A distinction was made between products derived from domestic forests and products derived from imports using the process outlined in the Domestic production volume within HWP consumption volume and stock volume section and Fig. 11.

#### *Material substitution effect*

For the material substitution effect for building construction, civil engineering, and furniture production, the volume of the life-cycle reduction in carbon emissions from fossil fuel consumption attributable to the substitution of HWPs for non-wooden materials was evaluated. However, it is necessary to determine a figure for the volume of HWP substitution for non-wooden materials. Thus, the BAU scenario was taken as a baseline, and the increase in HWP volumes in the Mod and the Agg scenarios as compared to the BAU scenario were taken to be the volume of substitution for non-wooden materials. For building construction, civil engineering, and furniture production, the annual carbon emissions reduction (tC year$^{-1}$) attributable to material substitution was evaluated by multiplying, respectively, the increase in the floor area of new wooden building construction (m$^2$ year$^{-1}$) and the increase in the consumption of each HWP (m$^3$ year$^{-1}$), by the carbon emissions reduction intensity (kgC m$^{-2}$, kgC m$^{-3}$). The carbon emissions reduction intensities are shown in Table 2.

In Table 2, the carbon emissions reduction intensity attributable to the substitution of wooden building for non-wooden building (kgC m$^{-2}$) was taken from MiLCA [55]. It was calculated by subtracting the emission intensity for wooden building (kgC m$^{-2}$) from the emission intensity for non-wooden building (reinforced concrete, steel reinforced concrete, steel and concrete blocks) (kgC m$^{-2}$), from the weighted average, based on 2013 building construction floor space (m$^2$). The emission reduction intensity attributable to the substitution of wooden piles for non-wooden piles in civil engineering was taken from [44]. The emission intensity for wooden piles (kgC m$^{-2}$) was subtracted from the average value (kgC m$^{-2}$) of the emission intensity per unit of improved area for cement and sand piles, to generate an emission reduction intensity per unit of HWP-use volume (kgC m$^{-3}$). The emission reduction intensity attributable to the substitution of

**Table 2 Carbon emissions reduction intensities attributable to fossil fuel consumption, as a result of HWP substitution for non-wooden materials**

| Material substitution | Unit | Value | References |
|---|---|---|---|
| Building construction: substitution of wooden buildings for non-wooden buildings | kgC m$^{-2}$ | 60.560 | [55] |
| Civil engineering: substitution of wooden piles for cement and sand piles | kgC m$^{-3}$ | 46.773 | [44] |
| Civil engineering: substitution of wooden guardrails for metal guardrails | kgC m$^{-3}$ | 64.477 | [43] |
| Furniture: substitution of wooden furniture for metal furniture | kgC m$^{-3}$ | 43.168 | [55] |

wooden guardrails for metal guardrails in civil engineering was taken from [43]. The emission intensity for wooden guardrails (kgC $m^{-1}$) was subtracted from the emission intensity for metal guardrails (kgC $m^{-1}$), to give a reduction intensity per unit of HWP-use volume (kgC $m^{-3}$). The reduction intensity attributable to the substitution of metal furniture with wooden furniture was taken from MiLCA [55]. The emission intensity for wooden furniture (kgC $item^{-1}$) was subtracted from the emission intensity for metal furniture (kgC $item^{-1}$), to generate a reduction intensity per unit of HWP-use volume (kgC $m^{-3}$).

### *Energy substitution effect*

The energy substitution effect was taken to be the substitution of logging residue, processing residue, and waste wood for heavy oil in energy generation, and the relevant annual reduction in the volume of carbon emissions was estimated using Eq. (22).

$$CRE_{t,i,j,k,q,e} = UWE_{t,i,j,k,q,e} \cdot SG_s \cdot LW \cdot EO/LO \tag{22}$$

where $CRE$ (tC $year^{-1}$) represents the volume of annual reduction in carbon emissions due to energy substitution, $LW$ (GJ $t^{-1}$) is the calorific value of wood (14.4) [44], $EO$ (tC $kl^{-1}$) is the carbon emissions volume accompanying the combustion of heavy oil (0.739) [44], and $LO$ (GJ $kl^{-1}$) is the calorific value of heavy oil (39.1) [44].

### *Climate change mitigation effect*

The climate change mitigation effect is the sum of the annual change in carbon storage volume outlined in the Carbon storage effect section, the volume of annual carbon emissions reduction owing to the material substitution outlined in the Material substitution effect section, and the volume of annual carbon emissions reduction owing to energy substitution outlined in the Energy substitution effect section.

#### Authors' contributions

CK designed the carbon balance model of this study, conducted the simulation, and wrote the manuscript. YT designed the model together with CK, and helped write the manuscript. MT built the research team that conducted the research project, participated in the design of this study, and helped write the manuscript. All authors have read and approved the final manuscript.

#### Author details

[1] Department of Environment Conservation, Graduate School of Agriculture, Tokyo University of Agriculture and Technology, 3-5-8 Saiwai-cho, Fuchu, Tokyo 183-8509, Japan. [2] Department of Wood Engineering, Forestry and Forest Products Research Institute, 1 Matsunosato, Tsukuba, Ibaraki 305-8687, Japan. [3] Shikoku Research Center, Forestry and Forest Products Research Institute, 2-915 Asakuranishi-cho, Kochi, Kochi 780-8077, Japan.

#### Acknowledgements

This research was supported by the Agriculture, Forestry and Fisheries Research Council Project, Development of Climate Change Mitigation Technologies in the field of Forests and Forestry, 2010–2014, JSPS KAKENHI Grant Number 26870181, MEXT KAKENHI Grant Number 15H02863, and the Policy Study Fund for Environmental Economics (the third period) of the Ministry of the Environment.

#### Competing interests

The authors declare that they have no competing interests.

#### References

1. Intergovernmental Panel on Climate Change (IPCC). Climate Change 2013, Fifth Assessment Report, The Physical Science Basis. 2013.
2. Intergovernmental Panel on Climate Change (IPCC). Climate Change 2014, Fifth Assessment Report, Mitigation of Climate Change. 2014.
3. Ellison D, Lundblad M, Petersson H. Carbon accounting and the climate policies of forestry. Environ Sci Policy. 2011;14:1062–78.
4. Hashimoto S, Nose M, Obara T, Moriguchi Y. Wood products: potential carbon sequestration and impact on net carbon emissions of industrialized countries. Environ Sci Policy. 2002;5:183–93.
5. Lim B, Brown S, Schlamadinger B. Carbon accounting for forest harvesting and wood products: review and evaluation of different approaches. Environ Sci Policy. 1999;2:207–16.
6. United Nations Framework Convention on Climate Change (UNFCCC). Report of the Conference of the Parties serving as the meeting of the Parties to the Kyoto Protocol on its seventh session, held in Durban from 28 November to 11 December 2011, Part Two: Action taken by the Conference of the Parties serving as the meeting of the Parties to the Kyoto. Protocol at its seventh session, Decisions adopted by the Conference of the Parties serving as the meeting of the Parties to the Kyoto Protocol. 2012.
7. Intergovernmental Panel on Climate Change (IPCC). 2013 Revised Supplementary Methods and Good Practice Guidance Arising from the Kyoto Protocol. 2014.
8. Pingoud K, Perälä AL, Soimakallio S, Pussinen A. Greenhouse gas impacts of harvested wood products, Evaluation and Development of methods, VTT Research Notes 2189. Espoo: VTT Technical Research Centre of Finland; 2003.
9. Pilli R, Fiorese G, Grassi G. EU mitigation potential of harvested wood products. Carbon Balance Manag. 2015;10:6.
10. Stockmann KD, Anderson NM, Skog KE, Healey SP, Loeffler DR, Jones G, Morrison JF. Estimates of carbon stored in harvested wood products from the United States forest service northern region, 1906–2010. Carbon Balance Manag. 2012;7:1.
11. Sikkema R, Junginger M, McFarlane P, Faaij A. The GHG contribution of the cascaded use of harvested wood products in comparison with the use of wood for energy—a case study on available forest resources in Canada. Environ Sci Policy. 2013;31:96–108.
12. Dymond CC. Forest carbon in North America: annual storage and emissions from British Columbia's harvest, 1965–2065. Carbon Balance Manag. 2012;7:8.
13. Dias AC, Louro M, Arroja L, Capela I. Carbon estimation in harvested wood products using a country-specific method: Portugal as a case study. Environ Sci Policy. 2007;10:250–9.
14. Raši R, Cienciala E, Priwitzer T, Palán S, Pavlenda P. Carbon balance in harvested wood products in Slovakia. Lesn Cas For J. 2015;61:101–6.
15. Winjum JK, Brown S, Schlamadinger B. Forest harvests and wood products: sources and sinks of atmospheric carbon dioxide. For Sci. 1998;44:272–84.
16. Holtsmark B. Quantifying the global warming potential of $CO_2$ emissions from wood fuels. GCB Bioenergy. 2013;. doi:10.1111/gcbb.12110.
17. Nepal P, Ince PJ, Skog KE, Chang SJ. Projection of US forest sector carbon sequestration under US and global timber market and wood energy consumption scenarios, 2010–2060. Biomass Bioenergy. 2012;45:251–64.
18. Woodbury P, Smith J, Heath L. Carbon sequestration in the US forest sector from 1990 to 2010. For Ecol Manag. 2007;241:14–27.
19. Chen J, Colombo SJ, Ter-Mikaelian MT, Heath LS. Carbon budget of Ontario's managed forests and harvested wood products, 2001–2100. For Ecol Manag. 2010;259:1385–98.
20. Liu G, Han S. Long-term forest management and timely transfer of carbon into wood products help reduce atmospheric carbon. Ecol Model. 2009;220:1719–23.
21. Smyth CE, Stinson G, Neilson E, Lemprière TC, Hafer M, Rampley GJ, Kurz WA. Quantifying the biophysical climate change mitigation potential of Canada's forest sector. Biogeosci. 2014;11:3515–29.
22. Knauf M, Köhl M, Mues V, Olschofsky K, Frühwald A. Modeling the

$CO_2$-effects of forest management and wood usage on a regional basis. Carbon Balance Manag. 2015;10:13.
23. Mathieu F, François N, Nicolas R, Frédéric M. Quantifying the impact of forest management on the carbon balance of the forest-wood product chain: a case study applied to even-aged oak stands in France. For Ecol Manag. 2012;279:176–88.
24. Karjalainen T, Pussinen A, Kellomäki S, Mäkipää R. Scenarios for the carbon balance of Finnish forests and wood products. Environ Sci Policy. 1999;2:165–75.
25. Werner F, Taverna R, Hofer P, Thürig E, Kaufmann E. National and global greenhouse gas dynamics of different forest management and wood use scenarios: a model-based assessment. Environ Sci Policy. 2010;13:72–85.
26. Lun F, Li W, Liu Y. Complete forest carbon cycle and budget in China, 1999–2008. For Ecol Manag. 2012;264:81–9.
27. Tsunetsugu Y, Tonosaki M. Quantitative estimation of carbon removal effects due to wood utilization up to 2050 in Japan: effects from carbon storage and substitution of fossil fuels by harvested wood products. J Wood Sci. 2010;56:339–44.
28. Hashimoto S, Moriguchi Y. Databook: Material and carbon flow of harvested wood in Japan, National Institute for Environmental Studies. 2004.
29. Kayo C, Tsunetsugu Y, Noda H, Tonosaki M. Carbon balance assessments of harvested wood products in Japan taking account of inter-regional flows. Environ Sci Policy. 2014;37:215–26.
30. Forestry and Forest Products Research Institute, Japan. Utilizing Forests under Climate Change: Ministry of Agriculture, Forestry, and Fisheries FY 2014 Commissioned Research Project. 2015.
31. Ministry of Agriculture, Forestry and Fisheries. Report on wood supply and demand 2013. 2014.
32. Verkerk H, Lindner M, Zanchi G, Zudin S. Assessing impacts of intensified biomass removal on deadwood in European forests. Ecol Indic. 2011;11(1):27–35.
33. Ministry of the Environment. Environmental Statistics 2013. 2014.
34. Forestry Agency. Annual Report on Trends in Forests and Forestry, Fiscal Year 2014. 2015.
35. Verkerk PJ, Mavsarb R, Giergiczny M, Lindner M, Edwards D, Schelhaas MJ. Assessing impacts of intensified biomass production and biodiversity protection on ecosystem services provided by European forests. Ecosyst Serv. 2014;9:155–65.
36. Böttcher H, Verkerk PJ, Gusti M, Havlik P, Grassi G. Projection of the future EU forest $CO_2$ sink as affected by recent bioenergy policies using two advanced forest management models. GCB Bioenergy. 2012;4:773–83.
37. Mubareka S, Jonsson R, Rinaldi F, Fiorese G, San-Miguel-Ayanz J, Sallnas O, Baranzelli C, Pilli R, Lavalle C, Kitous A. An integrated modelling framework for the forest-based bioeconomy. IEEE Earthzine. 2015;. doi:10.1101/011932.
38. Meyfroidt P, Rudel TK, Lambin EF. Forest transitions, trade, and the global displacement of land use. PNAS. 2010;107(49):20917–22.
39. Kastner T, Erb K, Nonhebel S. International wood trade and forest change: a global analysis. Glob Environ Change. 2011;21:947–56.
40. Tsunetsugu Y, Tonosaki M. Outlook study on the forest, forestry and forest industry: Data, theory and simulation. Forestry and Forest Products Research Institute, editor. Tokyo: Japan Forestry Investigation Committee; 2012. p 255–267.
41. Ministry of Land, Infrastructure, Transport and Tourism. Construction statistics 1990–2013. 1991–2014.
42. Tonosaki M. 2009 report on inter-disciplinary study on the increased use of wood in civil engineering; 2010. p 66.
43. Noda R, Kayo C, Yamanouchi M, Shibata N. Life cycle greenhouse gas emission of wooden guardrails—a study in Nagano Prefecture. J Wood Sci. 2015. (in press).
44. Kayo C, Hashimoto S, Numata A, Hamada M. Reductions in greenhouse gas emissions by using wood to protect against soil liquefaction. J Wood Sci. 2011;57(3):234–40.
45. National Institute of Population and Social Security Research. Population projections by region. 2014.
46. Ministry of Economy, Trade and Industry. Textiles and consumer goods statistics 1990–2013. 1991–2014.
47. Ministry of Finance. Trade statistics 1990–2013. 1991–2014.
48. Yano Research Institute Ltd. Inquiry into the provision of environmental information to the wood industry (wooden furniture and wooden sash edition): Inquiry report. 2003.
49. Ministry of Economy, Trade and Industry. Yearbook of paper, printing, plastic products and rubber products statistics 1990–2013. 1991–2014.
50. Government of Japan. Report on supplementary information on the LULUCF activities under Article 3, Paragraphs 3 and 4 of the Kyoto Protocol. 2009.
51. Forestry Agency. The current status of forestry and use of woody biomass. 2008. http://www.rinya.maff.go.jp/j/press/kaihatu/pdf/080522_1-05.pdf. Accessed 25 May 2015.
52. Ministry of the Environment. The current situation as regards reduction of wood generated by construction. 2013.
53. Ministry of Agriculture, Forestry and Fisheries. Reports on wood supply and demand 1990–2013. 1991–2014.
54. Ministry of Land, Infrastructure, Transport and Tourism. National net flow of cargo (logistics census) 1990–2011. 1991–2012. http://www.env.go.jp/council/former2013/03haiki/y0317-05/ref01-3.pdf. Accessed 10 May 2015.
55. Japan Environmental Management Association for Industry. MiLCA. 2014.

# 4

# Choice of satellite imagery and attribution of changes to disturbance type strongly affects forest carbon balance estimates

Vanessa S. Mascorro[1*], Nicholas C. Coops[1], Werner A. Kurz[2] and Marcela Olguín[3]

**Abstract**

**Background:** Remote sensing products can provide regular and consistent observations of the Earth´s surface to monitor and understand the condition and change of forest ecosystems and to inform estimates of terrestrial carbon dynamics. Yet, challenges remain to select the appropriate satellite data source for ecosystem carbon monitoring. In this study we examine the impacts of three attributes of four remote sensing products derived from Landsat, Landsat-SPOT, and MODIS satellite imagery on estimates of greenhouse gas emissions and removals: (1) the spatial resolution (30 vs. 250 m), (2) the temporal resolution (annual vs. multi-year observations), and (3) the attribution of forest cover changes to disturbance types using supplementary data.

**Results:** With a spatially-explicit version of the Carbon Budget Model of the Canadian Forest Sector (CBM-CFS3), we produced annual estimates of carbon fluxes from 2002 to 2010 over a 3.2 million ha forested region in the Yucatan Peninsula, Mexico. The cumulative carbon balance for the 9-year period differed by 30.7 million MgC (112.5 million Mg $CO_{2e}$) among the four remote sensing products used. The cumulative difference between scenarios with and without attribution of disturbance types was over 5 million Mg C for a single Landsat scene.

**Conclusions:** Uncertainty arising from activity data (rates of land-cover changes) can be reduced by, in order of priority, increasing spatial resolution from 250 to 30 m, obtaining annual observations of forest disturbances, and by attributing land-cover changes by disturbance type. Even missing a single year in the land-cover observations can lead to substantial errors in ecosystems with rapid forest regrowth, such as the Yucatan Peninsula.

**Keywords:** Carbon modeling, CBM-CFS3, Forest disturbance, MRV, Mexico, REDD+, Activity data

## Background

Monitoring forest cover and change is essential to quantify the amount of carbon that is stored in the vegetation and soils, and the corresponding greenhouse gas (GHG) emissions and removals [1, 2]. Forest ecosystems can mitigate the impacts of climate change by absorbing significant amounts of atmospheric carbon through plant photosynthesis [3, 4]. However, degradation of forests and conversion to non-forested lands is the second largest anthropogenic cause of carbon dioxide ($CO_2$) emissions into the atmosphere [5–8] which has led to initiatives such as Reducing Emissions from Deforestation and Forest Degradation (REDD+) [9].

Natural and anthropogenic disturbances are one of the main drivers that alter the forest structure over time, and understanding their impacts is therefore critical for quantifying carbon stock changes and associated emissions into and removals from the atmosphere [10–13]. Disturbances can range from fire, hurricane, insects and diseases, to timber harvesting, settlement expansion and other human activities. Since each disturbance type impacts the terrestrial carbon cycle in unique ways, annual observations of forest changes by disturbance type are necessary to accurately estimate carbon

*Correspondence: vanessa.mascorro@alumni.ubc.ca
[1] Faculty of Forestry, Department of Forest Resources Management, University of British Columbia, 2424 Main Mall, Vancouver, BC V6T 1Z4, Canada
Full list of author information is available at the end of the article

dynamics in the year of disturbance and the years after the disturbance [5, 10, 14].

Given the complexity of measuring and monitoring the terrestrial carbon cycle, a number of approaches have been developed [7]. Some techniques utilize data collected from ground-plot measurements and allometric equations that relate the physical attributes of trees to aboveground biomass in order to quantify forest carbon stocks (e.g., [15–17]). Other approaches use carbon budget models (e.g., [18]), remote sensing data (e.g. [19, 20]), or a combination of both with field measurements (e.g., [21–24]). Carbon budget modelling is a well-established approach to estimate carbon fluxes at regional to national-scales by integrating data from different spatial and temporal scales [25]. For example, Canada's National Forest Carbon Monitoring Accounting and Reporting System (NFCMARS) uses the Carbon Budget Model of the Canadian Forest Sector (CBM-CFS3) as the framework that integrates data from many sources [22, 26]. The CBM-CFS3 is also used as a decision support tool for forests managers to quantify the ecosystem carbon dynamics at the landscape level.

In compliance with the Intergovernmental Panel on Climate Change (IPCC) guidelines, the CBM-CFS3 quantifies carbon transfers among the five terrestrial carbon pools: above and belowground biomass, litter, dead wood and soil organic carbon, including atmospheric releases of $CO_2$ and non-$CO_2$ greenhouse gasses and transfers to the forest product sector [10]. The CBM-CFS3 offers the advantage of modeling the extent to which tree biomass and dead organic matter are affected by different disturbance types [27]. Following disturbance, it simulates subsequent forest transitions and successional dynamics and represents vegetation transfers to litter, coarse woody debris and soil organic carbon pools, and the decomposition of these pools. Based on the stage of stand development and ecological characteristics of the forest, the model accounts for growth and mortality using yield curves.

Carbon models, such as CBM-CFS3 require detailed spatio-temporal information about forest dynamics, including forest disturbance events, to accurately simulate carbon transfers, emissions and removals. Remote sensing (RS) is the primary source of land-cover data for forest monitoring due to its ability to monitor the Earth's surface on a regular and continuous basis, including areas otherwise difficult to access [28–30]. Recently, an increasing number of regional and global-scale RS studies and products have been developed specifically for forest carbon monitoring and REDD+ at different spatial and temporal scales [20, 31–33]. In a global study, Potter et al. [23] integrated large-scale Moderate Resolution Imaging Spectroradiometer (MODIS) satellite observations with the Carnegie Ames Stanford Approach (CASA), predicting a total of 0.51 Pg C year$^{-1}$ emissions from forest disturbance and biomass burning from 2000 to 2009. At a regional-scale, Masek and Collatz [21] integrated the CASA model with forest inventory data and higher spatial resolution satellite imagery to predict C fluxes from 1973 to 1999. They used Landsat time-series analysis to assess the effects of land-clearing disturbances (such as logging, harvesting and urbanization) on ecosystem productivity and quantified carbon emissions from biomass losses, decomposition and decay.

Significant variability among carbon stock estimates contributes to uncertainty in estimates of emissions and removals. Such variability is potentially due to differences inherent to the methods used, the spatial and temporal scales involved, and the definition of the components included [34]. A recent study by Achard et al. [33] estimated carbon emissions from deforestation in tropical countries for two decades (1990–2000, 2000–2010) by combining forest cover change maps derived from Landsat imagery and pan-tropical biomass maps. Contrasting their results with those obtained by Hansen et al. [35], they found discrepancies that can be explained by different approaches used in the satellite image analysis and their definitions of forest. A global study for tropical countries by Saatchi et al. [31] integrated inventory data and satellite light detection and ranging (LiDAR) data to quantify the carbon stored in the living biomass (247 Pg C) circa 2000. In a similar study, Baccini et al. [32] estimated carbon emissions from pan-tropical deforestation and land-use from 2000 to 2010, combining aboveground biomass with regional deforestation rates derived from LiDAR and MODIS, respectively. Comparing their results with the Forest Resource Assessment (FRA), they found their estimates (228.7 Pg C) were 21 % higher [32]. Likewise, Mitchard et al. [36] compared estimates from Saatchi et al. [31] and Baccini et al. [32] across the Amazon forest to a unique dataset of ground-plot data, and found that the regional differences over or underestimated the forest carbon estimates obtained from the ground data by >25 %.

It is critical to understand the potential role and implications of different RS observations for estimating ecosystem carbon dynamics following disturbance, in order to reduce uncertainties in measuring and monitoring carbon emissions and removals [37, 38]. In REDD+ studies, land-cover change observations are one of the main components (also called activity data) required for Monitoring Reporting and Verification (MRV) systems [39]. These activity data can be observations of changes in forest areas or land conversions between categories [40]. The design of efficient MRV systems requires an understanding of the implications of different attributes of remote sensing products on estimates of GHG emissions and removals.

This study assesses the impacts of using different RS products on estimates of terrestrial carbon dynamics. We examined the impacts of three attributes of RS products on the estimates of GHG emissions and removals: (1) the spatial resolution (30 vs. 250 m), (2) the temporal resolution (annual observations vs. multi-year observations), and (3) the attribution of forest cover changes to disturbance types. To do so we utilize four contrasting RS datasets—two land-cover change maps and two thematic maps—to compile activity data that we then use as spatially-explicit inputs of annual disturbance events to the CBM-CFS3. The RS data were derived from MODIS, Landsat, and a combination of Landsat-SPOT satellite imagery developed using different approaches, spatial and temporal resolutions. We derived two spatially-explicit layers for each of the RS products, one with land-cover change area, and the second attributing the observed change to its underlying disturbance cause. We then provided these activity data as inputs to CBM-CFS3 and estimated annual and cumulative greenhouse gas fluxes from 2002 to 2010 in the tropical dry forests of the Yucatan Peninsula, Mexico. We conclude with a discussion of key findings and implications of different activity data sources for future carbon budget analysis.

## Methods and data

### Study area

Our study focused on a ~3.2 million ha area located in the northwest side of the Yucatan Peninsula (YP), Mexico, covering parts mainly from Yucatan and Campeche states and a small area of Quintana Roo (see Fig. 1). The region's vegetation is dominated by secondary lowland dry tropical forests [41] of mixed ages that have regenerated after cycles of shifting agriculture and land abandonment. The region has a tropical climate, with rainy summers from June to October and dry winters from November to May. The mean average temperature is 26 °C with precipitation levels ranging from 900 to 1400 mm year$^{-1}$ [42]. The regional topography is characterized by flat limestone areas and low moderate hills, with an average slope of 7 % and a mean elevation of 116 m [18]. The main soil types are well-drained rendzinas and shallow rocky lithosols [43].

### Remote sensing data

Remote sensing observations are essential to track forest disturbance impacts and changes in land cover and forest

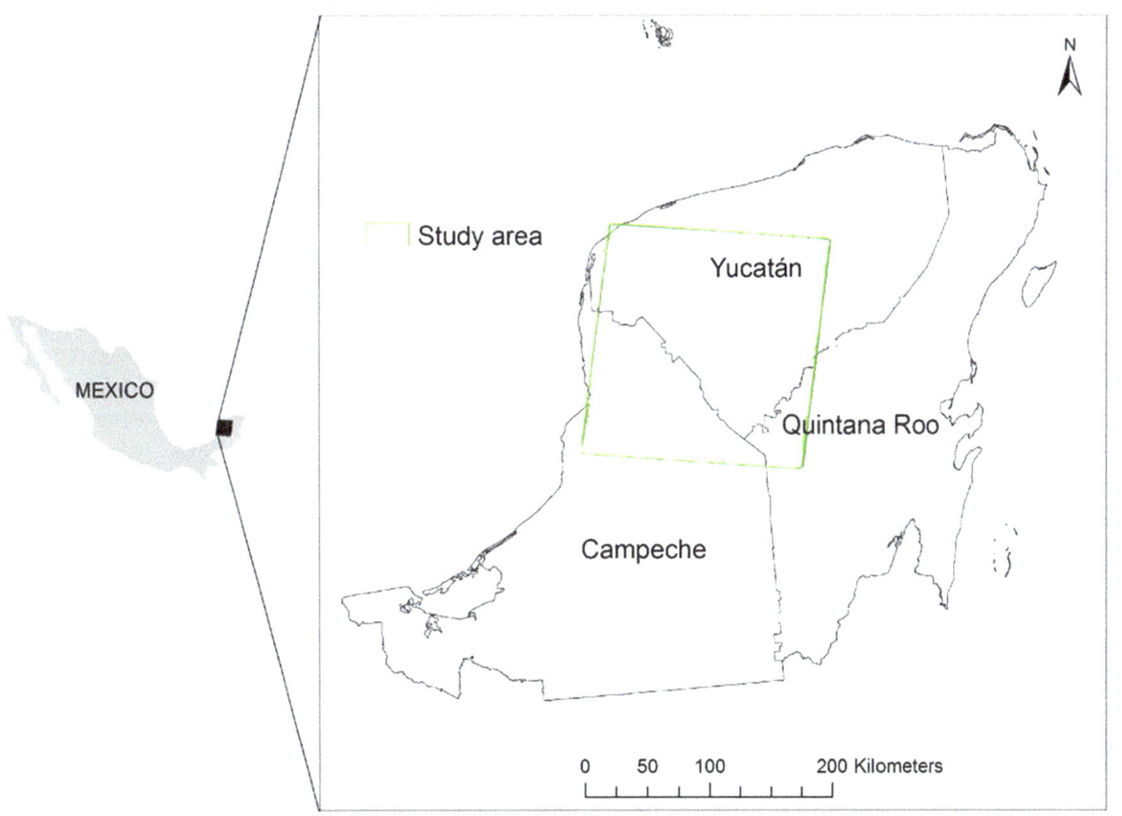


**Fig. 1** Study area: the Yucatan Peninsula, Mexico

carbon at a range of spatial scales [44, 45]. Land-cover changes are those remotely sensed observations of the Earth's surface that changed from one point in time over two or more time periods because of any disturbance [7], including both natural and human-induced causes. We compiled four remote sensing disturbance products from Landsat, MODIS and a combination of Landsat-SPOT imagery—two classification and two pixel-based maps—derived using a range of approaches to compare their impacts on carbon budget estimates (see Table 1). We use the term pixel-based approach to differentiate the two maps that are labeled pixel by pixel as change/no change, from the thematic or classification maps that are comprised by various pixels grouped together into larger areas sharing similar characteristics (land cover class).

### Landsat: VCT

The vegetation change tracker (VCT) is an automated algorithm that implements a disturbance index (DI, [46]) to reconstruct the history of the landscape disturbance on an annual or bi-annual basis (depending on data quality and availability) using Landsat 30 m time-series analysis [47]. This layer was generated and provided by the NASA Goddard Space Flight Center Biospheric Sciences Research branch. Each pixel of the spatial dataset was labeled according to the year in which the forest cover change was identified, from 1985 to 2010 (represented by 16 classes). From this map, we extracted disturbed areas for the years 2003, 2004, 2005, 2007, 2008 and 2010. Change data were unavailable due to cloud cover in 2002, 2006 and 2009.

### Landsat: Hansen

Annual land-cover change data from a global forest cover change map derived from Landsat imagery at 30 m spatial resolution were also available from Hansen et al. [35]. The coverage provided direct estimates of forest cover loss and gain from 2000 to 2012, including detailed information of annual forest loss. Assessments of forest cover loss included pixels that completely lost tree canopy cover from a stand-replacing disturbance. Here, we obtained annual cover loss areas for our study period (2002–2010).

### Landsat-SPOT: INEGI

We also retrieved three classification maps from the National Institute of Statistics and Geography of Mexico (in Spanish: INEGI): the vegetation and land-use cartography series developed for 2002, 2007 and 2011 at a scale 1:250,000 with approximately 70 classes (INEGI, SIII, SIV, and SV). The vegetation maps were generated using manual interpretation of Landsat and SPOT satellite images, validated with ground-plot data. In addition, a digital elevation model and ancillary datasets that describe the climate, geology, hydrology and topography of the Mexican ecosystems were used in their construction. These maps are a conventional source of information in Mexico for assessing the status and condition of the forested land base. The minimum mapping unit in all three vegetation and land-use cartography series is 25 ha [48]. We used the INEGI maps to generate two land-cover change maps: (1) changes detected from period observation one to two (2002–2007); and 2) changes detected between observation two and three (2007–2011; [49–51]). To simplify this process, we grouped the INEGI vegetation classes into twelve classes established by the National Commission for Knowledge and Use of Biodiversity of Mexico (CONABIO) in the Monitoring Activity Data for the Mexican REDD+ program (MAD-MEX; [48]) and assessed the change from one class to another.

### Modis

Finally, we retrieved six classification maps developed by CONABIO in the North American Land Change Monitoring System (NALCMS) project using MODIS satellite imagery at 250 m spatial resolution [52]. Annual spatially-explicit maps were developed from 2005 to 2011, using monthly composites of MODIS, multiple classifications (ensemble classifier) with decision trees and attributes from auxiliary datasets to characterize 15 land-cover classes in Mexico [53]. These attributes include a digital elevation map, aspect, slope, mean temperature, maximum precipitation, sampling data, and aerial photography at high-spatial resolution. From these maps, we assessed the annual transition from one land-cover class to another from 2005 to 2010.

**Table 1 Remote sensing products used as activity data inputs for carbon modeling with the CBM-CFS3 in the Yucatan Peninsula from 2002 to 2010**

| Map alias | Source | Type | Satellite | Spatial resolution | Temporality |
|---|---|---|---|---|---|
| VCT | NASA | Vegetation Change Tracker map | LANDSAT | 30 m | 2003, 2004, 2005, 2007, 2008 and 2010 |
| Hansen | University of Maryland | Forest cover loss map | LANDSAT | 30 m | 2002, 2003... 2010 |
| INEGI | INEGI | Classification map | LANDSAT-SPOT | 30 m (25 ha MMU) | 2002–2005, 2005–2007, 2007–2011 |
| MODIS | NALCMS/CONABIO | Classification map | MODIS | 250 m | 2005, 2006... 2011 |

*MMU* minimum mapping unit

#### Accuracy assessment of forest cover change estimates

We performed an independent assessment of the remote sensing disturbance datasets using ground-plot data obtained from the Mexican National Forest Inventory (NFI). The NFI is the primary data source of information for forest and ecosystem studies in Mexico [54]. Data are collected in a two-stage sampling scheme comprised of primary plots of one-hectare size and four sub-sampling sites of 400 $m^2$ each. More than 150 variables are measured on a 5-year cycle, including attributes of the overstory and understory vegetation, the soil, and environmental characteristics of the landscape. For more details concerning the variables, methods and sampling design, readers are referred to the NFI sampling manual [54].We obtained forest change data over the period 2004–2012 of basal area loss/no-loss per hectare computed by Mascorro et al. [55] for 647 plots that were located across the study area. We intersected these plots with the RS derived land-cover change areas to identify and compare the number of plots with forest loss to the proportion of changed areas detected by each RS product using a standard error matrix. The matrix performs a cross-tabulation between the numbers of observations mapped with RS that agree with what is observed on the ground to assess the level of accuracy of the mapped products [56, 57].

#### Disturbance attribution

We attributed the land-cover change estimates obtained from the RS products using a Multi-Scale, Multi-Source Disturbance (MS-D) approach developed by Mascorro et al. [55]. This approach integrates RS data, forest inventory and ancillary datasets to attribute the land-cover change observations to the most likely disturbance type (natural or anthropogenic). Using NFI data and historical records of forest disturbances in tabular format (both considered ground-truth data), Mascorro et al. [55] derived annual spatially-explicit layers of major forest disturbance types, and obtained the forest change observed in the ground-plots. A regression tree analysis was undertaken to identify which of the constraining variables (forest disturbance types) best explained the observed forest loss in the field. Once the most likely cause of change was identified, remote sensing maps were used to obtain pixels with land-cover change and overlaid these with the spatially-explicit layers of disturbances characterized by type. Finally, the change was attributed to the most likely disturbance cause according to the relevance resulting from the regression tree analysis.

We retrieved ancillary forest disturbance layers compiled by Mascorro et al. [55] and overlaid them with the VCT, Hansen, MODIS and INEGI RS observations. The disturbance datasets retrieved included annual, spatially-explicit, natural and anthropogenic disturbances from 2005 to 2010: fires, hurricanes, and forest management areas. To characterize the permanent conversion of forestland to non-forestland caused by settlement, we used road coverages retrieved from INEGI [58]. In addition, land-cover classes that changed to urban areas in the classification maps were attributed as settlement.

#### Pre-processing data

To prepare the input data and provide the parameters required by the CBM-CFS3 for each of the simulation runs, we used a software tool called "Recliner" developed and provided by the carbon accounting team of the Canadian Forest Service. The tool facilitates the processing of large volumes of spatially-explicit data and assists users with the preparation of required input data for each simulation. We customized Recliner and matched the RS disturbance types with pre-defined disturbance matrices stored in the CBM-CFS3. Disturbance matrices define the impacts on the carbon pools for each disturbance type and specify the amount of carbon that is transferred from the biomass to the dead organic matter pools or is released into the atmosphere during a particular disturbance type [10, 27]. Disturbance matrices also define the amount of carbon transferred from forest ecosystems to the forest product sector by harvesting activities, and the amount of carbon released to the atmosphere as methane ($CH_4$), carbon monoxide (CO) and carbon dioxide ($CO_2$). Fire disturbances were matched to "wildfires", settlement to "deforestation" and harvest to "clear-cut with slash-burn".

The CBM-CFS3 can simulate stand-replacing disturbance by restoring the stand age to the initiation stage, or simulate partial mortality. "Generic mortality" disturbance matrices ranging from 5 to 95 % of impact are among the default choices [27]. Since the CBM-CFS3 does not contain a specific matrix for hurricanes, we matched these events with a generic disturbance matrix of "10 % mortality", estimated by dividing the mean basal area loss explained by hurricanes in Mascorro et al. [55] studies, over the mean basal area estimated across the Yucatan Peninsula (1.71 $m^2ha^{-1}$/15.54 $m^2ha^{-1}$). With Recliner, we assigned the year of disturbance impact to the corresponding land-cover change detection year. In cases where annual data were not available, we distributed equally the disturbance events among the number of years contained in the period of detection (e.g. VCT disturbances detected in 2007 where assigned in equal proportions to 2006 and 2007, since no observations were available for 2006).

#### Carbon budget modeling

We provided annual disturbance events to the CBM-CFS3 to quantify carbon exchange from tree biomass mortality, plant detritus decay, soil organic carbon and forest regrowth after disturbance. Additional parameters and data required by the CBM-CFS3 for the simulations were provided by the carbon accounting team of the Canadian Forest Service, and researchers from the Mexican Forest Service. These parameters were kept constant in all simulations so that only the activity data differed among the eight scenarios.

Ancillary datasets included detailed information on forest inventory, forest type, forest status, ecological boundaries, age-class structure, and yield curves [59]. Stand age and yield curve data, in combination with the ecological characteristics of the site, are used by the CBM-CFS3 to simulate forest growth and carbon accumulation [10]. The stage of stand development is also used to simulate litterfall and decomposition rates for dead organic matter and soil carbon. For forest status, we used protected and non-protected conservations areas, which the model employs as a classifier to differentiate the conditions on the landscape. We also used as a classifier the ecoregions level-I retrieved from the Commission for Environmental Cooperation (CEC) [60]: tropical humid forests and tropical dry forests. Estimates of carbon sinks and sources were then generated with the CBM-CFS3 for each of the two spatially-explicit layers of the four RS products, i.e. with and without attribution of cover loss to disturbance types. For simulations without attribution to specific disturbance types we assumed that all forest cover changes were caused by stand-replacing harvest with slash-burning.

## Results

### Forest change detection

The spatial variation in the location and extent of changes in forest area detected by the VCT, Hansen, MODIS and INEGI maps can be observed in Figs. 2 and 3. Overall, MODIS products detected much less changes across the landscape than the Landsat products, likely due to its lower spatial resolution. As can be seen in Fig. 2b, d, landscape changes between the VCT and Hansen products exhibited similar patterns. With a 30 m spatial resolution, these products detected finer disturbance events throughout the study area. Undisturbed forest areas were primarily located in the center of the region, expanding to the northwest and southeast of the area. In contrast, the INEGI maps generated with a classification approach detected larger contiguous areas of change from land-cover class transitions, due to the various 25 ha MMU's of similar characteristics grouped together into a class (Fig. 3e, f).

The cumulative area of change detected in the study area by each RS product across the analysis period is shown in Fig. 4. It is apparent that the Landsat products detected the most change in all of the remote sensing approaches. However, the amount detected was less in the pixel-based approach than that derived from the classification maps (INEGI).

The range between the maximum (13 % by INEGI) and minimum (1 % by MODIS) cumulative percentages of area change detected, highlights the uncertainties resulting from the differences in spatial and temporal characteristics inherent in each RS product. Temporal trajectories of the land-cover changes detected in annual time-steps provided additional information on the inter-annual variation of disturbance events as shown in Fig. 5.

### Forest change accuracy assessment

Table 2, summarizes the results of intersecting the NFI plots describing forest cover loss, with the VCT, Hansen, and INEGI land-cover change products in an error matrix. Of the 647 NFI plots located in the study area, 119 showed a forest cover loss, and 528 had no evidence of any basal area reduction in the plot. The first row of each table contains the number of mapping units that each RS product detected as forest cover loss compared to what was observed on the ground with the NFI plots (the total sum of the first column). As can be seen in Table 2a and b, the overall accuracy of the non-classification maps, the VCT (83 %) and Hansen (82 %), maps produced using pixel-based image processing approaches was similar, whereas the INEGI accuracy was 79 % (Table 2c). However, we can observe that all three maps presented a low producer's accuracy, detecting only a small fraction of the changes observed in the NFI plots likely due to forest degradation. The largest uncertainties were found in the MODIS maps, as there was no correspondence between the MODIS changed pixels and the NFI plots.

### The impact of activity data on estimates of carbon fluxes

We used the CBM-CFS3 model to simulate the effects of activity data on emission and removal estimates during, and following disturbances from 2002 to 2010 (Fig. 6). Results show the annual net changes in ecosystem carbon stocks, with negative numbers indicating a decrease in carbon stocks (an emission of carbon to the atmosphere) and positive numbers an increase in carbon stocks (carbon uptake from the atmosphere). The estimates show both differences in the magnitude of changes and differences in trends due to differences in activity data.

The values from these simulations should not be considered as absolute values as research continues to improve yield curves and other ecological and modelling parameters that will likely affect future estimates of fluxes. What matters here is the impact of the activity

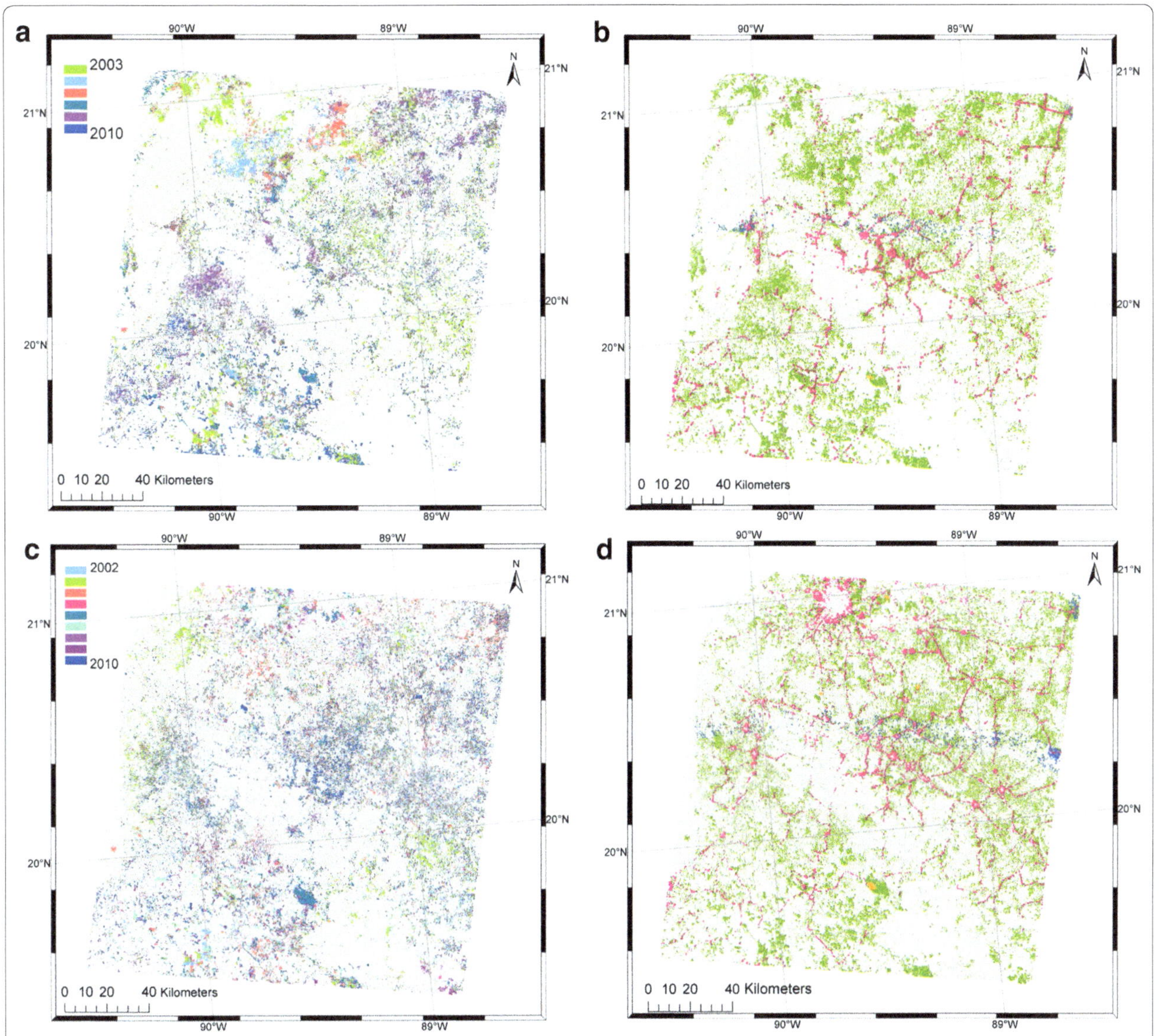


**Fig. 2** Land-cover change maps derived from the different remote sensing products. VCT map: (**a**, **b**); Hansen map: (**c**, **d**). *Left side* maps show annual non-attributed disturbances; *right side* maps show attributed disturbances: *green* harvest, *pink* settlement, *blue* hurricane, *orange* fire

data on the differences in the annual and cumulative estimates of carbon fluxes (Fig. 6).

The cumulative difference in the carbon balance estimates resulting from the four RS products for the 9-year period (2002–2010) was over 30.7 million Mg C (112.5 million Mg $CO_{2e}$) see Fig. 7, with MODIS-derived estimates suggesting a large sink while INEGI-derived estimates remained a source for the whole period (Fig. 6).

Since the ecological parameters and additional data inputs used in each of the simulations were kept constant, the observed differences in the estimates reflect only the effects of the spatial and temporal resolutions inherent to each of the RS observations, and the impacts of attributing cover changes to disturbance types with different impacts on carbon stocks.

Over the simulation period we observe the overriding trend of the MODIS disturbance detection underestimating cover changes and the associated carbon losses compared to the Landsat RS products. MODIS-based simulations shifted from a source to a sink in 2005 and remained a sink for the remainder of the simulations. Landsat-derived estimates in the other hand, followed a similar trend in 2002 and 2003, and began diverging in 2004. From 2005 to 2007, simulations based on VCT and Hansen predicted increasing carbon stocks. Consistent with the amount of land-cover change detected by each product

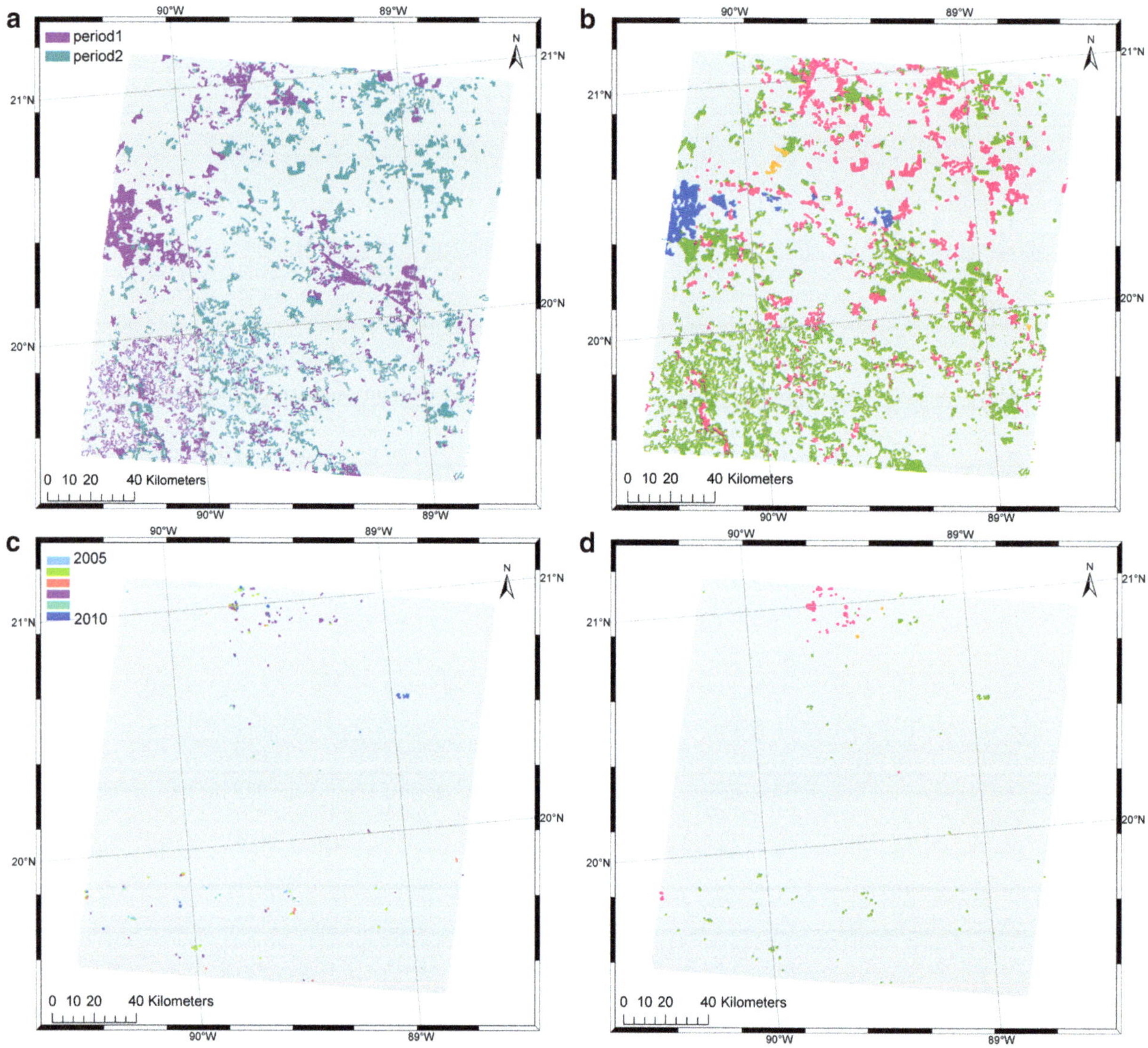


**Fig. 3** Land-cover change maps derived from the different remote sensing products (thematic). INEGI maps: (**a**, **b**); and the MODIS maps: (**c**, **d**). *Left side* maps show annual non-attributed disturbances; *right side* maps show attributed disturbances: *green* harvest, *pink* settlement, *blue* hurricane, *orange* fire

(Fig. 5), the rate of increase in carbon uptake estimated with VCT from 2005 to 2007 was higher than that simulated with Hansen. In 2007, Hansen removals were 234,188 Mg C, whereas the VCT estimates were 1,033,608 Mg C (Fig. 6). While some disturbance events were detected in 2005–2007, the amount of area affected was not enough to bring the sink back to a source (e.g., estimates in 2009 from Hansen). Fewer disturbances were observed by VCT in this period, which translated to fewer carbon emissions and more carbon accumulation due to forest growth without disturbance. Nevertheless, in 2008 estimates from these two data sources went in opposite directions.

In the Hansen simulation, the forest continued to sequester carbon in 2008 as the amount of affected area decreased by 37 % compared to 2007 (Fig. 5). This was translated by the CBM-CFS as less carbon emissions and more living biomass accumulation, sequestering 3 times more carbon than in 2007 (234,188 Mg C in 2007 vs 729,215 Mg C in 2008). Conversely, with the VCT, the amount of change area detected in 2008 was more than double compared to 2007 (10,409 vs 28,278 ha), releasing considerable amounts of carbon (1,033,964 Mg C) to the atmosphere from forest disturbance events that killed the vegetation and redistributed the

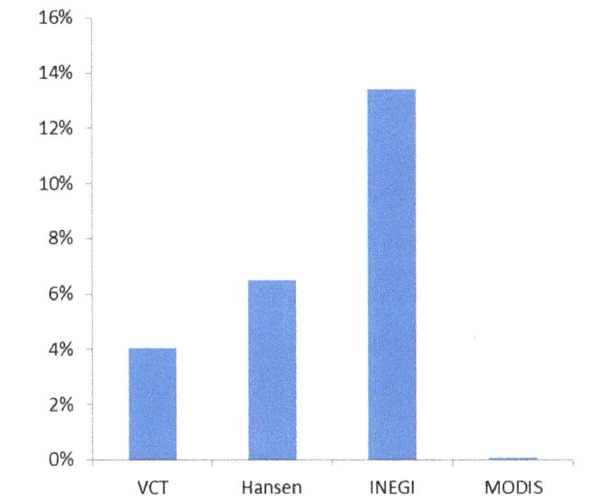


**Fig. 4** Cumulative area with change (as percentage of total area) detected in the study area from 2002 to 2010 with Landsat (VCT, Hansen and INEGI) and MODIS satellite imagery

carbon in the dead organic matter pools during that year.

Moreover, Mascorro et al. [55] found that 2009 was an intense year of disturbances in the Yucatan Peninsula. Accordingly, the impact of the disturbance events detected by Hansen, resulted in significant amounts of carbon emissions into the atmosphere (stocks dropped by 2,774,453 Mg C; Fig. 6). MODIS-derived estimates also showed a decrease in carbon content in 2009, but by much smaller amount due to its limited capacity to detect changes in the landscape. However, with the VCT no land-cover change observations were included in 2009 due to limitations of image acquisition and cloud coverage, and therefore the simulations show no decrease in carbon emissions in 2009, despite the fact that the change observed in 2010 was equally distributed to 2009 and 2010. It is likely that some of the 2009 disturbance events were not detected by the VCT in 2010 due to the fast regrowth rate of the tropical forests of the Yucatan Peninsula.

Overall, estimates from non-attributed changes presented more carbon emissions than the attributed estimates in all the simulations. In the INEGI simulations this difference was more evident. In the second period, emissions from the non-attributed disturbances (7 million Mg C) were more than double than the attributed ones (3.2 million Mg C). When converting the additional non-$CO_2$ GHG emissions from fires (see Fig. 8a–c) in the form of carbon monoxide (CO), methane ($CH_4$) and nitrous oxide ($N_2O$) with their global warming potential (GWP) to 100 years (1, 25 and 298 respectively), we can see the additional $CO_2$ equivalent ($CO_2e$) emissions contributed by non-$CO_2$ greenhouse gases (i.e. 3 million Mg $CO_2e$ more; see Fig. 8d).

## Discussion

This study analyzed the contribution of three main attributes from RS activity data on estimates of forest carbon dynamics: (1) the spatial resolution (30 vs. 250 m), (2) the temporal resolution (annual vs. multi-year observations), and (3) the attribution of forest cover changes to disturbance types. Annual carbon fluxes were simulated with the CBM-CFS3 from 2002 to 2010 over a 180 × 180 km study area of dry tropical forests in the Yucatan Peninsula, Mexico.

### Spatial resolution

The large difference in annual and cumulative carbon estimates generated with MODIS and Landsat-derived disturbance observations suggests that increased spatial resolution should be the first priority when deriving

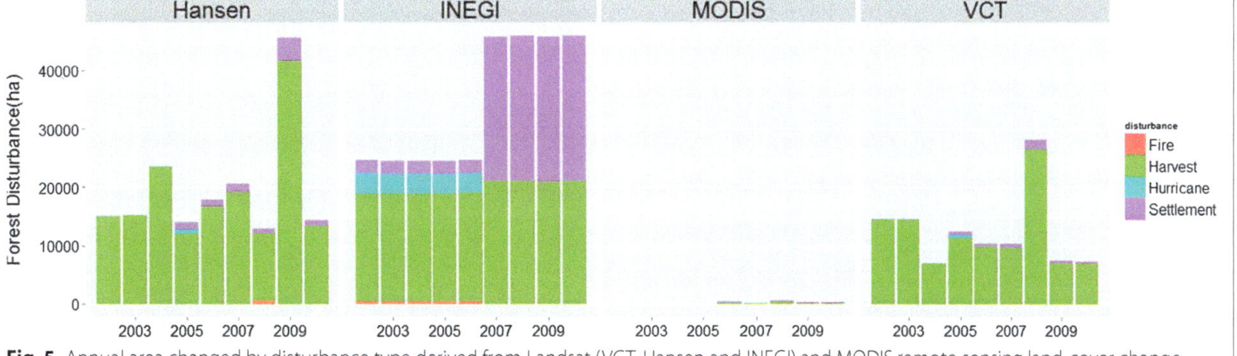


**Fig. 5** Annual area changed by disturbance type derived from Landsat (VCT, Hansen and INEGI) and MODIS remote sensing land-cover change products

**Table 2 Error matrix of landsat maps VCT (a), Hansen (b) and INEGI (c) identification of change and no change areas compared to the ground-plot data**

**(a) VCT**

| | Change | No change | Sum | Producer's |
|---|---|---|---|---|
| NFI plots | | | | |
| Change | 18 | 101 | 119 | 15 % |
| No change | 11 | 517 | 528 | 98 % |
| Sum | 29 | 618 | 647 | |
| User's | 62 % | 84 % | | |
| Overall accuracy = 83 % | | | | |

**(b) Hansen**

| | Change | No change | Sum | Producer's |
|---|---|---|---|---|
| NFI plots | | | | |
| Change | 27 | 92 | 119 | 23 % |
| No change | 22 | 506 | 528 | 96 % |
| Sum | 49 | 598 | 647 | |
| User's | 55 % | 85 % | | |
| Overall accuracy = 82 % | | | | |

**(c) INEGI**

| | Change | No change | Sum | Producer's |
|---|---|---|---|---|
| NFI plots | | | | |
| Change | 26 | 93 | 119 | 22 % |
| No change | 44 | 484 | 528 | 92 % |
| Sum | 70 | 577 | 647 | |
| User's | 37 % | 84 % | | |
| Overall accuracy = 79 % | | | | |

activity data for carbon modeling. Carbon flux estimates derived from Landsat at higher spatial resolution reflected the impacts of more accurate activity data. In contrast, the coarse spatial resolution of MODIS-derived products results in a substantial underestimate of change events and thus overestimates forest carbon sinks over the observation period. The latter occurred despite having MODIS observations on an annual basis from 2005 to 2010. While Mascorro et al. [55] were able to characterize major stand-replacing disturbances for the entire Yucatan Peninsula with MODIS, results from this study showed that this 250 m resolution satellite imagery has limitations for providing activity data for carbon modeling at finer spatial scales. Typically, satellite imagery from MODIS has focused on broad scale studies (e.g., [23, 29]).

Moreover, we found that the pixel-based image processing approach using VCT and Hansen was more accurate than change inferred from classification maps (i.e., INEGI, MODIS) (see Table 2). While the INEGI simulations of the first period observations (2002–2006) show a similar trend to VCT and Hansen, change observations for the period 2007–2011 are much higher than any of the others. This is likely due to the fact that INEGI maps were generated with a classification approach, detecting larger areas of change formed by clusters of pixels that share similar characteristics and attributes. Therefore the larger disturbances observed over the second period, resulted in higher estimates of change likely due to larger areas that transitioned from one class to another (as opposed to single-pixel changes). Some of the predicted changes using thematic maps can also result from inaccuracies inherent to the classification method [61, 62], and omission and commission errors [63]. Nevertheless, post-classification analysis—a change detection technique that compares the initial and final class condition between two classification maps in time- has become a popular method for change detection [64], that helped to better identify settlement expansion embedded in the land-cover classification approach as urban areas. Yet, RS products generated with a pixel-based approach presented some limitations in attributing the changes to the underlying disturbance type due to the lack of spatially-explicit ancillary datasets. Further research would be required to increase the accuracy of land-cover changes

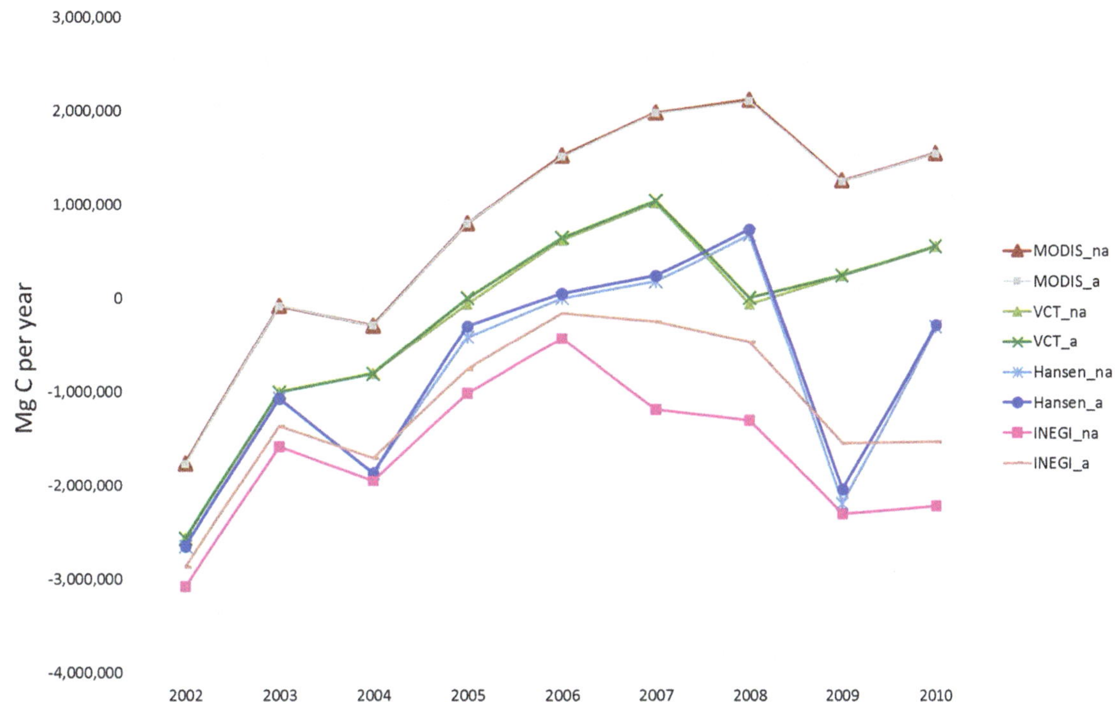


**Fig. 6** Annual carbon fluxes estimated with different sources of activity data in the Yucatan Peninsula from 2002 to 2010; *a* attributed, *na* no-attributed

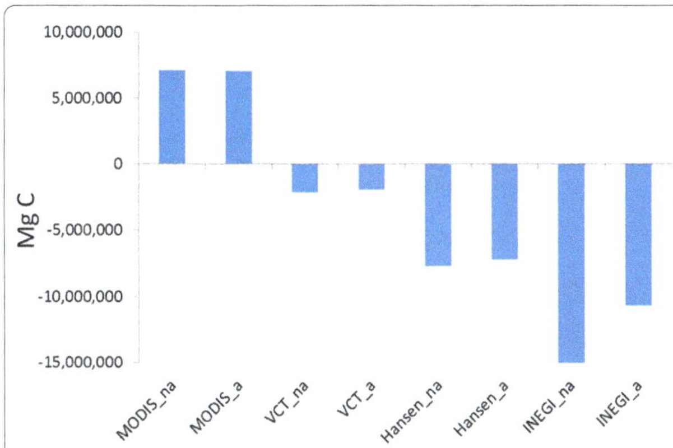


**Fig. 7** Cumulative carbon fluxes, estimated with different sources of activity data in the Yucatan Peninsula from 2002 to 2010; *a* attributed, *na* no-attributed

characterized by disturbance type using additional datasets.

### Temporal resolution

Reliable estimates of carbon dynamics require detailed observations of drivers of change on annual basis [14]. Our results showed that higher temporal resolution RS products improved the carbon dynamics estimates and captured the inter-annual variability in the forest. Consistent with the large disturbances detected in 2009, carbon fluxes predicted with the Hansen and MODIS products showed a decrease in total ecosystem carbon content. However, due to the lack of disturbance events detected with the VCT in 2009, corresponding decreases in the carbon stocks from the disturbance events were not reflected in the CBM-CFS3 estimates. This occurred despite the fact that the disturbance events observed over the period 2009–2010 were equally distributed in 2009 and 2010. This result showed that even missing a single year in the land-cover observations can lead to substantial errors; especially in ecosystems with rapid-regrowth forests, such as the Yucatan Peninsula.

### Attribution of land-cover changes by disturbance type

Natural and human-induced disturbances modify the stand age and succession dynamics of the landscape in a unique way, changing the live biomass and dead organic matter stocks and subsequent turnover [65, 66]. Results from this study showed the relevance of identifying the cause of change in the landscape. The cumulative difference between scenarios with and without attribution of disturbance types was over 5 million Mg C for a single Landsat scene over the 9-year-period. INEGI estimates over the second period (2007–2011) from the

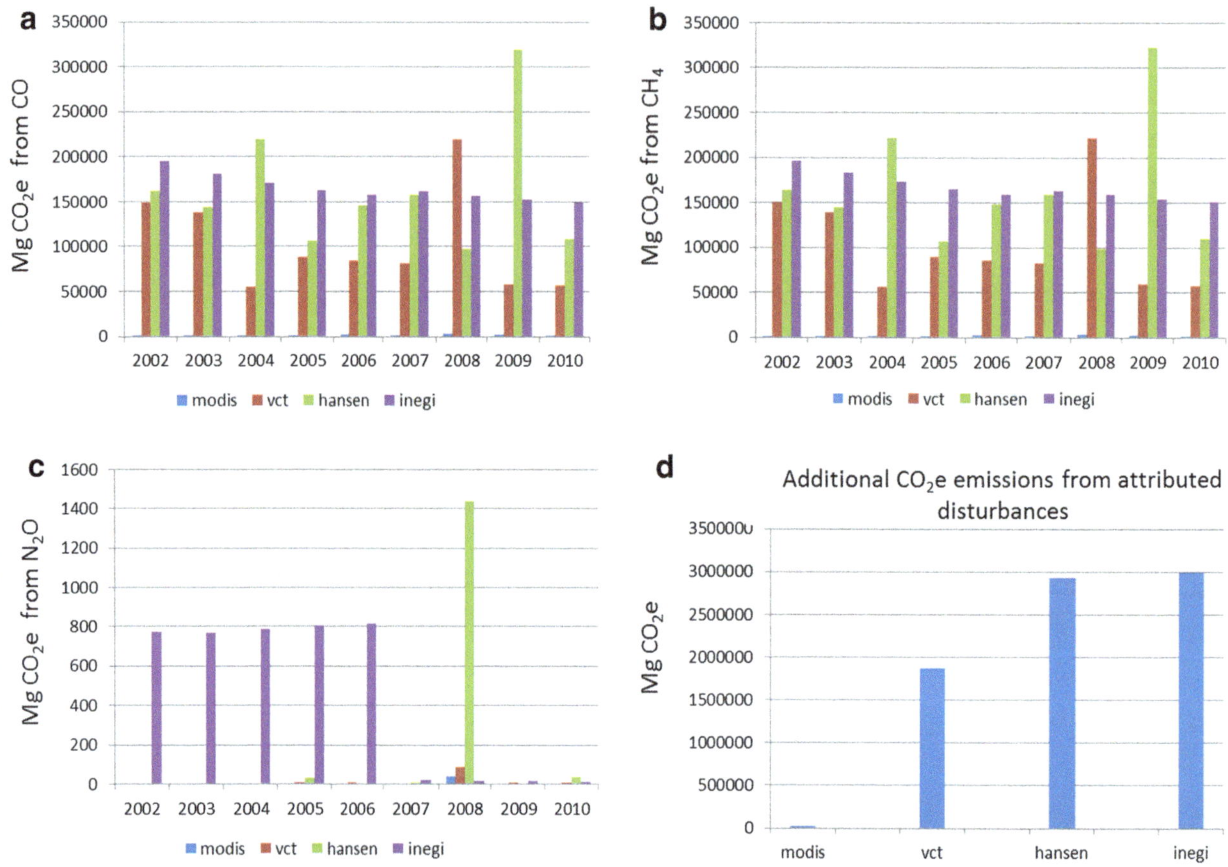


**Fig. 8** $CO_2$ equivalent GHG emissions from carbon monoxide CO (**a**), methane ($CH_4$) (**b**) and nitrous oxide $N_2O$ (**c**) with a Global Warming Potential (GWP) time horizon of 100 years, and their cumulative-additional $CO_2$e emissions from attributed disturbance types (**d**)

non-attributed disturbances, for example, dropped 84 % more than the attributed ones yielding higher emissions (3.2 million Mg C more). Since the default disturbance type associated with harvesting was clear-cut with slash-burn, it generated more emissions because it involved burning. Depending on the disturbance type, the CBM-CFS3 disturbance matrices determines how much biomass gets killed, re-distributing the carbon in the litter and dead organic matter pools accordingly; affecting the carbon changes more adequately. These differences between with and without attribution were even more pronounced when reported as $CO_2$ equivalent emissions (as required in REDD+ projects, see Fig. 8a–c), because fires cause additional non-$CO_2$ GHG emissions (in the form of $CH_4$ and $N_2O$), with higher global warming potentials than $CO_2$ (see Fig. 8d).

The carbon balance estimates obtained for this study are based on CBM-CFS3 simulations using preliminary inventory, age distribution and yield data for the study area [59]. Ongoing research as part of bigger collaboration of members from CONAFOR´s project "Reinforcing REDD+ and South–South Cooperation", SilvaCarbon program, Colegio de Postgraduados, and research supported by the CEC will refine and improve the CBM-CFS3 input data. The absolute values of the numerical estimates obtained in this study may thus change in the future, but the general conclusions about the impacts on carbon balance estimates of the spatial and temporal resolution of RS products, and the attribution of land-cover change to causes of disturbance will not be affected by future changes to other input data of the model.

## Conclusions

Systematic forest monitoring, reporting and verification (MRV) systems are required to aid in the successful development of national and regional strategies for REDD+, and to ensure long-term commitments to preserve forests [25, 40, 67]. This can only be achieved by implementing RS observations that will allow monitoring of large areas of land in a regular, consistent, and cost-efficient way and by developing modelling tools that can translate activity data derived from RS products into policy-relevant estimates of GHG emissions and removals. Results from this study provide an improved understanding of the role of RS disturbance observations on ecosystem carbon dynamics and the range of variability of carbon flux estimates following disturbance. Developing climate change mitigation scenarios, priorities, and initiatives requires further knowledge of the drivers of deforestation and forest degradation that can have a significant impact in the reduction of GHG emissions [7, 25, 68].

We conclude that spatial scale represents a serious mapping constraint when attempting to encompass large forest areas for ecosystem carbon accounting. Therefore, carbon monitoring decisions should consider direct trade-offs between the spatial detail of finer resolution products (e.g., Landsat, LiDAR, RapidEye) with more precision and less area covered per scene, versus moderate and broad spatial resolution observations derived at larger scales (e.g., MODIS). Our results support findings from the Global Forest Observations Initiative (GFOI) regarding MODIS imagery as "too large to be used for generating REDD+ activity data", suggesting it could be used in complimentary applications (e.g., monitoring near-real time forest change indicators) [69].

This research documents cumulative differences from four different RS products in estimates of carbon emissions of 30.7 million Mg C (112.5 million Mg $CO_2$e) over a 9-year period in a single Landsat scene in Mexico. While we did not attempt to extrapolate this to the national-scale implications, it is evident that with 135 Landsat scenes required to cover all of Mexico [48], the choice of satellite data sources to derive activity data can have a large impact on uncertainties in national estimates of forest carbon budgets. The magnitude of these differences highlights that efforts to improve the activity data can yield substantial reductions in uncertainty of GHG estimates at the regional or national scale. Uncertainties arising from activity data can be reduced by, in order of priority, increasing spatial resolution from 250 to 30 m, obtaining annual observations of forest disturbances, and by attributing land-cover changes by disturbance type. Results from this study can help justify the monitoring expenses required to reduce uncertainties in GHG emissions and removals estimates for countries like Mexico, with a variety of small scale-low impact disturbances and a strong interest in participating in REDD+ initiatives. Improved estimates also help meet the requirements of the Intergovernmental Panel on Climate Change (IPCC) to identify, quantify and reduce sources of uncertainty in estimates of GHG emissions and removals as far as is practicable.

**Abbreviations**

CEC: Commission for Environmental Cooperation; CBM-CFS3: carbon budget model of the Canadian Forest Sector; CONABIO: National Commission for Knowledge and Use of Biodiversity of Mexico; GHG: greenhouse gas; INEGI: National Institute of Statistics and Geography of Mexico; LiDAR: light detection and ranging; MODIS: moderate resolution imaging spectroradiometer; NFI: Mexican National Forest Inventory; REDD+: reducing emissions from deforestation and forest degradation; RS: remote sensing; VCT: vegetation change tracker algorithm; YP: Yucatan Peninsula.

**Authors' contributions**

All the authors have made substantial contribution towards the successful completion of this manuscript. They all have been involved in designing the study, drafting the manuscript and engaging in critical discussion. VSM, contributed with the processing, data analysis and write up. NCC and WAK contributed to the interpretation, quality control and revisions of the manuscript. MO contributed with the analysis of forest status, age distribution and yield data. All authors read and approved the final manuscript.

**Author details**

[1] Faculty of Forestry, Department of Forest Resources Management, University of British Columbia, 2424 Main Mall, Vancouver, BC V6T 1Z4, Canada. [2] Canadian Forest Service, Natural Resources Canada, 506 West Burnside Road, Victoria, BC V8Z 1M5, Canada. [3] Proyecto Mexico-Noruega, Comision Nacional Forestal, Coyoacán, D.F. Col. del Carmen Coyoacán, CP 04100 Mexico, Mexico.

**Acknowledgements**

This research was undertaken as part of the "Integrated Modeling and Assessment of North American Forest Carbon Dynamics and Climate Change Mitigation Options" project funded by the Commission for Environmental Cooperation (Montreal, Canada) representing Canada, Mexico and the USA. Additional funding was provided by the Secretary of Public Education of Mexico to Mascorro and the Natural Sciences and Engineering Research Council of Canada to Coops. We thank Shannon Franks and Jeff Masek from NASA for the provision of the VCT land cover time series and Rene Colditz from CONABIO for providing the MODIS maps. We thank Scott Morken and Max Fellows of the Canadian Forest Service for software development, training and support. We also thank Craig Wayson and Gregorio Ángeles for their help with developing the growth and age information that was used to run the model. Statements in this paper are by the authors and do not necessarily represent the positions of the sponsors or their governments.

**Competing interests**

The authors declare that they have no competing interests.

**References**

1. Bosworth D, Birdsey R, Joyce L, Millar C. Climate change and the nation's forests: challenges and opportunities. J For. 2008;106:214–21.
2. Turner M. Disturbance and landscape dynamics in a changing world. Ecology. 2010;91:2833–49.
3. Chapin FS III, McFarland J, David McGuire A, Euskirchen ES, Ruess RW, Kielland K. The changing global carbon cycle: linking plant-soil carbon dynamics to global consequences. J Ecol. 2009;97:840–50.
4. Pan Y, Birdsey RA, Fang J, Houghton R, Kauppi PE, Kurz WA, Phillips OL, Shvidenko A, Lewis SL, Canadell JG, Ciais P, Jackson RB, Pacala SW, McGuire AD, Piao S, Rautiainen A, Sitch S, Hayes D. A large and persistent carbon sink in the world's forests. Science. 2011;333:988–93.
5. Spalding D. The role of forests in global carbon budgeting. Forests an. In: Tyrrell ML, Ashton MS, Spalding D, Gentry B, Editors. Yale School of Forestry & Environmental Studies; 2009. p 223–53.
6. Hansen MC, Stehman SV, Potapov PV. Quantification of global gross forest cover loss. Proc Natl Acad Sci USA. 2010;107:8650–5.
7. Houghton RA, House JI, Pongratz J, van der Werf GR, DeFries RS, Hansen MC, Le Quéré C, Ramankutty N. Carbon emissions from land use and land-cover change. Biogeosciences. 2012;9:5125–42.
8. Le Quéré C, Moriarty R, Andrew RM, Peters GP, Ciais P, Friedlingstein P, Jones SD, Sitch S, Tans P, Arneth A, Boden TA, Bopp L, Bozec Y, Canadell JG, Chini LP, Chevallier F, Cosca CE, Harris I, Hoppema M, Houghton RA, House JI, Jain AK, Johannessen T, Kato E, Keeling RF, Kitidis V, Klein

Goldewijk K, Koven C, Landa CS, Landschützer P, et al. Global carbon budget 2014. Earth Syst Sci Data. 2015;7:47–85.
9. UNFCCC. Outcome of the work of the ad hoc working group on long-term cooperative action under the convention—policy approaches and positive incentives on issues relating to reducing emissions from deforestation and forest degradation in developing countries. United Nations Framework Convention on Climate Change COP 16; 2010.
10. Kurz WA, Dymond CC, White TM, Stinson G, Shaw CH, Rampley GJ, Smyth C, Simpson BN, Neilson ET, Trofymow JA, Metsaranta J, Apps MJ. CBM-CFS3: a model of carbon-dynamics in forestry and land-use change implementing IPCC standards. Ecol Modell. 2009;220:480–504.
11. Birdsey R, Pan Y, Houghton R. Sustainable landscapes in a world of change: tropical forests, land use and implementation of REDD+: Part II. Carbon Manag. 2013;4:567–9.
12. Pickell P, Gergel S, Coops N, Andison D. Monitoring forest change in landscapes under-going rapid energy development: challenges and new perspectives. Land. 2014;3:617–38.
13. Lorenz K, Lal R: Effects of disturbance, succession and management on carbon sequestration. In: Carbon sequestration in forest ecosystems. Dordrecht: Springer Netherlands; 2010. p. 103–57.
14. Kurz WA. An ecosystem context for global gross forest cover loss estimates. Proc Natl Acad Sci USA. 2010;107:9025–6.
15. Vargas R, Allen MF, Allen EB. Biomass and carbon accumulation in a fire chronosequence of a seasonally dry tropical forest. Glob Chang Biol. 2007;14:109–24.
16. Jaramillo VJ, Kauffman JB, Renteria-Rodriguez L, Cummings DL, Ellingson LJ. Biomass, carbon, and nitrogen pools in Mexican tropical dry forest landscapes. Ecosystems. 2003;6:609–29.
17. Orihuela-Belmonte DE, de Jong BHJ, Mendoza-Vega J, Van der Wal J, Paz-Pellat F, Soto-Pinto L, Flamenco-Sandoval A. Carbon stocks and accumulation rates in tropical secondary forests at the scale of community, landscape and forest type. Agric Ecosyst Environ. 2013;171:72–84.
18. Dai Z, Birdsey RA, Johnson KD, Dupuy JM, Hernandez-Stefanoni JL, Richardson K. Modeling carbon stocks in a secondary tropical dry forest in the Yucatan Peninsula, Mexico. Water Air Soil Pollut. 1925;2014:225.
19. De Fries RS, Houghton RA, Hansen MC, Field CB, Skole D, Townshend J. Carbon emissions from tropical deforestation and regrowth based on satellite observations for the 1980s and 1990s. Proc Natl Acad Sci USA. 2002;99:14256–61.
20. Asner GP, Powell GVN, Mascaro J, Knapp DE, Clark JK, Jacobson J, Kennedy-Bowdoin T, Balaji A, Paez-Acosta G, Victoria E, Secada L, Valqui M, Hughes RF. High-resolution forest carbon stocks and emissions in the Amazon. Proc Natl Acad Sci USA. 2010;107:16738–42.
21. Masek JG, Collatz GJ. Estimating forest carbon fluxes in a disturbed southeastern landscape: Integration of remote sensing, forest inventory, and biogeochemical modeling. J Geophys Res. 2006;111:G01006.
22. Stinson G, Kurz WA, Smyth CE, Neilson ET, Dymond CC, Metsaranta JM, Boisvenue C, Rampley GJ, Li Q, White TM, Blain D. An inventory-based analysis of Canada's managed forest carbon dynamics, 1990 to 2008. Glob Chang Biol. 2011;17:2227–44.
23. Potter C. Terrestrial ecosystem carbon fluxes predicted from MODIS satellite data and large-scale disturbance modeling. Int J Geosci. 2012;03:469–79.
24. Espírito-Santo FDB, Gloor M, Keller M, Malhi Y, Saatchi S, Nelson B, Junior RCO, Pereira C, Lloyd J, Frolking S, Palace M, Shimabukuro YE, Duarte V, Mendoza AM, López-González G, Baker TR, Feldpausch TR, Brienen RJW, Asner GP, Boyd DS, Phillips OL. Size and frequency of natural forest disturbances and the Amazon forest carbon balance. Nat Commun. 2014;5:3434.
25. Birdsey R, Angeles-Perez G, Kurz WA, Lister A, Olguin M, Pan Y, Wayson C, Wilson B, Johnson K. Approaches to monitoring changes in carbon stocks for REDD+. 2013:519–37.
26. Kurz WA, Apps MJ. Developing Canada's national forest carbon monitoring, accounting and reporting system to meet the reporting requirements of the Kyoto Protocol. Mitig Adapt Strateg Glob Chang. 2006;11:33–43.
27. Kull S, Kurz WA, Rampley GJ, Banfield G, Schivatcheva R, Apps M. Operational-Scale Carbon Budget Model of the Canadian Forest Sector (CBM-CFS3): Version 1.0, User's Guide. Edmonton, Alberta; 2011.
28. Coops NC, Wulder MA, White JC: Identifying and describing forest disturbance and spatial pattern: data selection issues and methodological implications; 2006. p 31–62.
29. Wulder MA, White JC, Gillis MD, Walsworth N, Hansen MC, Potapov P. Multiscale satellite and spatial information and analysis framework in support of a large-area forest monitoring and inventory update. Environ Monit Assess. 2010;170:417–33.
30. Hermosilla T, Wulder MA, White JC, Coops NC, Hobart GW. Remote sensing of environment an integrated landsat time series protocol for change detection and generation of annual gap-free surface reflectance composites. Remote Sens Environ. 2015;158:220–34.
31. Saatchi SS, Harris NL, Brown S, Lefsky M, Mitchard ETA, Salas W, Zutta BR, Buermann W, Lewis SL, Hagen S, Petrova S, White L, Silman M, Morel A. Benchmark map of forest carbon stocks in tropical regions across three continents. Proc Natl Acad Sci USA. 2011;108:9899–904.
32. Baccini A, Goetz SJ, Walker WS, Laporte NT, Sun M, Sulla-Menashe D, Hackler J, Beck PSA, Dubayah R, Friedl MA, Samanta S, Houghton RA. Estimated carbon dioxide emissions from tropical deforestation improved by carbon-density maps. Nat Clim Chang. 2012;2:182–5.
33. Achard F, Beuchle R, Mayaux P, Stibig H-J, Bodart C, Brink A, Carboni S, Desclée B, Donnay F, Eva HD, Lupi A, Raši R, Seliger R, Simonetti D. Determination of tropical deforestation rates and related carbon losses from 1990 to 2010. Glob Chang Biol. 2014;20:2540–54.
34. Chapin FS, Woodwell GM, Randerson JT, Rastetter EB, Lovett GM, Baldocchi DD, Clark DA, Harmon ME, Schimel DS, Valentini R, Wirth C, Aber JD, Cole JJ, Goulden ML, Harden JW, Heimann M, Howarth RW, Matson PA, McGuire AD, Melillo JM, Mooney HA, Neff JC, Houghton RA, Pace ML, Ryan MG, Running SW, Sala OE, Schlesinger WH, Schulze E-D. Reconciling carbon-cycle concepts, terminology, and methods. Ecosystems. 2006;9:1041–50.
35. Hansen MC, Potapov PV, Moore R, Hancher M, Turubanova SA, Tyukavina A, Thau D, Stehman SV, Goetz SJ, Loveland TR, Kommareddy A, Egorov A, Chini L, Justice CO, Townshend JRG. High-resolution global maps of 21st-century forest cover change. Science. 2013;342:850–3.
36. Mitchard ET, Feldpausch TR, Brienen RJW, Lopez-Gonzalez G, Monteagudo A, Baker TR, Lewis SL, Lloyd J, Quesada CA, Gloor M, ter Steege H, Meir P, Alvarez E, Araujo-Murakami A, Aragão LEOC, Arroyo L, Aymard G, Banki O, Bonal D, Brown S, Brown FI, Cerón CE, Chama Moscoso V, Chave J, Comiskey JA, Cornejo F, Corrales Medina M, Da Costa L, Costa FRC, Difiore A, et al. Markedly divergent estimates of Amazon forest carbon density from ground plots and satellites. Glob Ecol Biogeogr. 2014;23:935–46.
37. DeFries R, Achard F, Brown S, Herold M, Murdiyarso D, Schlamadinger B, de Souza C. Earth observations for estimating greenhouse gas emissions from deforestation in developing countries. Environ Sci Policy. 2007;10:385–94.
38. De Sy V, Herold M, Achard F, Asner GP, Held A, Kellndorfer J, Verbesselt J. Synergies of multiple remote sensing data sources for REDD+ monitoring. Curr Opin Environ Sustain. 2012;4:696–706.
39. GOFC-GOLD. A sourcebook of methods and procedures for monitoring and reporting anthropogenic greenhouse gas emissions and removals caused by deforestation, gains and losses of carbon stocks in forests remaining forests, and forestation; 2010.
40. Hewson J, Steininger M, Pesmajoglou S: REDD+ measurement, reporting and verification (MRV) manual. USAID-supported Forest Carbon, Markets and Communities Program; 2013.
41. Read L, Lawrence D. Recovery of biomass following shifting cultivation in dry tropical forests of the Yucatan. Ecol Appl. 2003;13:85–97.
42. Vandecar KL, Lawrence D, Richards D, Schneider L, Rogan J, Schmook B, Wilbur H. High mortality for rare species following hurricane disturbance in the southern Yucatan. 2011; 43:676–84.
43. Urquiza-Haas T, Dolman PM, Peres CA. Regional scale variation in forest structure and biomass in the Yucatan Peninsula, Mexico: effects of forest disturbance. For Ecol Manage. 2007;247:80–90.
44. Cohen WB, Yang Z, Kennedy R. Detecting trends in forest disturbance and recovery using yearly Landsat time series: 2. TimeSync—Tools for calibration and validation. Remote Sens Environ. 2010;114:2911–24.
45. Powers RP, Hermosilla T, Coops NC, Chen G. Remote sensing and object-based techniques for mapping fine-scale industrial disturbances. Int J Appl Earth Obs Geoinf. 2015;34:51–7.

46. Healey S, Cohen W, Zhiqiang Y, Krankina O. Comparison of Tasseled Cap-based Landsat data structures for use in forest disturbance detection. Remote Sens Environ. 2005;97:301–10.
47. Huang C, Goward SN, Masek JG, Thomas N, Zhu Z, Vogelmann JE. An automated approach for reconstructing recent forest disturbance history using dense Landsat time series stacks. Remote Sens Environ. 2010;114:183–98.
48. Gebhardt S, Wehrmann T, Ruiz M, Maeda P, Bishop J, Schramm M, Kopeinig R, Cartus O, Kellndorfer J, Ressl R, Santos L, Schmidt M. MAD-MEX: automatic wall-to-wall land cover monitoring for the Mexican REDD-MRV program using all Landsat data. Remote Sens. 2014;6:3923–43.
49. INEGI. Conjunto de Datos Vectoriales de La Carta de Uso Del Suelo Y Vegetación, Escala 1:250,000, Serie III (CONTINUO NACIONAL). Aguascalientes, Mexico: Instituto Nacional de Estadística y Geografía; 2003.
50. INEGI: Conjunto de Datos Vectoriales de La Carta de Uso Del Suelo Y Vegetación, Escala 1:250,000, Serie IV (CONTINUO NACIONAL). Aguascalientes, Mexico: Instituto Nacional de Estadística y Geografía; 2007.
51. INEGI. Conjunto de Datos Vectoriales de La Carta de Uso Del Suelo Y Vegetación, Escala 1:250,000, Serie V (CONTINUO NACIONAL). Aguascalientes, Mexico: Instituto Nacional de Estadística y Geografía; 2011.
52. Latifovic R, Homer C, Ressl R, Pouliot D, Hossain SN, Colditz RR, Giri OC, Victoria A. North american land change monitoring system; 2010.
53. Colditz RR, López Saldaña G, Maeda P, Espinoza JA, Tovar CM, Hernández AV, Benítez CZ, Cruz López I, Ressl R. Generation and analysis of the 2005 land cover map for Mexico using 250 m MODIS data. Remote Sens Environ. 2012;123:541–52.
54. INFyS. Inventario Nacional Forestal Y de Suelos Informe 2004–2009 (National Forest and Soils Inventory Report). Zapopan, Jalisco: Comision Nacional Forestal de Mexico; 2012.
55. Mascorro VS, Coops NC, Kurz WA, Olguín M. Attributing changes in land cover using independent disturbance datasets: a case study of the Yucatan Peninsula, Mexico. Reg Environ Chang. 2014.
56. Congalton RG. A review of assessing the accuracy of classifications of remotely sensed data. 1991;46(October 1990):35–46.
57. Olofsson P, Foody GM, Herold M, Stehman SV, Woodcock CE, Wulder MA. Good practices for estimating area and assessing accuracy of land change. Remote Sens Environ. 2014;148:42–57.
58. INEGI. Conjunto de Datos Vectoriales de Carreteras y Vialidades Urbanas Edición 1.0. 2013.
59. Olguin M, Wayson C, Kurz. W, Fellows M, Fellows, Ángeles G, Maldonado V, Carrillo O, López D. Input Data Improvements for Version 2 of State-Level Carbon Dynamics Runs Using the CBM-CFS3 Model in Mexico. Mexico City: Mexico-Norway Project of the National Forestry Commission of Mexico; 2014.
60. CEC. Ecological Regions of North America: Towards a Common Perspective. Montreal, Quebec: Commission for Environmental Cooperation; 1997.
61. Dal XL, Khorram S. Remotely sensed change detection based on artificial neural networks. Photogramm Eng Remote Sens. 1999;65:1187–94.
62. Fuller R, Smith G, Devereux B. The characterisation and measurement of land cover change through remote sensing: problems in operational applications? Int J Appl Earth Obs Geoinf. 2003;4:243–53.
63. Olofsson P, Foody GM, Herold M, Stehman SV, Woodcock CE, Wulder MA. Remote sensing of environment good practices for estimating area and assessing accuracy of land change. Remote Sens Environ. 2014;148:42–57.
64. Lu D, Mausel P, Brondízio E, Moran E. Change detection techniques. Int J Remote Sens. 2004;25:2365–401.
65. Lorenz K, Lal R. The Natural Dynamic of Carbon in Forest Ecosystems. 2010(C).
66. Franklin J, Mitchell R, Palik B. Natural disturbance and stand development principles for ecological forestry. USDA Forest Service 2007. p 44.
67. Herold M, Skutsch M. Monitoring, reporting and verification for national REDD+ programmes: two proposals. Environ Res Lett. 2011;6:014002.
68. Hosonuma N, Herold M, De Sy V, De Fries RS, Brockhaus M, Verchot L, Angelsen A, Romijn E. An assessment of deforestation and forest degradation drivers in developing countries. Environ Res Lett. 2012;7:044009.
69. GFOI. Integrating remote-sensing and ground-based observations for estimation of emissions and removals of greenhouse gases in forests. Geneva: Group on Earth Observations; 2014.

# 5

# Spatially explicit analysis of field inventories for national forest carbon monitoring

David C. Marvin* and Gregory P. Asner

**Abstract**

**Background:** Tropical forests provide a crucial carbon sink for a sizable portion of annual global $CO_2$ emissions. Policies that incentivize tropical forest conservation by monetizing forest carbon ultimately depend on accurate estimates of national carbon stocks, which are often based on field inventory sampling. As an exercise to understand the limitations of field inventory sampling, we tested whether two common field-plot sampling approaches could accurately estimate carbon stocks across approximately 76 million ha of Perúvian forests. A 1-ha resolution LiDAR-based map of carbon stocks was used as a model of the country's carbon geography.

**Results:** Both field inventory sampling approaches worked well in estimating total national carbon stocks, almost always falling within 10 % of the model national total. However, the sampling approaches were unable to produce accurate spatially-explicit estimates of the carbon geography of Perú, with estimates falling within 10 % of the model carbon geography across no more than 44 % of the country. We did not find any associations between carbon stock errors from the field plot estimates and six different environmental variables.

**Conclusions:** Field inventory plot sampling does not provide accurate carbon geography for a tropical country with wide ranging environmental gradients such as Perú. The lack of association between estimated carbon errors and environmental variables suggests field inventory sampling results from other nations would not differ from those reported here. Tropical forest nations should understand the risks associated with primarily field-based sampling approaches, and consider alternatives leading to more effective forest conservation and climate change mitigation.

**Keywords:** Field sampling, Forest carbon stocks, Tropical forest, Carnegie Airborne Observatory

## Background

More atmospheric carbon is absorbed and stored by tropical forests than any other terrestrial ecosystem on Earth [1]. This crucial ecosystem service provides a carbon sink larger than what is emitted by fossil fuel combustion across the entire European Union each year [2, 3]. Policies that monetize the amount of carbon stored annually by a hectare of tropical forest seek to incentivize forest conservation by making it more economical to leave the forest intact than to degrade or deforest the land. The resulting economic boon for landowners and countries that reduce deforestation and degradation also results in increased carbon sequestration in the form of woody biomass, reducing global net carbon emissions. For such policies to be successful, the uncertainty in standing carbon stocks and change (flux) must be reduced.

Accurate carbon flux calculations necessitate accurate estimates of standing carbon stocks at one or more time periods, unless carbon fluxes are measured using more direct methods (e.g., eddy covariance, atmospheric inversion). Both the price of carbon and the efficacy of climate change mitigation can be negatively affected by uncertainty in our understanding of carbon stocks, requiring the deployment of methods to make highly accurate spatial and temporal estimates of forest carbon. As forest carbon stock uncertainties increase, the monetary value of that carbon is decreased through a sliding scale discount [4,

*Correspondence: dmarvin@carnegiescience.edu
Department of Global Ecology, Carnegie Institution for Science, 260 Panama St., Stanford, CA 94305, USA

5], reducing investment opportunities and the economic benefits accrued by countries and landowners. Depending on the baseline levels of deforestation and carbon storage rates, Kohl et al. [4] found in a simulation study that many countries would generate no economic benefit with total errors in carbon estimates exceeding 5 % unless baseline deforestation rates are very high. Ultimately the power of these policies to increase forest conservation and mitigate climate change may rely on our ability to accurately quantify forest carbon in a spatially explicit manner, as opposed to generalized estimates for total carbon stocks of a landscape, habitat type, or eco-region.

Spatially explicit maps of forest carbon, or carbon geographies, allow for multi-stakeholder engagement at subnational levels [6]. Landholders and agencies within tropical countries control forest assets across a variety of spatial extents. Accordingly, knowledge of carbon stocks is needed at scales commensurate with the activities of these subnational stakeholders. Providing generalized carbon stock estimates for a given area, rather than a carbon geography, removes important spatial heterogeneity in forest carbon that could lead to over- or underestimates of carbon stocks, and consequently, the potential for poor land-use decisions. Both field inventories [7] and forest fragmentation [8] may introduce bias into forest carbon stock estimates; a bias that arises when local landscape heterogeneity is disregarded.

Forest carbon inventories typically rely (or are encouraged to rely) on a network of field plots installed either on a regular systematic grid or a random stratified grid [9–11]. Importantly, the IPCC guidelines for estimating and measuring carbon direct countries to use these sampling approaches [12, 13]. Although these guidelines are primarily designed for total national or average carbon density estimates, they are sometimes used for carbon geographies as well [14, 15]. Field plots are usually 1-ha in size (the generally accepted standard for a forest plot inventory [16]) but forest inventories that use smaller plot designs run the risk of the plots being even less representative of the surrounding forest [17]. The data collected in the field can then be combined with remote sensing data in two general ways: stratify-and-multiply, and model-linked. The stratify-and-multiply approach [18] uses remote sensing data to partition a country by land cover, climate, or other environmental (or biogeochemical) strata, or uses a regular (systematic) grid in place of partitioning by strata. Each unique stratum (or grid cell for systematic sampling) is assigned the field plot-estimated carbon stock and multiplied by the total area within the stratum. Alternatively in a model-linked approach, a model is calibrated to link field plot-estimated carbon stocks with multiple environmental variables at the location of the plots (or an intermediate remote sensing product such as LiDAR tree height), and the model applied to the entirety of the dataset for which there is no field data (or intermediate data product). The integration of remote sensing data with field data is preferable, but countries may not have the capacity to work with remote sensing data, due to a lack of either in-country expertise, funding to outsource the work, access to technology, or sufficient remotely sensed data [e.g., [19], [20]. In cases where field plot sampling cannot be supplemented by sufficient remote sensing data and/or modeling, spatial heterogeneity in forest carbon is necessarily disregarded because of the time and expense required for the massive field sampling needed to sufficiently capture such heterogeneity [7].

While field plot sampling alone is commonly used to scale ecosystem properties and processes from local-to-regional scales, the efficacy of this approach has never been assessed at the appropriate scale. As a result, we do not know whether field carbon inventories can be used to create accurate spatially-explicit maps of national carbon geography at one-ha resolution. Here, using Perú as an example country, we examine both systematic and stratified random field-based sampling designs with a LiDAR-estimated national carbon map spanning all 76 million ha of intact and recovering forest land (Fig. 1a) as a model of the country's true carbon stocks (see "Methods" section for more detail). Hereafter, we refer to this LiDAR-estimated carbon map as "the model carbon stock" or "the model aboveground carbon density (ACD)." For both sampling designs we examine the stratify-and-multiply approach (i.e., applying each sampled value to its corresponding unsampled population at the original spatial resolution of the map) and a model-linked approach using the random forest machine learning algorithm to create spatially explicit carbon geographies of the country. We assume perfect a priori knowledge of the country's stratification, and assume that each forest plot location is accessible. This is a theoretical exercise using no field inventory data; rather all values are extracted from the carbon map as if it reflected reality. We chose a 10 % interval (i.e., ±5 %) around the true carbon stock value as our threshold for an estimate to be considered accurate. We asked the following questions. (i) What sampling intensity is needed to accurately estimate the total national carbon stock? (ii) Can either sampling method accurately estimate the national carbon geography at 1-ha resolution? (iii) What topographic and climatic variables cause increased error in carbon stock estimates?

## Results and discussion

### Total national carbon stocks

#### *Systematic grid*

A systematic sampling grid with dimensions of ≤56 km (requiring a minimum of 236 field plots) resulted in

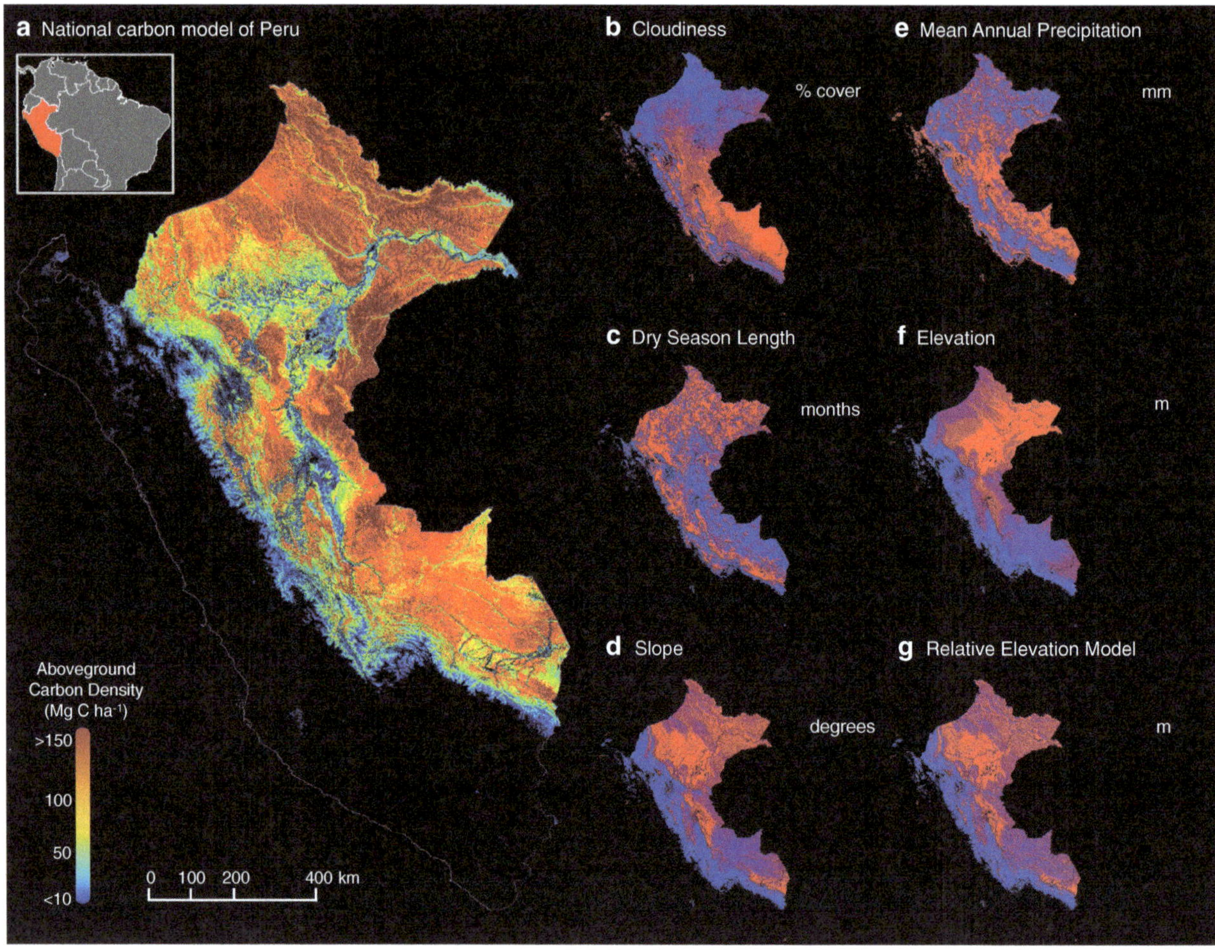


**Fig. 1** Maps used for the analyses. **a** The model aboveground carbon density (ACD) map of Perú at 1-ha resolution with all non-forested areas masked out (adapted from [6]), and the strata binned and *colored* by quintiles for **b** cloudiness, **c** dry season length, **d** slope, **e** mean annual precipitation, **f** elevation and **g** relative elevation (see "Methods" section)

<2.5 % difference between the total field plot-estimated and the total model national forest carbon stock (Fig. 2). Even for systematic grids of larger spatial grains, estimates tended to stay within the 10 % threshold, with just a few of the large grid sizes falling outside that range. This demonstrates that it is possible to estimate national total forest carbon stock quite accurately with minimal sampling effort using a systematic grid.

### *Stratified random*

Using a stratified random sampling approach, we find that only a single plot per substratum is needed to estimate the total carbon stock of Perú to within 10 % of the total model national carbon stock. A maximum of 5398 plots nationwide are needed to sample all combinations of strata representing more than 1000 ha. In fact, only the largest 15 % of the substrata (by area) need to be sampled with a single field plot to yield an accurate estimate of the total national carbon stock, requiring a minimum of 810 plots (Additional file 1: Fig. S1).

Both sampling methods result in accurate estimates of Perú's total carbon stock, usually falling within a few percent of the model total. However, the systematic grid approach requires less sampling intensity compared to the stratified random, needing only single plots spaced as far apart as 56 km over the entire country. Systematic grid sampling is the cheapest and most efficient way to accurately sample total national carbon stocks of even a country with highly heterogeneous forest carbon distributions.

While accurate estimates of total carbon at the national and regional scale are important for understanding broad geographic patterns of carbon distributions, they are very

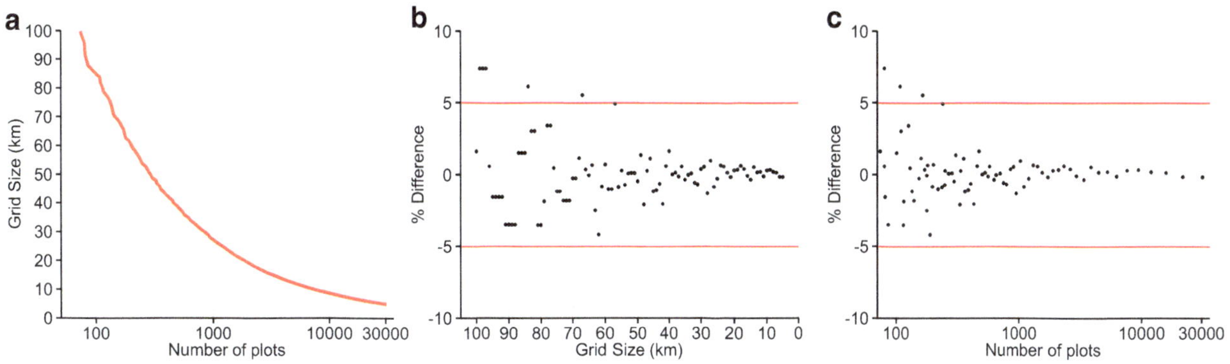


**Fig. 2** National total carbon stock estimation using systematic grid sampling showing **a** the relationship between the sampling grid dimensions and number of 1-ha field plots, **b** the percent difference between total field plot estimated and total model national carbon stocks by grid size and **c** same as in **b** but with the number of plots required by the grid sampling on the *x-axis* (*log scale*)

difficult to use in applications of carbon conservation and monetization, activities that are highly spatially dependent and vary at local, landscape, and sub-regional scales. Improved knowledge of spatial variation in carbon stocks, ideally at the hectare resolution, will enhance policies that seek to incentivize conservation and climate change mitigation through carbon valuation. Maps of national local carbon geographies must be developed with particular attention to accuracy, repeatability, and cost.

## Carbon geography

### *Systematic grid*

Using a systematic sampling grid to estimate ACD and then upscaling the result at 1-ha resolution using the stratify-and-multiply approach leads to a large percentage of inaccurate estimates across the country (i.e., field plot-estimated ACD failed to be within 10 % of the model ACD) (Fig. 3a, c). Regardless of the sampling grid size, this approach produces inaccurate estimates across approximately 78 % of the country (Fig. 4a, c). The inaccurate estimates are quite consistent, 82 % using a 100 km grid (requiring just 74 1-ha plots) vs. 71 % using a 5 km grid (requiring >30,000 1-ha plots), suggesting that this method performs poorly in creating an accurate map of Perú's carbon geography. Using a model-linked random forest approach, rather than stratify-and-multiply upscaling, improves the carbon map of Perú but still results in inaccurate estimates across more than half of the country (Fig. 5a, c). Using a systematic grid of 15 km (requiring approximately 3400 1-ha plots) as the training dataset for a random forest model, 63 % of the country is still inaccurately estimated. Even using the 5 km grid results in 56 % of the country inaccurately estimated using random forest upscaling, but still requires a field plot sampling intensity of 30,459 1-ha plots to produce the training dataset (Figs. 3e, f, 5a).

### *Stratified random*

The stratified random sampling stratify-and-multiply approach did not perform much better than the systematic grid stratify-and-multiply approach in estimating the carbon geography of Perú. The distribution of inaccurate estimates was similar to the systematic grid approach (Fig. 3b, d), and resulted in nearly the same amount of total inaccurate estimates across the country (Fig. 4b, d). Using only a single randomly placed plot per substrata (requiring around 5400 plots) leads to 79 % of the country inaccurately estimated. There is barely any improvement—with 76 % of the country inaccurately estimated—when the number of plots randomly placed inside a substrata is increased to 100 (requiring around 540,000 plots). Again the model-linked random forest upscaling approach did not substantially increase the area accurately mapped (Fig. 5b, c), with 63 % of the country inaccurately estimated regardless of the number of plots used per substratum (1, 5 or 100 plots).

Neither field plot sampling approach assessed here was able to accurately estimate the carbon geography at the 1-ha scale across more than 25–44 % of Perú. While field plot systematic sampling tended to equally under- and overestimate the model ACD across the country, the field plot stratified sampling consistently underestimated model ACD at the 1-ha scale (Fig. 4b, d). Chronically underestimating local carbon stocks will artificially deflate the per-hectare value of carbon, leading to reduced conservation incentives by landholders. Even when the total sampling is increased to an impractical number of plots (i.e., tens- to hundreds-of-thousands), the accuracy does not improve substantially. This demonstrates that neither sampling approach would be an appropriate choice for developing national maps of carbon geography, or changes in carbon geography via emissions or sequestration.

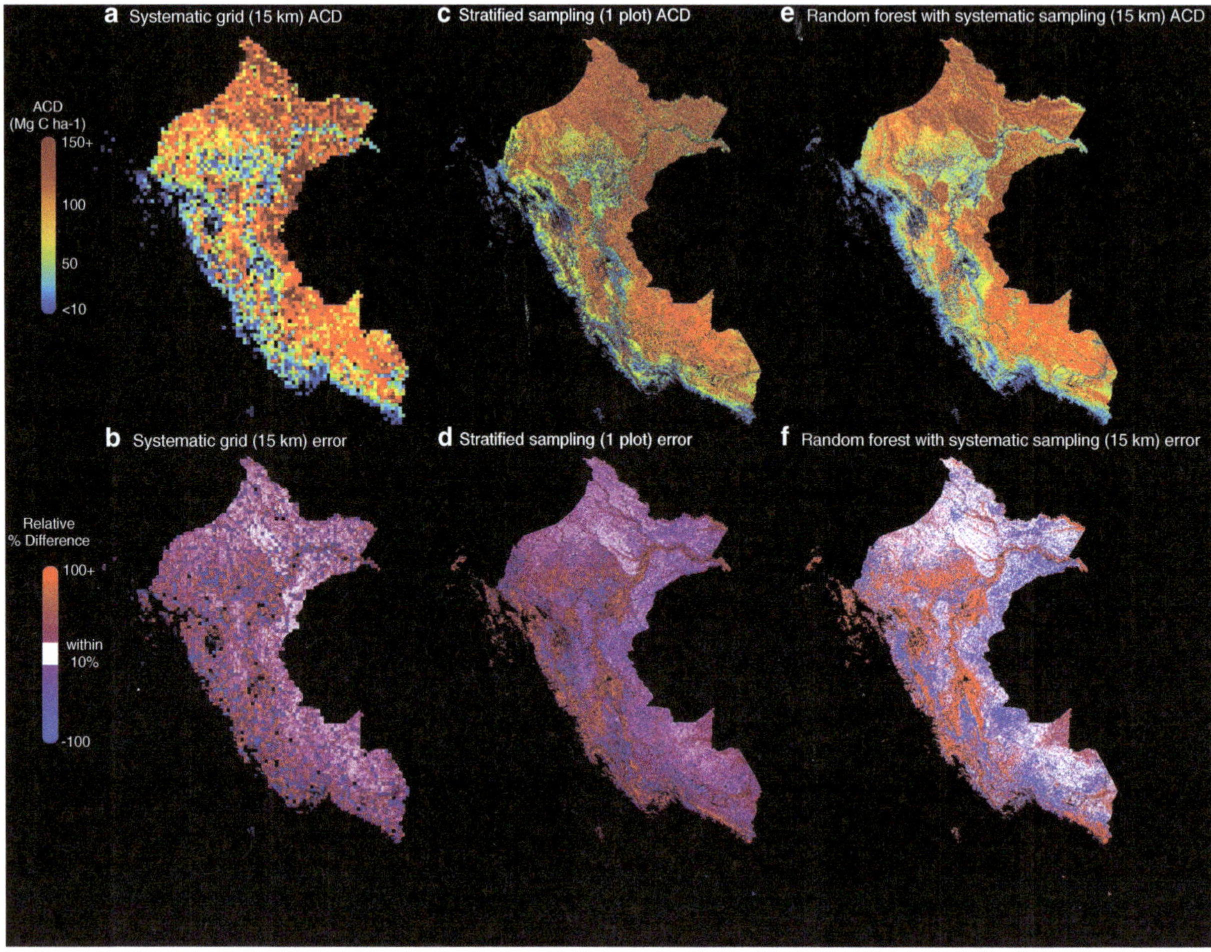


**Fig. 3** Results of the sampling and upscaling. **a** Estimated ACD using a 15 km systematic grid sampling approach, **b** percent relative difference between model ACD (from Fig. 1a) and **a**, **c** ACD using a 1-plot stratified sampling approach, **d** percent relative difference between model ACD and **c**, **e** ACD using random forest upscaling trained with 15 km systematic sampling grid and **f** percent relative difference between model ACD and (**e**)

### Underlying drivers of uncertainty

To understand whether there are particular strata or carbon values that lead to higher field plot estimated ACD errors, we used simulations to examine the number of field plots needed to reliably estimate (probability $\geq 0.9$) the mean ACD of all substrata (unique combinations of the 6 quintile-binned strata) (Fig. 1b–g) to within 10 % of the model ACD (Fig. 6a, b). While most substrata (81 %) can be accurately estimated at a certain sampling density, some never reach the 0.9 probability threshold at any sampling density, meaning they are so heterogeneous that field plot sampling is not feasible. The median number of field plots needed to reliably sample the mean model ACD of any substratum is 43 (mean = 61; Fig. 6b).

Further examination of the simulation outcomes by strata or carbon value shows a few strong patterns. Substrata that had higher mean model ACD values require fewer randomly placed plots to accurately sample the mean model ACD (Fig. 6c). This implies that the heteroskedasticity in the relationship used to develop the model ACD map (see "Methods" section) is not a major factor in the inability of the field plot sampling strategies to reproduce the model ACD. The strata with lower elevation and slope values require fewer plots to accurately sample the mean ACD (Fig. 6d, e), indicating that these areas are more likely to have lower heterogeneity in their carbon stocks. The other strata showed weaker or no patterns (Additional file 1: Fig. S2). Similar patterns were found when the same simulations were conducted with strata that were binned by equal range (Additional file 1: Figs. S3–S5).

We also extracted the relative percent error between the field plot sampling estimated ACD and the model ACD, and graphed this against each stratum on a hectare-by-hectare basis (Fig. 7). Both field-plot sampling methods produced similar patterns in their estimation errors

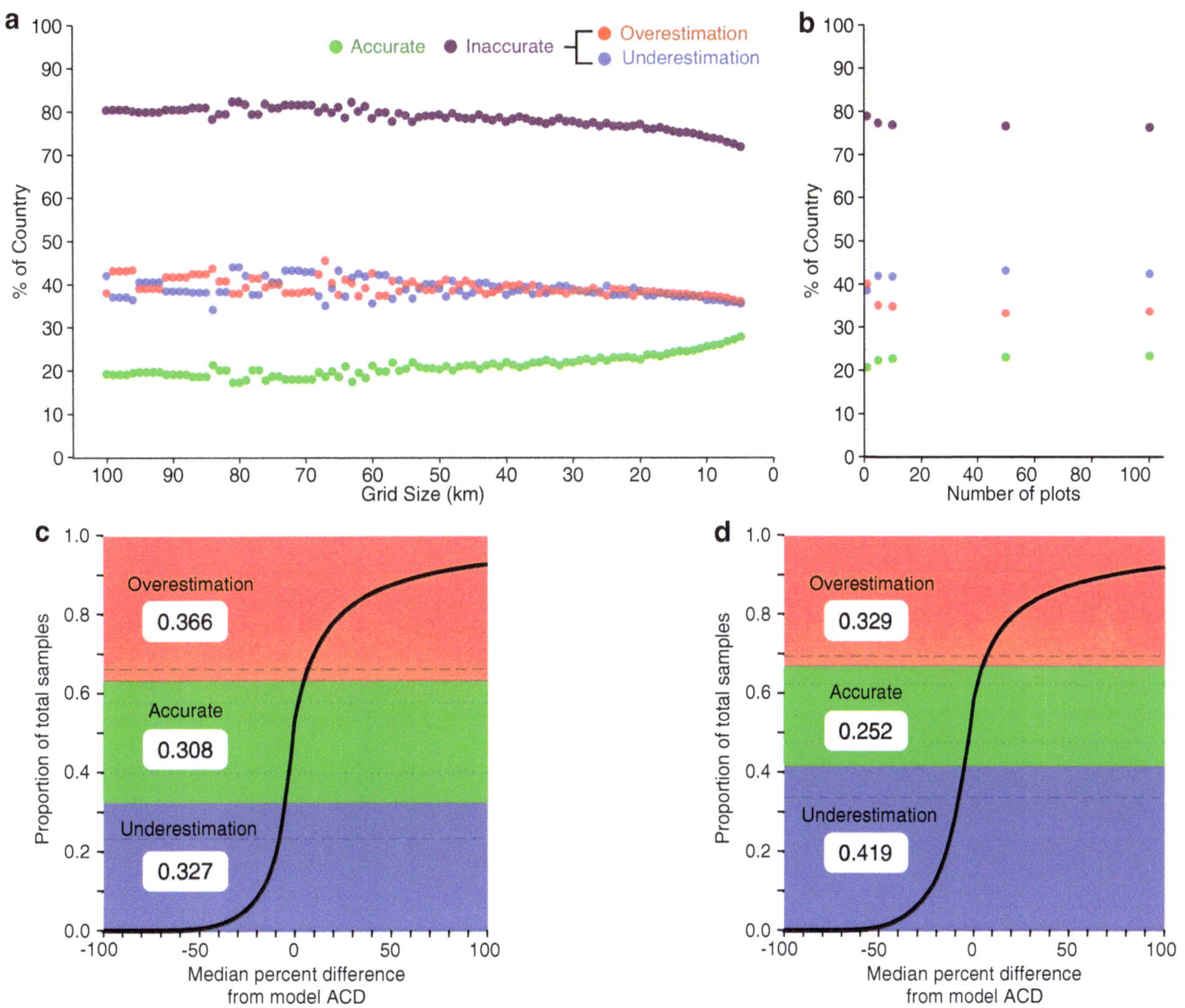


**Fig. 4** Accuracy assessment of plot sampling when using stratify-and-multiply upscaling to within at least 10 % of the model ACD using **a** systematic grid sampling and **b** stratified random sampling with same legend as in (**a**). The empirical cumulative distribution of the percent relative difference between the model ACD and the median estimated ACD error across **c** all systematic sampling grids and **d** all stratified random sampling plot sets (see "Methods" section). The *coloring* in **c**, **d** show the proportions that overestimate (*red*), underestimate (*blue*), and correctly estimate (*green*) within 10 % of the model ACD value. *Horizontal broken lines* in **c**, **d** show how the proportion correctly estimated would change with a more stringent (5 %, *dotted lines*) or less stringent (15 %, *dashed line*) error threshold

with each stratum. None of the strata show a particularly clear overall association with their ACD estimation error. Instead, generally the spread in error densities seem either restricted to low strata values in the case of elevation, slope, and relative elevation, or are more widespread across the strata values in the case of cloudiness, mean annual precipitation, and dry season length. This is probably more a reflection of the underlying data distributions than patterns in the relationship between estimation error and environmental variables (i.e., where there are more data you would expect a larger spread in the errors).

The lack of strong and clear trends between the percent relative error and the strata used in this analysis further reduces the utility of plot sampling for creating accurate, spatially explicit national carbon maps. If there were particular strata or subsets of strata that were underlying the resulting errors, then these could be isolated and sampled differently from the remaining areas of the country. The large errors within and between the strata indicate that widespread ACD estimation errors will be unavoidable when using a field-plot sampling strategy.

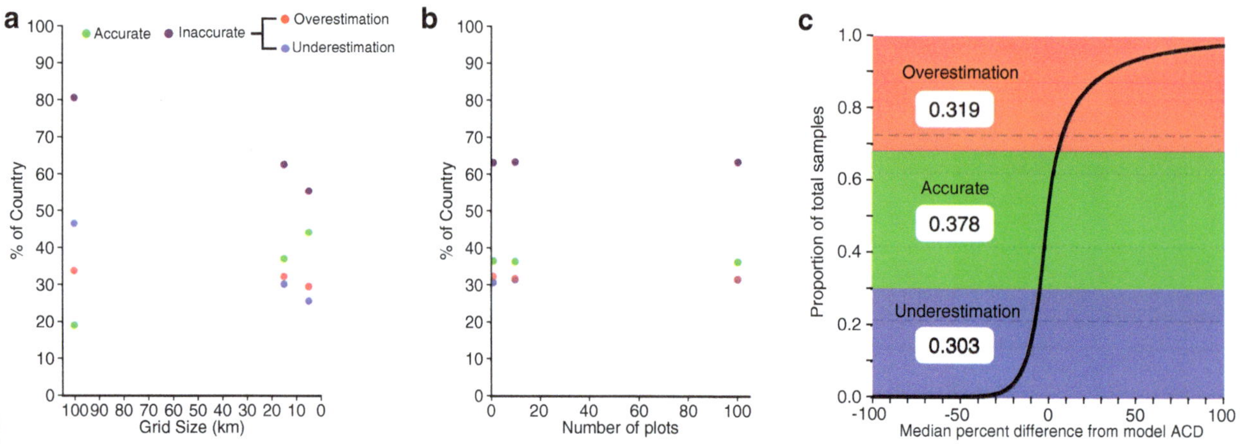


**Fig. 5** Accuracy assessment of plot sampling when using model-linked (random forest) upscaling to within at least 10 % of the model ACD using **a** systematic grid sampling and **b** stratified random sampling with same legend as in (**a**). The empirical cumulative distribution of the percent relative difference between the model ACD and the median estimated ACD error across **c** all systematic sampling grids and all stratified random sampling plot sets (see "Methods" section). The *coloring* in **c** show the proportions that overestimate (*red*), underestimate (*blue*), and correctly estimate (*green*) within 10 % of the model ACD value, and *horizontal broken lines* show how the proportion correctly estimated would change with a more stringent (5 %, *dotted lines*) or less stringent (15 %, *dashed line*) error threshold

## Conclusion

Creating a national carbon geography using field inventory plot sampling is unlikely to produce accurate results that can be deployed for use in spatially explicit actions to reduce carbon emissions. In this exercise we find that two common methods for large-scale field sampling fail to produce accurate (within a 10 % interval around the model ACD) carbon estimates for any more than 44 % of the total forested areas of Perú (approximately 76 million ha), based on a model of carbon geography. This holds true even when 12 layers of remote sensing imagery were used to upscale 30,459 field plot samples using a machine learning algorithm. While both sampling strategies can produce accurate estimates of the *total* carbon content of Perú using relatively few plots, the local carbon geographies of countries are far more important in a carbon valuation and carbon sequestration context. Without accurate, one-hectare resolution carbon geographies, land use decisions by subnational stakeholders (individual landowners and agencies) may be based on insufficient or biased information.

Perú is an ideal tropical country for this type of exercise because it hosts a wide range of topographic, biotic, geologic, and climatic variation resulting in highly heterogeneous landscape carbon distributions. While some countries may have very low carbon heterogeneity, most efforts to map national carbon stocks will face the issue of high sampling errors resulting from non-homogenous carbon distributions. This means our results are likely applicable to most tropical countries, or at least to substantial portions of any particular country.

Field inventories could be targeted toward more intensive sampling of areas likely to have higher carbon heterogeneity, potentially reducing the estimation uncertainties of any local carbon geography. This is the basic premise underlying any stratification approach [11], whether the basic stratification used here or those that incorporate remote sensing data. However, we did not find any variables driving the errors in local carbon estimates among the six topographic and climatic strata tested. These six strata are the environmental variables that best explain total variation in carbon across Perú [6]. Therefore, countries using field-based carbon mapping will find it difficult a priori to target sampling toward areas of potential high carbon estimation uncertainty. Moreover, even if a country were to have perfect knowledge of its strata, as in the stratification exercise presented here, field plot sampling still cannot reproduce the model ACD without incurring substantial errors across much of the country.

Countries seeking to value their forest carbon reserves for conservation and climate change mitigation must look beyond the sole use of field plot sampling. Of course, field inventory plots are critical for understanding local-scale ecological processes and for calibration/validation of remote sensing data. (We emphasize here the important distinction between the use of field plots as calibration/validation for remote sensing products, and using remote sensing products to scale field plot results to larger areas). While field plots are integral to remote sensing campaigns, they are not designed for, and do not perform well at, producing spatially explicit estimates of forest carbon stocks [21]—even when using satellite

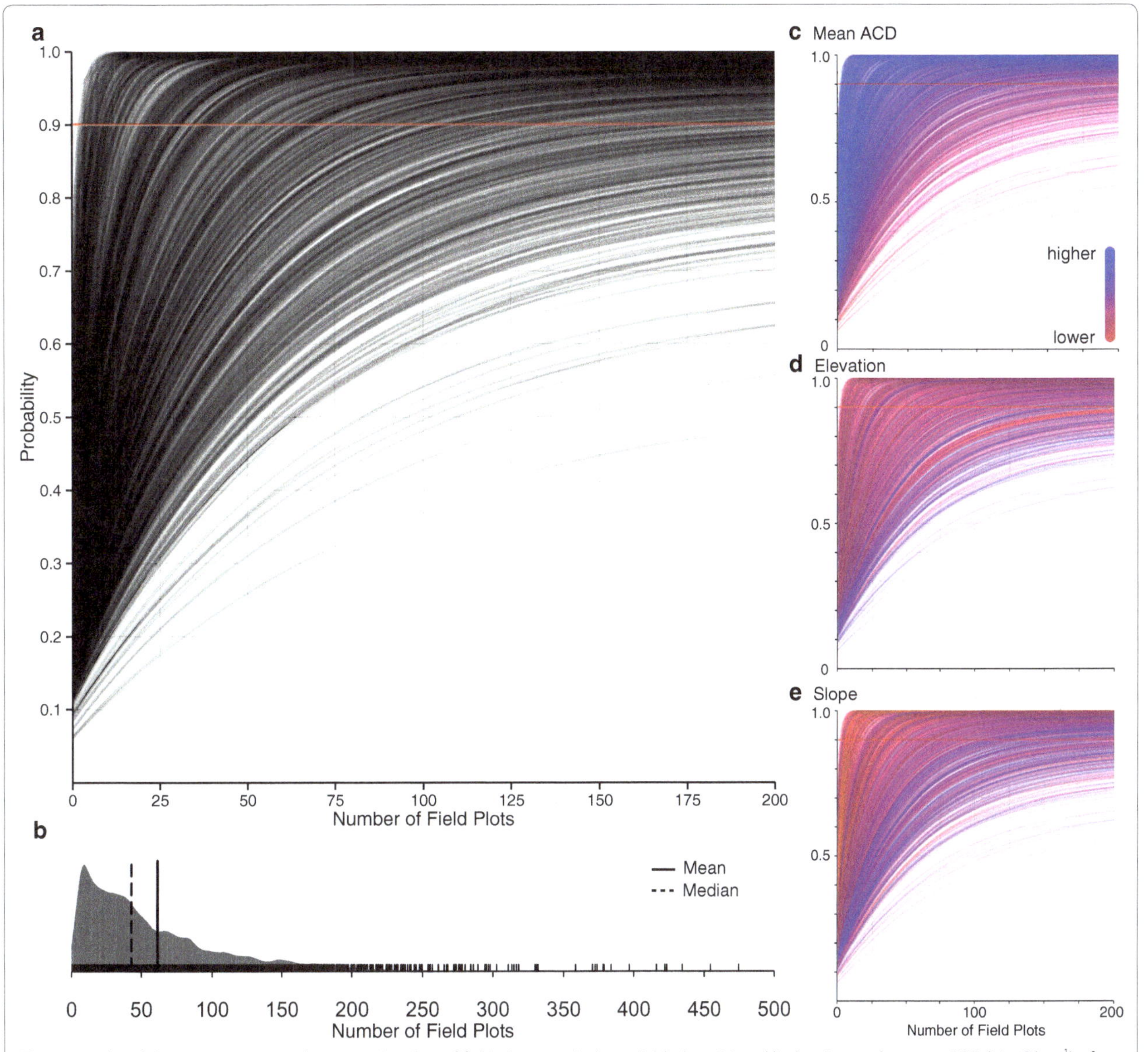


**Fig. 6** Results of the uncertainty simulations. **a** Number of field plots needed to reliably (pr=0.9, *red line*) estimate the mean ACD (Mg C ha$^{-1}$) of a substratum using quintile binning (note: a random selection of 1000 substrata are *plotted for clarity*). **b** Frequency distribution of the number of field plots needed to reliably estimate a substratum's mean ACD. Same as in **a** but *colored* by the substratum's **c** mean ACD, **d** elevation and **e** slope. Further topographic and climatic variables are shown in Additional file 1: Fig. S3

based sensors for upscaling. Field inventories are hard pressed to adequately capture the link between remotely sensed environmental variables and estimated carbon stocks. For example, 30,000 field plots spread across Perú—an extremely ambitious sampling effort for most countries—is still only 0.5 % of the total area (6.76 million ha) sampled by airborne LiDAR and used to create the model ACD map [6].

Instead tropical countries should look toward airborne and spaceborne sensors to fulfill the need for improved wall-to-wall aboveground carbon maps at hectare scales. The LiDAR sensor on the future ICESat-2 satellite is unlikely to accurately measure tropical forest carbon [22] and the GEDI sensor will be a short-term (1–2 year) mission on the International Space Station [23]. The lack of a long-term spaceborne sensor to measure global forest carbon suggests that we must rely on airborne platforms to carry out the bulk of national carbon sampling for at least the next decade. Airborne LiDAR can now successfully map forest carbon stocks at high-resolution at subnational to national scales at high accuracy and extremely low cost on a per-hectare basis if operated

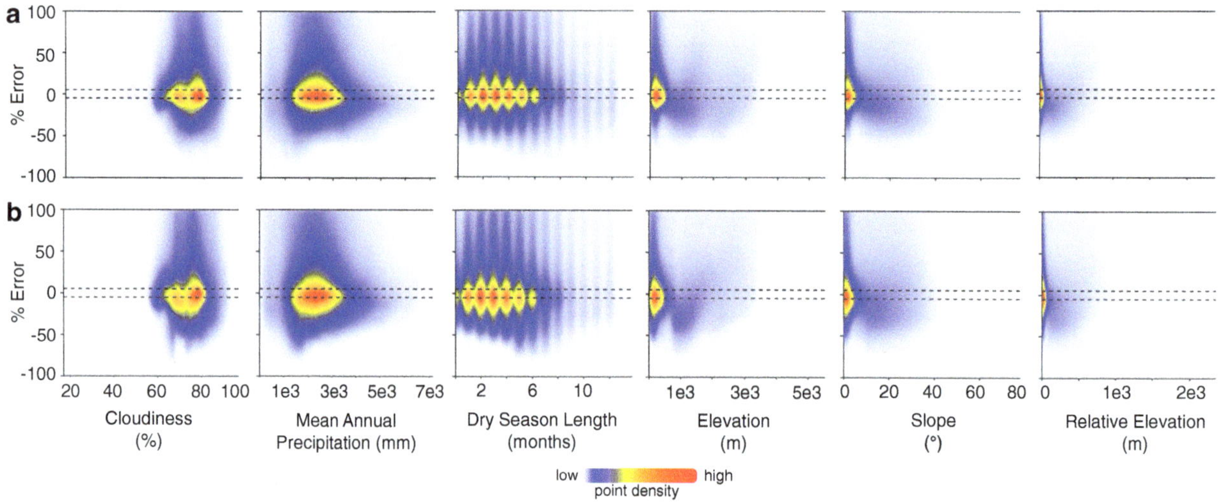


**Fig. 7** Underlying drivers of uncertainty. The percent relative difference between the model ACD and the median estimated ACD error across **a** all systematic sampling grids and **b** all stratified random sampling plot sets (see "Methods" section), plotted against each non-binned stratum at the 1-ha scale. *Plots* are composed of a randomly selected 10 % of the total dataset (approximately 7.6 million hectares), with the point density *color scale* on a square root transformation for plotting clarity. *Horizontal dashed black lines* show the 10 % accuracy range

non-commercially; hundreds of thousands of hectares can be mapped per day at accuracies that approach or match field plot based sampling [24, 25]. The economies of scale achieved by a single airborne LiDAR sensor leads to far more cost-effective and accurate carbon mapping that is easily repeatable—yielding estimates of carbon net changes over time—gives this method of mapping a drastic advantage over field plot sampling.

No doubt tropical forests and their substantial carbon sink capacities will play a major role during the implementation of agreements forged during the December 2015 UNFCCC climate conference in Paris. Tropical forest nations must be ready to respond with accurate assessments of their carbon stocks and reliably monitor their changes over time. While field inventory plots have long been the standard for meeting these challenges, with tens of millions of dollars spent on plot implementation and infrastructure, this should not preclude a move toward more accurate and cost-effective forest carbon mapping.

## Methods

### Study area

Perú is an ideal test case for investigating whether field plot sampling can accurately map carbon because its forests span a wide range of topographic, climatic, floristic, and geologic variables. Heterogeneity in aboveground carbon density (ACD) is driven by combinations of these variables, and higher ACD heterogeneity leads to less accurate field plot estimates of ACD [7]. Results from within the diverse environmental gradients of Perú can then be applied to other countries of similar conditions.

We used the 1-ha resolution aboveground carbon density map of Perú created by Asner et al. [6] as a model of the "true" carbon stocks of the country. We refer to this LiDAR-estimated carbon map throughout the paper as "the model carbon stock" or "the model aboveground carbon density (ACD)." The model ACD map was produced from a countrywide airborne LiDAR campaign with the Carnegie Airborne Observatory (CAO; cao.carnegiescience.edu). The LiDAR data were integrated with high-resolution satellite imaging, a large field plot sampling network (to calibrate/validate the LiDAR carbon mapping), and an advanced geospatial scaling algorithm [for more details see 6]. Please refer to Fig. S5 of [6] for model validation results, which compared 536,874 ha of LiDAR-measured forest TCH that were not used to train the model-estimated TCH of those same locations. The $R^2$ is 0.78 and the RMSE is 3.50 m. While the underlying relationship between LiDAR measured-TCH and estimated ACD may be heteroskedastic, with a higher ACD errors found at taller TCH, the relative ACD uncertainty that results is low (i.e., <20 % for high ACD (>120 MgC $ha^{-1}$) lowland forests).

We masked this map to exclude areas that are not forested, using a mean per hectare threshold of >70 % photosynthetic vegetation and >2 m top-of-canopy height (TCH). The photosynthetic vegetation map was created from a national-scale mapping of Perú using the CLASlite algorithm with mosaicked Landsat satellite imagery [26]. The TCH map is the underlying dataset used to produce the ACD map of Perú [6]. This left a forested area of 76,457,286 ha for the analysis. While this is 59.5 % of

Perú by land area, it contains 98.5 % of the total carbon of the entire country (6.798 Pg). All analyses were performed using R statistical software [27] with geospatial manipulations and analyses performed using the 'raster' package [28].

#### Environmental stratification layers

We compiled a set of six co-aligned, 1-ha resolution environmental stratification layers (hereafter 'strata') comprised of topographic and climatic variables covering all of Perú. Cloudiness (%) was created from long-term (2000–2010) NASA moderate resolution imaging spectroradiometer (MODIS) data and described in further detail by [6]. Mean annual precipitation (MAP) was calculated from NASA Tropical Rainfall Measuring Mission (TRMM) 2b31 monthly data (1998–2011), and dry season length (DSL) was calculated using the same TRMM data for all months with less than 100 mm of precipitation. Elevation, slope, and a relative elevation model (REM) were created from NASA Shuttle Radar Topography Mission (SRTM) at 90 m resolution. Relative elevation is a proxy for vegetation related water resources, and is developed by calculating the height of the ground above nearest water body [29]. Each stratum was masked by the same forest mask described above.

#### Field plot sampling designs

We tested two commonly used field plot sampling strategies: systematic sampling and stratified random sampling. This study is not meant to be an exhaustive review of spatial sampling techniques, rather to evaluate two methods that are commonly employed or recommended to map forest carbon over large geographic areas [9, 10, 12, 13]. Therefore, we did not consider other possible spatial sampling designs (e.g., simple random sampling, cluster sampling).

Systematic sampling uses sampling units (i.e., field inventory plots) placed at regular intervals according to an ordering scheme. For the country of Perú, we chose regularly spaced square grids with dimensions ranging from 5–100 km, increasing at 1-km intervals. Only those sampling units on the grid that fall within the bounds of the forest mask were retained (hereafter referred to as systematic grid sampling).

Stratified sampling places sampling units across a region according to pre-defined subregions that are more homogenous than the region as a whole, thereby reducing inherent sampling errors. The degree to which a stratified sampling technique accurately estimates the true population depends largely upon choosing homogenous subregions from which to conduct the subsampling [30]. We used unique combinations of the six strata described above to create these subregions (hereafter referred to as 'substrata'). First each stratum was binned by quintile, creating 5 categorical classes for each stratum (Fig. 1b–g). Then a map of all unique combinations of the six quintile-binned strata was produced, and any resulting substrata less than 1000 ha in area was excluded (hereafter referred to as 'quintile-binned substrata'). This resulted in 5398 unique substrata totaling 98.2 % of the forested area used for the analyses. We repeated the above steps, instead binning each stratum by 5 equal subsets of the total range of values of each stratum (hereafter referred to as 'range-binned substrata') (Additional file 1: Fig. S3). Again removing any substrata with less than 1000 ha, resulting in 447 substrata representing 99.9 % of the forested area used for the analyses. Sampling units were then randomly selected from within each substratum according to the simulations described below (hereafter referred to as 'stratified random sampling').

#### Estimating total national carbon stocks

##### *Systematic grid*

For each systematic sampling grid, the ACD value extracted from the centroid of each sampling grid cell was used as the ACD value for all of the 1-ha values in that sampling grid cell. The total number of grid cells in a systematic grid represents the number of field plots needed to create the estimated ACD map. We produced 96 different estimated ACD maps, each with a spatial resolution corresponding to the sampling grid dimensions used to create the map (5–100 km in 1 km increments). The ACD was summed across the entire country for each of the different estimated ACD maps, and compared to the total of the model ACD map.

##### *Stratified random*

We used Monte Carlo simulations to determine the probability of a randomly placed set of field inventory plots within each quintile-binned substratum to accurately estimate the total carbon stocks of Perú. For all of the 5398 substrata, a random sample of ACD values drawn from all ACD values of the substratum was selected and used to estimate the mean ACD of the substratum. The mean ACD of each substratum was multiplied by the total number of hectares in the substratum to get the estimated total ACD of that substratum. All estimates of total ACD from the substrata were then summed to get an estimated total ACD of the country, and compared to the total of the model ACD map. This was repeated 5000 times and the probability of correctly estimating the total ACD of Perú to within 10 % was determined. This simulation was run for sample sizes of 1 through 100 field plots in each substratum. We repeated this same simulation, but at each iteration progressively removing the smallest substrata (by area) in increments of 5 %. We used the mean ACD of the sampled substrata to estimate the total ACD of the substrata that were removed from the simulation. We did not test the range-binned substrata because of its poor performance in estimating the mean ACD of a substratum (see "Uncertainty simulations and extraction" section below).

### Estimating national carbon geographies

#### *Systematic grid*

Each 1-ha resolution systematic grid estimated ACD map (see above) represents a stratify-and-multiply upscaling approach whereby the sampled value from the systematic grid (which can be thought of as stratification in this context) is applied to the corresponding unsampled population (the rest of the 1-ha values in the systematic grid cell). Each estimated ACD map was co-aligned with the model ACD map of Perú, subtracted from the model ACD map, and the difference divided by the model ACD map to get the relative percent difference on a per hectare basis. We also performed a model-linked upscaling approach using random forest machine learning in addition to the stratify-and-multiply upscaling approach. We used the systematic sampling grid values as the training dataset for each random forest model. For the predictors of the model, we used the same 12 contiguous remote sensing layers that were used to create the model ACD map [see 6]. For each systematic grid, a random forest model was fit with the sampled values and their co-aligned values from the remote sensing layers. Each fitted model was then applied to the entire country using the 12 remote sensing layers to create a wall-to-wall random forest estimated ACD map [for more details on the random forest approach see [31]. This map was then compared to the model ACD map in the same manner as above.

#### *Stratified random*

We chose a subset {1, 5, 10, 50, 100} of field plot sets used to randomly sample each substratum to create a stratified random estimated ACD map using a stratify-and-multiply approach. For each substratum, a field plot set was randomly drawn from all ACD values of the substratum, and the mean of the set was mapped back onto all hectares of that substratum. This produced a set of maps of estimated ACD for the entire country of Perú at 1-ha resolution based on the stratified random sampling approach. For each map created from one of the five field plot sets, the stratified random estimated ACD map was subtracted from the model ACD map, and the difference divided by the model ACD map to get the relative percent difference. Again, we did not test the range-binned strata for the reason described in the preceding section. We also applied a model-linked upscaling approach using the random forest machine learning algorithm. Here the training datasets were composed of the stratified random sample values and their co-aligned values from the 12 remote sensing layers. Random forest models were fit and applied in the same way as described above.

### Uncertainty simulations and extractions

We ran Monte Carlo simulations to find the number of field plots it would take to accurately estimate the mean ACD of a substratum for the stratified random sampling approach. For each substratum, a random sample of ACD values (representing field plots) was selected from all possible ACD values of the substratum and the mean of the sample was compared to the mean of all ACD values of the substratum. This was repeated 5000 times to find the probability that the selected number of field plots would produce an estimate accurate to within 10 % of the mean of the substratum as estimated by the model ACD map. This simulation was run for sample sizes of 1 through 100 field plots. We then mapped the carbon value (either mean or total) and the environmental strata value of each substratum onto the results of these simulations to examine potential patterns.

We co-aligned the spatially explicit maps of estimated ACD created from both sampling approaches with the original (non-binned) strata. For each sampling approach, we created a median estimated ACD map across all of the sampling grids (systematic sampling, 5–100 km) and plot sets (stratified random, {1, 5, 10, 50, 100} plots). We then extracted the percent relative difference between the median estimated ACD map and the model ACD map. We also extracted the six climatic and topographic values associated with each mapped hectare.

## Additional file

**Additional file 1: Figure S1.** The fraction of stratified random substrata needed to estimate model total national carbon stocks. **Figure S2.** Number of field plots needed to reliably estimate the mean model substratum ACD using stratified sampling. **Figure S3.** Strata binned and colored by five equal ranges. **Figure S4.** Number of field plots needed to reliably estimate the mean model substratum ACD using equal range binning. **Figure S5.** Number of field plots needed to reliably estimate the mean model substratum ACD using equal range binning colored by environmental variables.

### Abbreviations

ACD: aboveground carbon density; CAO: Carnegie Airborne Observatory; IPCC: Intergovernmental Panel on Climate Change; LiDAR: Light Detection and Ranging; MAP: mean annual precipitation; MODIS: Moderate Resolution Imaging Spectroradiometer; REM: relative elevation model; SRTM: Shuttle Radar Topography Mission; TCH: top of canopy height; TRMM: Tropical Rainfall Monitoring Mission; UNFCCC: United Nations Framework Convention on Climate Change.

### Authors' contributions

DCM and GPA designed the study. GPA provided the data and DCM analyzed the data. DCM and GPA wrote the paper. All authors read and approved the final manuscript.

**Acknowledgements**
We thank P. Brodrick for helpful comments on the manuscript. This study was supported by the John D. and Catherine T. MacArthur Foundation and the endowment of the Carnegie Institution for Science. The Carnegie Airborne Observatory is made possible by the Gordon and Betty Moore Foundation, the John D. and Catherine T. MacArthur Foundation, Avatar Alliance Foundation, W.M. Keck Foundation, the Margaret A. Cargill Foundation, Grantham Foundation for the Protection of the Environment, Mary Anne Nyburg Baker and G. Leonard Baker Jr, and William R. Hearst III. We thank colleagues from the Perúvian Ministry of Environment (MINAM) for their long-term support and collaboration under the joint MINAM-Carnegie Institution program agreement.

**Competing interests**
The authors declare that they have no competing interests.

**References**
1. Pan Y, Birdsey RA, Fang J, Houghton R, Kauppi PE, Kurz WA, Phillips OL, Shvidenko A, Lewis SL, Canadell JG, Ciais P, Jackson RB, Pacala SW, McGuire AD, Piao S, Rautiainen A, Sitch S, Hayes D. A large and persistent carbon sink in the world's forests. Science. 2011;333:988–93.
2. LeQuere C, Moriarty R, Andrew RM, Peters GP, Ciais P, Friedlingstein P, Jones SD, Sitch S, Tans P, Arneth A, Boden TA, Bopp L, Bozec Y, Canadell JG, Chevallier F, Cosca CE, Harris I, Hoppema M, Houghton RA, House JI, Jain A, Johannessen T, Kato E, Keeling RF, Kitidis V, KleinGoldewijk K, Koven C, Landa CS, Landschützer P, Lenton A, et al. Global carbon budget 2014. Earth Syst Sci Data Discuss. 2014;7:521–610.
3. CDIAC: Global carbon project: Full global carbon budget. US. Department of Energy; 2012.
4. Köhl M, Baldauf T, Plugge D, Krug J. Reduced emissions from deforestation and forest degradation (REDD): a climate change mitigation strategy on a critical track. Carbon Balance Manag. 2009;4:10.
5. REDD Offset Working Group: California, Acre and Chiapas: partnering to reduce emissions from tropical deforestation. Green Technology Leadership Group; 2012.
6. Asner GP, Knapp DE, Martin RE, Tupayachi R, Anderson CB, Mascaro J, Sinca F, Chadwick KD, Higgins M, Farfan W, Llactayo W, Silman MR. Targeted carbon conservation at national scales with high-resolution monitoring. Proc Natl Acad Sci USA. 2014;111:E5016–22.
7. Marvin DC, Asner GP, Knapp DE, Anderson CB, Martin RE, Sinca F, Tupayachi R. Amazonian landscapes and the bias in field studies of forest structure and biomass. Proc Natl Acad Sci USA. 2014;111:E5224–32.
8. Chaplin-Kramer R, Ramler I, Sharp R, Haddad NM, Gerber JS, West PC, Mandle L, Engstrom P, Baccini A, Sim S, Mueller C, King H. Degradation in carbon stocks near tropical forest edges. Nat Commun. 2015;6:10158.
9. Tomppo E, Gschwantner T, Lawrence M, McRoberts RE. National forest inventories. Dordrecht: Springer Science & Business Media; 2009.
10. GOFC-GOLD. A sourcebook of methods and procedures for monitoring and reporting anthropogenic greenhouse gas emissions and removals associated with deforestation, gains and losses of carbon stocks in forests remaining forests, and forestation. GOFC-GOLD land cover project office, The Netherlands: Wageningen University; 2014.
11. Maniatis D, Mollicone D. Options for sampling and stratification for national forest inventories to implement REDD+ under the UNFCCC. Carbon Balance Manag. 2010;5:9.
12. IPCC: 2013 Revised Supplementary Methods and Good Practice Guidance Arising From the Kyoto Protocol. Switzerland: IPCC; 2014.
13. IPCC: Guidelines for national greenhouse gas inventories. Japan: IGES; 2006.
14. Blackard J, Finco M, Helmer E, Holden G, Hoppus M, Jacobs D, Lister A, Moisen G, Nelson M, Riemann R. Mapping U.S. forest biomass using nationwide forest inventory data and moderate resolution information. Remote Sens Environ. 2008;112:1658–77.
15. Cartus O, Kellndorfer J, Walker W, Franco C, Bishop J, Santos L, Fuentes J. A national, detailed map of forest aboveground carbon stocks in Mexico. Remote Sens. 2014;6:5559–88.
16. Phillips OL, Baker T, Feldpausch T, Brienen R: RAINFOR field manual for plot establishment and remeasurement. R Soc. 2016:1–22.
17. Réjou-Méchain M, Muller-Landau HC, DETTO M, Thomas SC, Le Toan T, Saatchi SS, Barreto-Silva JS, Bourg NA, Bunyavejchewin S, Butt N, Brockelman WY, Cao M, Cárdenas D, Chiang JM, Chuyong GB, Clay K, Condit R, Dattaraja HS, Davies SJ, Duque Á, Esufali S, Ewango C, Fernando RHS, Fletcher CD, Gunatilleke IAUN, Hao Z, Harms KE, Hart TB, Hérault B, Howe RW, et al. Local spatial structure of forest biomass and its consequences for remote sensing of carbon stocks. Biogeosci Discuss. 2014;11:5711–42.
18. Goetz SJ, Baccini A, Laporte NT, Johns T, Walker W, Kellndorfer J, Houghton RA, Sun M. Mapping and monitoring carbon stocks with satellite observations: a comparison of methods. Carbon Balance Manag. 2009;4:2.
19. Romijn E, Herold M, Kooistra L, Murdiyarso D, Verchot L. Assessing capacities of non-annex I countries for national forest monitoring in the context of REDD. Environ Sci Policy. 2012;19–20:33–48.
20. Romijn E, Lantican CB, Herold M, Lindquist E, Ochieng R, Wijaya A, Murdiyarso D, Verchot L. Assessing change in national forest monitoring capacities of 99 tropical countries. For Ecol Manage. 2015;352:109–23.
21. Saatchi S, Mascaro J, Xu L, Keller M, Yang Y, Duffy P, Espírito Santo F, Baccini A, Chambers J, Schimel D. Seeing the forest beyond the trees. Glob Ecol Biogeogr. 2015;24:606–10.
22. Luccio M. Improving global carbon estimates with LiDAR. Sensors and systems. 2014. Accessed 2 May 2016.
23. Goetz S, Dubayah R. Advances in remote sensing technology and implications for measuring and monitoring forest carbon stocks and change. Carbon Manag. 2011;2:231–44.
24. Mascaro J, Detto M, Asner GP, Muller-Landau HC. Evaluating uncertainty in mapping forest carbon with airborne LiDAR. Remote Sens Environ. 2011;115:3770–4.
25. Asner GP, Mascaro J. Mapping tropical forest carbon: calibrating plot estimates to a simple LiDAR metric. Remote Sens Environ. 2014;140:614–24.
26. Asner GP, Knapp DE, Balaji A, Paez-Acosta G. Automated mapping of tropical deforestation and forest degradation: cLASlite. J Appl Remote Sens. 2009;3:033543.
27. R Development Core Team: R: A language and environment for statistical computing. 2014.
28. Package "raster" [cran.r-project.org/web/packages/raster].
29. Asner GP, Clark JK, Mascaro J, Vaudry R, Chadwick KD, Vieilledent G, Rasamoelina M, Balaji A, Kennedy-Bowdoin T, Maatoug L, Colgan MS, Knapp DE. Human and environmental controls over aboveground carbon storage in Madagascar. Carbon Balance Manag. 2012;7:2.
30. Wang J, Haining R, Cao Z. Sample surveying to estimate the mean of a heterogeneous surface: reducing the error variance through zoning. Int J Geogr Inf Sci. 2010;24:523–43.
31. Mascaro J, Asner GP, Knapp DE, Kennedy-Bowdoin T, Martin RE, Anderson C, Higgins M, Chadwick KD. A tale of two "forests": random forest machine learning aids tropical forest carbon mapping. PLoS One. 2014;9:e85993–9.

6

# Estimating carbon sequestration in the piedmont ecoregion of the United States from 1971 to 2010

Jinxun Liu[1,6*], Benjamin M. Sleeter[1], Zhiliang Zhu[2], Linda S. Heath[3,4], Zhengxi Tan[5], Tamara S. Wilson[1], Jason Sherba[1] and Decheng Zhou[7]

**Abstract**

**Background:** Human activities have diverse and profound impacts on ecosystem carbon cycles. The Piedmont ecoregion in the eastern United States has undergone significant land use and land cover change in the past few decades. The purpose of this study was to use newly available land use and land cover change data to quantify carbon changes within the ecoregion. Land use and land cover change data (60-m spatial resolution) derived from sequential remotely sensed Landsat imagery were used to generate 960-m resolution land cover change maps for the Piedmont ecoregion. These maps were used in the Integrated Biosphere Simulator (IBIS) to simulate ecosystem carbon stock and flux changes from 1971 to 2010.

**Results:** Results show that land use change, especially urbanization and forest harvest had significant impacts on carbon sources and sinks. From 1971 to 2010, forest ecosystems sequestered 0.25 Mg C $ha^{-1}$ $yr^{-1}$, while agricultural ecosystems sequestered 0.03 Mg C $ha^{-1}$ $yr^{-1}$. The total ecosystem C stock increased from 2271 Tg C in 1971 to 2402 Tg C in 2010, with an annual average increase of 3.3 Tg C $yr^{-1}$.

**Conclusions:** Terrestrial lands in the Piedmont ecoregion were estimated to be weak net carbon sink during the study period. The major factors contributing to the carbon sink were forest growth and afforestation; the major factors contributing to terrestrial emissions were human induced land cover change, especially urbanization and forest harvest. An additional amount of carbon continues to be stored in harvested wood products. If this pool were included the carbon sink would be stronger.

**Keywords:** Land-use change, Carbon change, Piedmont ecoregion, IBIS model

## Background

Increasing concentrations of carbon dioxide ($CO_2$) in the atmosphere is a major cause of global warming. Major terrestrial $CO_2$ emissions have been found where humans have disturbed the land by deforestation and agricultural practices [1, 2]. Because both forest and agricultural ecosystems are critical components of terrestrial C sequestration, many intensive observation and modeling studies have been undertaken to quantify ecosystem C change and C sequestration potential. Existing research shows forest ecosystems in the United States have been acting as C sinks, varying from 0.3 to 4 Mg C $ha^{-1}$ $yr^{-1}$ [3–11]. While forest harvest and natural disturbance lower forest C sequestration potential, forest rotation processes and natural recovery could make a forest system C neutral or a C sink if given enough time for recovery [12–14]. Studies of agricultural systems in the United States suggest that land-use changes caused severe soil organic carbon (SOC) loss from 1850 to 1960; but since the 1960s, improved farming practices (e.g., no-till) and increased C return to the soil have caused SOC to stabilize or possibly increase in some areas [15, 16]. Simulations of forest and agricultural ecosystems have produced large uncertainties regarding spatial and temporal variability of

*Correspondence: jxliu@usgs.gov
[1] Western Geographic Science Center, USGS, Menlo Park, CA 94025, USA
Full list of author information is available at the end of the article

carbon dynamics, and identification of the driving forces of change [17–22]. Most uncertainties originate from difficulties in quantifying the impacts of disturbances and environmental variables. Land-use and land-cover change (LUCC) is a major disturbance factor, which strongly influences carbon budget calculations [2, 21, 23–25]. However, it has been a challenge to detect and quantify the dynamic nature of LUCC over large areas [21, 26, 27]. In the past, LUCC information in large-scale carbon sequestration modeling was not well developed, mainly due to the lack of consistent data describing changes in land use and land cover.

Several LUCC-oriented carbon studies have been conducted based on reconstructed LUCC histories [20, 23, 28–32]. However, these land cover change histories were usually averaged at a coarse spatial scale. Additionally, remote sensing is often used to detect tree cover loss at the time of disturbance whereas detection of regeneration following harvest is delayed. For agricultural ecosystems, previous research was usually at a local scale and under experimental control [33]. Quantifying the magnitude and spatial variation of regional carbon sources or sinks was found to be difficult because of the high spatial variability in site conditions and the diversity of human management.

More recently, high-resolution land-change datasets, such as the US Geological Survey's Land Cover Trends (LCT) dataset have become available [34, 35]. The LCT data is the longest temporal record of consistent, empirically-derived, high resolution LUCC data available for the US at present. This ecoregion-based assessment of land-use change was guided by a nationally consistent study design including mapping, statistical methods, field studies, and analysis [26, 34, 36]. The sequential LUCC maps for the Piedmont ecoregion have a 60-m spatial resolution, a much finer resolution than any previously used in C accounting for the conterminous United States [19, 20, 37].

In this study, we report the use of the Integrated Biosphere Simulator (IBIS) in simulating carbon dynamics of forest and agricultural ecosystems in the Piedmont ecoregion from 1971 to 2010. The 60-m resolution 1973–2000 LCT data were used to generate 960-m annual land cover change maps from 1971 to 2010 (see more details in the "Methods" section). We focused on the effective use of the annual maps in analyzing land-change effects on biomass and soil C, as well as harvested C trends related to forest cover change.

## Methods

### The Piedmont ecoregion and LUCC detection

The Piedmont is a hilly, transitional ecoregion between the flatland near the Atlantic coast and the mountainous Appalachian ecoregions of the eastern United States. It has an area of 165,460 $km^2$, as delineated by EPA level III ecoregions [38]. Annual precipitation ranges from 1100 to 1400 mm. Average annual minimum temperature ranges from 7 to 12 °C, and maximum temperature ranges from 20 to 25 °C.

The Piedmont was an important farming region in the 19th century, but during the 20th century, forestry and land development became more competitive uses of land. In recent decades, the Piedmont has had a relatively fast land conversion rate compared with other ecoregions in the eastern US. Nearly 15 % of land in the Piedmont converted to a different cover type between 1973 and 2000 [34]. The LCT land cover change quantification was based on interpretation of land change across eleven 20-km by 20-km sample blocks randomly selected using a stratified randomsampling design [26, 36]. Imagery from the Landsat archive was manually interpreted for each sample block at five dates (1973, 1979, 1986, 1992, 2000). The classified maps include ten land use and land cover categories: water, developed, mechanically disturbed (clear cutting), mining, naturally barren (vegetation cover less than 10 %), forest, grassland/shrubland, agricultural land, wetland, and non-mechanically disturbed (i.e., forest cover loss attributed to fire, flooding, or disease) (For more details, see [26, 34]. The sample block locations are shown in Fig. 1. An example of land transition rates between 1992 and 2000 is listed in Table 1. Overall, from 1973 to 2000, the percentage of forest area decreased from 61 to 56 %, agricultural area decreased from 25 to 24 %, and the combined developed and mining areas increased from 11 to 17 % of the ecoregion area.

### IBIS model framework and calibration

The integrated biosphere simulator (IBIS) [40, 41] is a physically consistent modeling framework that follows basic rules of physics, plant physiology, and biogeochemistry. The original model combined features of a mechanistic model of canopy photosynthesis [42], a semi-mechanistic model of stomatal conductance [43], an algorithm on phenology [44], and several soil biogeochemical models [45–47] in a single application. IBIS has the ability to simulate major land surface processes, canopy physiology, vegetation phenology, long-term vegetation dynamics, ecosystem productivity, and carbon cycling.

A modified version of IBIS included nitrogen (N) controls on the carbon cycle [48], land-use and land-cover change, and wildland fire effects [49], and Methane ($CH_4$) emission [50]. Figure 2 shows the major C pools, fluxes and flow pathways simulated by the IBIS model. The input C flux of the ecosystem is the net primary productivity (NPP), which is calculated from climate, soil

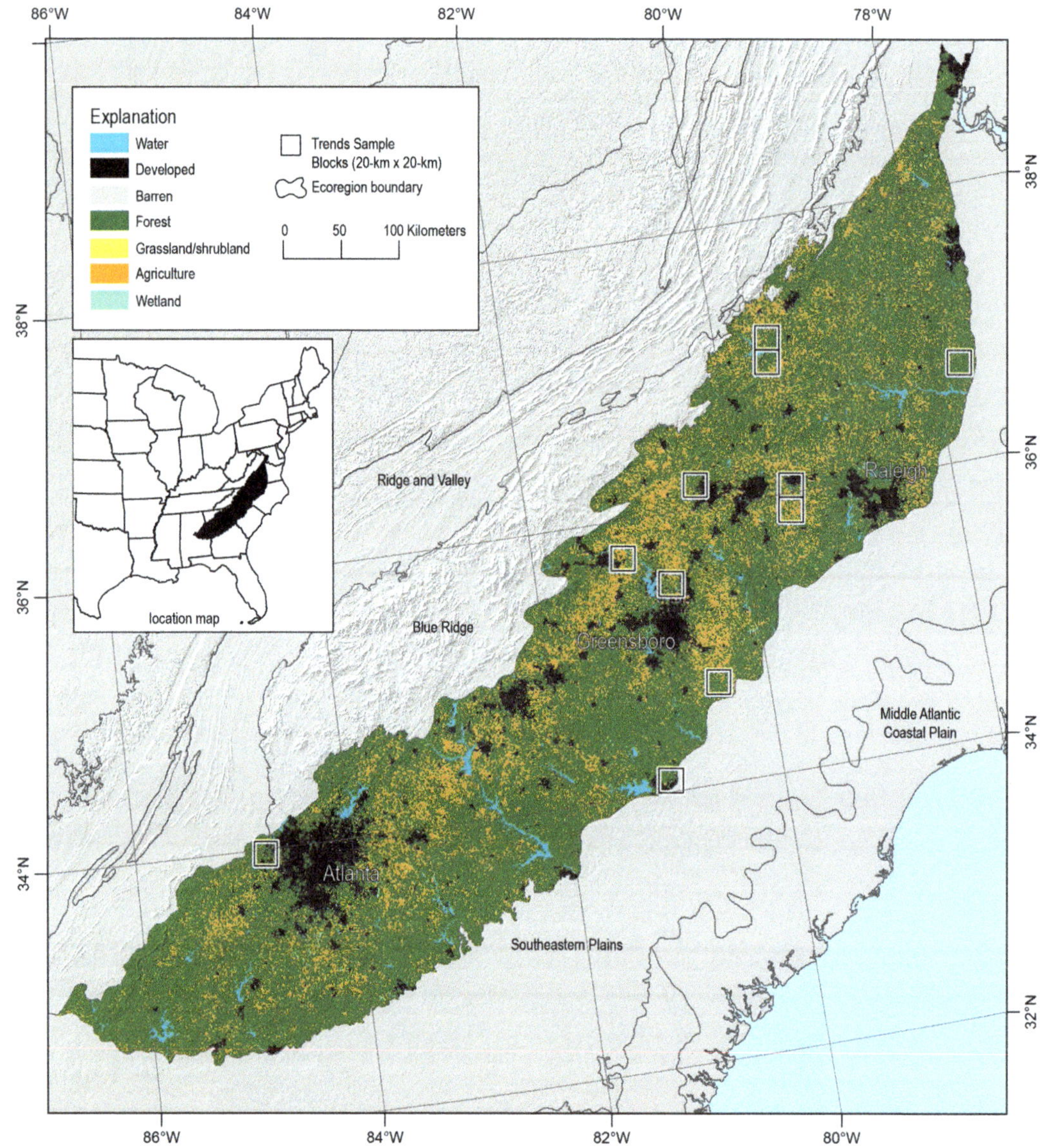


**Fig. 1** The Piedmont ecoregion from the 1992 National Land Cover Dataset (Vogelmann et al. [39]) modified to the ten land use and land cover classes of the USGS Land Cover Trends Project. *Square boxes* indicate the eleven 20 $km^2$ land cover change sampling blocks at 60 m spatial resolution

and vegetation conditions. The output C flux includes woody biomass harvest, crop harvest, fire combustion, heterotrophic respiration and C leaching. Climate drivers mainly determine the ecosystem NPP and respiration calculations while disturbance (LUCC) events mainly affect the C removal and reset the vegetation C pools.

Process-based biogeochemical models like IBIS usually simulate over an hourly time step. Therefore decade- and century-long simulations for a large region would take a significant length of time (weeks to months) if running at a very high spatial resolution (e.g. 60-m). Therefore, IBIS has been recently enhanced in several aspects, including: (1) program recoded to support parallel computing using Message Passing Interface (MPI) on super-computers; (2) treatment of fractional vegetation cover within a single land pixel in order to use the newly available higher resolution LUCC products; (3) spatial scalars for tree biomass growth and crop grain production at county level to deal with diverse geography. The IBIS conceptual land pixel includes multiple plant functional types, each competing for light, moisture and nutrient during simulation (Fig. 3A). The existing forest and agricultural cover fraction maps for the

**Table 1 Piedmont ecoregion land cover transition rates between 1992 and 2000**

| | Water | Developed | M.disturbed | Mining | Nat.bare | Forest | Gra/shr | Agri. | Wetland | N.M.dist. | Total |
|---|---|---|---|---|---|---|---|---|---|---|---|
| Water | *3576.9* | 4.0 | 21.4 | 0.0 | 0.0 | 210.2 | 0.0 | 93.2 | 31.3 | 0.0 | 3937.1 |
| Developed | 11.2 | *24,013.1* | 172.9 | 9.3 | 0.1 | 2262.8 | 1.2 | 653.1 | 4.3 | 0.0 | 27,128.1 |
| M.disturbed | 0.0 | 0.0 | *92.8* | 0.0 | 0.0 | 3176.9 | 0.5 | 16.6 | 16.5 | 0.0 | 3303.2 |
| Mining | 0.0 | 0.3 | 0.9 | *315.0* | 0.0 | 64.7 | 0.0 | 25.3 | 0.0 | 0.0 | 406.3 |
| Nat.bare | 0.0 | 0.0 | 0.0 | 0.0 | *3.9* | 0.0 | 0.0 | 0.0 | 0.0 | 0.0 | 3.9 |
| Forest | 23.9 | 2.8 | 3669.0 | 1.1 | 0.0 | *87,101.2* | 40.2 | 313.5 | 0.3 | 0.0 | 91,151.9 |
| Grass/shrub | 0.0 | 0.0 | 30.1 | 0.0 | 0.0 | 8.6 | *19.1* | 10.1 | 0.0 | 0.0 | 68.0 |
| Agriculture | 0.7 | 0.0 | 154.3 | 0.0 | 0.3 | 557.7 | 0.0 | *37,509.3* | 0.1 | 0.0 | 38,222.3 |
| Wetland | 0.0 | 0.0 | 14.8 | 0.0 | 0.0 | 0.0 | 0.0 | 0.0 | *1224.2* | 0.0 | 1239.0 |
| N.M.dist. | 0.0 | 0.0 | 0.0 | 0.0 | 0.0 | 0.0 | 0.0 | 0.0 | 0.0 | *0.0* | 0.0 |
| Total | 3612.7 | 24,020.3 | 4156.2 | 325.4 | 4.3 | 93,382.1 | 61.1 | 38,621.2 | 1276.6 | 0.0 | 165,459.9 |

Areas summed by row indicate total land area in 2000; Areas summed by column indicate total land area in 1992. Area units: $km^2$. Total forest area in 1992 was 93,382.1 $km^2$. Forest to developed land (urban) conversion between 1992 and 2000 was 2262.8 $km^2$. Forest remaining as forest was 87,101.2 $km^2$. Forest to agriculture conversion was 557.7 $km^2$. Total forest area in 2000 was 91,151.9 $km^2$, of which 313.5 $km^2$ was converted from agriculture

Values italicized within the table represent land use and land cover remaining constant

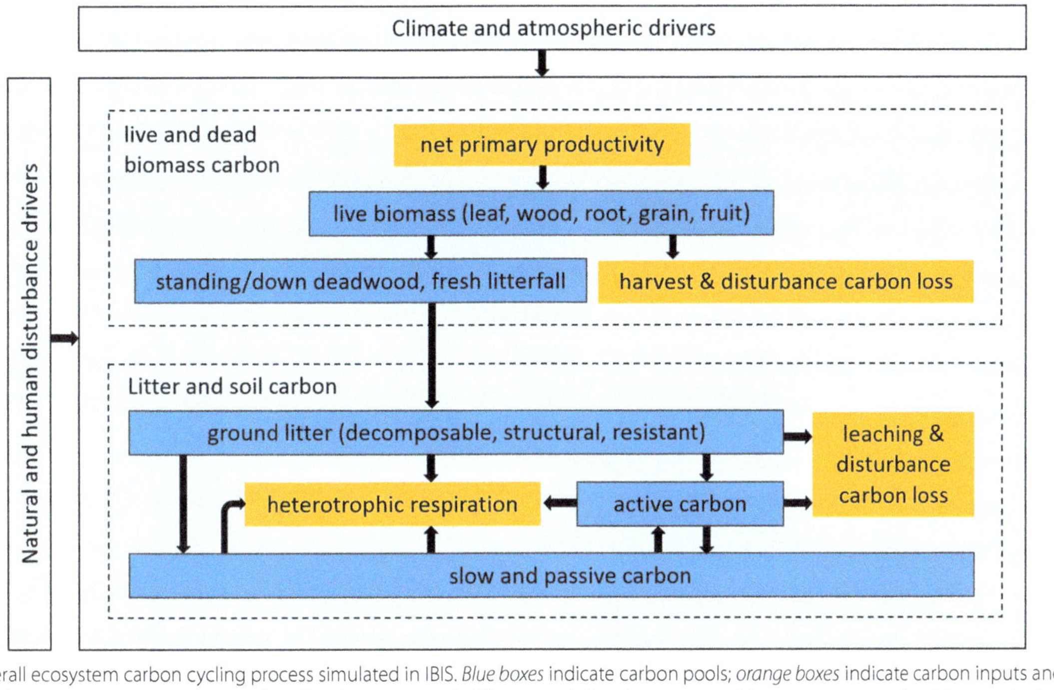


**Fig. 2** Overall ecosystem carbon cycling process simulated in IBIS. *Blue boxes* indicate carbon pools; *orange boxes* indicate carbon inputs and outputs to the ecosystem; *arrows* show carbon flow between pools. Climate and disturbance control the ecosystem productivity, respiration, and carbon removals

Piedmont ecoregion were built from the 30-m vegetation canopy maps from the LandFire Project and aggregated to 960-m (Fig. 3). Most of the 960-m pixels are a mixture of several land cover types. The modified IBIS tracks the percent area of each land cover type within each land pixel. When a LUCC or disturbance event (e.g., reforestation, deforestation, urbanization, etc.) occurs, cover fractions are transferred between relevant land cover types.

The current IBIS version deals with 11 types of disturbances: (1) fire, (2) logging, (3) deforestation to grass/shrub, (4) deforestation to cropland, (5) afforestation from grass/shrub, (6) afforestation from agriculture, (7) urbanization from forest, (8) urbanization from grass/shrub, (9) urbanization from cropland, (10) agricultural expansion (grass/shrub to cropland), (11) agricultural contraction (cropland to grass/shrub). Logging and fire events may

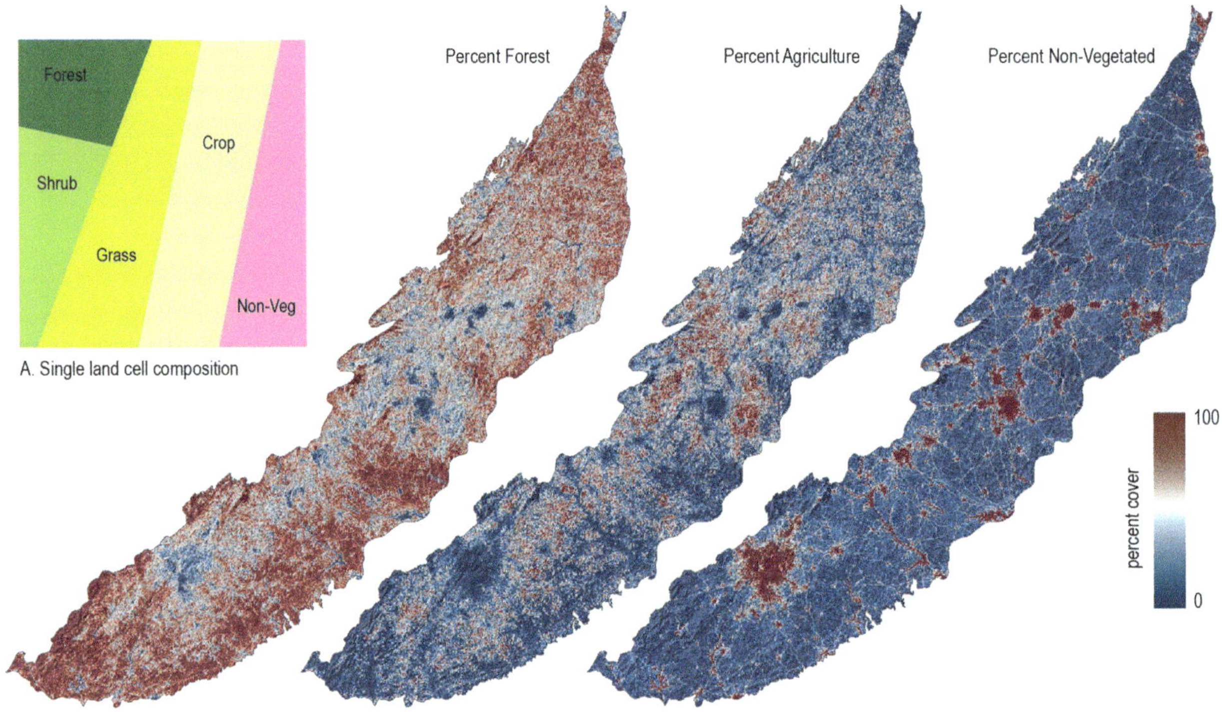


**Fig. 3** Conceptual IBIS single land pixel composition and the distribution and percentages of forest land, agricultural land and non-vegetation cover in the Piedmont ecoregion based on data obtained from the LandFire Project aggregated to 960-m from 30-m

only trigger C removal and additional tree mortality; forest cover fraction will remain unchanged, allowing for forest regrowth. Other types of disturbance will remove carbon from the landscape and also alter land-cover fractions. For example, forest to cropland transition (deforestation) will re-allocate previous forest cover fraction to the cropland cover fraction, and remove all forest carbon from the landscape. As a result, the following simulation year will have no forest productivity, but more crop productivity due to crop area fraction increase.

In addition to disturbances detectable through remote sensing methods, we also consider the less easily detectable events like forest thinning activities. Forest thinning rate is calculated using recent annualized forest inventory data collected by the US Forest Service, Forest Inventory and Analysis Program [51, 52]. Thinning activity is loosely defined as the cutting-related biomass carbon loss of less than 50 % during two consecutive observation periods (around 5 years) in order to make the overall thinned area percentage (i.e., 61 % of the total forest cutting area) in agreement with earlier estimates [53, 54]. For the Piedmont ecoregion, the annual thinning rates in terms of total live aboveground biomass carbon range from 0.31 to 1.16 % (average 0.81 %) in different counties. Forest thinning removes an amount of tree carbon which is not usually a detectable change in forest cover fraction.

For this study, the following LUCC were considered: logging, deforestation (forest to agriculture conversion), afforestation (agriculture to forest conversion), agriculture contraction (agriculture to grassland conversion), agriculture expansion (grassland to agriculture conversion), and urbanization (forest to urban, grassland to urban, and agriculture to urban).

Carbon output variables of the IBIS model include live and dead biomass, soil organic carbon, carbon losses from disturbance, as well as net primary productivity (NPP) and net biome productivity (NBP). In this study, we used the dominant vegetation cover (i.e., forest, agriculture, shrub, and grass) to summarize carbon variables because most land pixels are mixed with more than one cover type. Statistics for forest land pixels usually include a certain amount of other vegetation types. Similarly, agricultural land summaries may also contain a small amount of forest and other vegetation covers.

IBIS uses biome level plant functional types (PFT) to represent major vegetation groups, which are coarsely defined in the model based on climate conditions. Some stand or landscape level carbon control factors for forest systems are not considered in the current version of IBIS, such as tree species, age class and stem density. Similarly, the modeled crop system only includes two generic crop PFTs (C3 and C4 crops). This makes site level model calibration difficult.

To validate the model, a county level calibration procedure was developed. For forest, 1-km MODIS forest NPP for 2001–2005 was averaged at the county level and compared with IBIS simulated forest NPP. Then, an adjustment scalar was introduced. The scalar was assumed to help in dealing with unknown environmental factors (e.g., species, age, stem density). For crops, without considering details like crop species, irrigation, fertilization, and double cropping in IBIS, we used the county level USDA NASS crop yield statistics and IBIS grain yield to generate a county level grain yield scalar, which partly reflects the human activity difference by geolocation. The scalars were used to modify the Maximum Rubisco-limited rate of carboxylation (Vmax) of related PFTs (forest or crop) in a new IBIS run. In addition to NPP and grain yield calibration, simulated forest live biomass at 100 years of age was also calibrated. Forest growth curves from the Carbon On Line Estimator (COLE) [55, 56] were used as the general forest growth references to be compared with IBIS growth curves.

### Data sources

Land-change information from the LCT project, wildland fire scar and burn severity data from the Monitoring Trends in Burn Severity project [57], and vegetation canopy percentage and vegetation height information from the Landfire project [58] were the key variables for calculating vegetation fraction and biomass on each land pixel, as well as the effects of logging, deforestation, afforestation, urbanization, agricultural expansion and contraction, and wildland fire on C changes. These 30- and 60-m datasets were aggregated to 960-m resolution for this study.

We extended LUCC mapping to include 1971–1972 and 2001–2010, using the LCT land conversion rates of 1973–1979 and 1992–2000, respectively. We used the LUCAS model [59] to create an annual time-series of land use and land cover maps for the period 1971–2010. The 2001 National Land Cover Database (NLCD) [60] was used as the starting point, and changes were backcasted into the historical period based on (1) rates of change between the four temporal periods from the LCT data, and (2) adjacency rules which prescribed change to occur adjacent to existing cells (also see Daniel et al. [61]). For the period 2001–2010, the LUCAS model was run forward in time using the LCT rates from the 1992–2001 period. All simulations were done on an annual timestep, at a spatial resolution of 1-km.

Global atmospheric $CO_2$ concentration trends were based on observed data [62]. Spatially heterogeneous atmospheric $CO_2$ measurements from the Scanning Imaging Absorption Spectrometer for Atmospheric Cartography (SCIMACHY) from ENVISAT (0.5° resolution, 2003–2009) were converted to monthly surface $CO_2$ concentration to produce an average monthly $CO_2$ difference to global average $CO_2$ map [63], which was used to adjust the $CO_2$ fertilization calculation.

The PRISM 4-km resolution monthly precipitation and temperature data from 1971 to 2010 were used as the main climate driver. Other climate variables, such as relative humidity and wind speed, were included as monthly normals across the 1961–1990 time series. The SSURGO soil carbon and texture (960 m resolution, ~2000) dataset was used for initial soil conditions [64]. The soil profiles contain up to six depth layers (to 7, 15, 25, 50, 100, 200 cm depths) and include sand, silt, and clay fractions for each.

For the forest growth calibration, summary results for each FIA survey unit were downloaded from the COLE website (http://www.ncasi2.org/COLE/). The 100 year average forest biomass growth values were used to calibrate IBIS simulated 100 year total tree biomass growth. For agricultural ecosystems, county-level grain yield statistics were downloaded from the USDA National Agricultural Statistics Service (NASS) website (https://www.quickstats.nass.usda.gov/) and used to calibrate simulated cropland grain yield.

## Results

### Land cover change trends and annual maps

Statistics showed net decreases in forest (−4.76 %) and agricultural (−1.27 %) land area and net increases in urban (4.49 %) and disturbance areas (1.14 %) between 1973 and 2000. Grassland, wetland, and water bodies had much lower change rates (<0.3 %) during this period. The calculated annual land change amounts for the IBIS C simulation (1971–2010) were mapped at 960 m spatial resolution. Rotational logging accounted for the most area over the study period. The percentage of forest area logged versus total forested area increased from 1.43 to 3.62 % over the simulation period, indicating an intensification in the rotational logging cycle, in addition to increasing rates of deforestation due primarily to urbanization. The overall land change statistics and mapping are shown in Fig. 4.

### Regional ecosystem carbon trends

Changes in the major C pools of forest and agricultural ecosystems are shown in Fig. 5. For forest ecosystems (forest dominant land pixels), live biomass C increased from 5.61 to 6.72 kg C $m^{-2}$ over the 40-year simulation period. SOC increased from 6.73 to 7.06 kg C $m^{-2}$. Dead wood and litter carbon pools increased slightly from 2.87 to 3.13 kg C $m^{-2}$. Overall, the forest ecosystem was a net C sink of 0.025 kg C $m^{-2}$ $yr^{-1}$. The agricultural ecosystem was a small C sink during the simulation period. Live

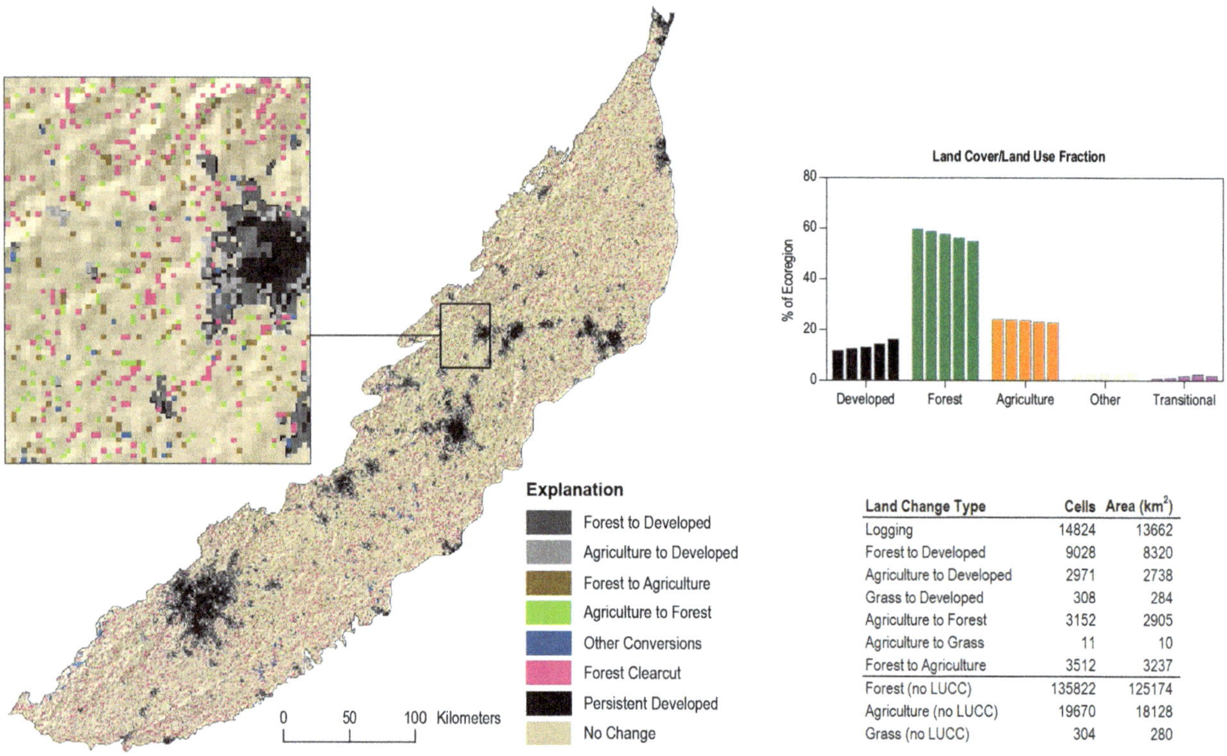


| Land Change Type | Cells | Area ($km^2$) |
|---|---|---|
| Logging | 14824 | 13662 |
| Forest to Developed | 9028 | 8320 |
| Agriculture to Developed | 2971 | 2738 |
| Grass to Developed | 308 | 284 |
| Agriculture to Forest | 3152 | 2905 |
| Agriculture to Grass | 11 | 10 |
| Forest to Agriculture | 3512 | 3237 |
| Forest (no LUCC) | 135822 | 125174 |
| Agriculture (no LUCC) | 19670 | 18128 |
| Grass (no LUCC) | 304 | 280 |

**Fig. 4** Total land cover change between 1971 and 2010 in the Piedmont ecoregion. Map is at 960-m resolution. Changing pixels indicate the simulated locations of land cover change

biomass C remained stable around 2.38 kg C $m^{-2}$; SOC increased from 5.60 to 6.02 kg C $m^{-2}$, and litter and dead biomass increased from 1.25 to 1.36 kg C $m^{-2}$ over the 40-years. The agricultural lands are not made up of pure land cover pixels and still include some forest cover. The average annual C sink was only 0.003 kg C $m^{-2}$ $yr^{-1}$.

In terms of ecosystem productivity, both forest and agricultural ecosystems showed some inter-annual fluctuations of NPP and NBP over the study period (Fig. 6). Average forest NPP increased from 0.61 to 0.71 kg C $m^{-2}$ $yr^{-1}$. Agricultural ecosystem NPP increased from 0.39 to 0.46 kg C $m^{-2}$ $yr^{-1}$.

Logging removal (rotational clear-cutting) from forest ecosystems averaged 0.012 kg C $m^{-2}$ $yr^{-1}$, with the 1970s exhibiting the lowest harvest rates (0.006 kg C $m^{-2}$ $yr^{-1}$), and the late 1980s and early 1990s exhibiting the highest (0.020 kg C $m^{-2}$ $yr^{-1}$). The estimated forest thinning removal averaged 0.048 kg C $m^{-2}$ $yr^{-1}$ during the study period. Carbon removal following forest to agriculture conversion and forest to urban conversion was about 0.011 kg C $m^{-2}$ $yr^{-1}$. Due to the non-pure land pixels at the 960 m spatial resolution, there was also agricultural removal (grain and straw) in forest areas, which increased from 0.017 to 0.021 kg C $m^{-2}$ $yr^{-1}$. The forest NBP averaged 0.025 kg C $m^{-2}$ $yr^{-1}$. Carbon removal from the agricultural system (grain and straw) increased from 0.075 to 0.118 kg C $m^{-2}$ $yr^{-1}$ over the study period. An estimated 0.001 kg C $m^{-2}$ $yr^{-1}$ was lost from logging removals in agricultural ecosystems due to non-pure land pixels. The agricultural ecosystem NBP averaged about 0.003 kg C $m^{-2}$ $yr^{-1}$.

For the whole ecoregion, the 40 year average NBP was 3.32 Tg C on a valid calculation area of 157,415 $km^2$, with forest dominant ecosystems accounting for approximately 3.32 Tg C (~133,380 $km^2$, averaged between 1971 and 2010) and agricultural dominant ecosystems accounting for approximately 0.06 Tg C (~19,522 $km^2$). Newly established urban land lost carbon at a rate of 0.014 kg C $m^{-2}$ $yr^{-1}$, and totalled 0.05 Tg C per year (~4497 $km^2$, averaged between 1971 and 2010). The spatial distribution of the 40 year average NPP and NBP are also displayed in Fig. 6.

## Land cover change impact

We analysed the simulated NBP map series to show the consequences of LUCC on the C cycle (Fig. 7). Piedmont forest ecosystems had an average NBP of 25 g C $m^{-2}$ $yr^{-1}$ over the 40 year period. Forest dominant ecosystems without land conversions had an average NBP of 36 g C $m^{-2}$ $yr^{-1}$, while forests that had undergone

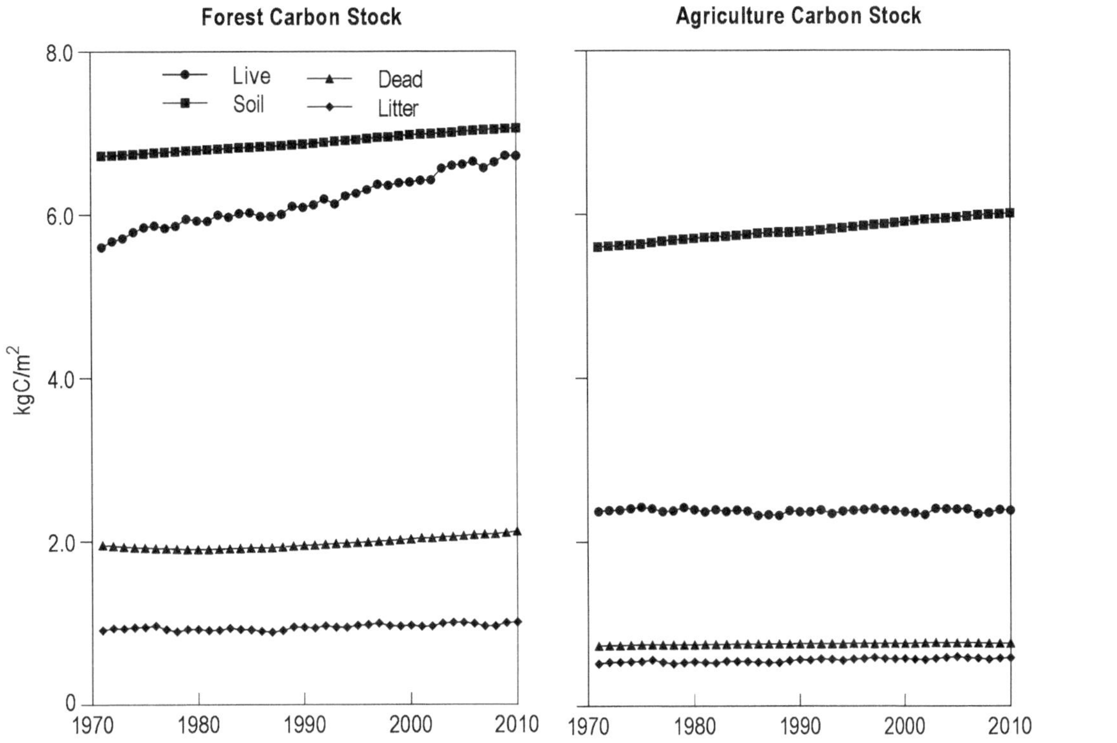


**Fig. 5** Trends of major carbon pools in forest and agricultural ecosystems in the Piedmont ecoregion between 1971 and 2010. The major pools are living biomass ("live"), soil organic carbon ("soc"), dead biomass ("dead"), and litter

rotational clear-cutting, deforestation to agriculture land, or clearing for urbanization showed C losses at 41, 117 and 128 g C $m^{-2}$ $yr^{-1}$, respectively. On the other hand, afforestation locations gained 112 g C $m^{-2}$ $yr^{-1}$ on average. Cropland had an average NBP of 3 g C $m^{-2}$ $yr^{-1}$ over the 40 year period. Without land conversions, cropland had an average NBP of 12 g C $m^{-2}$ $yr^{-1}$. Agricultural land converting to grassland led to C loss at a rate of 12 g C $m^{-2}$ $yr^{-1}$. Additionally, conversions of agricultural land to urban also led to C loss, 107 g C $m^{-2}$ $yr^{-1}$, mainly due to removal of trees on non-pure agricultural lands. The regional total NBP for each LUCC type were also displayed on Fig. 7.

### Carbon budget of Piedmont ecoregion

The NBP of forest and agricultural lands in the Piedmont ecoregion varied annually as indicated in Fig. 6. The Piedmont ecoregion has 170,806 simulated 960 m resolution land pixels and with 1 kg C $m^{-2}$, that equals 157.4 Tg C (1 Tg $= 10^{12}$ g $=$ 1 million tons) for the whole region. Figure 8 shows the average annual C budget of the Piedmont over the 40 year period. The four aggregated carbon pools show that Piedmont soil had the largest carbon storage and also had the largest carbon increase. Live biomass carbon increase was small due to removal and mortality. The major carbon emissions were litter and soil respiration. Overall, NBP for the Piedmont Ecoregion was a small net sink at a rate of 3.3 Tg C per year.

### Model validation and uncertainty

This study used county level observations and model results from other sources to adjust model outputs. We used FIA COLE data and USDA NASS grain yield data as field truth datasets. The MODIS NPP product (based on remote sensing and modelling) was used once to generally adjust the NPP range because the IBIS model does not consider tree species, local management, etc., which were potentially well captured by remote sensing. We avoided overfitting because IBIS has its own NPP algorithm. Figure 9 shows the adjustment.

In general, IBIS NPP matched with MODIS NPP. Since this was not a pixel level comparison, IBIS forest NPP still maintained its spatial variability driven by climate, soil, vegetation and disturbance. Forest biomass and crop yield output fit closely to field observation data because multiple iterations were run on the related spatial scalars. Again, the calibration was done at the county level, therefore the output maintained spatial heterogeneity.

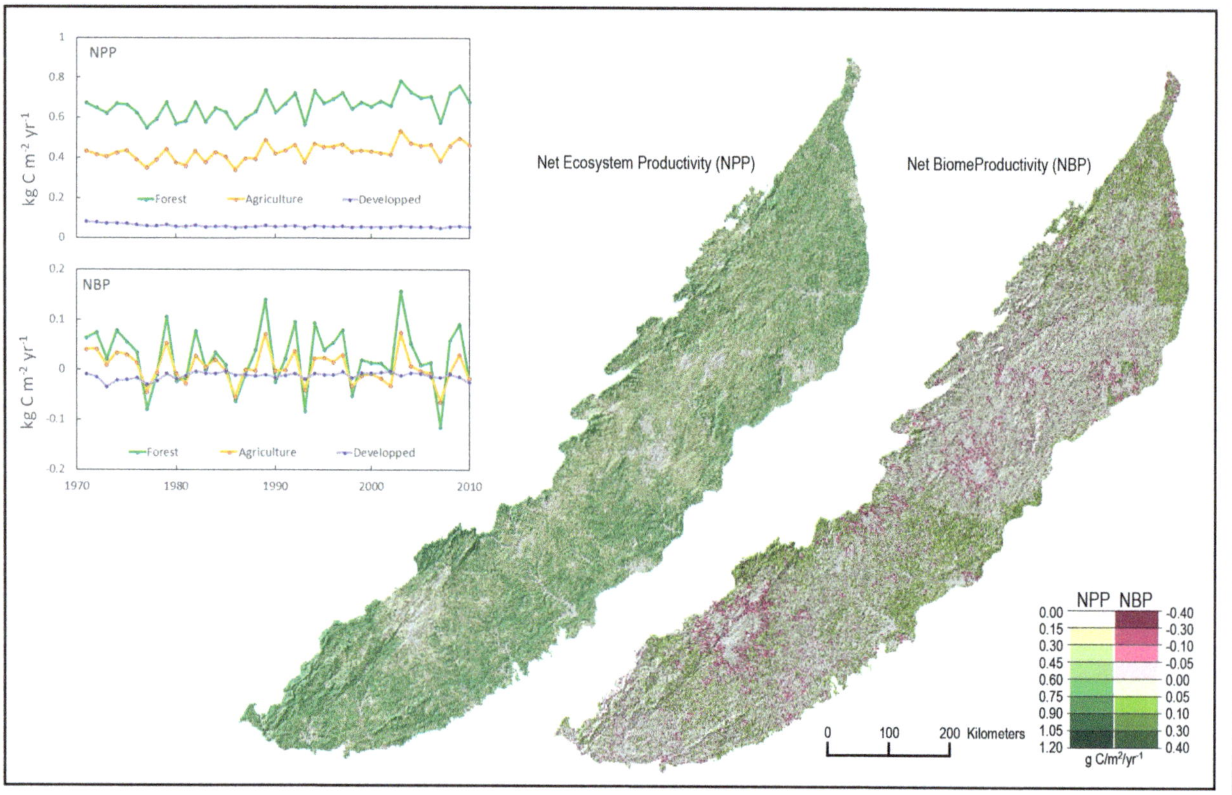


**Fig. 6** Trends and spatial distribution of net primary productivity (NPP) and net biome productivity (NBP) in the Piedmont ecoregion between 1971 and 2010. The maps show the average over 40 years, expressed in kg C $m^{-2}$ $yr^{-1}$

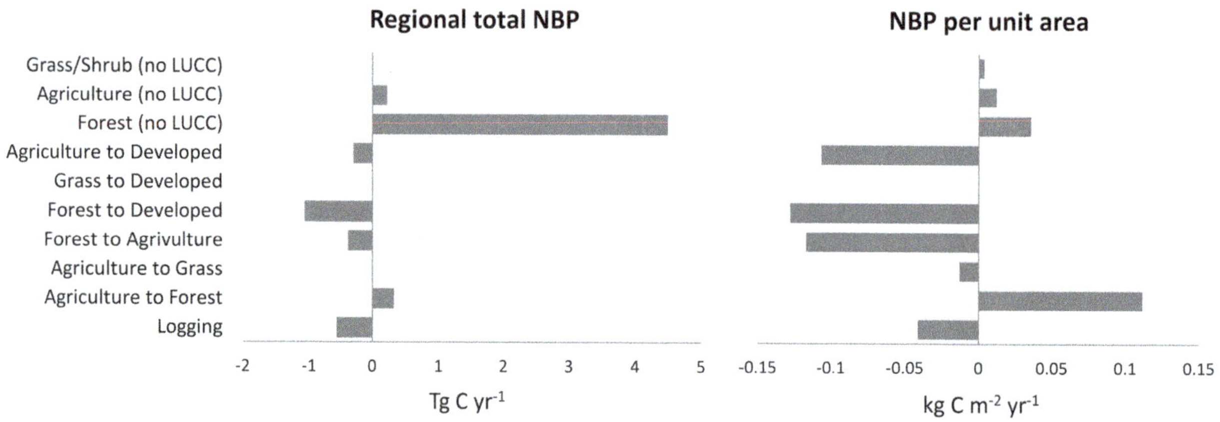


**Fig. 7** Land cover change impact on net biome productivity (NBP) in Piedmont ecosystems

## Discussion

### Total ecosystem C budgets

Our current estimate of NPP for the Piedmont forests was 656 g C $m^{-2}$ $yr^{-1}$ (6.56 Mg C $ha^{-1}$ $yr^{-1}$). This result is higher than a MODIS NPP estimate of 550 g C $m^{-2}$ $yr^{-1}$ (5.5 Mg C $ha^{-1}$ $yr^{-1}$) for temperate forests [65]. However, our forest NPP estimation is close to MODIS NPP estimation in the Piedmont region as indicated in our calibration result (Fig. 9). In our simulations, the average annual forest biomass C growth (before harvesting) was about 1.2 Mg C $ha^{-1}$ $yr^{-1}$. Assuming 70 % of this growth is above-ground biomass C growth (0.84 Mg C $ha^{-1}$ $yr^{-1}$), then it is very close to Caspersen's [66] estimate of average US forest above-ground biomass

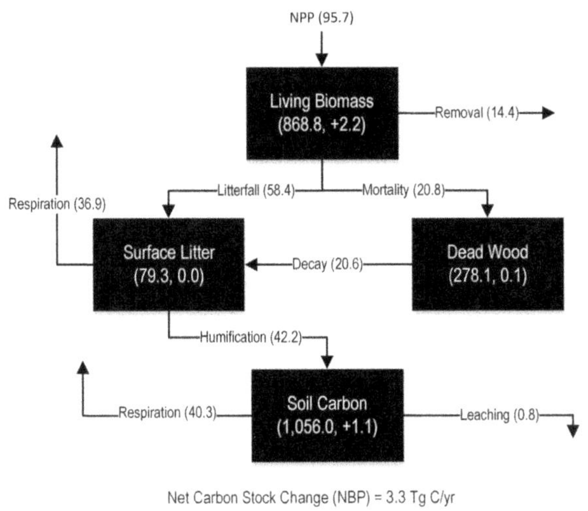


**Fig. 8** Overall average annual Piedmont ecoregion carbon budget from 1971 to 2010 expressed in teragrams (Tg) of C per year. *Blocks* are carbon pools, with average pool size and average annual change amount. All other *names* and *numbers* are carbon fluxes

C increase (0.8 Mg C $ha^{-1}$ $yr^{-1}$). However, after adding the carbon removal amount, our Piedmont forest NBP from 1971 to 2000 was only 0.25 Mg C $ha^{-1}$ $yr^{-1}$. This estimate is significantly lower than that reported by Hurtt et al. [20], who estimated a forest C sink, in the 1980s, for all the United States, of about 0.9 Mg C $ha^{-1}$ $yr^{-1}$. It is also lower than the report of Tan et al. [67] focusing on US federal lands, which estimated a rate of about 0.6 Mg C $ha^{-1}$ $yr^{-1}$. It is much lower than a study of the nearby Appalachian forests, where NBP was estimated at approximately 1.8 Mg C $ha^{-1}$ $yr^{-1}$ [24], mainly due to more intensive forest harvesting in the Piedmont forests compared to the Appalachian forests. As previously indicated, forest thinning was considered in our simulations, which accounted for two times the amount of C removal compared to forest clear-cutting (including deforestation). Logging, thinning and deforestation were the major causes of Piedmont's low C sink strength.

Agricultural ecosystem NPP averaged 425 g C $m^{-2}$ $yr^{-1}$ (4.25 Mg C $ha^{-1}$ $yr^{-1}$), and NBP was 3 g C $m^{-2}$ $yr^{-1}$ (0.03 Mg C $ha^{-1}$ $yr^{-1}$). Our NBP estimate was lower than controlled site level studies (e.g., [68], which reported a sink of 36 g C $m^{-2}$ $yr^{-1}$ for a complex crop rotation receiving both manure and chemical fertilizers). However, it is already understood that regional simulation results are typically lower than site-level results [69]. Especially when fractional cropland areas are considered, lower NPP and NBP is not unexpected. In addition, we applied a straw removal rate of 50 % in the study. This ratio was another factor influencing the C sink level on agricultural lands.

### Impact of LUCC

Although deforestation commonly led to a C source and reforestation led to a C sink as indicated in Fig. 7, the influence on the regional C budget depends on the total area affected. The total forest area of the US has been very stable at 2.98 million $km^2$ during the past several decades, with a 0.1 % average fluctuation [70, 71]. During the 1990s and 2000s, US total forest land had increased by approximately 1 % of total US land area [72]. However, net forest area decrease (accounting for both deforestation and afforestation), in the Piedmont ecoregion, from 1971 to 2000, was about 5 % of the total land area and

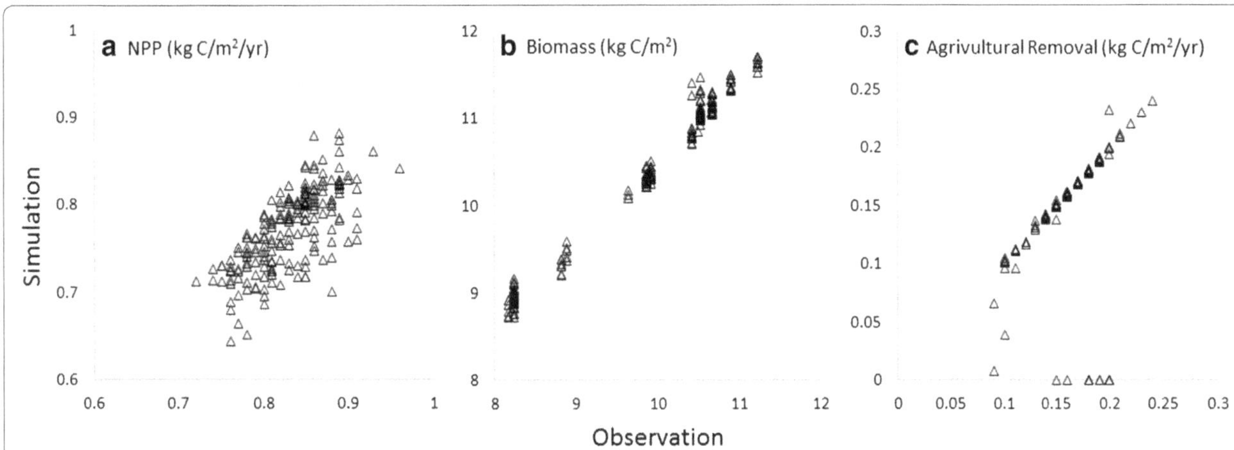


**Fig. 9** IBIS model validation against remote sensing and field survey data at the county level. **a** Simulated forest NPP against MODIS NPP product; **b** Simulated 100 year forest tree biomass against FIA based COLE database; **c** Simulated agricultural harvest carbon removal (grain + hay + straw) against USDA NASS crop harvest data. Each data point represents a county in the Piedmont ecoregion (total 198 counties)

about 10 % of the total forested area. This would have a direct influence on the overall C budget. In this study, we applied a simple rule for vegetation regrowth following urbanization related land cover changes, i.e. allowing up to 15 % vegetation cover on newly generated urban lands. Urban lands on average were a C source of about 14 g C $m^{-2}$ $yr^{-1}$ (0.14 Mg C $ha^{-1}$ $yr^{-1}$).

The intensive stand-replacing disturbances occurring in the Piedmont ecoregion were a key factor leading to below-average NBP levels compared to other regions. Disturbed sites may need at least 20 years to regain the lost C [4, 14, 73]. In this study, the logging land pixels had an average NBP of −41 g C $m^{-2}$ $yr^{-1}$ over a 40 year period (average logging was about 20 years), indicating the recovery process may take even longer. On the other hand, forest thinning led to the largest amount of C removal in our simulation and represents an important source of uncertainty in modeling regional carbon dynamics.

For agricultural ecosystems, LUCC effects were linked to land conversions only. At present, we didn't model other land management activities on cropland, such as fertilization and irrigation.

Combined, an estimated 14.4 Tg C was removed from ecosystems every year, of which 69 % (~9.9 Tg C) were from forest thinning and clear-cut. Some of the removed carbon continues to be stored in harvested wood products (HWP). However, dynamics of this important carbon pool were not considered in this study. If carbon storage in HWP were included the carbon sink would be stronger.

#### LUCC data across scales

Currently, research by the USGS LandCarbon project [74] is underway to produce 30-m, annual, wall-to-wall land cover change maps at the national scale based on the Landsat 40 year data archive. The up-coming highly relevant LUCC map product will allow carbon models to make much better C estimations and better differentiate drivers from combined climate and land change interactions. However, using these high resolution land cover change data can be difficult for large regional C assessments. Aggregating the high resolution data to a coarser resolution would help complicated process-based models to use the LUCC information effectively. For even larger scale C modelling work, such as global C simulations, the LUCC information could be more inaccurate at coarse spatial resolution (such as 1–2°). New LUCC products such as the Global Change Assessment Model (GCAM) already focus on building a mixed land cover product. Therefore, using mixed land cover data in C models is a new and necessary approach for carbon accounting, especially for large region C assessments.

#### Statistics of mixed land pixels

Although it is meaningful to use fractional land cover for carbon simulation in large regions, there exists the challenge of correctly interpreting LUCC and its effects. The use of forest dominant or agriculture dominant lands can potentially lead to confusion. For example, the Piedmont ecoregion has about 122,590 $km^2$ of forest dominant land and 18,472 $km^2$ of cropland dominant land. Yet this does not mean forest cover is 6.6 times larger than agricultural land cover. The Land Cover Trends data (pure land pixels) actually shows that forest area is about 2.4 times agricultural land area in the Piedmont. Therefore, alternative summary methods should be explored.

### Conclusion

The Piedmont ecoregion in the eastern US was estimated as a weak C sink with an average C gain of 3.3 Tg C $yr^{-1}$. The overall per unit area C sink was 0.025 kg C $m^{-2}$ $yr^{-1}$, which is much smaller than the rates in the Appalachians and other eastern ecoregions. The major cause was the rapid human induced land-cover and land-use changes, especially forest logging, thinning, and urbanization. The method used in this study helps to quantify the overall human land use effect on the carbon budget and would be suitable at national and global scales when more detailed and consistent LUCC data becomes available.

#### Authors' contributions

JL carried out the IBIS modeling work. BS carried out the land cover change data interpretation and analysis studies. ZZ was involved in the whole study design with JL and BS. LH guided the understanding and use of the COLE database. TW, ZT and JS helped to draft the manuscript. DZ helped with forest thinning quantification. All authors read and approved the final manuscript.

#### Author details

[1] Western Geographic Science Center, USGS, Menlo Park, CA 94025, USA. [2] USGS, 12201 Sunrise Valley Drive, Reston, VA 20192, USA. [3] USDA Forest Service, Northern Research Station, Durham, NH 03824, USA. [4] Present Address: USDA Forest Service, Washington Office R&D, Washington DC 20250, USA. [5] Arctic Slope Regional Corporation, Contractor to USGS EROS, Sioux Falls, SD 57198, USA. [6] San José State University Research Foundation, San José, CA 95112, USA. [7] Jiangsu Key Laboratory of Agricultural Meteorology, Nanjing University Of Information Science And Technology, Nanjing 210044, China.

#### Acknowledgements

This study is supported by the USGS LandCarbon Project. This research used resources of the Argonne Leadership Computing Facility, which is a DOE Office of Science User Facility supported under Contract DE-AC02-06CH11357.

#### Competing interests

The authors declare that they have no competing interests.

#### References

1. Intergovernmental Panel on Climate Change (IPCC). Climate change—the IPCC scientific assessment. In: Houghton JT, Jenkins GJ, Ephraums JJ, editors. Cambridge: Cambridge University Press; 1990.

2. Intergovernmental Panel on Climate Change (IPCC). Land use, land-use change, and forestry—a special report of the intergovernmental panel on climate change. In: Watson RT, Noble IR, Bolin B, Ravindranath NH, Verardo DJ, Dokken DJ, editors. Cambridge: Cambridge University Press; 2000.
3. Birdsey RA, Heath LS. Carbon changes in US forests. In: Joyce LA, editor. Climate change and the productivity of America's forests (Gen. Tech. Rep. RM-GTR-271). Fort Collins: USDA Forest Service, Rocky Mountain Research Station; 1995.
4. Delcourt HR, Harris WF. Carbon budget of the southeastern US biota: analysis of historical change in trend from source to sink. Science. 1980;210:321–3.
5. Fan S, Gloor M, Mahlman J, Pacala S, Sarmiento J, Takahashi T, Tans P. A large terrestrial carbon sink in North America implied by atmospheric and oceanic carbon dioxide data and models. Science. 1998;282:442–6.
6. Goodale CL, Apps MJ. Forest carbon sinks in the Northern hemisphere. Ecol Appl. 2002;12:891–9.
7. Heath LS, Smith JE. Soil carbon accounting and assumptions for forestry and forest-related land use changes. In: Joyce LA, Birdsey R, editors. The impact of climate change on America's forests (Gen. Tech. Rep. RMRS-GTR-59). Fort Collins: USDA Forest Service, Rocky Mountain Research Station; 2000. p. 89–101.
8. Heath LS, Birdsey RA, Williams DW. Methodology for estimating soil carbon for the forest carbon budget model of the United States. Environ Pollut. 2002;116:373–80.
9. Turner DP, Koerper G, Harmon M, Lee J. A carbon budget for forests of the conterminous United States. Ecol Appl. 1995;5:421–36.
10. Brown SL, Schroeder PE. Spatial patterns of aboveground production and mortality of woody biomass for eastern US forests. Ecol Appl. 1999;9:968–80.
11. Heath LS, Birdsey RA. Carbon trends of productive temperate forests of the coterminous United States. Water Air Soil Pollut. 1993;70:279–93.
12. Covington HR, Harris WF. Changes in the forest floor organic mater and nutrient content following clear cutting in northern hardwoods. Ecology. 1981;62:41–8.
13. Federer CA. Organic matter and nitrogen content of the forest floor in even-aged northern hardwoods. Can J For Res. 1984;14:763–7.
14. Gholz HL, Fisher RF. Organic matter production and distribution in slash pine (*Pinus elliottii*) plantations. Ecology. 1982;63:1827–39.
15. Flach KW, Barnwell TO, Crosson P. Impacts of agriculture on atmospheric carbon dioxide. In: Paul EA, Paustian K, Elliott ET, Cole CV, editors. Soil organic matter in temperate agroecosystems. Boca Raton: CRC Press; 1997. p. 3–13.
16. Liu S, Bliss N, Sundquist E, Huntington TG. Modeling carbon dynamics in vegetation and soil under the impact of soil erosion and deposition. Global Biogeochem Cycles. 2003;17:1074–97.
17. Ciais P, Peylin P, Bousquet P. Regional biospheric carbon fluxes as inferred from atmospheric $CO_2$ measurements. Ecol Appl. 2000;10:1574–89.
18. Dixon RK, Brown S, Houghton RA, Solomon AM, Trexler MC, Wisniewski J. Carbon pool and flux of global forest ecosystems. Science. 1994;263:185–90.
19. Houghton RA, Hackler JL, Lawrence KT. The US carbon budget: contributions from land-use change. Science. 1999;285:574–8.
20. Hurtt GC, Pacala SW, Moorcroft PR, Caspersen J, Shevliakova E, Houghton RA, Moore B III. Projecting the future of the US carbon sink. Proc Natl Acad Sci USA. 2002;99:1389–94. doi:10.1073/pnas.012249999.
21. Liu S, Loveland TR, Kurtz RM. Contemporary carbon dynamics in terrestrial ecosystems in the Southeastern Plains of the United States. Environ Manag. 2004;33:S442–56.
22. Pacala SW, Houghton RA, Birdsey RA, Heath L, Sundquist ET, Stallard RF, Ciais P, Moorcroft P, Caspersen JP, Shevliakova E, Moore B, Kohlmaier G, Holland E, Gloor H, Harmon ME, Fan SM, Sarmiento JL, Goodale CL, Schimel D, Field CB, Hurtt GC, Baker D, Peylin P. Consistent land- and atmosphere-based US carbon sink estimates. Science. 2001;292:2316–20.
23. Foley JA, DeFries R, Asner GP, Barford C, Bonan G, Carpenter SR, Chapin FS, Coe MT, Daily GC, Gibbs HK, Helkowski JH, Holloway T, Howard EA, Kucharik CJ, Monfreda C, Patz JA, Prentice IC, Ramankutty N, Snyder PK. Global consequences of land use. Science. 2005;309:570–4.
24. Liu J, Liu S, Loveland TR. Temporal evolution of carbon budgets of the Appalachian forests in the US from 1972 to 2000. For Ecol Manag. 2006;222:191–201.
25. Shao G, Qian T, Liu Y, Martin B. The role of urbanisation in increasing atmospheric $CO_2$ concentrations: think globally, act locally. Int J Sustain Dev World Ecol. 2008;15(4):302–8.
26. Loveland TR, Sohl TL, Stehman SV, Gallant AL, Sayler KL, Napton DE. A strategy for estimating the rates of recent United States land-cover changes. Photogramm Eng Remote Sens. 2002;68:1091–9.
27. Shao G, Wu J. On the accuracy of landscape pattern analysis using remote sensing data. Landscape Ecol. 2008;23:505–11.
28. Carter MR, Parton WJ, Rowland IC, Schultz JE, Steed GR. Simulation of soil organic carbon and nitrogen changes in cereal and pasture systems of Southern Australia. Aust J Soil Res. 1993;31:481–91.
29. Howard PJA, Loveland PJ, Bradley RI, Dry FT, Howard DM, Howard DC. The carbon content of soil and its geographical distribution in Great Britain. Soil Use Manag. 1995;11:9–15.
30. Kelly RH, Parton WJ, Crocker GJ, Grace PR, Klir J, Korschens M, Poulton PR, Richter DD. Simulating trends in soil organic carbon in long-term experiments using the century model. Geoderma. 1997;81:75–90.
31. Parton WJ, Rasmussen PE. Long-term effects of crop management in wheat/fallow: II. century model simulations. Soil Sci Soc Am J. 1994;58:530–6.
32. Parton W, Tappan G, Ojima D, Tschakert P. Ecological Impact of historical and future land-use patterns in Senegal. J Arid Environ. 2004;59:605–24.
33. Paul EA, Paustian K, Elliott ET, Cole CV, editors. Soil organic matter in temperate agroecosystems. Boca Raton: CRC Press; 1997.
34. Sleeter BM, Sohl TL, Loveland TR, Auch RF, Acevedo W, Drummond MA, Sayler KL, Stehman SV. Land-cover change in the conterminous United States from 1973–2000. Glob Environ Change. 2013;23:733–48. doi:10.1016/j.gloenvcha.2013.03.006.
35. Soulard CE, Acevedo W, Auch RF, Sohl TL, Drummond MA, Sleeter BM, Sorenson DG, Kambly S, Wilson TS, Taylor JL, Sayler KL, Stier MP, Barnes CA, Methven SC, Loveland TR, Headley R, Brooks MS. Land cover trends dataset, 1973–2000, US Geological Survey Data Series 844. 2014. http://pubs.usgs.gov/ds/844/. Accessed 31 Aug 2014.
36. Stehman SV, Sohl TL, Loveland TR. Statistical sampling to characterize recent UnitedStates land-cover change. Remote Sens Environ. 2003;86:517–29.
37. Members VEMAP. Vegetation/ecosystem modeling and analysis project: comparing biogeography and biogeochemistry models in a continental-scale study of terrestrial ecosystem responses to climate change and $CO_2$ doubling. Global Biogeochem Cycles. 1995;4:407–37.
38. US Environmental Protection Agency. Level III ecoregions of the continental United States, (1:7,500,000-scale map). Corvalis: US Environmental Protection Agency, National Health and Environmental Effects Research Laboratory; 1999.
39. Vogelmann JE, Howard SM, Yang L, Larson CR, Wylie BK, Van Driel JN. Completion of the1990's National Land Cover Data Set for the conterminous United States. Photogramm Eng Remote Sens. 2001;67(6):650–62.
40. Foley JA, Colin PI, Ramankutty N, Levis S, Pollard D, Sitch S, Haxeltine A. An integrated biosphere model of land surface processes, terrestrial carbon balance, and vegetation dynamics. Global Biogeochem Cycles. 1996;10:603–28.
41. Kucharik CJ, et al. Testing the performance of a dynamic global ecosystem model: water balance, carbon balance, and vegetation structure. Global Biogeochem Cycles. 2000;14:795–825.
42. Farquhar GD, von Caemmerer S, Berry JA. A biochemical model of photosynthetic $CO_2$ assimilation in leaves of C3 species. Planta. 1980;149:78–90. doi:10.1007/BF00386231.
43. Ball JT, Woodrow IE, Berry JA. A model predicting stomatal conductance and its contribution to the control of photosynthesis under different environmental condition. In: Biggings J, editor. Progress in photosynthesis research. Leiden: Martinus Nijhoff; 1986. p. 221–4.
44. Botta A, Viovy N, Ciais P, Friedlingstein P, Monfray P. A global prognostic scheme of leaf onset using satellite data. Glob Chang Biol. 2000;6(7):709–25.
45. Parton W, Schimel D, Cole C, Ojima D. Division s-3-soil microbiology and biochemistry. Soil Sci Soc Am J. 1987;51:1173–9.
46. Parton WJ, et al. Observations and modeling of biomass and soil organic matter dynamics for the grassland biome worldwide. Global Biogeochem Cycles. 1993;7(4):785–809.
47. Verberne E, Hassink J, De Willigen P, Groot J, Van Veen J. Modelling organic matter dynamics in different soils. Neth J Agric Sci. 1990;38(3):221–38.

48. Liu J, Price DT, Chen J. Nitrogen controls on ecosystem carbon sequestration: a model implementation and application to Saskatchewan, Canada. Ecol Model. 2005;186:178–95.
49. Liu J, Vogelmann JE, Zhu Z, Key CH, Sleeter BM, Price DT, Chen JM, Cochrane MA, Eidenshink JC, Howard SM, Bliss NB, Jiang H. Estimating California ecosystem carbon change using process model and land cover disturbance data: 1951–2000. Ecol Model. 2011;222:2333–41.
50. Zhu Q, Liu J, Peng C, Chen H, Fang X, Jiang H, Yang G, Zhu D, Wang W, Zhou X. Modelling methane emissions from natural wetlands by development and application of the TRIPLEX-GHG model. Geosci Model Dev. 2014;7(981–999):2014.
51. Zhou D, Liu S, Oeding J, Zhao, S. Forest cutting and impacts on carbon in the eastern United States. 2013. Scientific Reports 3, Article number: 3547.
52. US Department of Agriculture Forest Service. 2015. Forest inventory and analysis data and tools webpage. http://www.fia.fs.fed.us/tools-data/. Accessed 23 Nov 2015.
53. Masek JG, Cohen WB, Leckie D, Wulder MA, Vargas R, de Jong B, Healey S, Law B, Birdsey R, Houghton R. Recent rates of forest harvest and conversion in North America. J Geophys Res. 2011. doi:10.1029/2010JG001471.
54. Oswalt SN, Smith WB. US forest resource facts and historical trends. FS-1035. (US Department of Agriculture, Forest Service). 2012.
55. Van Deusen P, Heath LS. COLE web applications suite. NCASI and USDA Forest Service, Northern Research Station. 2015. http://www.ncasi2.org/COLE/. Accessed 23 Nov 2015.
56. Van Deusen PC, Heath LS. Weighted analysis methods for mapped plot forest inventory data: tables, regressions, maps and graphs. For Ecol Manag. 2010;260:1607–12.
57. Eidenshenk J, Schwind B, Brewer K, Zhu Z, Quayle B, Howard S. A project for monitoring trends in burn severity. Fire Ecol Spec Issue. 2007;3(1):3–21.
58. LANDFIRE. Existing vegetation type layer, LANDFIRE 1.1.0, US Department of the Interior, Geological Survey. 2008. http://landfire.cr.usgs.gov/viewer/. Accessed Oct 2010.
59. Sleeter BM, Liu J, Daniel C, Frid L, Zhu Z. An integrated approach to modeling changes in land use, land cover, and disturbance and their impact on ecosystem carbon dynamics: a case study in the Sierra Nevada Mountains of California. AIMS Environ Sci. 2015;2(3):577–606.
60. Homer C, Dewitz J, Fry J, Coan M, Hossain N, Larson C, Herold N, McKerrow A, VanDriel JN, Wickham J. Completion of the 2001 National land cover database for the conterminous United States. Photogramm Eng Remote Sens. 2007;73(4):337–41.
61. Daniel C, Frid L, Sleeter B, Fortin M-J. State-and-transition simulation models: A frameworkfor forecasting landscape change. Methods Ecol Evol. 2016. doi:10.1111/2041-210X.12597.
62. Keeling CD et al. Exchanges of atmospheric $CO_2$ and $13CO_2$ with the terrestrial biosphere and oceans from 1978 to 2000. In: Global aspects, SIO Reference Series, No. 01–06, San Diego: Scripps Institution of Oceanography; 2001.
63. Zhang Z, Jiang H, Liu J, Ju W, Zhang X. Effect of heterogeneous atmospheric $CO_2$ on simulated global carbon budget. Global Planet Change. 2013;101:33–51.
64. Soil Survey Staff, Natural Resources Conservation Service, United States Department of Agriculture. StateSoil Survey (SSURGO) database, Soil Data Mart. 2009. http://soildatamart.nrcs.usda.gov/Default.aspx. Accessed 30 Dec 2009.
65. Turner DP, Ritts WD, Cohen WB, Gamon JA. Site-level evaluation of satellite-based global terrestrial gross primary production and net primary production monitoring. Glob Change Biol. 2005;11:666–84.
66. Caspersen JP, Pacala SW, Jenkins JC, Hurtt GC, Moorcroft PR, Birdsey RA. Contributions of land-use history to carbon accumulation in US forests. Science. 2000;290:1148–51.
67. Tan Z, Liu S, Sohl T, Wu Y, Young C. Ecosystem carbon stocks and sequestration potential of federal lands across the conterminous United States. Proc Natl Acad Sci. 2015;112(41):12723–8.
68. Jarecki MK, Lal R. Soil organic carbon sequestration rates in two long-term no-till experiments in Ohio. Soil Sci. 2005;170:280–91.
69. Hollinger SE, Bernacchi Carl J, Meyers Tilden P. Carbon budget of mature no-till ecosystem in North Central Region of the United States. Agric For Meteorol. 2005;130:59–69.
70. Heath LS, Smith JE. Land use change and forestry and related sections. In: US EPA. Inventory of US greenhouse gas emissions and sinks: 1990–1997. EPA 236-R-99-003. 1999.
71. Powell DS, Faulkner JL, Darr DR, Zhu Z, MacCleery DW. Forest resources of the United States, 1992. Gen. Tech. Rep. RM-234. Fort Collins: US Forest Service, Rocky Mountain Forest and Range Experiment Station; 1993.
72. US Environmental Protection Agency. Inventory of US greenhouse gas emissions and sinks: 1990–2014. Washington, DC. EPA 430-R-16-002. 2016.
73. Morrison IK, Foster NW, Hazlett PW. Carbon reserves, carbon cycling, and harvesting effects in three mature forest types in Canada. New Zealand J For Sci. 1993;23:403–12.
74. Zhiliang Z, Bergamaschi B, Bernknopf R, Clow D, Dye D, Faulkner S, Forney W, Gleason R, Hawbaker T, Liu J, Liu S, Prisley S, Reed B, Reeves M, Rollins M, Sleeter B, Sohl T, Stackpoole S, Stehman S, Striegl R, Wein A, Zhu, Z. A method for assessing carbon stocks, carbon sequestration, and greenhouse-gas fluxes in ecosystems of the United States under present conditions and future scenarios: US Geological Survey Scientific Investigations Report 2010–5233. 2010.

# 7

# A methodological framework to assess the carbon balance of tropical managed forests

Camille Piponiot[1,9], Antoine Cabon[2], Laurent Descroix[3], Aurélie Dourdain[4], Lucas Mazzei[5], Benjamin Ouliac[6], Ervan Rutishauser[7], Plinio Sist[8] and Bruno Hérault[4*]

**Abstract**

**Background:** Managed forests are a major component of tropical landscapes. Production forests as designated by national forest services cover up to 400 million ha, i.e. half of the forested area in the humid tropics. Forest management thus plays a major role in the global carbon budget, but with a lack of unified method to estimate carbon fluxes from tropical managed forests. In this study we propose a new time- and spatially-explicit methodology to estimate the above-ground carbon budget of selective logging at regional scale.

**Results:** The yearly balance of a logging unit, i.e. the elementary management unit of a forest estate, is modelled by aggregating three sub-models encompassing (i) emissions from extracted wood, (ii) emissions from logging damage and deforested areas and (iii) carbon storage from post-logging recovery. Models are parametrised and uncertainties are propagated through a MCMC algorithm. As a case study, we used 38 years of National Forest Inventories in French Guiana, northeastern Amazonia, to estimate the above-ground carbon balance (i.e. the net carbon exchange with the atmosphere) of selectively logged forests. Over this period, the net carbon balance of selective logging in the French Guianan Permanent Forest Estate is estimated to be comprised between 0.12 and 1.33 Tg C, with a median value of 0.64 Tg C. Uncertainties over the model could be diminished by improving the accuracy of both logging damage and large woody necromass decay submodels.

**Conclusions:** We propose an innovating carbon accounting framework relying upon basic logging statistics. This flexible tool allows carbon budget of tropical managed forests to be estimated in a wide range of tropical regions.

**Keywords:** Carbon cycle, Selective logging, Error propagation, Amazonia, Production forests

## Background

There is a growing interest of the international community in tropical forests and their key role in the global carbon cycle. In particular, tropical managed forests, i.e. tropical forests managed for timber production, are important in the context of climate change mitigation as they might either act as carbon source or sink, depending on their management type and cutting cycle length [1]. With half of tropical humid forests, i.e. more than 400 million ha, designated by National Forest Services (NFS) as production forests [2], forest management will play a decisive role in future global carbon cycle. However, while there is extensive scientific literature on undisturbed tropical forests, our knowledge on carbon balance (i.e. the net carbon exchange with the atmosphere) in managed forests remains to be improved. With international programs like REDD+ (Reducing Emissions from Deforestation and Degradation) aiming at mitigating carbon emissions, notably through improved forest management, it is essential to have a clear understanding of the carbon balance of tropical managed forests in order to develop unified accounting methods for carbon emissions.

The second "D" of REDD+, forest degradation, refers to the reduction of carbon stocks in areas that have retained sufficient canopy cover to be still considered as forests [3]. In tropical forests, selective logging is generally

*Correspondence: bruno.herault@cirad.fr
[4] Cirad, UMR EcoFoG (AgroParisTech, CNRS, Inra, Université de la Guyane, Université des Antilles), Campus agronomique, 97310 Kourou, French Guiana
Full list of author information is available at the end of the article

considered an important factor of forest degradation. Selective logging consists in harvesting only a few trees and leaving the rest of the forest to natural regeneration until the next logging event. Selective logging practices result in long-lasting carbon emissions to the atmosphere [4]. In the sole Brazilian Amazon, forestry-related carbon emissions are estimated to be equivalent to 60–123 % of deforestation emissions [5]. Emissions come from (i) the extracted wood, (ii) logging damage, i.e. trees killed (purposely or incidentally) during logging operations and (iii) skid trails, logging roads and decks used for log yarding. Logging operations (e.g. tree felling, log yarding or skidding) induce incidental damage to surrounding trees, proportional to logging intensity in conventional [6] or reduced impact logging [7]. Injured or smashed trees generally die soon after logging operation [8, 9], and, along with log residuals such as crowns or stumps, are left over in the forest. The quite slow decay of these woody debris and harvested logs is a long-term source of carbon to the atmosphere.

However selective logging does not induce carbon emissions only. Logging gaps and incidental mortality release local competition for key resources, such as light [10], and enhance carbon accumulation in remnant trees [11]. Some local studies suggest that carbon emissions peak soon after logging (5–10 years), while tree growth and recruitment counterbalance tree mortality after a decade, resulting in positive carbon accumulation in logged forests until an equilibrium is reached [12]. If not further disturbed, forests recover most of their initial carbon stock at the end of the cutting cycle, depending on logging intensity and damage [13, 14]. While skid trails and logging decks generally close rapidly after logging, logging roads may be reopened and persist for decades in the landscape [15].

Most studies on the carbon balance of selectively logged forests have estimated pre- and post-logging carbon stocks with remote sensing methods [16], allowing large-scale estimation of logging extent and quantification of canopy damage [14, 17, 18]. However optical remote sensing methods require further field calibration to avoid estimation biases [17, 19]. Accurate estimation of carbon loss and recovery based on evolution of canopy openness is questionable, as ground damage remain poorly assessed with remote sensing and canopy rapidly closes with crown expansion of survivors. [20]. Another problem lies in estimating carbon recovery rates when canopy gaps rapidly close with crown expansion of survivors and growth of recruits [21]. LiDAR and RADAR are a promising tools to overcome those limitations but remain expensive to be applied over large areas [22]. The International Panel on Climate Change [23] recommends basic methods based on activity data, e.g. the annual volume of extracted roundwood (in $m^3 \cdot year^{-1}$) or the annual logged area (in $ha \cdot year^{-1}$). So far, only a few studies have estimated the carbon balance of tropical logged forests from activity data. Griscom et al. [24] estimated carbon emissions at concession level with detailed maps of skidtrails, logged trees, damage and logging decks. But such maps are generally not available at regional level impeding the wide-use of this approach. Pearson et al. [25] developed emissions factors for the main logging compartments to be multiplied to extracted volumes, as a surrogate for carbon emissions from selective logging. For sake of simplicity, most studies have estimated committed emissions, i.e. all the carbon lost by logging is assumed to return instantaneously to the atmosphere. This choice is highly questionable as carbon emissions are spread over decades [4]. Some studies have indeed adopted a more dynamic approach, considering carbon emissions through time in managed forests [26]. However to date there is no method assessing all carbon fluxes (carbon accumulation as well as all emissions) from tropical managed forests through time.

In this study we developed a methodological framework to estimate the above-ground carbon fluxes from selective logging. Carbon fluxes are assessed first at the logging unit scale, i.e. the elementary management unit of the forest estate, and then integrated in space and time to estimate the regional carbon balance over time. The annual carbon net change of a given logging unit is estimated by aggregating carbon fluxes of three sub-models: (1) emissions from extracted wood, (2) emissions from woody debris (logging residues, damage, road and logging deck openings) left in the forest, and (3) sequestration from forest regrowth. Some submodels come from the literature while others were specifically developed and parameterised in a Bayesian framework for this study. Uncertainty over the global model are propagated through a Markov Chain Monte Carlo algorithm.

Our case study is French Guiana, a French territory located in the Guiana shield (northeastern Amazonia) with a total area of 84,000 $km^2$. Above 95 % of its area (8Mha) is occupied by tropical rainforest: timber extraction by selective logging is the 3rd economic sector. Selective logging occurs exclusively in the Permanent Forest Estate (2.4 Mha) (Fig. 1), an area managed by the French National Forest Service (NFS). We used 38-year NFS statistics on 1155 logging units (Fig. 2) to estimate the carbon balance of managed forests at countrywide scale.

## Methods

### The model framework

*Methods overview* We created a model to assess annual above-ground carbon fluxes at the logging unit scale,

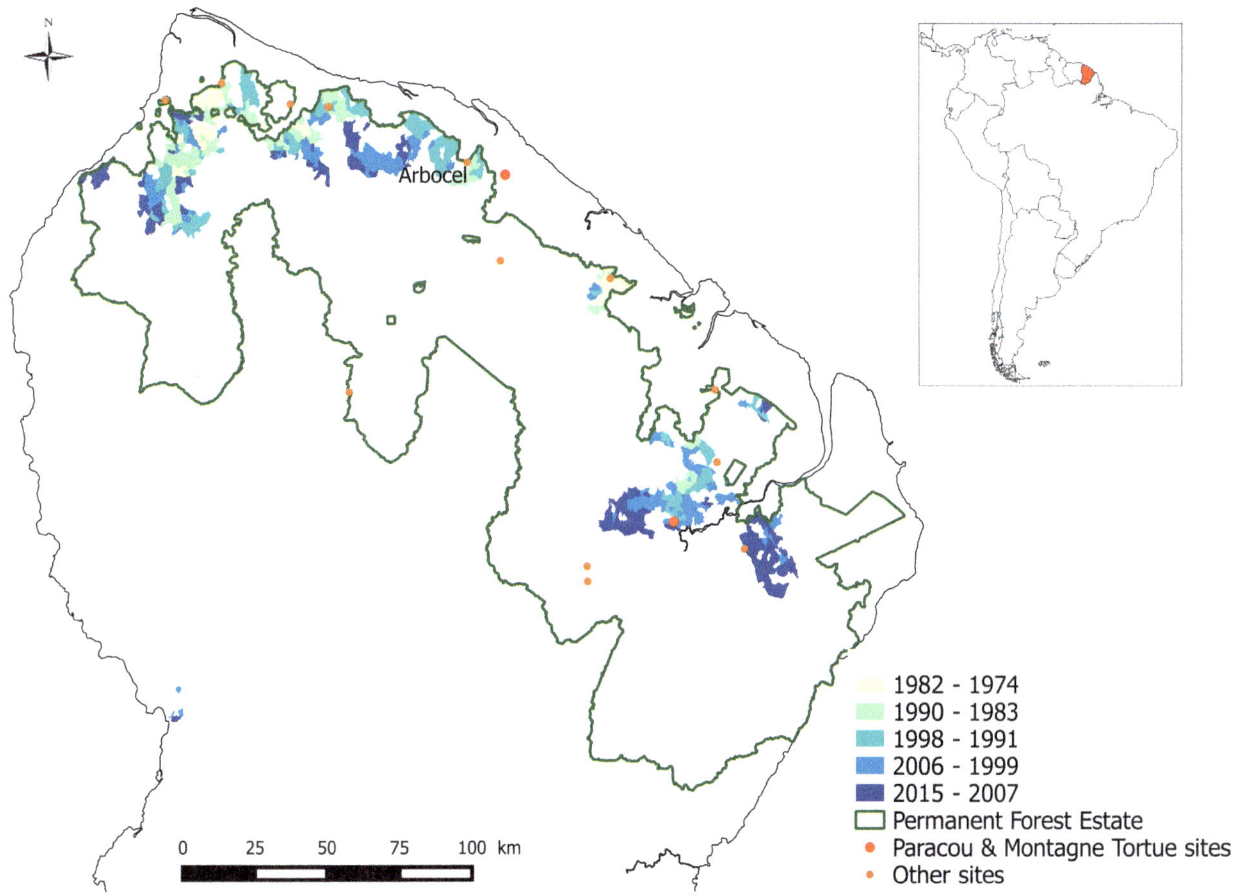


**Fig. 1** Map of the Permanent Forest Estate in French Guiana. Logging units are *coloured* according to the year of logging: from *white and light blue* for most anciently logged plots to *dark blue* for most recently logged plots. *Dots* are experimental sites from the Guyafor network: *red dots* have both logged and control plots (Paracou and Tortue sites); *smaller orange dots* have only control plots. The borders of the Permanent Forest Estate are the *bold green lines*

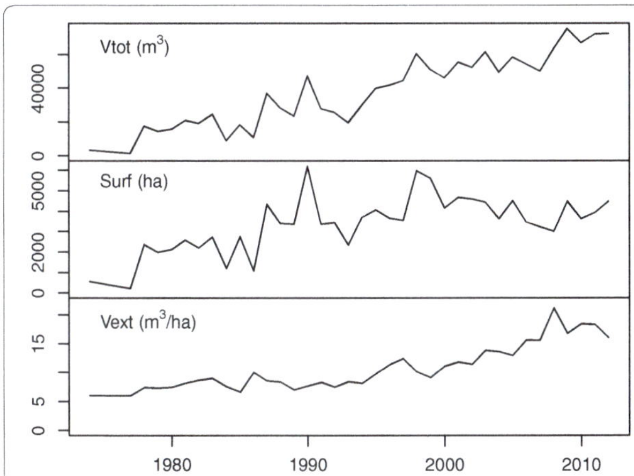


**Fig. 2** National Forest Service statistics on selective logging in the French Guiana Permanent Forest Estate between 1974–2012. *Top graph* annual extracted volumes of roundwood *Vtot* ( $m^3$) ; *middle graph* annual logged areas *Surf* (ha); *bottom graph* annual logging intensities *Vext* ($m^3 \cdot ha^{-1}$)

from the logging year up to the last year of NFS statistics, and then integrated these results over time and space. Conventionally, fluxes from carbon emissions are positive and fluxes from carbon accumulation are negative.

We modeled separately (1) emissions from extracted wood decay (Sawmill); (2) emissions from woody debris decay (Dead wood), divided into (2.a) logging damage and (2.b) deforestation on logging roads, main skid trails and logging decks; (3) sequestration from forest regrowth (3.a) in logging gaps and (3.b) on abandoned skid trails (Live biomass) (Fig. 3). For each submodel we first assessed the evolution of post-logging carbon stocks (in $MgC \cdot ha^{-1}$) in each logging unit. Annual carbon fluxes at time $t$ (in $MgC \cdot year^{-1}$) are obtained by calculating the difference between stocks at time $t$ and $t-1$, and multiplied by the logged area (in ha). Model parameters are considered as uncertain quantities, and the error is propagated with Monte Carlo methods. Doing so, we were

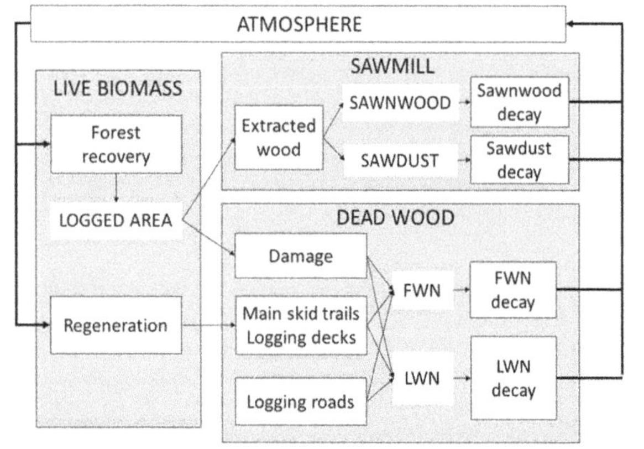


**Fig. 3** Overall scheme of the carbon balance and submodels articulation. *Thick arrows* are carbon fluxes. *Grey boxes* are the three carbon compartments: Sawmill, Dead Wood and Live Biomass. Submodels are in surrounded white boxes. FWN and LWN are fine and large woody necromass respectively

able to track the annual carbon budget in each logging unit, and to conduct simultaneously a sensitivity analysis to identify the main sources of uncertainties in the global model.

*Calibration data* In some cases, submodels come from the literature. In other cases, and especially when there was no existing model, submodels were developed and parametrised with data from Amazonian permanent forest plots. We used data from control plots to assess pre-logging forest structure variables and logged plots to assess post-logging dynamics. Two networks of permanent forest plots were used in this study: (i) Guyafor in French Guiana (regional scale) (Fig. 1); (ii) TmFO, a network of logged plots spread across Amazonia (continental scale) [27]. From Guyafor, we used data from 11 sites, all containing control plots and two of them (Montagne Tortue, Paracou) also containing logged plots. From the TmFO network, we used data from four Amazonian sites (Jari, Tapajos, Tortue, Paracou) that have logged plots where extracted volumes have been measured. To calibrate the model of forest regeneration on logging decks and main skid trails, we used data from a 25-ha plot in French Guiana, Arbocel (see Fig. 1): the plot was clearcut in 1976 and then left to natural regeneration [28]. Some submodels and/or calibration data are specific to the context of selective logging in French Guiana: adapting this framework to other areas, countries or for upscaling purposes may require new data or new submodels to be regenerated. However, the general structure and the articulation of the different submodels is generic enough to be applied elsewhere.

*Input data* To run the model, we need basic NFS statistics available for every logging unit $p$: (i) the logged surface $Surf_p$ (ha), corresponding in French Guiana to on average 50 % of the total area, (ii) the total extracted volume $Vtot_p$ ($m^3$), (iii) the year of logging $t0_p$ and (iv) the initial above-ground carbon stock $ACS_p$ ($MgC{\cdot}ha^{-1}$).

For our study, statistics on selective logging in the Permanent Forest Estate over the 1974–2012 time period were provided by the French NFS. According to NFS expertise, between 1974 and 1994 logging intensities are assumed to be equal to 6 $m^3{\cdot}ha^{-1}$ in the western forests, and equal to 10 $m^3{\cdot}ha^{-1}$ in the rest of the Permanent Forest Estate. After 1994, logging statistics are highly reliable and logging intensities $Vext_p$ are calculated as follows:

$$Vext_p = \frac{Vtot_p}{Surf_p} \tag{1}$$

Initial above-ground carbon stocks were extracted from the recently developed carbon map of French Guiana [29].

**Sawmill**

The carbon $Ext_p$ ($MgC{\cdot}ha^{-1}$) of logs extracted from the logging unit $p$ is defined as :

$$Ext_p = Vext_p \times dext \times 0.5 \tag{2}$$

where $Vext_p$ is the logging intensity ($m^3{\cdot}ha^{-1}$), the biomass is assumed to be 50 % carbon [30] and $dext = 0.736$ is the mean wood density of timber species in French Guiana. $dext$ has been estimated as the mean wood density of *Dicorynia guianensis* ($d = 0.76$), *Qualea rosea* ($d = 0.725$) and *Sextonia rubra* ($d = 0.65$) [31], weighted by their relative contribution to annual timber extraction [32].

The carbon stored in extracted wood ($Ext_p$) is then transformed into sawnwood with an efficiency of $f_{SW} = 0.33$ [33], the remaining 67 % consisting of sawdust. Sawnwood and sawdust then decompose with different decay rates. Sawdust is considered to decompose entirely within the first year after logging [34]. The quantity of sawdust $DecSD_p$ ($MgC{\cdot}ha^{-1}$) in the logging unit $p$ at the year of logging $t0_p$ is:

$$DecSD_p = (1 - f_{SW}) \times Ext_p \tag{3}$$

Sawnwood decay is approximated with an exponential decay. The carbon in sawnwood $DecSW_{p,t}$ ($MgC{\cdot}ha^{-1}$) from logging unit $p$ at time $t$ is :

$$DecSW_{p,t} = f_{SW} \times Ext_p \times exp\Big(-\lambda_{SW} \times (t - t0_p)\Big) \tag{4}$$

with $Ext_p$ the carbon in extracted wood, $t0_p$ the year of logging and $\lambda_{SW}$ the decay rate of sawnwood in the tropics estimated as

$$\lambda_{SW} = \frac{ln(2)}{t_{0.5}} \tag{5}$$

with

$$t_{0.5} \sim N_{t0}(30, 15^2) \quad (6)$$

where $N_{t0}(30, 15^2)$ is the normal distribution truncated at zero with mean equal to 30 and standard deviation equal to 15, according to IPCC guidelines [30]. So far, no life cycle analysis has yet been done for woody products in the tropics and adapting the framework developed in Europe and North America in tropical countries remains challenging [35].

### Dead wood

Both logging damage and deforestation on roads, main skid trails and logging decks are sources of dead wood that is left in the forest.

#### *Logging damage*

The logging damage $Dam_p$ (MgC·ha$^{-1}$) within the logging unit $p$ consists in above-ground carbon in trees harvested or killed (purposely or incidentally) during logging operations (tree felling and log yarding or skidding) minus carbon of extracted logs ($Ext_p$). Logging damage were assessed with data from 4 TmFO sites in the Eastern Amazon, where we have logging intensities $Vext_p$ and logging damage $Dam_p$ (Fig. 4). Logging gaps and secondary skid trails are accounted for as logging damage, whereas main skid trails and logging roads are located outside of calibration plots and are considered as deforestation.

The relationship between $Vext_p$ and $Dam_p$ was modelled as follows:

$$Dam_p \sim lnN(\theta_0 + \theta_1 \times Vext_p; \sigma^2_{dam}) \quad (7)$$

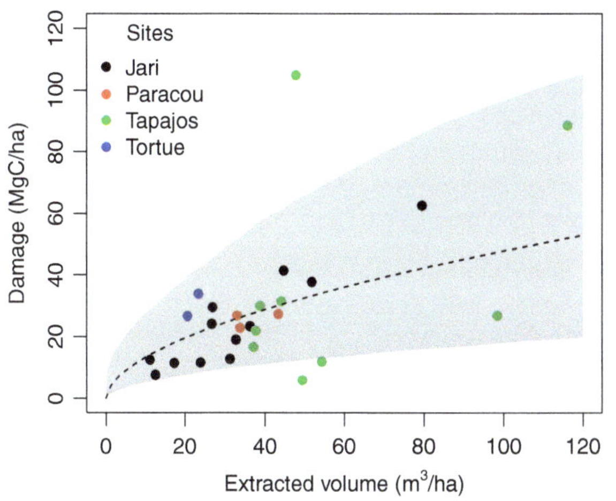


**Fig. 4** Modelling logging damage (MgC·ha$^{-1}$) from extracted wood (m$^3$·ha$^{-1}$). The *dashed line* is the prediction with maximum likelihood, the shaded area is the 95 % credibility interval on the prediction. Data are taken from four Amazonian sites: Jari (*black*), Paracou (*red*), Tortue (*green*), Tapajos (*blue*)

where $lnN(\mu, \sigma^2_{dam})$ is the lognormal distribution of mean $\mu$ and standard deviation $\sigma_{dam}$. We chose the lognormal distribution to have errors proportional to the extracted volume, i.e. damage is less predictable for higher logging intensities.

The parameters $(\theta_0, \theta_1, \sigma_{dam})$ were estimated with a Bayesian inference, using a Markov Chain Monte Carlo method to obtain the posterior joint distribution. Uninformative priors were chosen. Each observation (one plot) was given a weight equal to the area of the plot.

#### *Deforestation*

Some basic infrastructure (logging roads, main skid trails, logging decks) is generally set up to store boles and get them out of the forest. As it has to be easily reached from every part of a logging unit $p$, the total area cleared for infrastructure is assumed to be proportional to the total logged area $Surf_p$. The length of logging roads may depend on the topography but, in the French Guiana Permanent Forest Estate, the topography is quite homogeneous.

*Logging roads* According to NFS expertise (personal communication):

$$\delta_p \sim U(0.12; 0.25) \quad (8)$$

where $\delta_p$ is the entire width of clearing for logging roads in logging unit $p$ (expressed in hectometers hm). Based on the study of Guitet et al. [36], we fit the length of logging roads $\eta_p$ (hm·ha$^{-1}$) as:

$$\eta_p \sim exp(17.95) \quad (9)$$

The carbon removed for road clearing $Road_p$ (MgC·ha$^{-1}$) and left to decay in the logging unit $p$ is :

$$Road_p = ACS_p \times \delta_p \times \eta_p \quad (10)$$

where $ACS_p$ is the initial above-ground carbon stock (MgC·ha$^{-1}$).

*Main skid trails* Main skid trails are built to allow the skidders to get in the logging units. According to Guitet et al. [36]:

$$SurfST_p \sim U(0.04; 0.10) \quad (11)$$

where $SurfST_p$ is the area (ha) of main skid trails per hectare logged. As large trees are generally avoided by skidders during yarding, we assumed that only trees with $DBH$ (Diameter at Breast Height) $\leq$50 cm were killed. For a logging unit $p$, the above-ground carbon removed on skid trails $STrail_p$ (Mg·ha$^{-1}$) is:

$$STrail_p = ACS_p \times SurfST_p \times f_{50} \quad (12)$$

where $ACS_p$ is the initial above-ground carbon stock of the logging unit $p$ (Mg·ha$^{-1}$) and $f_{50}$ is the proportion of $ACS_p$ in trees $DBH$ <50 cm. $f_{50}$ was estimated with data from the control plots of the 11 Guyafor sites. For one site $j$, the proportion $f_{50,j}$ was calculated as follows:

$$f_{50,j} = \frac{\sum_{DBH_k \leq 50} ACS_{j,k}}{\sum_k ACS_{j,k}} \tag{13}$$

where $\sum_{DBH \leq 50} ACS_{j,k}$ is the above-ground carbon of all trees with $DBH \leq 50$ cm and $\sum_k ACS_{j,k}$ is the total above-ground carbon of the site. We report the obtained distribution in Table 1.

*Logging decks* Logging decks are cleared to store harvested boles. Based on detailed mapping of 11 logging decks in logging units HKO95 (283 ha) and HKO96 (466 ha) and with the information that one logging deck is set up for every 70 ha (NFS expertise, personal communication), we used:

$$\frac{LDeck_p}{ACS_p} \sim lnN(-6.12, 0.58^2) \tag{14}$$

with $ACS_p$ the above ground carbon stock of the logging unit $p$, and $LDeck_p$ ($MgC{\cdot}ha^{-1}$) is the carbon removed in logging decks.

### *Dead wood decay*

Assuming that all woody debris are left out for decay in situ, dead wood was divided into two pools with different decay rates: (i) fine woody necromass (FWN), i.e. leaves, twigs, and branches with diameter <10 cm, and (ii) large woody necromass (LWN), i.e. logs and branches with diameter $\geq$10 cm.

*Fraction of Large Woody Necromass* We used Chambers et al. [37] model to estimate the fraction of large woody necromass $LWN_i$ at tree scale ($i$).

For a tree $i$ : If $DBH_i < 100$ cm:

**Table 1 Parameters of the carbon balance model**

| Submodel | Parameter | Distribution | Justification | Data source |
|---|---|---|---|---|
| Extracted wood | $dext$ | 0.736 | 3 main commercial species | [31, 32][a] |
| | $f_{SW}$ | 0.33 | No uncertainty reported | [33][b] |
| Sawnwood decay | $\lambda_{SW}$ | $N_{t0}(30, 15^2)$ | Positive normal distribution | [30][c] |
| Damage | $\theta_0$ | $N(1.28, 0.48^2)$ | Normal approximation of the posterior | TmFO sites[b] |
| | $\theta_1$ | $N(0.56, 0.14^2)$ | Normal approximation of the posterior | TmFO sites[b] |
| | $\sigma_{dam}$ | $N(0.37, 0.04^2)$ | Normal approximation of the posterior | TmFO sites[b] |
| Logging roads | $\delta$ | $U(0.12, 0.25)$ | Personal communication | NFS expertise[a] |
| | $\eta$ | $exp(17.95)$ | Decreasing monotonic distribution | [36][a] |
| Main skid trails | $SurfST$ | $U(0.04, 0.10)$ | Only range of values is reported | [36][a] |
| | $f_{50}$ | $Beta(15.10, 15.32)$ | Proportion | Guyafor[a] |
| Logging decks | $LDeck$ | $ACS \times lnN(-6.12, 0.58)$ | Positive distribution | Original data[a] |
| LWN decay | $fLWN$ | $Beta(1115, 169)$ | Proportion | Guyafor[a] |
| | $\lambda_1$ | $N(0.069, 0.028^2)$ | Normal approximation of the posterior | Guyafor[a] |
| | $\lambda_2$ | $N(0.198, 0.074^2)$ | Normal approximation of the posterior | Guyafor[a] |
| | $\pi_1$ | $N(0.512, 0.032^2)$ | Normal approximation of the posterior | Guyafor[a] |
| | $\sigma_{LWN}$ | $N(0.015, 0.0035^2)$ | Normal approximation of the posterior | Guyafor[a] |
| FWN decay | $\lambda_{FWN}$ | $N(0.19, 0.026^2)$ | Reported distribution | [37][c] |
| Forest recovery | $\alpha$ | $N(1.100, 0.03^2)$ | Reported distribution | [13][c] |
| Regeneration | $\mu_0$ | $N(-4.505, 0.124^2)$ | Maximum likelihood estimate | Arbocel[a] |
| | $\mu_1$ | $N(1.085, 0.038^2)$ | Maximum likelihood estimate | Arbocel[a] |
| | $\varphi$ | 490.7 | Maximum likelihood estimate | Arbocel[a] |

Parameters are grouped by submodel in which they appear. *FWN* fine woody necromass decay; *LWN* large woody necromass. *ACS* initial above-ground carbon stock ($MgC{\cdot}ha^{-1}$). Parameters are: *dext* mean density of extracted roundwood in French Guiana; $f_{SW}$ efficiency of wood transformation in sawmills; $\lambda_{SW}$ decay rate of sawnwood; $(\theta_0, \theta_1, \sigma_{dam})$ parameters of the relationship between extracted wood and logging damage; $(\eta, \delta)$ length ($hm{\cdot}ha^{-1}$) and width (hm) of logging roads; *SurfST* main skid trails area (ha); $f_{50}$ proportion of above-ground carbon in trees DBH <50 cm; *LDeck* carbon loss in logging decks; $f_{LWN}$ fraction of necromass carbon in large woody necromass; $(\lambda_1, \lambda_2, \pi_1, \sigma_{LWN})$ parameters of large woody necromass decay; $\alpha$ parameter of the recovery model; $\mu_0$, $\mu_1$ and $\varphi$: parameters of the regeneration beta model

[a] Valid in French Guiana

[b] Valid in Amazonia

[c] Valid in the tropics

$$LWN_i = 0.774 + 0.0018 \times DBH_i \tag{15}$$

else $LWN_i = 0.95$.

This fraction of large woody necromass has been estimated for all individuals in control plots of the 11 Guyafor sites. For one site $j$, the fraction of large woody necromass is:

$$fLWN_j = \frac{\sum_i (LWN_{i,j} \times ACS_{i,j})}{\sum_i ACS_{i,j}} \tag{16}$$

We report the obtained distribution of $fLWN_j$ in Table 1.

At $t = t0_p$ ($t0_p$: year of logging) in the logging unit $p$:

$$FWN_{p,t0_p} = (1 - f_{LWN}) \times (Dam_p + Road_p + STrail_p + LDeck_p) \tag{17}$$

$$LWN_{p,t0_p} = f_{LWN} \times (Dam_p + Road_p + STrail_p + Ldeck_p) \tag{18}$$

where $FWN_{p,t0_p}$ and $LWN_{p,t0_p}$ (MgC·ha$^{-1}$) are the carbon in fine and large woody necromass, respectively, left on the logging unit $p$ at $t0_p$.

*Decay of Fine Woody Necromass* Fine woody necromass decay in logging unit $p$ at time $t$ is estimated by the following model [38]:

$$DecFWN_{p,t} = FWN_{p,t0_p} \times exp\Big(-\lambda_{FWN} \times (t - t0)\Big) \tag{19}$$

where $DecFWN_{p,t}$ (MgC·ha$^{-1}$) is the fine woody necromass left in the logging unit $p$ at time $t$ and $\lambda_{FWN}$ is the decay rate of fine woody necromass (Table 1).

*Decay of Large Woody Necromass* There was no published model for decay of large woody necromass at the logging unit level and derived individual tree large woody necromass (LWN) decay from [4]:

$$LWN_{i,t} = LWN_{i,t0} \times exp\Big(-\frac{\gamma_1}{WD_i^{\gamma_2} \times (circ_i)^{\gamma_3}} \times (t - t0)\Big) \tag{20}$$

where $LWN_{i,t}$ is the large woody necromass of the individual $i$ at time $t$, $WD_i$ the wood density of the individual $i$ and $circ_i$ the circumference at breast height of the individual $i$, and $(\gamma_k)_{1,2,3}$ are parameters of the model [4]. For all 11 Guyafor sites, the model was applied to all individuals $i$ of the control plots. Individual trajectories are then aggregated to have the large woody necromass left at time $t$ on the site $j$:

$$LWN_{j,t} = \sum_i LWN_{i,j,t} \tag{21}$$

where $LWN_{j,t}$ is the large woody necromass left at time $t$ on the site $j$; $LWN_{i,j,t}$ is the large woody necromass at time $t$ of the individual $i$ from the site $j$.

A new model was developed at the logging unit level: to better fit the data, a double-decay model was chosen instead of a one-parameter decay model, because it drastically reduced the bias in our predictions. The double-decay model works as if there were two pools of large woody necromass (i.e. sap- and heart-wood) with two different decay rates.

$$\frac{DecLWN_{p,t}}{LWN_{p,t0}} = \pi_1 \times exp\Big(-\lambda_1 \times (t - t0_p)\Big) + (1 - \pi_1) \times exp\Big(-\lambda_2 \times (t - t0_p)\Big) + \varepsilon_{LWN,p} \tag{22}$$

with

$$\pi_1 > 0.5$$

and

$$\varepsilon_{LWN,p} \sim N(0; \sigma_{LWN})$$

where $\frac{DecLWN_{p,t}}{LWN_{p,t0_p}}$ is the fraction of the initial large woody necromass left on logging unit $p$ at time $t$; $\lambda_1$ and $\lambda_2$ are the two decay rates and $\pi_1$ the fraction of large woody necromass decomposing with the rate $\lambda_1$; $\varepsilon_{LWN,p}$ is the additive error for the logging unit $p$, taken in $N(0; \sigma^2_{LWN})$, the centred normal distribution of standard deviation $\sigma_{LWN}$.

Parameters $(\lambda_1; \lambda_2; \pi_1; \sigma_{LWN})$ were estimated as follows: (i) 100 sets of parameters $(\gamma_k)_{1,2,3}$ of the individual decay model were taken in their distribution; (ii) for each set of individual parameters $(\gamma_k)_{1,2,3}$, the trajectories of large woody necromass $LWN_{j,t}$ for each site $j$ in the 11 Guyafor sites was estimated; (iii) parameters $(\lambda_1; \lambda_2; \pi_1; \sigma_{LWN})$ were then estimated using a Bayesian inference with a Markov Chain Monte Carlo method. Uninformative priors were chosen. One set of parameters $(\lambda_1; \lambda_2; \pi_1; \sigma_{LWN})$ was stored for every set of individual parameters $(\gamma_k)_{1,2,3}$. 100 sets of parameters $(\lambda_1; \lambda_2; \pi_1; \sigma_{LWN})$ were then stored to be used in the full model (the normal approximations of the posterior distributions are in Table 1).

### Live biomass

#### *Carbon recovery in logged areas*

The assumption of a linear carbon recovery was made, i.e. the rate at which a logged area stores above-ground carbon after logging is constant and becomes null at recovery time, when the above-ground carbon has reached its original level. This assumption is arguable but is consistent with available data on post-logging dynamics in the region [13]. Under this assumption, the recovery time (*RT*) can be estimated with the sole percentage of above-ground carbon loss caused by logging:

$$RT_p = \left( \frac{100 \times (Ext_p + Dam_p)}{ACS_p} \right)^{\alpha} \tag{23}$$

with

$$\alpha \sim \mathcal{N}(1.100, 0.013^2) \tag{24}$$

with $RT_p$ (yr) the recovery time of the logging unit $p$, $Ext_p + Dam_p$ (MgC·ha$^{-1}$) the above-ground carbon loss per ha logged in the logging unit $p$ and $ACS_p$ the initial above-ground carbon stock (MgC·ha$^{-1}$).

The above-ground carbon recovered $Recov_{p,t}$ (MgC·ha$^{-1}$) at time $t$ in the logged area of the logging unit $p$ is then :

$$Recov_{p,t} = \min\left( Ext_p + Dam_p; (t - t0_p) \times \frac{Ext_p + Dam_p}{RT_p} \right) \tag{25}$$

When $t - t0_p \leq RT_p$ (i.e. before recovery time), the logged area stores carbon with a rate $\frac{Ext_p + Dam_p}{RT_p}$. When $t - t0_p > RT_p$, the logged area has recovered all the above-ground carbon lost $Ext_p + Dam_p$ and the carbon storage stops.

#### *Regeneration on abandoned main skid trails and logging decks*

The fate of the carbon dynamics on abandonned skid trails and logging decks was modelled taking advantage of four 1.65-ha plots clearcut 38 years ago (Fig. 5), and regularly measured since [28]. Carbon stocks regeneration on skid trails with compacted soils may be expected to be slower than on large clear-cuts but no data on skid trails reforestation have been produced in French Guiana. The following model was inferred:

$$Rege_{p,t} = ACS_p \times Y_{p,t} \tag{26}$$

with

$$Y_{p,t} \sim Beta\left(\mu_0 + \mu_1 \times log(t - t0_p + 1); \varphi\right) \tag{27}$$

where $Rege_{p,t}$ (MgC·ha$^{-1}$) is the above-ground carbon gain from forest regeneration on plot $p$ at time $t$, $ACS_p$ (MgC·ha$^{-1}$) is the initial above-ground carbon stock of the plot $p$ (i.e. before the clear cut), $Y_{p,t}$ is the proportion of the initial above-ground carbon stock recovered on the plot $p$ at time $t$. $Y_{p,t}$ follows a beta distribution of mean $\mu_0 + \mu_1 \times log(t - t0_p + 1)$ and of precision $\varphi$ (parameters ($\mu_0$, $\mu_1$, $\varphi$) estimated distributions are in Table 1). The beta distribution is defined on [0, 1] and takes two shape parameters.

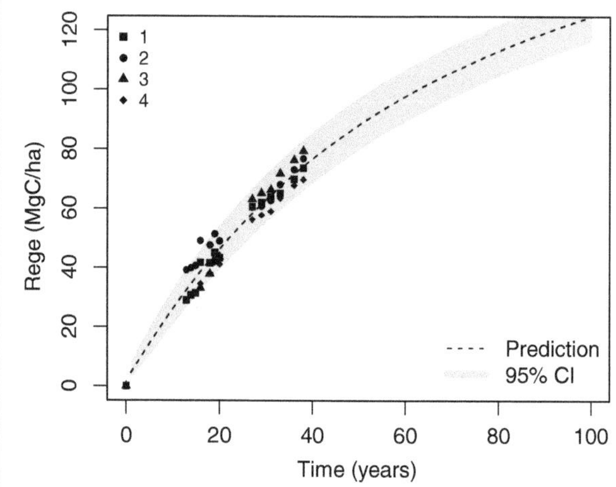


**Fig. 5** Beta model of carbon regeneration (*Rege*, MgC·ha$^{-1}$) on four clearcut plots. The *dashed line* is the prediction with maximum likelihood, the shaded area is the 95 % credibility interval on the prediction. Data are taken from four 1.65-ha plots in Arbocel site (French Guiana), clearcut 38 year ago and let to natural regeneration

### Accounting for uncertainties

#### *Uncertainty propagation*

The model was built and parametrised with MCMC algorithms in order to track uncertainties. Uncertainty propagation was done with the following steps: (i) every parameter is randomly taken in its distribution; (ii) the model is applied over all logging units of the region with these parameters values; (iii) results are integrated in space and time and stored. These three steps are repeated 5000 times and summary statistics are then calculated.

#### *Sensitivity analysis*

To assess the uncertainty induced by each submodel, we carried out a sensitivity analysis where the parameters of each submodel are set to their maximum likelihood values one after the other, making the submodel deterministic while the rest of the model remains stochastic. The results for each deterministic submodel are then compared to the total model: the "shrinkage" of their summary statistics reflects the uncertainty linked to the parameters of a specific submodel, more than the model and modelling choices themselves.

## Results and discussion

### Methodological aspects

In this study we developed a new methodology to assess the above-ground carbon balance of tropical managed forests. The model is based on a book-keeping approach: it aggregates post-logging carbon fluxes of every logging unit over a given area and period of time to obtain the regional carbon balance. The main advantage of our methodology compared to other published methods estimating emissions from selectively logged concessions [24, 25] lies in the fact that both post-logging carbon sequestration and necromass respiration are explicit functions of time. This framework is flexible enough to be adapted to new situations and areas, but potential users should keep in mind that applying this model to other areas will require in-depth analysis on the suitability of the parameters value, variables distribution and submodels choices. The framework can also be adapted to integrate new information on post-logging forest dynamics or to take into account new logging technique, e.g. new techniques of reduced impact logging (RIL) likely to decrease the quantity of damage for the same logging intensity [39] but used in less than 5 % of selectively logged areas so far [1]. Part of logging damage may also be used for new purposes, e.g. burned in energy plants as wood fuel [40]. The model can be adapted and re-parametrised very quickly to take into account such changes.

We aknowledge that French Guiana is a special case in the tropics in regards to data quality and availability. It is therefore easier to develop a methodology there than in other countries where both input and calibration data are scarse. Submodels were either taken from the literature or specifically developed for this study with data at regional and continental scales. For each submodel, parameters are considered as uncertain quantities making good use of the Bayesian inference framework. Depending on spatial extent of the data used to calibrate the corresponding submodel, parameters distribution have different geographic validity (Table 1). Submodels from the literature and our submodel of logging damage (*Dam*) are considered to be valid over Amazonia. The estimated distribution of corresponding parameters could therefore be used for assessing the carbon balance of any Amazonian region. The other models developed in this study [deforestation on roads, main skid trails and decks, large woody necromass (LWN) decay and forest regeneration] were calibrated with data from a regional plot network, Guyafor or with NFS data and should be used with caution elsewhere. In order to study the carbon balance of other regions, data from permanent plots in the region of interest should be used to validate, and if necessary recalibrate submodels.

The method is simple and only requires two sets of input data : logged areas in space and time and the intensities at which these forest areas were logged. These data can come either from official logging statistics or from remote sensing methods. In many countries, logging operations occur almost exclusively inside logging concessions for which management plans do exist. For instance, in the Congo Basin, 45 Mha of forest are managed under the concession system [41]. In these cases, the logging statistics needed can be taken directly from management plans. In other countries where management plans do not exist or are not fully applied [42], remote sensing tools can be useful to map logged areas. For instance, Asner et al. [5] tracked logged areas in five States of the Brazilian Amazon during three consecutive years using Landsat imagery and CLAS methodology. Logging intensities are more difficult to obtain with remote sensing methods: a way to overcome this problem is to use region-wide statistics on commercialized wood that are generally quite robust because wood boles from either legal or illegal logging eventually need authorizations to be transported and commercialized. The simplest way to estimate logging intensities is then to use an average logging intensity, or to use local expertise to locally refine this average value. Bayesian approaches are quite appropriate to mix *a priori* knowledge with quantitative datasets. For both logged areas and and logging intensities, a close attention should be paid to uncertainties and their propagation into the complete model.

### Carbon balance at the logging unit scale

To fully understand the theoretical behaviour of our model we focused on two logging units both logged in 1996: one with a low logging intensity (1.1 $m^3 \cdot ha^{-1}$) and one with a high logging intensity (33 $m^3 \cdot ha^{-1}$). For those two logging units, we estimated carbon fluxes ($MgC \cdot ha^{-1} \cdot year^{-1}$) during an 65-year period, which corresponds to the official cutting cycle in French Guiana (Fig. 6).

#### *Annual fluxes*

At low logging intensity, the two main sources of carbon are both logging damage and deforestation (on roads and main skid trails). At higher logging intensities, emissions from deforestation become negligible compared to emissions from damage, consistent with previous remote-sensing results in the Brazilian Amazon [43]. According to our model, carbon fluxes reach a steady state before the end of the 65-year period for both logging units (Fig. 6).

The different trends of the net carbon change are mostly explained by the interplay of two components: emissions from logging damage and accumulation from recovery in logging gaps. On one hand, higher logging intensities obviously cause more logging damage, resulting in higher carbon emissions from damage decay. On the other hand, the annual carbon flux from recovery in logging gaps is less dependent on logging damage (Fig. 6): a logging unit that has undergone high logging intensities will thus take longer to recover.

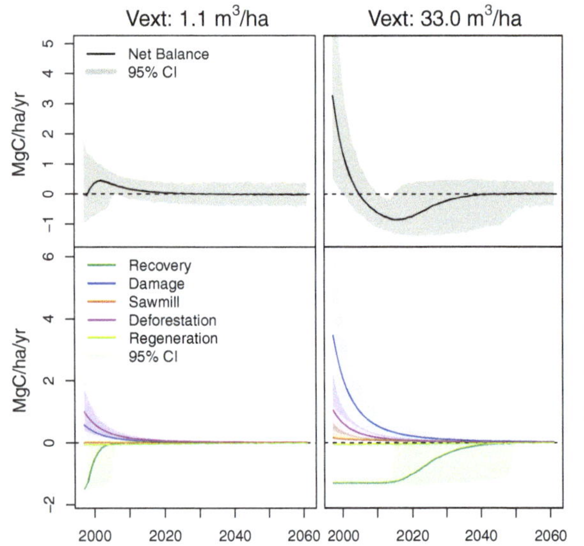


**Fig. 6** Annual above-ground carbon fluxes from logging on a 65-year period in two logging units logged in 1994 with different logging intensities. The first logging unit underwent a low-intensity logging (*left*) with a mean extracted volume $V_{ext} = 1.1\,m^3{\cdot}ha^{-1}$whereas the second logging unit underwent a high-intensity logging (*right*) with a mean extracted volume $V_{ext} = 33.0\,m^3{\cdot}ha^{-1}$. For each plot, the net carbon flux (*top*) is the result of four components: emissions from logging damage (*blue*), emissions from road, deforestation on main skid trail and logging deck (*purple*), emissions from extracted timber (*red*), accumulation from post-logging recovery on the plot (*forest green*) and accumulation from regeneration on main skid trails and logging decks (*light green*). The *shaded areas* correspond to the 95 % credibility intervals

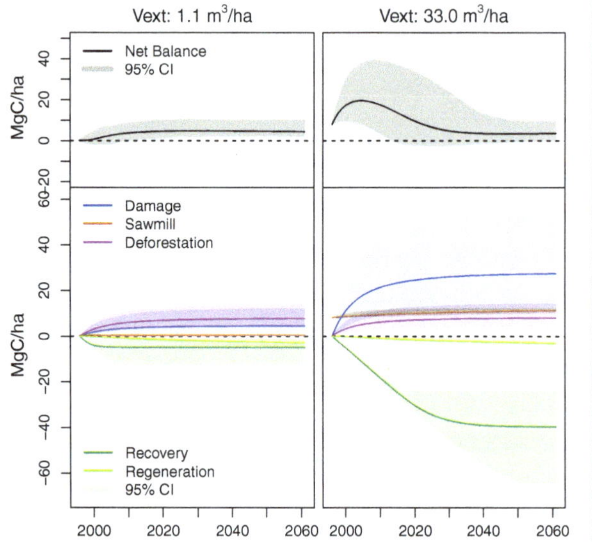


**Fig. 7** Above-ground carbon balance from logging on a 65-year period in two logging units logged in 1994 with different logging intensities. The carbon balance at time *t* is the integral of annual carbon fluxes, i.e. the sum of annual carbon fluxes from the year of logging (1996) to *t*. The first logging unit underwent a low-intensity logging (*left*) with a mean extracted volume $V_{ext}$=1.1 $m^3{\cdot}ha^{-1}$ whereas the second logging unit underwent a high-intensity logging (*right*) with a mean extracted volume $V_{ext} = 33.0\,m^3{\cdot}ha^{-1}$. For each plot, the net carbon balance (*top*) is the result of fourcomponents: emissions from logging damage (*blue*), emissions from road, deforestation on main skid trail and logging deck (*purple*), emissions from extracted timber (*red*), accumulation from post-logging recovery on the plot (*forest green*) and accumulation from regeneration on main skid trails and logging decks (*light green*). The *shaded areas* correspond to the 95 % credibility intervals

Even under high logging intensity (33 $m^3{\cdot}ha^{-1}$), the time to null carbon fluxes does not exceed 50 year. Previous results from satellite data have shown that in the Brazilian Amazon, where logging intensities usually range between 20 and 30 $m^3{\cdot}ha^{-1}$, carbon emissions last for 2–3 decades [14]. This last study also showed that carbon forest recovery may last for up to a century, twice as much as our highest prediction (Fig. 6). This difference in recovery times may arise from the model of forest recovery chosen in our study. We chose the model proposed in Rutishauser et al. [13], that assumes a constant recovery rate. With lack of insight in long-term dynamics of recovering forests, we made the conservative assumption that the carbon accumulation stops when initial carbon stock is recovered. However it is possible that recovering forests behave differently when the carbon stock approaches its initial value. The carbon storage could have a logistic behaviour, decelerating when approaching the initial carbon stock, or could oscillate around an equilibrium value (i.e. the initial carbon stock) and stabilize only after a longer time period [44].

Under the assumptions of our model, the carbon fluxes of a logged forest return to their equilibrium state before the end of the cutting cycle. At second harvest cycle, the carbon stocks will likely be close to their initial value, except on logging roads and skid trails. This does not mean that the forest will have returned to its original state as selective logging is known to have long-lasting effects on forest structure, timber volume and biodiversity [45].

### *Cumulative fluxes*

To assess the carbon balance of our two logging units, we estimated the cumulative carbon fluxes (Fig. 7). The cumulative carbon flux at time *t* is the sum of annual carbon fluxes between the logging year and *t*.

The carbon balance of the first logging unit (with low logging intensity) is not different from zero in the first years because the recovery rate is quite similar to emissions from dead wood decay and deforestation. At the

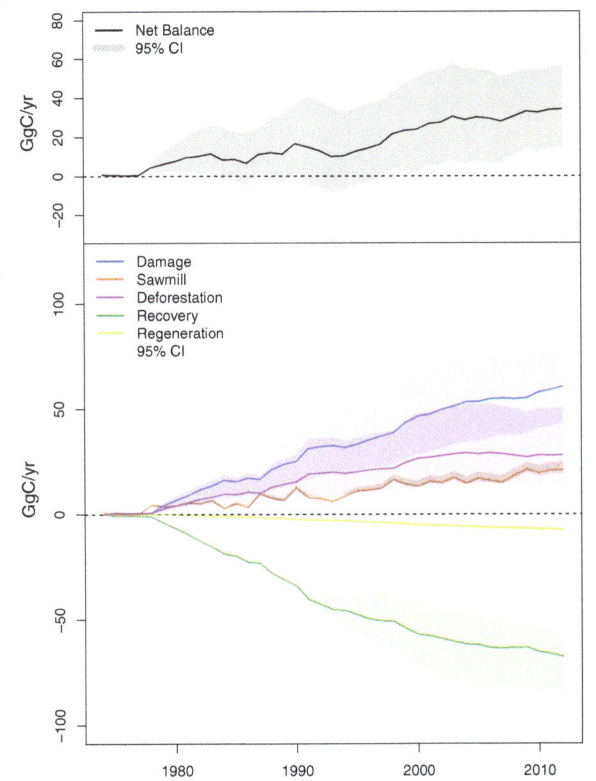


**Fig. 8** Annual above-ground carbon fluxes from logging in the Permanent Forest Estate between 1974–2012. The net carbon flux (*top*) is the result of four components: emissions from logging damage (*blue*), emissions from road, deforestation on main skid trail and logging deck (*purple*), emissions from extracted timber (*red*), accumulation from post-logging recovery on the plot (*forest green*) and accumulation from regeneration on main skid trails and logging decks (*light green*). The *shaded areas* correspond to the 95 % credibility intervals

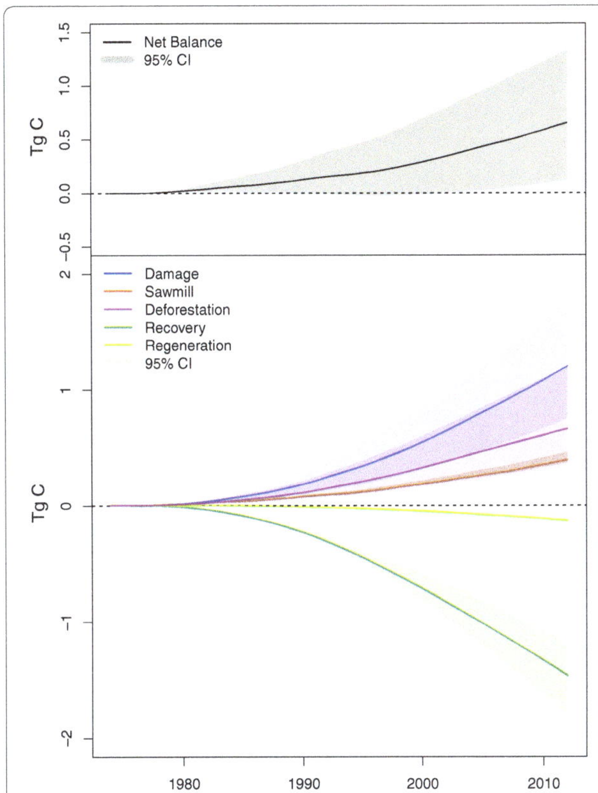


**Fig. 9** Above-ground carbon balance of the French Guiana Permanent Forest Estate between 1974–2012. The carbon balance at time *t* is the integral of annual carbon fluxes, i.e. the sum of annual carbon fluxes from the first year (1974) to *t*. The net carbon balance (*top*) is the result of four components: emissions from logging damage (*blue*), emissions from road, deforestation on main skid trail and logging deck (*purple*), emissions from extracted timber (*red*), accumulation from post-logging recovery on the plot (*forest green*) and accumulation from regeneration on main skid trails and logging decks (*light green*). The *shaded areas* correspond to the 95 % credibility intervals

end of the cutting cycle, the carbon balance of the logging unit is about equal to the carbon emissions from deforestation (on roads and skid trails). The second logging unit (with high logging intensity) has high emissions from damage decay in the first 8 years after logging, that exceed carbon accumulation from forest recovery: the carbon balance of the logging unit increases. Between 8 and 40 years after logging, carbon accumulation from forest recovery exceeds carbon emissions: the carbon balance decreases. After 40 years, the carbon balance of the logging unit stabilizes at a value about equal to the emissions from deforestation.

In the first years after logging, the net carbon balance is mostly driven by the logging damage, directly related to logging intensity. High logging intensities cause more damage, and thus induce higher emissions from dead wood decay. At the end of the cutting cycle, both logging units have the same carbon balance, corresponding to the carbon emissions from deforestation on logging roads and main skid trails. Indeed, there is little regeneration in main skid trails, and no regeneration in deforested roads whereas carbon recovery in logging gaps compensates the carbon emissions from extracted wood and logging damage. In terms of forest management, the carbon balance of a logging unit at the end of a cutting cycle thus depends mostly on the total surface of logging roads and skid trails. To mitigate carbon emissions, logging should be planned to avoid road and skid trail opening. This can be done by intensifying timber extraction per logged area and thus lowering the amount of infrastructure needed to get the logs out of the area.

### Carbon balance at the regional scale

After assessing carbon fluxes for each logged logging unit, results were aggregated over space to assess the carbon fluxes over the whole Permanent Forest Estate, and

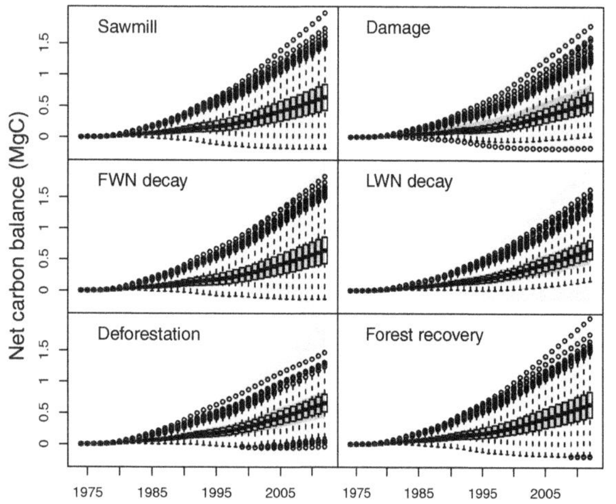


**Fig. 10** Estimating the uncertainty due to each submodel into the net carbon balance. The box plots represent the distribution of the net carbon balance when parameters of each submodel are set to their maximum likelihood value. The shaded area is the range of the net carbon balance for the total model, i.e. when all parameters are taken in their statistical distribution

over time to assess the carbon balance between 1974 and 2012 (Figs. 8, 9).

### *Annual fluxes*

As for the logging unit level, the largest carbon emissions arise from the decay of logging damage (Fig. 8): this is consistent with results from field data on other Amazonian countries [25]. Carbon accumulation is mainly explained by forest recovery in the logged area, forest regeneration on main skid trails and logging decks being negligible.

Net carbon fluxes are positive: the Permanent Forest Estate in French Guiana behaves like a carbon source during the 1974–2012 period. However zero is included in the 95 % credibility interval until 2000: the positiveness of carbon fluxes is not significant between 1974 and 2000. In the mid 1990' there is a sudden increase in net carbon fluxes, and after 2000 the lower bound of the 95 % credibility interval is positive: between 2000 and 2012, logging in the Permanent Forest Estate significantly emitted carbon to the atmosphere. This change in the trend of net carbon flux corresponds to increased logging intensities in the mid 1990s (Fig. 2).

### *The carbon balance at the regional scale*

Selective logging countrywide has induced significant carbon emissions since 2000 (Fig. 8), resulting in a positive net carbon balance in 2012 (Fig. 9). The net above-ground carbon balance in 2012 is estimated to be comprised between 0.12 and 1.33 Tg C, with a median value of 0.64 Tg C. With 1.36 Pg C stored in above-ground carbon [29], selective logging has emitted between 0.01 and 0.10 % of the French Guiana forest stock over the 1974–2012 period. Such ratio may look quite low when compared to the above-ground carbon stock of all the total forested area, but we have to keep in mind that the units logged over the 1974–2012 in the Permanent Forest Estate cover 0.12 Mha, i.e. 5 % of the total forested area in French Guiana (Fig. 1). Moreover, logging intensities in French Guiana are low (12.7 $m^3 \cdot ha^{-1} \cdot year^{-1}$, Fig. 2) compared to the Brazilian Amazon where logging intensities average 23.2 $m^3 \cdot ha^{-1} \cdot year^{-1}$ [5]. Applying our model to the rest of Amazonian managed forests should thus raise higher carbon emissions.

## Possible model improvements

### *Deforestation, damage and LWN decay are key submodels*

The sensitivity analysis reveals three main sources of uncertainty in our model: deforestation, large woody necromass decay and logging damage (Fig. 10). Despite relatively low values of logging areas that are annually deforested, the key role of the deforestation submodel in generating the net carbon balance uncertainties (Fig. 10) may look quite surprising. Because the carbon loss on permanently-deforested logging roads will never be recovered, carbon emissions accumulate over time, and their uncertainties as well. At the end of the 40-year period, the cumulative uncertainties then become non-negligible. The quantity of damage intervenes in two carbon fluxes: emissions from logging damage decay and carbon recovery (Fig. 3). If the objective is to estimate one of these two fluxes, efforts should be made to improve the prediction of logging damage. Until now the only explanatory variable of our logging damage model is the logging intensity *Vext* ($m^3 \cdot ha^{-1}$). A great proportion of the variability remains unexplained, and other factors may help improve the prediction of logging damage. For instance, including further information such as the logging techniques applied (conventional or reduced-impact logging), the local topography, or the forest structure would certainly have improved damage predictions. Nevertheless, the main objective is to estimate the net carbon balance, improving the estimation of logging damage should not be a priority. Indeed, recovery dynamics are very fast, lasting for a few decades at the most (Fig. 6). At the regional scale, emissions from damage decay are almost instantaneously mitigated by recovery mechanisms (Figs. 8, 9), so the variability of logging damage has little influence on the overall variability of the carbon balance (Fig. 10). On the contrary, the variability of large woody necromass decay explains a great part of the overall uncertainty of the total carbon balance (Fig. 10). In this submodel, uncertainties come from the individual-based model [4] and from the error propagation during the passage to the logging unit scale. To reduce uncertainties over the large woody necromass decay model, direct field measurements should be made to estimate the decay rate directly at the logging unit scale.

### *Below-ground carbon*

This study remains restricted to aboveground carbon, the main carbon pool in tropical forests [46], but below-ground carbon (roots, soil) could also be impacted by selective logging and result in carbon emissions. Some studies suggest that soil carbon stocks may decrease after selective logging [47], but carbon sequestration in roots and in the woody debris that are not respired and that return to soil could also be expected [48]. In this study we do not consider below-ground carbon fluxes due to lack of robust information. Gathering more information and adding below-ground carbon dynamics into the model would further improve the model.

## Conclusions

We here provide a comprehensive methodology to estimate above-ground carbon fluxes in managed tropical forests at both local and regional scales, easily replicable in other tropical regions. This is a further contribution towards more accurate assessment of the carbon budget of tropical managed forests. The framework is also flexible and easy to adapt to new models: in particular, new logging techniques or new management practices (*e.g. fuel wood*) could be added to the model. This can be useful when investigating the carbon balance of upcoming opportunities and of future management choices for tropical production forests.

#### Authors' contributions

Study conception and design: CP, AC, BH. Acquisition of data: LD, AD, BO, LM, ER, PS. Analysis and interpretation of data: CP, AC, LD, AD, BO, BH. Drafting of manuscript: CP, BH. Critical revision: CP, AC, LD, AD, BO, LM, ER, PS, BH. All authors read and approved the final manuscript.


#### Author details

[1] Université de la Guyane, UMR EcoFoG (AgroParisTech, CNRS, Inra, Université des Antilles, Cirad), Campus agronomique, 97310 Kourou, French Guiana. [2] Centre Tecnològic Forestal de Catalunya, Crta. de Sant Llorenç de Morunys, Km.2, 25280 Solsona, Spain. [3] ONF-Guyane, Réserve de Montabo, 97307 Cayenne, French Guiana. [4] Cirad, UMR EcoFoG (AgroParisTech, CNRS, Inra, Université de la Guyane, Université des Antilles), Campus agronomique, 97310 Kourou, French Guiana. [5] EMBRAPA Amazônia Oriental, Trav. Dr. Enéas Pinheiro, Belém, Brazil. [6] Guyane Energie Climat, 16 rue Victor Schoelcher, 97300 Cayenne, French Guiana. [7] CarboForExpert, 1248 Hermance, Switzerland. [8] Cirad UR Forêts et Sociétés, 34398 Montpellier, France. [9] CNRS, UMR EcoFoG (AgroParisTech, Inra, Université de la Guyane, Université des Antilles, Cirad), Campus Agronomique, 97310 Kourou, French Guiana.



#### Acknowledgements

We are grateful to CIRAD and the Observatoire du Carbone project (French Guiana Region and European structural fundings) for financial support. This study was partially funded by an Investissement d'Avenir Grant of the ANR (CEBA: ANR-10-LABEX-0025).


#### Competing interests

The authors declare that they have no competing interests.

#### References


1. Putz FE, Zuidema PA, Pinard MA, Boot RG, Sayer JA, Sheil D, et al. Improved tropical forest management for carbon retention. PLoS Biol. 2008;6(7):1368–9.
2. Blaser J, Sarre A, Poore D, Johnson S. Status of tropical forest management 2011, vol 38. ITTO; 2011. p. 27–31.
3. Mertz O, Müller D, Sikor T, Hett C, Heinimann A, Castella JC, et al. The forgotten D: challenges of addressing forest degradation in complex mosaic landscapes under REDD+. Geografisk Tidsskrift-Danish J Geogr. 2012;112:63–76.
4. Hérault B, Beauchêne J, Muller F, Wagner F, Baraloto C, Blanc L, et al. Modeling decay rates of dead wood in a neotropical forest. Oecologia. 2010;164(1):243–51.
5. Asner GP, Knapp DE, Broadbent EN, Oliveira PJC, Keller M, Silva JN. Selective logging in the Brazilian Amazon. Science. 2005;310(5747):480–2.
6. Picard N, Gourlet-Fleury S, Forni É. Estimating damage from selective logging and implications for tropical forest management. Can J For Res. 2012;42(3):605–13.
7. Sist P, Ferreira FN. Sustainability of reduced-impact logging in the Eastern Amazon. For Ecol Manag. 2007;243(2–3):199–209.
8. Sist P, Mazzei L, Blanc L, Rutishauser E. Large trees as key elements of carbon storage and dynamics after selective logging in the Eastern Amazon. For Ecol Manag. 2014;318:103–9.
9. Shenkin A, Bolker B, Peña-Claros M, Licona JC, Putz FE. Fates of trees damaged by logging in Amazonian Bolivia. For Ecol Manag. 2016;2015(357):50–9.
10. Rutishauser E, Hérault B, Petronelli P, Sist P. Tree height reduction after selective logging in a tropical forest. Biotropica. 2016;48(3):285–9.
11. Hérault B, Ouallet J, Blanc L, Wagner F, Baraloto C. Growth responses of neotropical trees to logging gaps. J Appl Ecol. 2010;47(4):821–31.
12. Blanc L, Echard M, Herault B, Bonal D, Marcon E, Chave J, et al. Dynamics of aboveground carbon stocks in a selectively logged tropical forest. Ecol Appl. 2009;19(6):1397–404.
13. Rutishauser E, Hérault B, Baraloto C, Blanc L, Descroix L, Sotta ED, et al. Rapid tree carbon stock recovery in managed Amazonian forests. Curr Biol. 2015;25(18):R787–8.
14. Huang M, Asner GP. Long-term carbon loss and recovery following selective logging in Amazon forests. Glob Biogeochem Cycles. 2010;24(3):1–15.
15. Kleinschroth F, Gourlet-Fleury S, Sist P, Mortier F, Healey JR. Legacy of logging roads in the Congo Basin: how persistent are the scars in forest cover? Ecosphere. 2015;6(4):1–17.
16. Goetz SJ, Baccini A, Laporte NT, Johns T, Walker W, Kellndorfer J, et al. Mapping and monitoring carbon stocks with satellite observations: a comparison of methods. Carbon Balance Manag. 2009;4:2.
17. Asner GP, Keller M, Pereira R, Zweede JC. Remote sensing of selective logging in Amazonia. Remote Sens Environ. 2002;80(3):483–96.
18. Asner GP, Powell GVN, Mascaro J, Knapp DE, Clark JK, Jacobson J, et al. High-resolution forest carbon stocks and emissions in the Amazon. Proc Natl Acad Sci. 2010;107(38):16738–42.
19. Réjou-Méchain M, Muller-Landau HC, Detto M, Thomas SC, Le Toan T, Saatchi SS, et al. Local spatial structure of forest biomass and its consequences for remote sensing of carbon stocks. Biogeosci Discuss. 2014;11(4):5711–42.
20. Peres CA, Barlow J, Laurance WF. Detecting anthropogenic disturbance in tropical forests. Trends Ecol Evol. 2006;21(5):227–9.
21. Pereira R, Zweede J, Asner GP, Keller M. Forest canopy damage and recovery in reduced-impact and conventional selective logging in eastern Para, Brazil. For Ecol Manag. 2002;168:77–89.
22. Meyer V, Saatchi SS, Chave J, Dalling JW, Bohlman S, Ga Fricker, et al. Detecting tropical forest biomass dynamics from repeated airborne lidar measurements. Biogeosciences. 2013;10(8):5421–38.
23. IPCC. 2006 IPCC guidelines for national greenhouse gas inventories, chapter 4. 2006, p. 4.14.
24. Griscom B, Ellis P, Putz FE. Carbon emissions performance of commercial logging in East Kalimantan, Indonesia. Glob Change Biol. 2014;20(3):923–37.
25. Pearson TRH, Brown S, Casarim FM. Carbon emissions from tropical forest

degradation caused by logging. Environ Res Lett. 2014;9(3):034017.
26. Khun V, Sasaki N. Cumulative carbon fluxes due to selective logging in Southeast Asia. Low Carbon Econ. 2014;5:180–91.
27. Sist P, Rutishauser E, Peña-Claros M, Shenkin A, Hérault B, Blanc L, et al. The tropical managed forests observatory: a research network addressing the future of tropical logged forests. Appl Veg Sci. 2014;18:171–4.
28. Toriola D, Chareyre P, Buttler A. Distribution of primary forest plant species in a 19-year old secondary forest in French Guiana. J Trop Ecol. 1998;14(3):323–40.
29. Guitet S, Hérault B, Molto Q, Brunaux O, Couteron P. Spatial structure of above-ground biomass limits accuracy of carbon mapping in rainforest but large scale forest inventories can help to overcome. PLoS One. 2015;10:e0138456.
30. Penman J, Gytarsky M, Hiraishi T, Krug T. Good practice guidance for land use, land-use change and forestry. Intergovernmental Panel on Climate Change (IPCC), chapter 3, Appendix 3.A.1.3. UNEP; 2003. p. 268–70.
31. Ollivier M, Baraloto C, Marcon E. A trait database for Guianan rain forest trees permits intra- and inter-specific contrasts. Ann For Sci. 2007;64(7):781–6.
32. Fargeon H, Aubry-kientz M, Brunaux O, Descroix L, Guitet S, Rossi V, et al. Vulnerability of commercial tree species to water stress in logged forests of the Guiana shield. Forests. 2016;7(105):1–21.
33. Keller M, Asner GP, Silva JNM, Palace M. Sustainability of selective logging of upland forests in the Brazilian Amazon: carbon budgets and remote sensing as tools for evaluation of logging effects. In: Working forests in the tropics: conservation through sustainable management? New York: Columbia University Press; 2004. p. 41–63.
34. Asner GP, Keller M, Lentini M, Merry F, Souza CJ. Selective logging and its relation to deforestation. Geophys Monogr Ser. 2009;186:25–42.
35. Eshun JF, Potting J, Leemans R. LCA of the timber sector in Ghana: preliminary life cycle impact assessment (LCIA). Int J Life Cycle Assess. 2011;16(7):625–38.
36. Guitet S, Pithon S, Brunaux O, Jubelin G, Gond V. Impacts of logging on the canopy and the consequences for forest management in French Guiana. For Ecol Manag. 2012;277:124–31.
37. Chambers JQ, Higuchi N, Teixeira LM, dos Santos J, Laurance SG, Trumbore SE. Response of tree biomass and wood litter to disturbance in a Central Amazon forest. Oecologia. 2004;141(4):596–611.
38. Chambers JQ, Higuchi N, Schimel JP, Ferreira LV, Melack JM. Decomposition and carbon cycling of dead trees in tropical forests of the central Amazon. Oecologia. 2000;122:380–8.
39. Putz FE, Sist P, Fredericksen T, Dykstra D. Reduced-impact logging: challenges and opportunities. For Ecol Manag. 2008;256(7):1427–33.
40. Berndes G, Hoogwijk M, van den Broek R. The contribution of biomass in the future global energy supply: a review of 17 studies. Biomass Bioenerg. 2003;25(1):1–28.
41. Karsenty A, Vermuelen C. Toward "Concessions 2.0": articulating inclusive and exclusive management in production forests in Central Africa. Int For Rev. 2016;18:1.
42. Monteiro A, Cardoso D, Conrado D, Veríssimo A, Souza C. Forest management transparency report—State of Pará (2012–2013). Imazon; 2013. http://imazon.org.br/publicacoes/forest-management-transparency-reportstate-of-para-2011-to-2012/?lang=en.
43. Asner GP, Keller M, Silva J. Spatial and temporal dynamics of forest canopy gaps following selective logging in the eastern Amazon. Glob Change Biol. 2004;10(5):765–83.
44. Holling CS. Resilience and stability of ecological systems. Annu Rev Ecol Syst. 1973;4(1):1–23.
45. Putz FE, Pa Zuidema, Synnott T, Peña-Claros M, Ma Pinard, Sheil D, et al. Sustaining conservation values in selectively logged tropical forests: the attained and the attainable. Conserv Lett. 2012;5(4):296–303.
46. Pan Y, Birdsey RA, Phillips OL, Jackson RB. The structure, distribution, and biomass of the World's forests. Annu Rev Ecol Evol Syst. 2013;44:593–622.
47. Chiti T, Perugini L, Vespertino D, Valentini R. Effect of selective logging on soil organic carbon dynamics in tropical forests in central and western Africa. Plant Soil. 2015;399:283–94.
48. Chambers JQ, Schimel JP, Nobre CA. Respiration from coarse wood litter in central Amazon forests. Biogeochemistry. 2001;52:115–31.

# 8

# Modelling forest carbon stock changes as affected by harvest and natural disturbances. II. EU-level analysis

Roberto Pilli[1*], Giacomo Grassi[1], Werner A. Kurz[2], Jose V. Moris[3] and Raúl Abad Viñas[1]

**Abstract**

**Background:** Forests and the forest sector may play an important role in mitigating climate change. The Paris Agreement and the recent legislative proposal to include the land use sector in the EU 2030 climate targets reflect this expectation. However, greater confidence on estimates from national greenhouse gas inventories (GHGI) and more comprehensive analyses of mitigation options are needed to seize this mitigation potential. The aim of this paper is to provide a tool at EU level for verifying the EU GHGI and for simulating specific policy and forest management scenarios. Therefore, the Carbon Budget Model (CBM) was applied for an integrated assessment of the EU forest carbon (C) balance from 2000 to 2012, including: (i) estimates of the C stock and net $CO_2$ emissions for forest management (FM), afforestation/reforestation (AR) and deforestation (D), covering carbon in both the forest and the harvest wood product (HWP) pools; (ii) an overall analysis of the C dynamics associated with harvest and natural disturbances (mainly storms and fires); (iii) a comparison of our estimates with the data reported in the EU GHGI.

**Results:** Overall, the average annual FM sink (−365 Mt $CO_2$ year$^{-1}$) estimated by the CBM in the period 2000–2012 corresponds to about 7 % of total GHG emissions at the EU level for the same period (excluding land use, land-use change and forestry). The HWP pool sink (−44 Mt $CO_2$ year$^{-1}$) contributes an additional 1 %. Emissions from D (about 33 Mt $CO_2$ year$^{-1}$) are more than compensated by the sink in AR (about 43 Mt $CO_2$ year$^{-1}$ over the period). For FM, the estimates from the CBM were about 8 % lower than the EU GHGI, a value well within the typical uncertainty range of the EU forest sink estimates. For AR and D the match with the EU GHGI was nearly perfect (difference $<\pm 2$ % in the period 2008–2012). Our analysis on harvest and natural disturbances shows that: (i) the impact of harvest is much greater than natural disturbances but, because of salvage logging (often very relevant), the impact of natural disturbances is often not easily distinguishable from the impact of harvest, and (ii) the impact of storms on the biomass C stock is 5–10 times greater than fires, but while storms cause only indirect emissions (i.e., a transfer of C from living biomass to dead organic matter), fires cause both direct and indirect emissions.

**Conclusions:** This study presents the application of a consistent methodological approach, based on an inventory-based model, adapted to the forest management conditions of EU countries. The approach captures, with satisfactory detail, the C sink reported in the EU GHGI and the country-specific variability due to harvest, natural disturbances and land-use changes. To our knowledge, this is the most comprehensive study of its kind at EU level, i.e., including all the forest pools, HWP and natural disturbances, and a comparison with the EU GHGI. The results provide the basis for possible future policy-relevant applications of this model, e.g., as a tool to support GHGIs (e.g., on accounting for natural disturbances) and to verify the EU GHGI, and for the simulation of specific scenarios at EU level.

*Correspondence: roberto.pilli@jrc.ec.europa.eu
[1] European Commission, Joint Research Centre, Directorate D – Sustainable Resources - Bio-Economy Unit, Via E. Fermi 2749, 21027 Ispra, VA, Italy
Full list of author information is available at the end of the article

**Keywords:** EU, Carbon Budget Model, Forest management, Afforestation, Deforestation, Harvest, Natural disturbances, GHGI

## Background

An effective role of forests in climate change mitigation requires a comprehensive assessment, from scientific and policy perspectives. From a scientific point of view, recent studies demonstrate the relevance of biophysical aspects of the forest-climate interactions, that may be important locally or in specific time frames [1]. Even if the recent paper by Naudts et al. [2], casting doubts on the role of European forests in mitigating climate change over the last centuries, there is increasing and largely consistent scientific evidence that forests in Europe are currently making a relevant and positive contribution to climate change mitigation (see [3, 4]). From a policy perspective, it is relevant to understand how this contribution may be efficiency translated into different mitigation options, including the sink in the forest, the sink outside the forest (in harvested wood products, HWP) and the use of wood for energy and material substitution [5]. Given the heterogeneity of the European forest system, assessing the specific regional circumstances, opportunities and challenges is fundamental [6]. At the same time, maximizing the sum of these mitigation options requires an integrated, dynamic modeling framework to quantify in a robust way the unavoidable trade-offs (e.g. between the forest sink and the bioenergy), which are often not appropriately considered [7]. Furthermore, for such a modeling framework to be directly policy-relevant, the policy context such as the rules for reporting and accounting emissions and removals from forests need to be taken into account.

In particular, the current rules under the Kyoto Protocol (KP) significantly changed for the second commitment period (CP2, 2013–2020) (see, [8–10]). According to these new rules, as reflected in the latest IPCC guidance [11]: (i) the reporting and accounting of forest management (FM, i.e., land in the forest land use category in 1989) are now mandatory (through a 'forest reference level'), in addition to the already mandatory accounting of afforestation/reforestation and deforestation (AR and D, i.e., forest land-use changes since 1990); (ii) the accounting of FM shall include the carbon (C) stock changes in the HWP pool; and (iii) emissions and subsequent removals from natural disturbances may be excluded from the accounting under certain conditions. While some further change is foreseen under the proposed post-2020 EU regulation on land use and forestry [12], most of these rules are expected to continue (e.g. forest reference level, HWP, natural disturbances).

The greenhouse gas inventories (GHGIs) represent the basis to assess the compliance of any climate mitigation target. The GHGI of the EU, submitted annually to the United Nations Framework Convention on Climate Change (UNFCCC) and its KP, is the sum of the inventories of 28 Members States (MS), which include about 158 Mha of forests [13]. The species composition, the current and past management practices, the amount of natural disturbances, and the quality and type of information available on the forest resources, differ among countries. Moreover, conceptual and methodological differences in countries' GHGIs produce discrepancies in the resulting estimates that are currently not entirely addressed and require further work to achieve reliable and consistent estimates throughout Europe [14, 15].

The complex EU forest sector can be represented using a process-based approach (e.g. [1, 16]) or using, for each country, empirical forest-inventory based models (e.g. [17]). Traditionally, process-based models have mainly been used to simulate the long-term evolution of forest C dynamics at large scales, including the potential effects of climate change [18], but they generally do not include a detailed analysis of forest management practices. Therefore, empirical models still remain the primary tool to simulate the detailed effects of different management options on short-term forest C dynamics [19, 20] at small to medium spatial scales (e.g. from forest stands to countries). When compiled from regional or country level scales to a continental scale, the empirical model results can be compared with the data produced through process-based approaches [21, 22], and can provide additional information on the main drivers of forest carbon dynamics at the EU scale. Different forest-inventory based models were used in the European context, to estimate the future forest C sink under different policies and management scenarios (i.e., [20]) or the impact of natural disturbances on the forest C stock [23] or the realizable potential supply of woody biomass [24]. None of these studies, however, considered in a comprehensive way the overall EU forest C sink consistently with the current international reporting and accounting regulations, i.e., including FM, AR and D, HWP and natural disturbances.

In a recent study, the Carbon Budget Model (CBM), developed by the Canadian Forest Service [25] was applied to 26 EU MSs to model the forest C dynamics from FM at the country level for the period 2000–2012, including the impact of the major natural disturbances [26]. In that study, after having validated the CBM results for a representative country, the country-specific results were evaluated against the individual 2014 GHGIs submitted to the

UNFCCC by each EU MS. This evaluation is an essential pre-requisite to analyze the overall EU forest C balance and the level of confidence on the EU GHGI. Achieving this confidence is key to allow the forest sink to be included in the EU climate target [27]. The goal of this paper is to provide a tool for verifying the whole EU GHGI and for simulating specific policy and forest management scenarios at EU level. In particular, with the present paper, focused at the EU level and largely based on the same methodological assumptions used by Pilli et al. [26], we aim to: (i) estimate the C stock and the $CO_2$ emissions and removals for FM, covering the carbon both in the forest pools (total living biomass, dead organic matter, mineral soil) and in HWP pool; (ii) estimate the $CO_2$ emissions and removals for forest land-use changes (i.e., AR and D); (iii) provide an overall EU-level analysis of the impacts of harvest and natural disturbances (mainly storms and fires); and (iv) compare our estimates with the estimates reported in the EU GHGI and other continental-scale studies.

## Results and discussion

The aggregated results at the EU level for the forest-related activities defined by the KP are reported for FM (i.e., forest existing in 1989) AR and D (i.e., forest and land-use changes since 1990) in "Forest C stock (2008–2012)" section. Here, the results obtained using the CBM are compared with the data reported in the 2014 EU GHGI[1] [13]. According to the EU GHGI, the two MSs not considered in this study, Cyprus and Malta, provide a negligible contribution to the EU forest sink (0.02 %).

The resulting forest C dynamics are described in "Forest C dynamics (2000–2012)" section. Here, the data are reported from an atmospheric perspective, where negative values represent a sink ($CO_2$ removals) and positive values a source ($CO_2$ emissions). Results cover only $CO_2$ and exclude organic soils. Even if emissions from drained organic soils and non-$CO_2$ emissions from forest fires may be relevant in some countries (e.g., [28, 29]), at the EU level they account for 5 and 2 % (in terms of $CO_2$-eq.), respectively, of net annual forest sink [13]. In "Main drivers determining forest C dynamics" section, we discuss the main drivers determining the forest C sink dynamic, further distinguished between harvest ("Harvest" section) and natural disturbances ("Natural disturbances" section).

### Forest C stock (2008–2012)

The average C stock per hectare estimated for FM by the model for the period 2008–2012 (i.e., the First Commitment Period, CP1, of the KP), is equal to 142.3 Mg C ha$^{-1}$ at the EU level, including 68.4, 19.3 and 54.5 Mg C ha$^{-1}$, for living biomass, DOM and mineral soil, respectively (Table 1). According to Pilli et al. [30], the total C stock of HWP is about 1921 Tg C (average for the CP1), which compares the total C stock in living biomass of 9437 Tg C in this study.

Overall, our estimates on living biomass are in good agreement with the data reported by most of other reports or studies. The Global Forest Resource Assessment 2010 [31] reports for 2010 in Europe, excluding the Russian Federation but including some non-EU countries (i.e., for a total forest area that is about 8 Mha larger than the forest area considered by our study), an average C stock equal to 63.9, 18.6 and 96.6 Mg C ha$^{-1}$, for the living biomass, dead wood plus litter and soil (including peat), respectively. The State of Europe's Forests 2015 [32] reports for EU-28 and assuming a forest area slightly larger than our study (+9 %) a biomass C stock in 2010 equal to 58.8 Mg C ha$^{-1}$.

Providing estimates for dead wood and litter C stock is difficult [33]. Verkerk et al. [34] applied the EFISCEN model to 24 EU MS (i.e., the same considered by our study, except Croatia and Greece) and estimated an average amount of deadwood of 12.3 t ha$^{-1}$ in 2005. Assuming a C content of 0.5, this equals an average C stock of 6.1 t C ha$^{-1}$, about 40 % lower than our estimate.

An accurate assessment of the soil C stock is also difficult due to the range of available model and inventory results [33, 35]. Our estimate of the total C stock in forest mineral soils (7524 Tg C) is intermediate between the values reported by other studies: about 5000 Tg C reported by Liski et al. [36] and 13,700 Tg C (including the O-layer) estimated by Goodale et al. [37]. Since the soil C stock is affected by both natural and anthropogenic factors [38, 39], comparing the average C stock estimated by our model, equal to 54.5 Mg C ha$^{-1}$, with other studies is even more difficult. Shulp et al. [40] report, for the Netherlands, a mean C stock in mineral soil between 53 and 97 Mg C ha$^{-1}$, with significant statistical differences mainly due to the species composition. The average C stock in mineral soil estimated by the last Italian National Forest Inventory (NFI) (concluded in 2009) varies between 68 and 96 Mg C ha$^{-1}$, depending by the species composition [41], while the Swedish NFI estimated, for 2000, an average C stock for coniferous forest soil, equal to 73 ± 10 Mg C ha$^{-1}$ [42].

### Forest C dynamics (2000–2012)

The sum of the net $CO_2$ removals from all land-use activities and carbon pools considered (FM + AR + D + HWP) is, on average, equal to −409 Mt $CO_2$ year$^{-1}$ between 2000 and 2012 (Fig. 1, panel c). This

[1] Since this paper reports the results of a long-term study, based on data available until 2014, further comparisons with the data reported by the 2015 GHGI is not appropriate.

**Table 1 Average C stock estimated for FM by the CBM model for the period 2008–2012 (KP-CP1) at EU level**

| Average historical (2008–2012) C stock | | DOM | | Soil | Living biomass | | | Total ecosystem |
|---|---|---|---|---|---|---|---|---|
| | | Deadw. | Litter | | Abovegr. | Belowgr. | Tot Liv. Biom. | |
| C stock | Av. (Mg C $ha^{-1}$) | 10.8 | 8.5 | 54.5 | 56.0 | 12.4 | 68.4 | 142.3 |
| | Total (Tg C) | 1493 | 1177 | 7524 | 7720 | 1717 | 9437 | 19,632 |

corresponds to about 8 % of the total GHG emissions in the EU for the same period (without LULUCF). This amount may be further distinguished between different land-use activities (FM, AR and D) and pools. About 90 % of the total C sink is due to FM (Fig. 1, panel a, including HWP), while AR (Fig. 1, panel b) contributes to the remaining 10 %, with an increasing fraction due to the ageing of the new forest area. Deforestation is a source by about 33 Mt $CO_2$ $year^{-1}$ between 2000 and 2012. About 80 % of the total C sink (after subtracting emissions from D) is due to the living biomass pool (70 % accounted as FM and 10 % as AR), 10 % is due to DOM (mainly accounted as FM) and the remaining 10 % is related to the HWP pool.

Within FM, the highest inter-annual variations (due to harvest and natural disturbances) were estimated for the living biomass pool, varying from −237 Mt $CO_2$ $year^{-1}$ in 2000 to −311 Mt $CO_2$ $year^{-1}$ in 2012 (Fig. 1, panel a). As expected, the DOM pool in the CBM has the opposite trend, because natural disturbances such as storms, fires and insect attacks transfer carbon from biomass to DOM pools (see for example 2000, 2005 and 2007) from where the carbon will be released to the atmosphere through subsequent decomposition. For this pool, we estimated an average C sink, for the entire period, equal to −43 Mt $CO_2$ $year^{-1}$. For the mineral soil we estimated a modest and rather stable C sink over the entire period, equal on average to −3 Mt $CO_2$ $year^{-1}$. This trend is consistent with increasing biomass of the EU forests, which means increasing inputs from litter and dead wood to the soil pool, and with the short time horizon considered, i.e., the process of soil C accumulation is typically a slow process. This process is simulated by the CBM through a series of biomass annual turnover rates and transfer rates [25]. Similarly to other soil models [43] the results provided by CBM may be influenced by uncertainty in the model initialization that may directly affect the estimate of the C stock change of this pool [44].

In terms of C stock change, we estimate average values, for the entire period, equal to 0.01, 0.08 and 0.60 Mg C $ha^{-1}$ $year^{-1}$ respectively for living biomass, DOM and soil.

The $CO_2$ sink of the FM living biomass pool estimated by the CBM is about 12 % lower than the data reported in the EU GHGI[2] (see Fig. 1, plot a) and is in line with most of other studies with similar area and time frames (e.g. [45, 46]). The largest differences with the GHGI, in 2000, 2005 and 2007, are related to the different assumptions about the impact of natural disturbances (mainly storms that occurred in central and northern European countries). Indeed, the effect of natural disturbances on the redistribution of C among pools (from living biomass to DOM), well represented in the CBM, is often not evident in the GHGIs [22, 26]. The difference between the CBM and the EU GHGI is reduced to 8 % when DOM and mineral soil are also considered. Since the uncertainty of $CO_2$ estimates for "forest land remaining forest land" at the EU level is around 18 % [13],[3] and given the fact that methods to estimate emissions/removals by the CBM are largely independent from those of the EU countries [26], we consider the match between the CBM results and the EU GHGI to be satisfactory. The increasing discrepancy in more recent years is mainly due to few countries (mainly Poland and France), which will deserve further investigation, e.g. inappropriate data or assumption used by the CBM model or problems with the GHGIs. Indeed, at country level, where updated NFI data are available and the model's assumptions on harvest and natural disturbances are consistent with the countries' input data, the estimates provided by the CBM are generally consistent (both in the trend and in the amount) with the GHGI data [26].

For AR (Fig. 1, panel b), the living biomass C sink gradually increases until 2003, due to the constant annual rate of AR prior to 2008 (see Fig. 7 in the "Methods" section). We estimated a very small source for the atmosphere from DOM and soil, due to the effect of afforestation on the soil pool during the first years [47]. As explained in the methods, the biomass C sink is directly related to the values reported by the yield tables applied by model. In most cases, due to the young age of the forests that were afforested since 1990, we assumed that no silvicultural treatment was applied to broadleaved species stands younger than 15 years and to coniferous stands younger

[2] In the area-based comparison, the FM area of the EU GHGI was decreased by the same amount decreased for CBM runs, i.e. excluding unproductive forests and overseas territories, see "Methods" for further details.

[3] The uncertainty raises to 25–50 % when analyzed at country level.

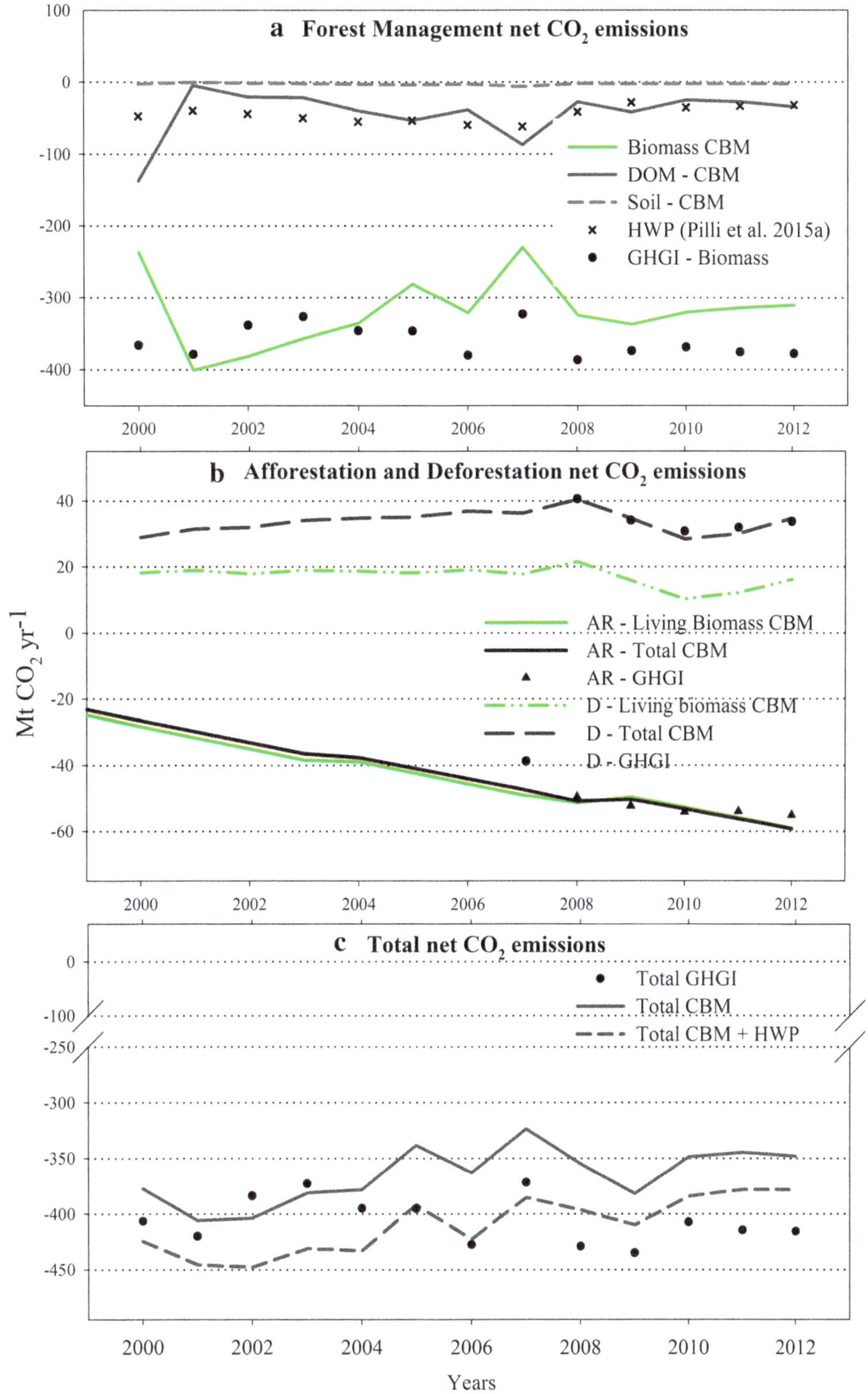
a Forest Management net CO2 emissions
Biomass CBM
DOM - CBM
Soil - CBM
HWP (Pilli et al. 2015a)
GHGI - Biomass
b Afforestation and Deforestation net CO2 emissions
AR - Living Biomass CBM
AR - Total CBM
AR - GHGI
D - Living biomass CBM
D - Total CBM
D - GHGI
c Total net CO2 emissions
Total GHGI
Total CBM
Total CBM + HWP
Mt CO2 yr-1
Years
100
0
-100
-200
-300
-400
40
20
-20
-40
-60
-250
-350
-450
2000
2002
2004
2006
2008
2010
2012

(See figure on previous page.)
**Fig. 1** EU-level net $CO_2$ emissions (in Mt $CO_2$ $year^{-1}$) for: **a** Forest management (FM), as estimated by the CBM for the living biomass, DOM and soil pools, by Pilli et al. [30] for the HWP pool and as reported for the living biomass pool in the EU GHGI (see [13] and the "Methods" for further details); **b** afforestation/reforestation (AR) and deforestation (D) since 1990, as estimated by the CBM for living biomass and all the pools (total) and as reported (all pools) in the KP CRF tables of the EU GHGI; **c** sum of FM, AR and D, as estimated for all the pools by the CBM (with and without HWP) and as reported in the EU GHGI (without HWP). Organic soils are always excluded from GHGIs, to allow a more consistent comparison with CBM. All the values are reported from an atmospheric perspective, i.e., negative values represent a sink and positive a source

than 20 years. The only exception was for Portugal's Eucalyptus plantations, where we assumed a minimum rotation length of 12 years. Due to the effect of these treatments, the biomass sink has a first step in 2003–2004 (due to the harvest on Eucalyptus plantations), and a second step in 2009–2010, due to the first harvest applied to coniferous plantations. Overall, the total amount of harvest provided by AR is negligible, equal to about 6.3 $Mm^3$ in 2012, i.e., about 1.2 % of the total amount of harvest obtained from FM at the EU level. The total annual C sink (−54 Mt $CO_2$ $year^{-1}$) provided by AR for the CP1 (2008–2012) is equal to about 1.2 % of the total GHGs in the EU for the same period (without LULUCF).

For deforestation (Fig. 1, panel b), the CBM estimates the loss of C from the living biomass, DOM and soil pools based on the areas subject to deforestation (taken from countries' GHGIs). Overall, for the CP1 (2008–2012), these emissions equal about 0.7 % of the total GHGs in the EU for the same period (without LULUCF).

Our estimates for AR and D compare very well with the EU GHGI for CP1 (2008–2012) [48]. While emissions from drained organic soils may be important in some MS (e.g. Finland, Sweden, Ireland) at the EU level the impact of organic soils for AR and D is small (about 2–3 Mt $CO_2$ $year^{-1}$). Overall, this good match with the GHGIs is expected, because the CBM uses the same rates of AR and D areas as reported in the GHGIs and because of the good agreement for the estimates of biomass carbon densities. For AR, however, a certain degree of independence between CBM and GHGIs arises from the choice of the yield tables and the harvest assumptions.

## Main drivers determining forest C dynamics

The forest C sink is essentially the difference between the net increment and the losses, i.e., harvest and natural disturbances. Forest growth, and the evolution of net annual increment[4] over time is estimated during the model run, by combining, for each country and time step (i.e., year), a yield table library based on the NFI annual increment with the forest inventory and its age class distribution (see [25, 44]). Given that the net increment typically changes relatively slowly and in recent years appears rather stable at EU level [49] here we focus on harvest (Fig. 2, panel a) and natural disturbances, such as storms and ice (panel b) and fire (panel c).

### *Harvest*

The CBM represents the amounts used for fuelwood (FW) and for industrial roundwood (IRW) (Fig. 2, panel a). According to the IPCC [11] and the UNFCCC [50], for the Second Commitment Period of the KP (KP-CP2) the C in the FW has to be accounted as a direct $CO_2$ emission into the atmosphere, while the C in the IRW products has to be further quantified to estimate the C stock changes in the HWP pool, including product categories. End-of life disposal of HWP in landfills is considered an instantaneous oxidation.

According to our estimates using the CBM model, the direct emissions related to the FW continuously increased, from about 82 Mt $CO_2$ $year^{-1}$ in 2000 to about 100 Mt $CO_2$ $year^{-1}$ in 2012 (i.e., +2 % $year^{-1}$). Based on the country-specific assumptions applied by CBM, the FW may be provided by: (i) direct harvest removals, i.e., specific silvicultural treatments applied to forest stands (e.g., clearcuts on coppices or commercial thinnings on high forests); (ii) indirect harvest of branches, other wood components and snags during other silvicultural treatments (i.e., thinnings and clearcut where the merchantable biomass is used as IRW); and (iii) the salvage logging after disturbance events (mainly fires).

The amount of IRW that was moved from living biomass to the HWP pool was equal, on average, to 309 Mt $CO_2$ $year^{-1}$ between 2000 and 2012. The two peaks reported in 2005 and 2007 are due to salvage logging after the storms in 2005 and 2007 (see Fig. 2, panel b). Interestingly, no major peak is reported by the statistics after the big storm that occurred in 1999/2000 and after the storms occurred in some countries (i.e., Austria, Estonia and, above all, France) in 2009, suggesting that salvage logging either was not very relevant or it was spread over more years. According to Rüter [51], the analysis of IRW data to obtain C stock changes in the HWP pools involves the service life of products (i.e., the annual decay rate), the estimate of the domestic production, the balance between C inflow and outflow from the HWP pool, the exclusion of harvest from deforestation, and many

[4] According to FAO, the net annual increment is defined as the average periodic increment of volume of all trees measured at a certain minimum Dbh, including the increment of trees which have been felled and excluding the increment of trees which have died during the reference period [62].

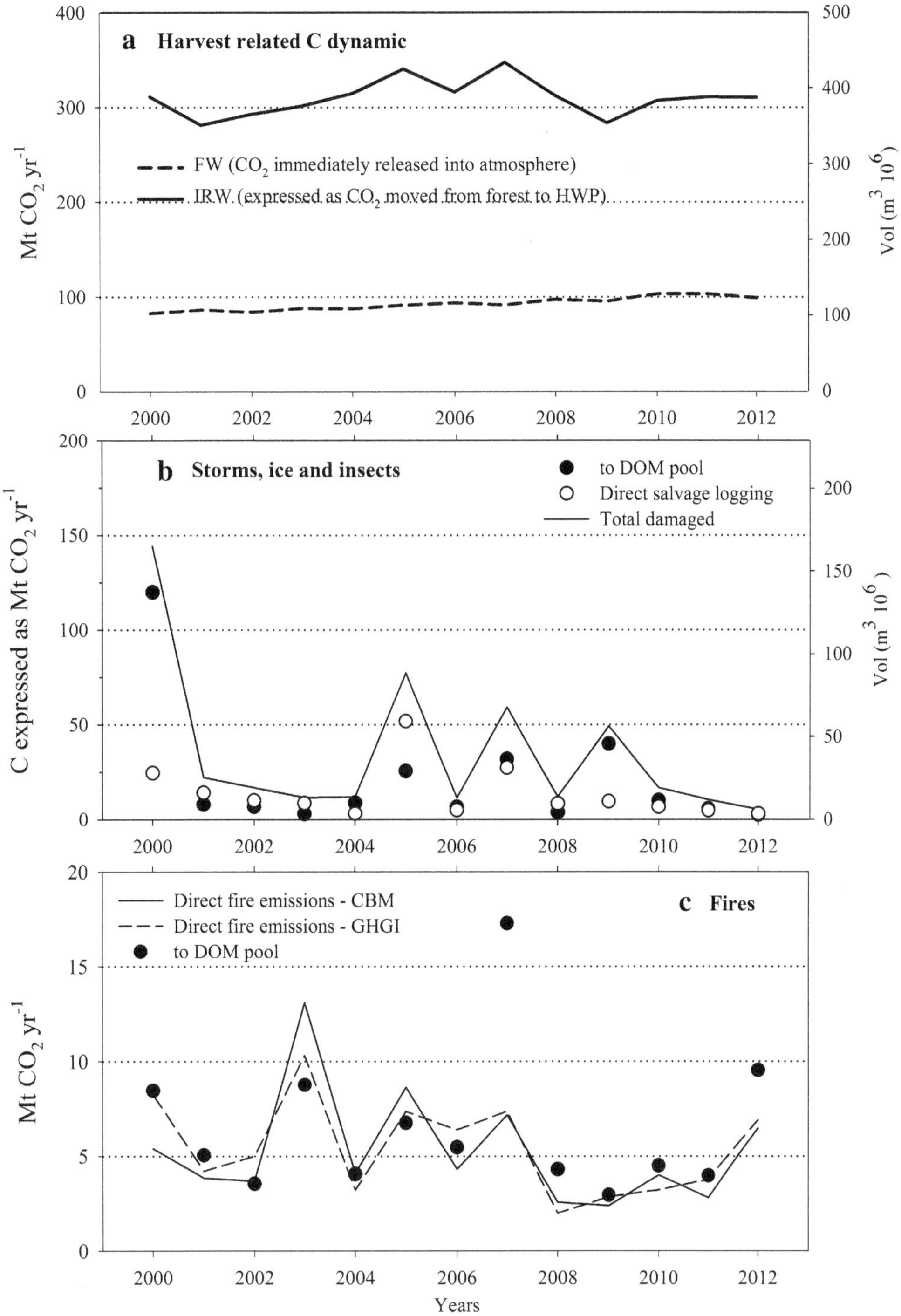


**Fig. 2** **a** Harvest-related C dynamics (in Mt $CO_2$ year$^{-1}$, reported on the *left axis*) and volume removed according to our model (in million m$^3$, on the *right axis*), in terms of fuelwood (FW) immediately released into the atmosphere and industrial roundwood (IRW) moved from forest to HWP; **b** indirect total $CO_2$ emissions (in Mt $CO_2$ year$^{-1}$, reported on the *left axis*) due to storm and insect attacks (reported as volume, in million m$^3$, on the *left axis*), further distinguished between the amount of biomass moved from living biomass to DOM pool (*black dots*) due to the disturbances and directly recovered as salvage logging (*white dots*); **c** $CO_2$ emissions (in Mt $CO_2$ year$^{-1}$) due to fires, distinguished between direct—i.e. immediately released into the atmosphere (according to our estimates and compared with the data reported in the EU GHGI [13])—and indirect emissions, i.e. moved from biomass to DOM pool (*black dots*), from where it will decompose in subsequent years

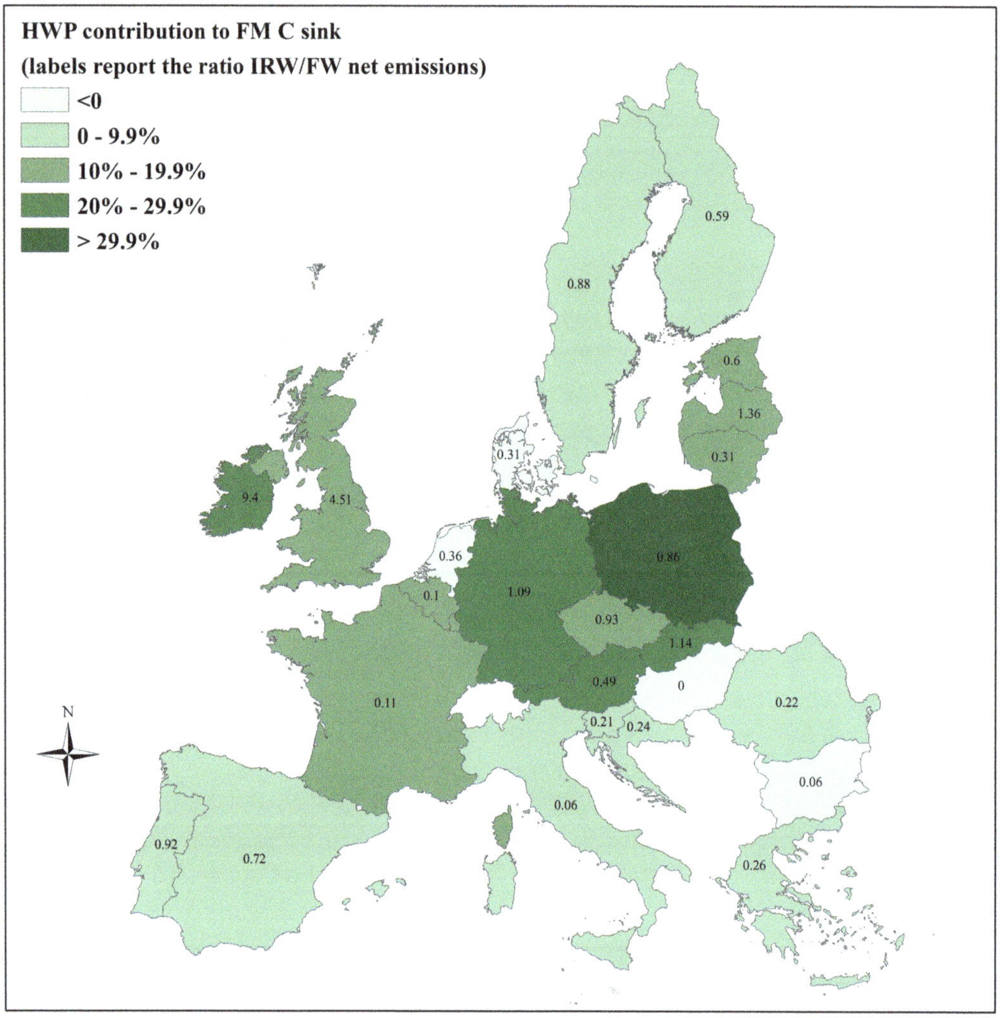


**Fig. 3** HWP mitigation contribution to the forest management (FM) C sink (*colors* as per legend). The numbers report the ratio between the industrial roundwood (IRW) C sink and the fuelwood (FW) direct emissions to the atmosphere

other factors. Taking into account all these factors and applying the same harvest rate used in the present study, Pilli et al. [30] have applied the IPCC Tier 2 production approach [11] to estimate the HWP mitigation potential at the EU level. The resulting net $CO_2$ sink in the HWP pool is equal on average to −44 Mt $CO_2$ year$^{-1}$ between 2000 and 2012. The ratio between the IRW C sink and the FW direct emissions to the atmosphere (on average 93 Mt $CO_2$ year$^{-1}$ between 2000 and 2012) is equal, on average, to 0.48 at EU level but it varies between countries, as highlighted by the labels in Fig. 3.

The sum of the net $CO_2$ emissions of the forest pools plus the HWP (as estimated by [30]) is, on average, equal to −409 Mt $CO_2$ year$^{-1}$ between 2000 and 2012 (Fig. 1, panel c). Our estimates on HWP at the EU level are very similar to the data submitted by the countries to the KP [52], and indicate that the HWP mitigation contribution is currently equal to about 10 % of the total forest net $CO_2$ emissions at the EU level. Pan et al. [53] estimated that at the global level, the C sequestration in HWP accounted for 8 % of the total C sink in established forests. As expected, at the EU level this percentage is higher and in five out of 26 countries the contribution of the HWP pool to the total FM C sink is >20 %, for the historical period 2000–2012 (see Fig. 3). However, because these estimates are based on the IPCC production approach, the C sink in HWP for countries with large exports is attributed to the country of harvest, i.e. where the wood originated and which may not be the country where the wood is in use.

As highlighted by Pilli et al. [30], in the future, the current HWP sink can be maintained either by further increasing (on average by 1 % per year) the current harvest, or by shifting more of the harvest to long-lived products [54]. In some countries, this contribution is negligible compared to the total forest C sink. In four countries where the IRW pool is a C source, the HWP pool has negative impacts on the overall C sink. This highlights the need to consider the specific national circumstances, when analyzing the possible contribution of the HWP C pool as potential mitigation tool.

### *Natural disturbances*

*Storms and ice* The overall C dynamics related to storms and ice are shown in Fig. 2, panel B. These disturbances do not produce any direct emission of $CO_2$ to the atmosphere, but they cause a transfer of C from the living biomass to both the DOM pool and the HWP pool (due to direct salvage logging). This process, reported in detail at the country level in Fig. 4, is simulated by the CBM through disturbance matrices for each disturbance type applied to each country (according to available information from the literature). Disturbance matrices quantify the proportion of C that is moved: (i) from the living biomass to DOM and (ii) from the living biomass to the HWP pool (see Fig. 5). Indeed, a consistent fraction of the living biomass damaged by these events is removed as salvage logging immediately after the disturbance (i.e., in the same year) or few years later (see [26, 55, 56]). The storm that occurred in 1999/2000 (reported as 2000 in Figs. 2, 4 and 5), with a total C stock transfer from the living biomass to DOM of about 150 Mt $CO_2$, caused opposing peaks in the living biomass and DOM pools (Fig. 1, panel a). The same effect is clearly reported for the other major disturbances (e.g., 2005, 2007 and 2009). On average, we estimated that, between 2000 and 2012, about 34 Mt $CO_2$ $year^{-1}$ were moved from the living biomass to DOM and to HWP due to the effects of storms; excluding the amount of biomass directly removed as salvage logging, this amount decreases to 21 Mt $CO_2$ $year^{-1}$.

The main storms at the EU level (the area affected by ice is negligible compared with storms) occurred in 1999/2000 and 2005 (see Figs. 2, 8, panel b in the Material). In the first case (which we count as 2000), the so called storms "Lothar" and "Martin", occurred on 27th–28th December 1999, and affected mainly France and Germany (Fig. 4). According to Gardiner et al. ([55], Appendix 3) between 184 and 204 $Mm^3$ were directly damaged by these events. Based on our model run, about 170 $Mm^3$ were damaged. In 2005, different storms affected many European countries, including "Gudrun" and "Erwin", which damaged about 75 million $m^3$ in northern Europe (see [55], Appendix 3).

Overall, between 2000 and 2012, about 272,000 ha $year^{-1}$ (based on the data collected by our study) were affected by storms in EU countries and, according to our estimates, on average 36.5 $Mm^3$ $year^{-1}$ were damaged. The direct salvage of storm residues simulated by CBM was equal, on average, to 13 $Mm^3$ $year^{-1}$ but with high interannual variability (see Fig. 5). Indeed, when large disturbances occur, we cannot expect that all the biomass affected by storms will be removed during the same year (this is the case for the two main disturbances that occurred in 1999–2000 and 2005). A fraction of this biomass will be removed during the following years (see for example the case of Germany in Fig. 5 between 2000 and 2003). A further amount of salvage logging may be recovered through the normal silvicultural practices (i.e., thinnings, clear-cuts, etc.) applied by the model at the country level but this amount cannot be directly estimated. On average, excluding the largest disturbance events (i.e., 2000 and 2005), we estimated that about 40 % of the total biomass damaged by storms (including branches and other wood components) was directly removed as salvage logging between 2000 and 2012.

*Fire* The second major disturbance type considered by our study is wildfire and slash burning (if relevant). Average annual direct emissions to the atmosphere due to the burning of biomass and dead organic matter are equal to 5.27 Mt $CO_2$ $year^{-1}$ between 2000 and 2012 (Fig. 2, panel c). As for the storms, strong inter-annual variations are related to the amount of area burned with peaks in 2003, 2005 and 2007, mainly concentrated on the Mediterranean countries (see Figs. 6, 8, panel c). In some cases (Slovak Republic and Lithuania), the emissions reported in Fig. 6 are mainly due to burning of harvesting residues.

Based on our assumptions, the total emissions due to fires are equal, on average, to 15 % of the total indirect losses due to the effect of storms. Indeed, while storms have a clear effect on the forest living biomass pool (Fig. 1, panel a), the effect of fires on this pool appears, at the EU level, much less evident. However, while storms do not contribute to direct $CO_2$ emissions and a considerable amount of wood is recovered through salvage logging, fires do cause direct emissions to the atmosphere and the amount of salvage logging varies considerably among countries. For example, in Portugal a substantial amount of wood may be recovered as salvage logging after fire disturbances, while this practice is negligible in Greece and Italy [26]. This explains the peak on transfers of living biomass to DOM reported in 2007 (see Fig. 2, panel c), when fires mainly occurred in Italy and Greece. Overall, our estimates are consistent with the total $CO_2$ emissions reported in the EU GHGI, except for few a years (e.g. 2000, 2003 and 2007). A full comparison between the CBM and the EU GHGI however is difficult because of the different assumptions and methodologies used by some MS (see [13, 14]).

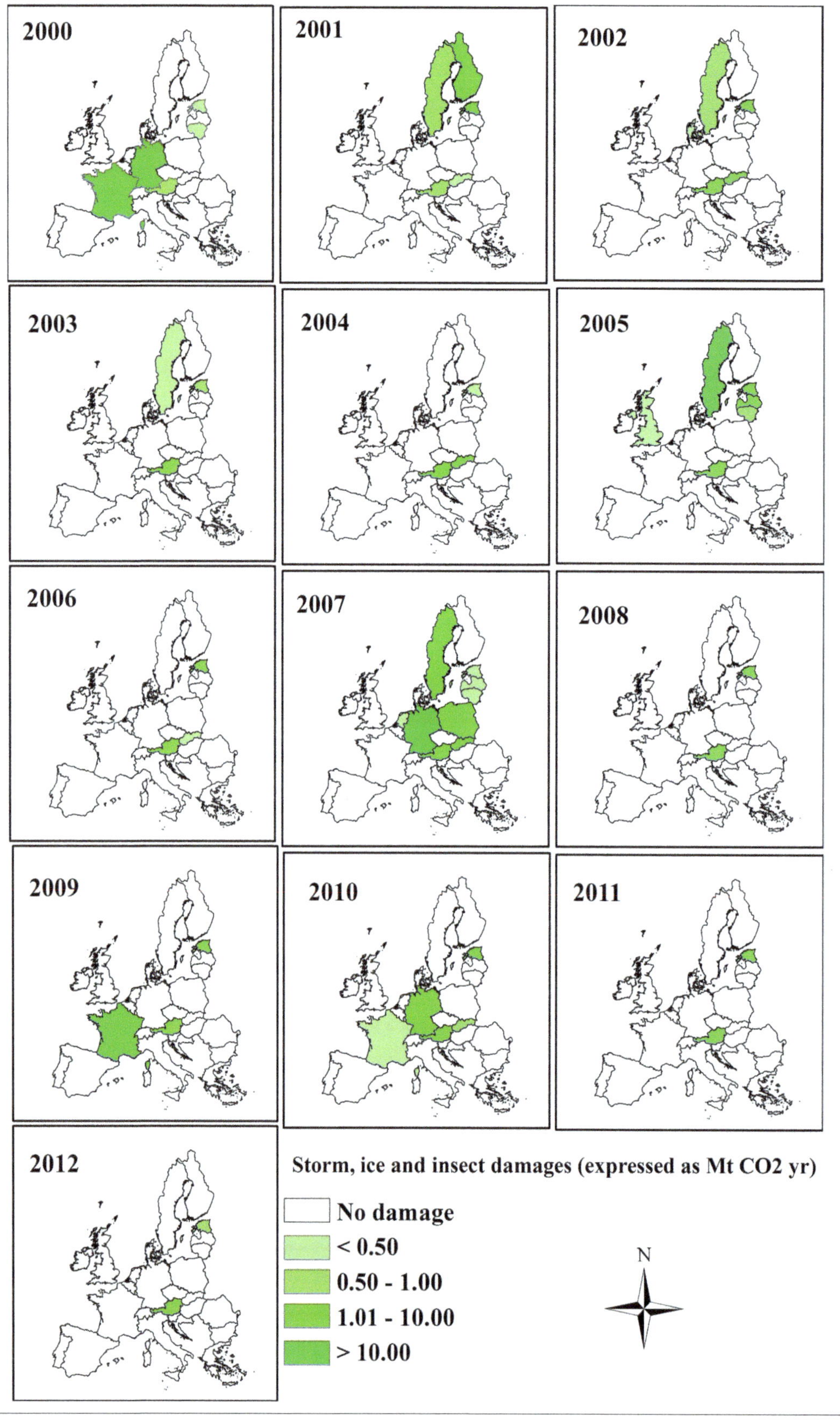
2000
2001
2002
2003
2004
2005
2006
2007
2008
2009
2010
2011
2012
Storm, ice and insect damages (expressed as Mt CO2 yr)
No damage
< 0.50
0.50 - 1.00
1.01 - 10.00
> 10.00
N

(See figure on previous page.)
**Fig. 4** The map reports, for each year and country, the amount of living biomass C stock (expressed as Mt $CO_2$ year$^{-1}$, even if these are not direct $CO_2$ emissions to the atmosphere) damaged by storm, ice and insect disturbances, as estimated by our model

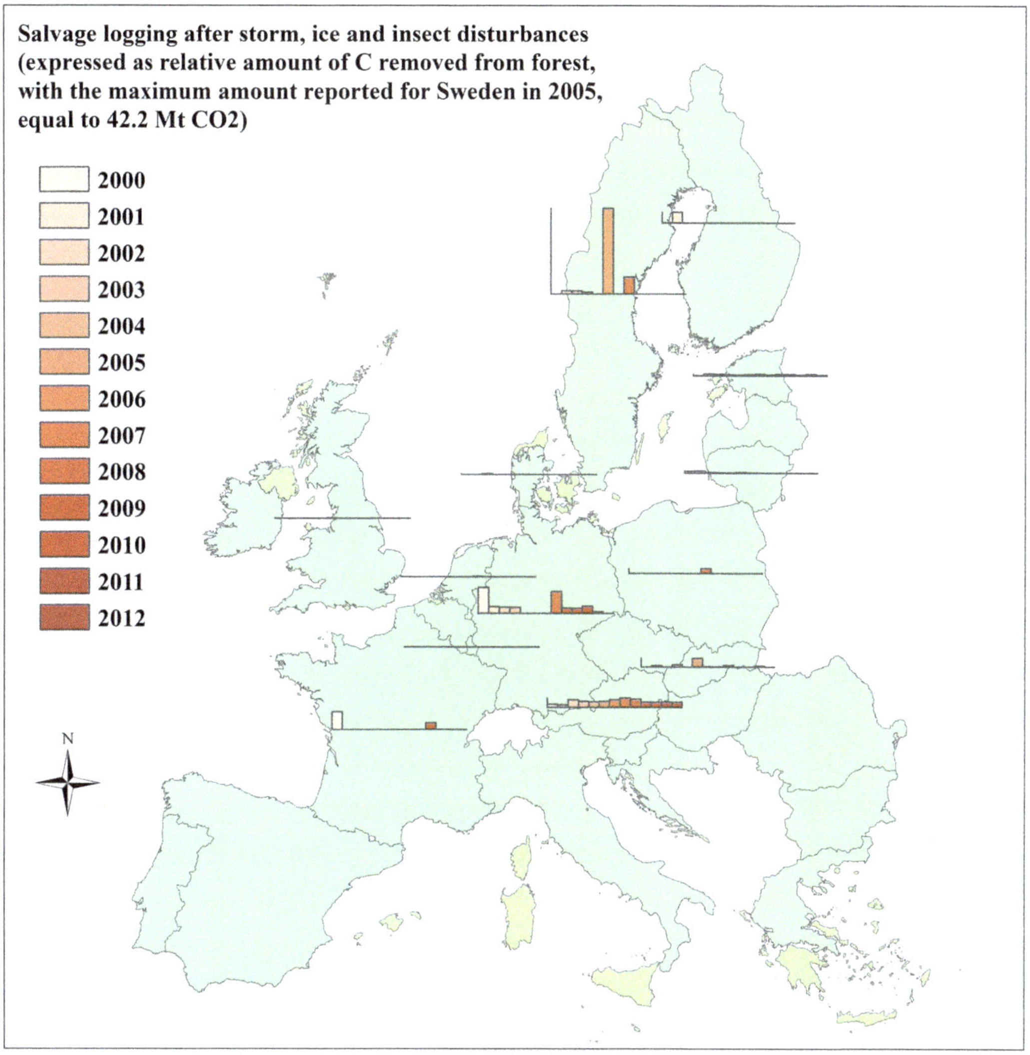


**Fig. 5** Salvage logging after storms, ice and insect disturbances, as estimated by the CBM between 2000 and 2012, reported as relative amount of C removed from forest, with the maximum amount reported for Sweden (2005) equal to 42.2 Mt $CO_2$

## Conclusions

This paper analyzed the CBM results at the EU level, in terms of C emissions and removals for all forest activities foreseen by the Kyoto Protocol: forest management (i.e. land in the forest land-use category in 1989), afforestation/reforestation and deforestation (, i.e. forest related land-use changes since 1990). We considered all forest carbon pools (including HWP for FM) and we analyzed the main drivers of forest C stock and sink dynamics over the period 2000–2012 (i.e., harvest, natural disturbances and land-use changes).

Overall, the sink estimated by the CBM from FM, AR and D during 2000–2012 corresponds to about 7 % of total GHGs at the EU level for the same period. The sink from the HWP pool contributes an additional 1 %. The CBM results for FM, AR and D are very close to the reported values from the EU GHGI, i.e. 8 % lower for FM (in the CBM vs. GHGI) and almost identical for AR and D.

In absolute terms, the impact of harvest is much greater than natural disturbances but, because of salvage logging, the impact of natural disturbances is often not

2000
2001
2002
2003
2004
2005
2006
2007
2008
2009
2010
2011
2012
Fires emissions (Mt CO2 yr)
No fire emissions
<0.0100
0.0101 - 0.0500
0.0501 - 0.1000
> 0.1000
N

(See figure on previous page.)
**Fig. 6** The map reports, for each year and country, the direct $CO_2$ emissions to the atmosphere (in Mt $CO_2$ year$^{-1}$) due to fire disturbances, as estimated by our model. Where no direct fire emissions is reported, they were assumed as negligible, according to the information reported by each country on the National Inventory Reports submitted to UNFCCC

easily distinguishable from the impact of harvest. Over the period analyzed in our study, the sum of these two drivers often caused year-by-year variations of about 10–15 % of the FM sink at the EU level. Between 2000 and 2012, land-use changes are also important: the annual sink from AR reached 15 % of the FM sink, while annual emissions from D were on average equal to nearly 10 % of the FM sink.

Our analysis on natural disturbances showed that at the EU level the impact of storms on the C stock balance is quantitatively far more important than the impact of fires, i.e. 5–10 times greater when both direct and indirect emissions are considered. While other studies quantified the impact of these events on the European forest carbon balance (e.g. [23, 57]), the detailed analysis of the C dynamics of natural disturbances presented in this study is useful both to identify climate change mitigation options and to allow better process understanding for the purpose of accounting emissions and removals under the KP and the post-2020 EU regulations. In particular, while storms cause a transfer of C from biomass to DOM pools (from where it will decompose unless salvage logged), fires cause an approximately equal amount of direct emission (of $CO_2$ to the atmosphere) and indirect emission (i.e. C moved from living biomass to DOM from where it will decompose). A consistent fraction of the total amount of harvest may be provided by salvage logging after disturbance events, and in several cases the impact of natural disturbances is not visible in the GHGIs either because the dynamics described above are not taken into account (e.g. Italy, [44]) or because the methods used to estimate GHGs do not detect interannual variations (e.g. Germany, [26]). In the light of the rules under the KP [11], our results suggest that: (i) different management strategies (e.g., salvage logging) applied after natural disturbances may significantly affect the total C balance, and (ii) modelling the detailed C dynamics associated to natural disturbances is an essential prerequisite to apply the provision of natural disturbances under the KP (i.e. excluding the corresponding emissions and subsequent removals from the accounting under certain circumstances).

In conclusion, this study presents the application of a consistent methodological approach, based on an inventory-based model, adapted to the forest management conditions of 26 different EU countries. The approach captures, with satisfactory detail, the C sink reported in the EU GHGI and the country-specific variability due to harvest, natural disturbances and land-use changes. These results provide the basis for possible policy-relevant future applications of the CBM, e.g., as a tool to support GHGIs (e.g. on accounting for natural disturbances) and to verify GHGIs, and for the simulation of specific scenarios at EU level. Applying the same model from a regional [58], to country [26, 44] and at the EU level (this study), may help a consistent assessment of different forest sector mitigation strategies appropriate to the specific regional circumstances, and in evaluating the overall contribution of the forest sector towards EU emissions reduction targets.

To our knowledge, this is the most comprehensive study of its kind at EU level. Indeed, even if the EU forest C sink was previously analyzed by many other studies, none of them provided an overall assessment disaggregated between FM, AR and D, including all the forest pools, HWP and natural disturbances, and a comparison with the EU GHGI. Such comprehensive assessments and comparisons are increasingly needed to help improving the quality of GHGIs and ultimately increasing the credibility of the forest sink as a potential mitigation option within the EU [27] and the international frameworks [10].

As part of an integrated modeling framework, further possible developments include linking the CBM to models predicting land-use changes, the impact of climate change on primary productivity, the harvest demand and the material substitution effects of the industrial roundwood products.

## Methods

This study used the Carbon Budget Model (CBM, [25]) to estimate forest C stock and the net $CO_2$ emissions in 26 EU countries, covering a total forest area of about 146 Mha, disaggregated in 189 forest types (FTs) distributed over 178 administrative regions and 35 climatic units. The area included about 138 Mha of FM (at time step zero of the model runs) and, in 2012, 8 Mha of AR, and 2.8 Mha of D, since 1990. Due to the D area, the area of FM slightly decreased during the model runs, but this area decrease was compensated by an increase in AR area. Unproductive forests (according to countries' GHGIs) and overseas territories were not included in this study [26].

The spatial framework applied by the CBM conceptually follows IPCC Reporting Method 1 [11] in which the spatial units are defined by their geographic boundaries and all forest stands are geographically referenced to a spatial unit (SPU). For FM, the Carbon Budget Model and the general methodological assumptions applied to each country were described previously [26]. Further details can be found elsewhere for the model itself (e.g. [25]), for its applications to European countries [26, 44] and for the representation of natural disturbances [26]. A summary of the main NFI input data used for each country is reported in Table 2. We considered 26 administrative units (i.e., European countries, as reported by Table 2) and 35 climatic units (CLUs, as defined by [59]), with mean annual temperatures (mainly affecting the DOM turnover rate), ranging from −7.5 to +17.5 °C.

Within a SPU, each forest stand is characterized by age, area and seven classifiers that provide administrative and ecological information, the link to the appropriate yield curves, and parameters defining the silvicultural system such as the forest composition (defined according to different FTs), the management type (MT), and the main use of the harvest provided by each SPU, as fuelwood or industrial roundwood. For each country, these parameters were mainly derived by national forest inventories. According to country-specific information, MTs may

**Table 2 Summary of the main parameters applied by CBM model for each country**

| Country | Original NFI year | Time step 0 (years) | CBM FM area (Mha)[b] | Harvest rate (av. 2000–2012, Mm3) | County specific biomass equations |
|---|---|---|---|---|---|
| Austria | 2008 | 1998 | 3.2 | 22.9 | X |
| Belgium | 1999 | 1999 | 0.7 | 4.3 | |
| Bulgaria | 2000 | 2000 | 3.2 | 5.3 | |
| Croatia | 2006[a] | 1996 | 2.0 | 4.6 | |
| Czech Republic | 2000 | 2000 | 2.6 | 17.0 | X |
| Denmark | 2004 | 1994 | 0.5 | 2.3 | |
| Estonia | 2000 | 2000 | 2.1 | 7.9 | |
| Finland | 1999 | 1999 | 21.7 | 55.0 | |
| France | 2008 | 1998 | 14.6 | 54.9 | |
| Germany | 2002 | 1992 | 10.6 | 74.7 | X |
| Greece | 1992[a] | 1992 | 1.2 | 1.6 | |
| Hungary | 2008 | 1998 | 1.6 | 6.2 | X |
| Ireland | 2005 | 1995 | 0.5 | 2.8 | |
| Italy | 2005 | 1995 | 7.4 | 10.2 | X |
| Latvia | 2009 | 1999 | 3.2 | 15.8 | X |
| Lithuania | 2006 | 1996 | 2.0 | 7.7 | |
| Luxembourg | 1999 | 1999 | 0.1 | 0.3 | |
| Netherlands | 1997 | 1997 | 0.3 | 1.2 | |
| Poland | 1993 | 1993 | 8.9 | 37.8 | |
| Portugal | 2005 | 1995 | 3.6 | 12.2 | X |
| Romania | 1985 | 1985 | 6.6 | 17.2 | X |
| Slovakia | 2000 | 2000 | 1.9 | 9.0 | |
| Slovenia | 2000 | 2000 | 1.1 | 3.3 | |
| Spain | 2002 | 1992 | 12.6 | 16.8 | |
| Sweden | 2006 | 1996 | 22.6 | 79.5 | |
| United Kingdom | 1997 | 1997 | 2.5 | 9.8 | |
| EU | | | 137.9 | 480.7 | 8 countries |

The table reports the NFI original reference year; the year since the model was applied; the FM area used by CBM at time step 0; the average harvest rate used; the countries where specific equations to convert the merchantable volume into aboveground biomass were selected. Two countries were not modeled: Cyprus (no NFI data available) and Malta (very small forest area, mainly covered by shrub lands)

[a] Analysis based on data from forest management plans

[b] FM area used by CBM at time step 0. According to KP rules, FM is the area of forest in 1990, decreased with any subsequent deforestation. The FM area is taken from the official submissions made by countries to UNFCCC/Kyoto Protocol, giving priority to data from KP-CRF tables when available (i.e., if FM had been elected during the first KP commitment period), or alternatively taking data from the convention CRF tables (using 'forest land remaining forest land' in 1990 as a proxy for FM). To obtain FM area at time step 0, the D area reported by all countries under the Kyoto Protocol was used. Please note that CBM runs did not include forests reported as "not productive" (e.g., 0.4 Mha in Austria, 0.02 Mha in Bulgaria, 5 Mha in Sweden) and overseas territories (8.2 Mha in France)

include even-aged high forests, uneven-aged high forests, coppices and specific sylvicutural systems such as clear-cuts (with different rotation lengths for each FT), thinnings, shelterwood systems, partial cuttings, etc. (detailed information on the main model's assumptions for five representative countries are reported in the Supplementary Information of [26]).

To assess the FM area, data from KP reporting were used when available (for 18 countries in the period 2008–2012); alternatively "forest land remaining forest land" data (from UNFCCC reporting) were used, i.e. for all countries before 2008 and for those counties that did not elect FM during 2008–2012. Countries' data for AR and D for 2008–2012 always came from KP reporting.

The following carbon pools were considered: living biomass (aboveground and belowground), dead organic matter (DOM), mineral soil and harvested wood products (HWP). Even if the CBM estimates $CH_4$ and $N_2O$ emissions, this study includes only $CO_2$ and excludes organic soils.

In the CBM, species-specific, stand-level equations [60] convert merchantable volume production into aboveground biomass, partitioned into merchantable stemwood, other (tops, branches, sub-merchantable size trees) and foliage components. Where additional information provided by NFIs or by literature was available (see last column in Table 2), country-specific equations were selected to convert the merchantable volume into aboveground biomass.

The CBM starts the initialization process with all DOM pools containing zero C stocks and then simulates multiple iterations of growth and stand-replacing disturbances, gradually increasing the size of the DOM pools [25]. The rotations continue until the slowly-decaying C pools at the end of two successive rotations meet a difference tolerance of 0.1 %. Once this criterion has been met, the CBM applies a user-selected last disturbance event which affects the amount of C in the DOM pools, and then links DOM dynamics to biomass dynamics. Inputs from biomass to DOM pools, during the model run, result from biomass litterfall and turnover as well as natural and human-caused disturbances. The DOM parameters were first calibrated and validated on some specific study at country and regional level [44, 58] and, if necessary, further modified for specific countries [26].

We use two sets of yield tables (YT) in these analyses [26, 44]. Historical YTs derived from the standing volumes per age class reported by the NFI represent the impacts of growth and partial disturbances during stand development. Current YTs derived from the current annual increment reported in country NFIs represent the stand-level volume accumulation in the absence of natural disturbances and management practices.

For 21 countries, we also evaluated the impact of natural disturbance events (a summary at EU level is reported in Fig. 8, below), including storms and ice (15 countries), fires (11 countries) and insect attacks (i.e., bark beetles attacks, for 2 countries). Specific information on the assumptions on natural disturbances are reported by [26],

The following sections provide specific information on the application of CBM to areas affected by land-use changes (AR, D lands) and a summary, at the EU level, of the main assumptions on harvest applied by our study.

**AR and D assumptions**

We used for the analyses of AR and D areas the same 26 administrative units (i.e., European countries) and 35 CLUs applied for FM. For both AR and D, we used the area reported up to 2012 by each country in the GHGIs submitted the KP (Fig. 7). Before 2008, only the cumulative values since 1990 are available [13] and therefore we used an average annual rate of AR and D for the period 1990–2007.

AR was modelled through country-specific model runs, always starting in 1990. The total amount of AR per year was distributed between different FTs using the same proportions of FTs observed in the FM area. Based on a preliminary model assessment, we generally used the current YT library for AR [61]. This library was derived from the increment data reported by each country. These values represent the gross volume yield of each stand (while the YTs derived by the standing volume include the impact of past silvicultural treatment) and therefore are more suitable for young stands, generally younger than 20 years (i.e. for AR), where in general no silvicultural treatments are applied.

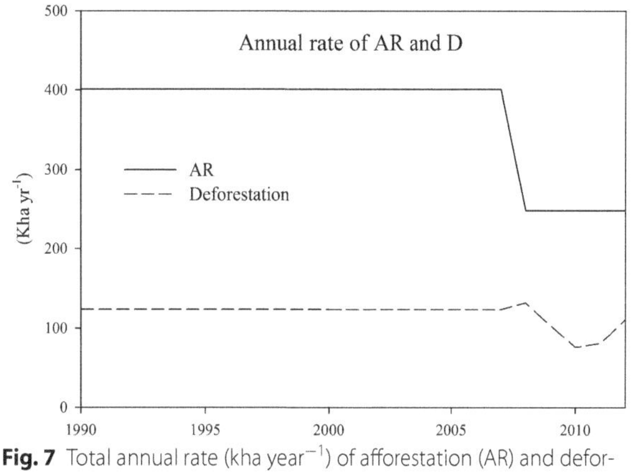


**Fig. 7** Total annual rate (kha year$^{-1}$) of afforestation (AR) and deforestation (D), used by CBM at the EU level, based on countries' GHGIs under the Kyoto Protocol. Data from 1990 to 2007 are the cumulative area of AR and D in 2008 converted to annual average rates of land-use change

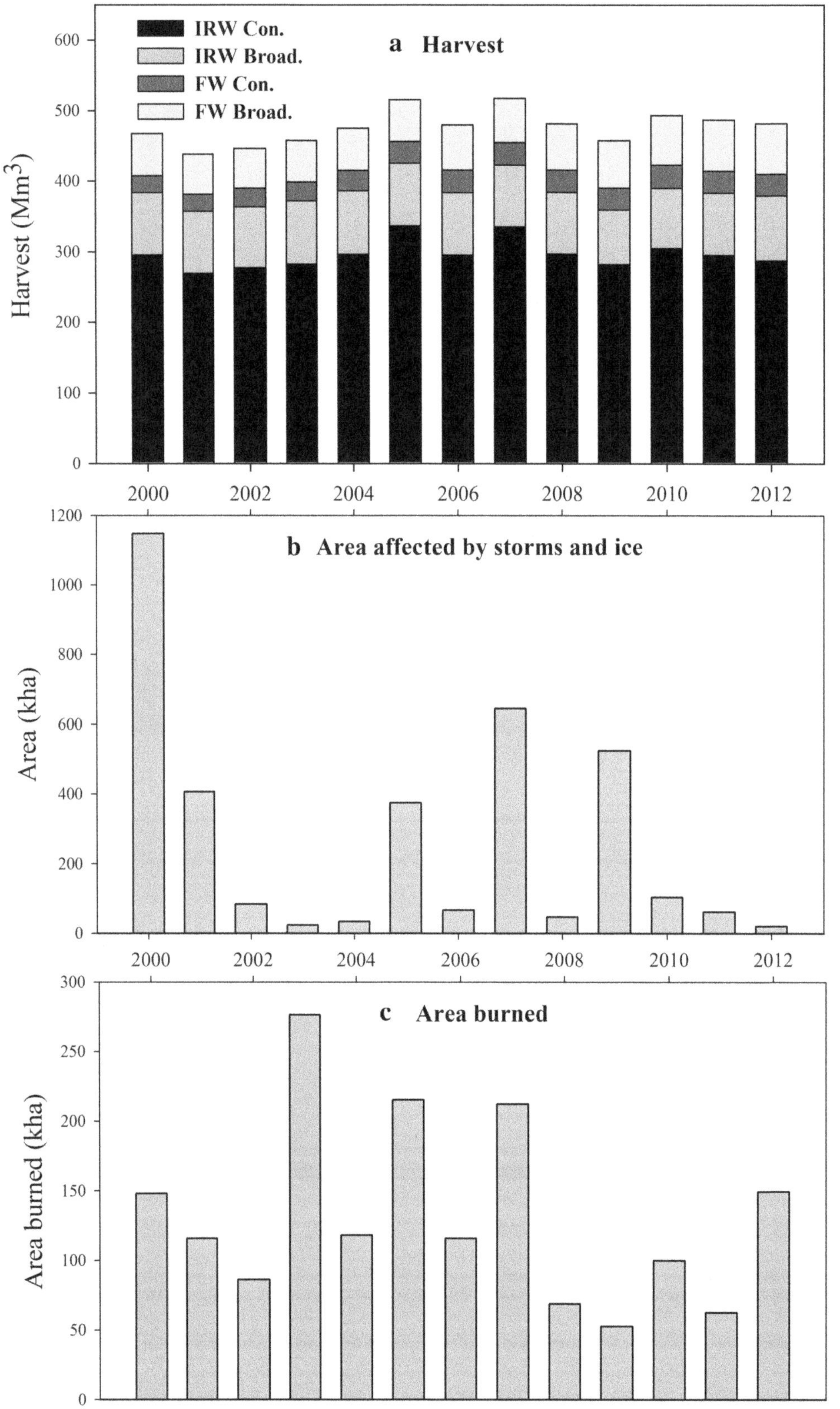


**Fig. 8** **a** Total harvest demand (in millions m$^3$) applied by CBM at the EU level, further distinguished between industrial roundwood (IRW) and fuelwood (FW) and between coniferous and broadleaved species; **b** total area affected by storms and ice (in kha); **c** total area affected by fire (in kha)

We assumed that the harvest rate was entirely satisfied by the FM area (with Portugal the only exception, see [26]), apart from a modest amount of harvest provided by D. The possible amount of harvest provided by AR (i.e. post-1990 forest) was generally very small [61].

For AR we estimated the maximum potential (and theoretical) amount of harvest provided assuming a common set of silvicultural practices for all the countries, with a 15 % commercial thinning applied to broadleaved forests 15-years or older and a 20 % commercial thinning applied to coniferous forests 20-years or older [61]. The only case where the harvest from AR was relevant was Portugal.

## EU summary of the main input data

A summary, at the EU level, of the main input data applied by our study is reported in Fig. 8, including: the total harvest rate applied by CBM at the EU level, further distinguished between industrial roundwood (IRW) and fuelwood (FW) and between coniferous and broadleaved species (panel a); total area affected by storms and ice (panel b); total area affected by fire (panel c). The harvest rate was mainly derived by a specific study on the HWP at EU level [30] where, for each country, FAOSTAT data (further distinguished between IRW and FW) were collected, compared with other data sources (i.e., Forest Resource Assessment, National Inventory Report, NFIs, ecc.) and eventually corrected in order to account for possible inconsistencies (i.e., due to the bark fraction or different methodological approaches).

Since the amount of IRW used in the present study is consistent with [30], we derived the IRW C sink by this last study, based on the IPCC production approach [11]. In this method, estimates of net emissions are derived from a stock change calculations applied to products derived from domestic harvest, i.e., imported HWP are excluded in the national estimates.

### Abbreviations

AR: afforestation and reforestation; C: carbon; CBM: Carbon Budget Model; CP1: First Commitment Period; CP2: Second Commitment Period; D: deforestation; Dbh: diameter at breast height; DOM: dead organic matter; EU: European Union; FM: forest management; FT: forest type; FW: fuelwood; GHG: greenhouse gas; GHGI: greenhouse gas inventory; HWP: harvested wood product; IPCC: Intergovernmental Panel on Climate Change; IRW: industrial Roundwood; KP: Kyoto Protocol; LULUCF: land Use, Land-Use Change and Forestry; MT: management type; NFI: National Forest Inventory; NIR: National Inventory Report; UNFCCC: United Nations Framework Convention on Climate Change; YT: yield table.

### Authors' contributions

RP carried out the data analysis, in collaboration with JWM. GG and WAK helped in the design of the study and the interpretation of results and together with RP wrote the manuscript, in collaboration with RAV. All authors read and approved the final manuscript.

### Author details

[1] European Commission, Joint Research Centre, Directorate D – Sustainable Resources - Bio-Economy Unit, Via E. Fermi 2749, 21027 Ispra, VA, Italy. [2] Natural Resources Canada, Canadian Forest Service, Victoria, BC V8Z 1M5, Canada. [3] Department of Agriculture, Forest and Food Sciences, University of Torino, Via Leonardo Da Vinci 44, 10095 Grugliasco, TO, Italy.

### Acknowledgements

This paper was prepared in the context of the Contract n. 31502, Administrative Arrangement 070307/2009/539525/AA/C5 between JRC and DG CLIMA. Further information was collected in the context of the AA 071201/2011/611111/CLIMA.A2. The analysis performed for each country was in general based on data publicly available and on additional information collected at country level, in collaboration with many colleagues and experts for each country. We especially thank our colleagues, Giulia Fiorese, Viorel Blujdea and Tibor Priwitzer, who provided useful comments and suggestions.

We also thank four anonymous reviewers, who provided useful comments and suggestions to improve the manuscript.

The views expressed are purely those of the authors and may not in any circumstances be regarded as stating an official position of the European Commission or Natural Resources Canada.

### Competing interests

The authors declare that they have no competing interests.

### References

1. Alkama R, Cescatti A. Biophysical climate impacts of recent changes in global forest cover. Science. 2016;351:600–4.
2. Naudts K, Chen Y, McGrath M, Ryder J, Valade A, Otto J, Lussayert S. Europe's forest management did not mitigate climate warming. Science. 2016;351:597–600.
3. Lundmark T, Bergh J, Hofer P, Lundström A, Nordin A, Poudel BC, Sathre R, Taverna R, Werner F. Potential roles of Swedish forestry in the context of climate change mitigation. Forests. 2014;5:557–78.
4. URL (last access April 2016). http://science.sciencemag.org/content/351/6273/597.e-letters.
5. Knauf M, Köhl M, Mues V, Olschofsky K, Frühwald A. Modeling the CO2-effects of forest management and wood usage on a regional basis. Carbon Balance Manag. 2015;10:13.
6. Nabuurs G-J, Delacorte P, Ellison D, Hanewinkel M, Lindner M, Nesbit M, Ollikainen M, Savaresi A. A new role for forests and the forest sector in the EU post-2020 climate targets. From Science to Policy 2. European Forest Institute; 2015.
7. Mubareka S, Jonsson R, Rinaldi F, Fiorese G, San Miguel J, Sallnas O, Baranzelli C, Pilli R, Lavalle C, Kitous A. An integrated modelling framework for the forest-Based bioeconomy. Earthzine, 2014. URL (last access March 2015). http://earthzine.org/2015/01/19/an-integrated-modelling-framework-for-the-forest-based-bioeconomy/.
8. UNFCCC. Decision 2/CMP.7 Land use, land-use change and forestry, 2012. URL (last access March 2016). http://unfccc.int/resource/docs/2011/cmp7/eng/10a01.
9. Grassi G, den Elzen MGJ, Hof AF, Pilli R, Federici S. The role of the land use, land use change and forestry sector in achieving annex I reduction pledges. Clim Change. 2012;115(3–4):873–81.
10. UNFCCC. Adoption of the Paris Agreement. Report No. FCCC/CP/2015/L.9/Rev.1, 20155. URL (last access March 2016): http://unfccc.int/resource/docs/2015/cop21/eng/l09r01.pdf. (UNFCCC, 2015).
11. IPCC. In: Hiraishi T, Krug T, Tanabe K, Srivastava N, Baasansuren J, Fukuda M, Troxler TG, editors. Intergovernmental panel on climate change. Revised supplementary methods and good practice guidance arising from the Kyoto Protocol, 2013. Switzerland; 2014.
12. Proposal for a Regulation of the European Parliament and the Council on the inclusion of greenhouse gas emissions and removals from land use, land use change and forestry into the 2030 climate and energy

framework. URL (last access July 2016). http://ec.europa.eu/transparency/regdoc/rep/1/2016/EN/1-2016-479-EN-F1-1.PDF.
13. EU NIR. Annual European Community greenhouse gas inventory 1990–2012 and inventory report 2014. Submission to the UNFCCC Secretariat. European Environment Agency, Technical report No 09/2014. URL (last access February 2016). https://www.google.it/search?q=Annual+European+Community+greenhouse+gas+inventory+1990%E2%80%932012+and+inventory+report+2014&ie=utf-8&oe=utf-8&gws_rd=cr&ei=fwi7VoSoJIXAOoCTroAF.
14. Neumann M, Moreno A, Mues V, Härkönen S, Mura M, Bouriaud O, Lang M, Achten MJ, Thivolle-Cazat A, Broisz K, Merganič J, Decuyper M, Alberdi I, Astrup R, Mohren F, Hasenauer H. Comparison of carbon estimation methods for European forests. For Ecol Man. 2016;361:397–420.
15. Blujdea VNB, Abadviñas R, Federici S, Grassi G. The EU greenhouse gas inventory for the LULUCF sector: I. Overview and comparative analysis of methods used by EU member states. Carbon Manag. 2016;6:5–6.
16. Ciais P, Schelhaas MJ, Zaehle S, Piao SL, Cescatti A, Liski J, Luyssaert S, Le-Maire G, Schulze E-D, Bouriaud O, Freibauer A, Valentini R, Nabuurs GJ. Carbon accumulation in European forests. Nat Geosci. 2008;1:425–92.
17. Karjalainen T, Pussinen A, Liski J, Nabuurs G-N, Eggers T, Lapveteläinen T, Kaipainen T. Scenario analysis of the impacts of forest management and climate change on the European forest sector carbon budget. For Policy Econ. 2003;5:141–55.
18. Pretzsch H, Grote R, Reineking B, Rötzer TH, Seifert ST. Models for forest ecosystem management: a European perspective. Ann Bot-London. 2008;101:1065–87.
19. Nabuurs GJ, Schelhaas MJ, Pussinen A. Validation of the European forest information scenario model (EFISCEN) and a projection of Finnish forests. Silva Fenn. 2000;34(2):167–79.
20. Böttcher H, Kurz WA, Freibauer A. Accounting of forest carbon sink and sources under a future climate protocol-factoring out past disturbance and management effects on age-class structure. For Ecol Man. 2008;11:669–86.
21. Ťupek B, Zanchi G, Verkerk G, Churkina G, Viovy N, Hughes J, Lindner M. A comparison of alternative modelling approaches to evaluate the European forest carbon fluxes. For Ecol Man. 2010;260:241–51.
22. Groen T, Verkerk PJ, Böttcher H, Grassi G, Cienciala E, Black K, Fortin M, Köthke M, Lehtonen A, Nabuurs G-J, Petrova L, Blujdea V. What causes differences between national estimates of forest management carbon emissions and removals compared to estimates of large-scale models? Environ Sci Policy. 2013;33:222–32.
23. Seidl R, Schelhaas MJ, Rammer W, Verkerk PJ. Increasing forest disturbances in Europe and their impact on carbon storage. Nat Climate Change. 2014;4:806–10.
24. Verkerk PJ, Anttila P, Eggers J, Lindner J, Asikainen A. The realizable potential supply of woody biomass from forests in the European Union. For Ecol Man. 2011;261:2007–15.
25. Kurz WA, Dymond CC, White TM, Stinson G, Shaw CH, Rampley G, Smyth C, Simpson BN, Neilson E, Trofymow JA, Metsaranta J, Apps MJ. CBM-CFS3: a model of carbon-dynamics in forestry and land-use change implementing IPCC standards. Ecol Model. 2009;220:480–504.
26. Pilli R, Grassi G, Kurz WA, Viñas RA, Guerrero N. Modelling forest carbon stock changes as affected by harvest and natural disturbances. I. Comparison of model results for forest management with EU countries' estimates. Carbon Balance Manag. 2016;11:5.
27. Climate action, consultation on addressing greenhouse gas emissions from agriculture and LULUCF in the context of the 2030 EU climate and energy framework, 2016. URL (last access July 2016). http://ec.europa.eu/clima/consultations/articles/0026_en.htm.
28. Gardi C, Jeffery S, Saltelli A. An Estimate of potential threats levels to soil biodiversity in EU. Glob Change Biol. 2013;19:1538–48.
29. Panagos P, Hiederer R, Van Liedekerke M, Bampa F. Estimating soil organic carbon in Europe based on data collected through an European network. Ecol Ind. 2013;24:439–50.
30. Pilli R, Fiorese G, Grassi G. EU mitigation potential of harvested wood products. Carbon Balance Manag. 2015;10:6.
31. FAO. Global forest resources assessment 2010. Main Report. FAO Forestry Paper, 2010, 163. URL (last access February 2016). http://www.fao.org/docrep/013/i1757e/i1757e.pdf.
32. Forest Europe, UNECE and FAO. State of Europe's Forests 2015. Status and Trends in Sustainable Forest Management in Europe, 2015. URL (last access March 2016). http://www.foresteurope.org/docs/fullsoef2015.pdf.
33. Didion M, Blujdea V, Grassi G, Hernández L, Jandl R, Kriiska K, Lehtonen A, Saint-André L. Models for reporting forest litter and soil C pools in national greenhouse gas inventories: methodological considerations and requirements. Carbon Manag. 2016. doi:10.1080/17583004.2016.1166457.
34. Verkerk PJ, Lindner M, Zanchi G, Zudin S. Assessing impacts of intensified biomass removal on deadwood in European forests. Ecol Indic. 2009;11:27–35.
35. Baritz R, Seufert G, Montanarella L, Van Ranst E. Carbon concentrations and stocks in forest soils of Europe. For Ecol Man. 2010;260:262–77.
36. Liski J, Perruchoud D, Karjalainen T. Increasing carbon stocks in the forest soils of Western Europe. For Ecol Man. 2002;169:163–79.
37. Goodale CL, Apps MJ, Birdsey RA, Field CB, Heath LS, Houghton RA, Jenkins JC, Kohlmaier GH, Kurz WA, Liu SR, Nabuurs GJ, Nilsson S, Shvidenko AZ. Forest carbon sinks in the Northern Hemisphere. Ecol Appl. 2002;12:891–9.
38. Lal R. Forest soils and carbon sequestration. For Ecol Man. 2005;2020(1–3):242–58.
39. Achat DL, Fortin M, Landmann G, Ringeval B, Augusto L. Forest soil carbon is threatened by intensive biomass harvesting. Sci Rep. 2015;5:15991.
40. Schulp C, Nabuurs G-J, Verburg PH, de Waal R. Effect of tree species on carbon stocks in forest floor and mineral soil and implications for soil carbon inventories. For Ecol Man. 2008;256(3):482–90.
41. Gasparini P, Di Cosmo L. Forest carbon in Italian forests: stocks, inherent variability and predictability using NFI data. For Ecol Man. 2015;337:186–95.
42. Ortiz C, Liski J, Gärdenäs A, Lehtonen A, Lundblad M, Stendahl J, Ågren G, Karltun E. Soil organic carbon stock changes in Swedish forest soils—a comparison of uncertainties and their sources through a national inventory and two simulation models. Ecol Model. 2013;251:221–31.
43. UNECE/FAO. The European forest sector outlook study II. 2010–2030. United Nations, Geneva, 2011. URL (last access February 2016). https://www.unece.org/fileadmin/DAM/timber/publications/sp-28.pdf.
44. Pilli R, Grassi G, Kurz WA, Smyth CE, Bluydea V. Application of the CBM-CFS3 model to estimate Italy's forest carbon budget, 1995–2020. Ecol Model. 2013;266:144–71.
45. Böttcher H, Verkerk PJ, Gusti M, Havlík P, Grassi G. Projection of the future EU forest CO2 sink as affected by recent bioenergy policies using two advanced forest management models. GCB Bioenergy. 2012;4(6):773–83.
46. European Union. EU Reference Scenario 2016. Energy, transport and GHG emissions. Trends to 2050. European Union; 2016.
47. Paul KI, Polglase PJ, Richard GP. Predicted change in soil carbon following afforestation or reforestation, and analysis of controlling factors by linking a C accounting model (CAMFor) to models of forest growth (3PG), litter decomposition (GENDEC) and soil C turnover (RothC). For Ecol Manag. 2003;177:485–501.
48. KP CRF Tables, Submission 2014. URL (last access June 2014). http://unfccc.int/national_reports/annex_i_ghg_inventories/national_inventories_submissions/items/7383.php.
49. Nabuurs G-J, Lindner M, Verkerk PJ, Gunia K, Deda P, Michalak R, Grassi G. First signs of carbon sink saturation in European forest biomass. Nat Climate Change. 2013;3:792–6.
50. UNFCCC, decision 2/CMP.7 on land use land use change and forestry, 2011. URL (last access February 2016). http://unfccc.int/resource/docs/2011/cmp7/eng/10a01.pdf.
51. Rüter S. Projections of net-emissions from harvested wood products in European Countries. Work report of the institute of wood technology and wood biology, Johann Heinrich von Thünen-Institute (vTI), Hamburg; 2011. Report No: 2011/1.
52. AWG-KP, Ad Hoc working group on further commitments for Annex I Countries under Kyoto Protocol. Submissions for Forest Management Reference Level, 2011. URL (last access March 2015). http://unfccc.int/bodies/awg-kp/items/5896.php.
53. Pan Y, Birdsey RA, Fang J, Houghton R, Kauppi P, Kurz WA, Phillips OL, Shvidenko A, Lewis SL, Canadell JG, Ciais P, Jackson RB, Pacala SW, McGuire AD, Piao S, Rautiainen A, Sitch S, Hayes D. a large and persistent carbon sink in the world's forests. Science. 2011;333:988–93.

54. Smyth CE, Stinson G, Neilson E, Lemprière TC, Hafer M, Rampley GJ, Kurz WA. Quantifying the biophysical climate change mitigation potential of Canada's forest sector. Biogeosciences. 2014;11:3515–29.
55. Gardiner B, Blennow K, Carnus JM, Fleischer P, Ingemarson F, Landmann G, Lindner M, Marzano M, Nicoll B, Orazio C, Peyron JL, Reviron MP, Schelhaas MJ, Schuck A, Spielmann M, Usbeck T. Destructive storms in European forests: past and forthcoming impacts. Final report to European commission—DG Environment, 2010. URL (last access March 2015). http://www.efiatlantic.efi.int/portal/databases/forestorms/.
56. Bottero A, Garbarino M, Long JN, Motta R. The interacting ecological effects of large-scale disturbances and salvage logging on montane spruce forest regenereation in the western European Alps. For Ecol Man. 2013;292:19–28.
57. Lindner M, Maroschek M, Netherer S, Kremer A, Barbati A, Garcia-Gonzalo J, Seidl R, Delzon S, Corona P, Kolström M, Lexer MJ, Marchetti M. Climate change impacts, adaptive capacity, and vulnerability of European forest ecosystems. For Ecol Man. 2010;259:698–709.
58. Pilli R, Grassi G, Cescatti A. Historical analysis and modeling of the forest carbon dynamics using the carbon budget model: an example for the Trento Province (NE, Italy). Forest@. 2014;11:20–35.
59. Pilli R. Calibrating CORINE Land Cover 2000 on forest inventories and climatic data: an example for Italy. Int J Appl Earth Obs. 2012;9:59–71.
60. Boudewyn P, Song X, Magnussen S, Gillis MD. Model-based, Volume-to-Biomass Conversion for Forested and Vegetated Land in Canada. Canadian Forest Service, Victoria, Canada, 2007 (Inf. Rep. BC-X-411). URL (last access, March 2015). http://cfs.nrcan.gc.ca/publications/?id=27434.
61. Pilli R, Grassi G, Moris JV, Kurz WA. Assessing the carbon sink of afforestation with the carbon budget model at the country level: an example for Italy. iFor Biogeosci For. 2014;8:410–21.
62. FAO, FRA 2015—terms and definitions. Forest resources assessment working paper 180, 2015. URL (last access July 2016). http://www.fao.org/docrep/017/ap862e/ap862e00.pdf.

# Shade tree diversity and aboveground carbon stocks in *Theobroma cacao* agroforestry systems: implications for REDD+ implementation in a West African cacao landscape

Evans Dawoe[1*], Winston Asante[1,2], Emmanuel Acheampong[1] and Paul Bosu[3]

**Abstract**

**Background:** The promotion of cacao agroforestry is one of the ways of diversifying farmer income and creating incentives through their inclusion in REDD+ interventions. We estimated the aboveground carbon stocks in cacao and shade trees, determined the floristic diversity of shade trees and explored the possibility of implementing REDD+ interventions in cacao landscapes. Using replicated multi-site transect approach, data were collected from nine 1-ha plots established on 5 km long transects in ten cacao growing districts in Ghana West Africa. Biomass of cacao and shade trees was determined using allometric equations.

**Results:** One thousand four hundred and one (1401) shade trees comprising 109 species from 33 families were recorded. Total number of species ranged from 34 to 49. *Newbouldia laevis* (Bignoniacea) was the most frequently occurring specie and constituted 43.2 % of all shade trees. The most predominant families were Sterculiaceae and Moraceae (10 species each), followed by Meliaceae and Mimosaceae (8 species each) and Caesalpiniacaea (6 species). Shannon diversity indices (H', $H_{max}$ and J') and species richness were low compared to other similar studies. Shade tree densities ranged from 16.2 ± 3.0 to 22.8 ± 1.7 stems $ha^{-1}$ and differed significantly between sites. Carbon stocks of shade trees differed between sites but were similar in cacao trees. The average C stock in cacao trees was 7.45 ± 0.41 Mg C $ha^{-1}$ compared with 8.32 ± 1.15 Mg C $ha^{-1}$ in the shade trees.

**Conclusions:** Cacao landscapes in Ghana have the potential of contributing to forest carbon stocks enhancement by increasing the stocking density of shade trees to recommended levels.

**Keywords:** Shade trees, *Theobroma cacao*, Species richness, Diversity indices, Carbon stocks, REDD+

## Background

The cacao sector in Ghana plays a significant role in the fight against poverty and the development of the economy of Ghana as a whole. Globally, Ghana is the world's third largest producer (after Cote d'Ivoire and Indonesia) and has maintained its position as the second largest exporter of cacao beans after Cote d'Ivoire for the period 2005–2011 [1]. Cacao is Ghana's leading cash crop and the highest export crop earner accounting for 8.2 % of the country's gross domestic product (GDP) and 30 percent of total export earnings in 2010 [2]. In terms of employment, the livelihood of about six million people (25–30 % of the population) depends on the cacao sector [3].

Ghana's cacao production is characterized by small-scale farming with an average productive cacao area per household of approximately 2–3 ha [1, 4]. Approximately 800,000 families grow cacao on 1.6 million hectares of

*Correspondence: elkdawoe.canr@knust.edu.gh; edawoe@gmail.com
[1] Faculty of Renewable Natural Resources (FRNR), Kwame Nkrumah University of Science and Technology (KNUST), Kumasi, Ghana
Full list of author information is available at the end of the article

land with the Western region having the highest production value (over 50 % of total production), followed by the Ashanti region (accounting for about 16 % of total production). The Eastern and Brong-Ahafo regions together account for about 19 % of total production [1].

There is a general belief that cultivation of cacao, as traditionally practiced in Africa, has the potential to restore carbon stocks to levels comparable to that in the native forest which they replaced. Cacao is an understory species and so cultivation is traditionally done under tree canopy. It is therefore intuitive that carbon stocks in cacao farms with dense tree canopy cover will be much similar to that in the original native forest than in farms with lower or no canopy shade. Indeed, several studies have revealed that a larger proportion of the carbon stocks in most cacao agroforests are contributed by the shade trees [5–8]. The question is whether carbon stocks in these emerging cacao plantations will be within reasonable proportion of the original native forests they have replaced. Understanding carbon accumulation dynamics in cacao agro-ecosystems is important to the long-term management of cacao farms that inure to climate change mitigation and the associated socioeconomic effects.

Although multi-strata shaded-cacao systems still occur in Ghana, there has been an increasing move towards intensification of cacao management with shade tree removal and monoculture practices [9, 10]. In the Western Region of Ghana, the majority of cacao farms are now predominantly being managed with low shade or no shade [11, 12]. The need to introduce approaches that combine supporting local cacao livelihoods for poverty alleviation and reducing ecosystem degradation in cacao production systems in Ghana is imperative. This is expected to be achieved through actions that improve business development skills of farmers, increase cacao productivity, enhance natural resources management, conserve biodiversity, and reduce emissions associated with smallholder cacao farming systems. It is expected that through this initiative, smallholder cacao farmers in Ghana can qualify and earn additional income through the REDD+ (Reduced Emissions from Deforestation and Forest Degradation) mechanism and related incentive packages associated with carbon mitigation programmes. Thus, information relating to carbon stocks especially from shade trees, is vital to the design of any mitigation intervention. However, though some research efforts have focused on carbon stocks and tree diversity within cacao farms, most of these data sets show results of selected farms [13, 14], with no clear indication of landscape level play out. There is also no information on the potential role of the cacao systems at the landscape level in future climate change mitigation mechanisms based on the prevailing site level conditions in the cacao systems, though these future mitigation mechanisms are being conceived at the landscape level for implementation.

The overall objective of this study was to quantify the baseline carbon stocks, tree diversity and shade tree crown cover variation associated with smallholder cacao agroforestry systems in the cacao landscape of Ghana. Specifically, we estimated the baseline aboveground carbon stocks in shade and cacao trees, explored the relationships between tree parameters and carbon storage and determined floristic diversity of shade trees. We hypothesized that cacao landscapes in Ghana have the potential of contributing positively to the country's REDD+ efforts.

## Methods

### Study area

The study was conducted in the mid-western cacao growing areas of Ghana covering three of the five cacao-growing regions namely, Western, Ashanti, Brong-Ahafo and regions, located between longitude 01° 30′ and 30 5′ W and latitude 06° 0′ and 07° 0′ N (Fig. 1). The area lies within the dry and moist semi-deciduous forest zones, characterized by the wet semi-equatorial climatic conditions with a double maxima rainfall (April–July and September–November) ranging between 1700 and 1850 mm per annum. The main dry season lasts from December to mid-March Average ambient temperatures are uniformly high throughout the year and average 26.0 °C. The predominant vegetation in the study districts is the semi-deciduous forest type, which is characterized by predominantly *Celtis-Triplochiton* association as described by [15], with some of the trees in the upper and middle layers shedding their leaves in the dry season. The cacao agroforestry systems in the study districts (political/administrative area where the sites are located) are mostly mixed stands of cacao and with variable proportions of naturally generated upper canopy shade trees such as *Terminalia superba* Engl. & Diels, *Triplochiton scleroxylon* K. Schum., *Alstonia boonei* de Wild and *Ceiba pentandra* (L.) Gaertn. Fruit trees such as orange (*Citrus sinensis* L.) Osbeck, Avocado (*Persea americana*) and mango (*Mangifera indica* L.), for shade, food and other purposes may be planted. The soils of the study area are from weathered phyllites and dominated by ochrosols (to the north) and ochrosol-oxisol intergrade (Rhodic Ferralsol) to the south [16]. They are generally deep, moderately well drained and brashy with a silty-loam humus texture in the 0–15 cm soil layer, which gives it a high moisture retention capacity [17].

Land tenure arrangements in the study area can be classified into farmers who own their land, family lands, rented lands and lands under sharecropping arrangements. Nearly half of all cacao farmers in the area own

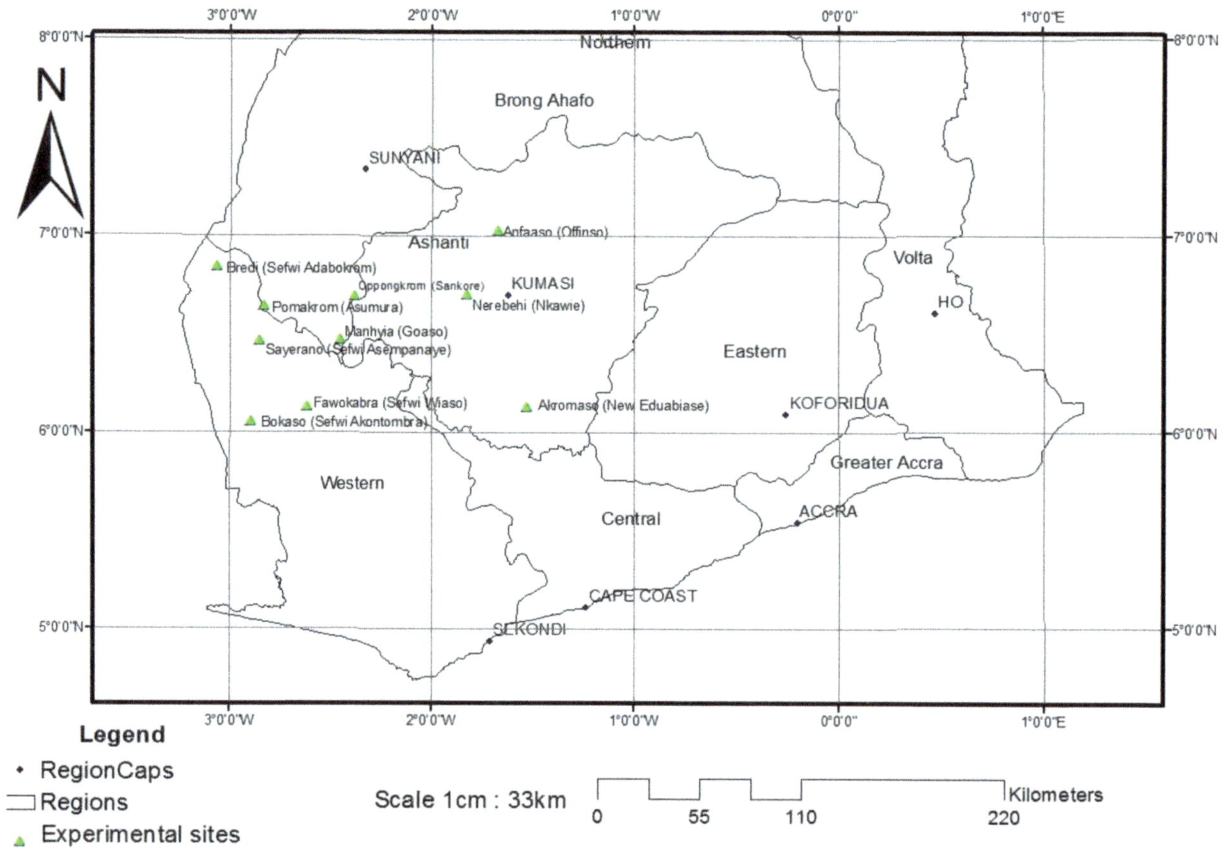

**Fig. 1** Locations of study communities and their administrative regions

the cacao farmlands themselves with about 22 % being family lands and 33 % managed under sharecropping arrangements [18].

### Study site selection

Selection of study sites (the location/situate of transects) was based on the biophysical and socioeconomic diversity of the landscape. Ten (10) cacao growing districts were selected, three each in the Ashanti and Brong-Ahafo Regions, and four in the Western Region after an initial reconnaissance of the study area. In each district, one of several communities (nearest town/village within whose immediate environs the transects were laid) with contiguous cacao plantations was randomly selected as a focal or reference community (Fig. 1).

### Plot installation and data collection

In each study community, three transects (laid out in a Y-shaped fashion and roughly aligned at 120° from each other) were established in a contiguous cacao landscape (Fig. 2). Each transect was at least 5 km long. Three plots, each measuring 100 m × 100 m, were established along each transect at intervals of at least 500 m. There were nine plots per site giving a total of ninety plots across all sites. In cases where the recording plot fell within a shade land use, the transect continued till the next cacao establishment. Within each 100 m × 100 m plot, subplots measuring 25 m × 25 m were demarcated. The subplots were placed at alternate points along the main transect line (Fig. 2). Plot level assessments included measurements of cacao (on 25 m × 25 m subplots) and shade canopy trees (100 m × 100 m main plots). All shade trees with diameter at breast height (DBH = 1.3 m) of 5 cm and above were identified (scientific and local names) by an experienced botanist. Subsequently tree botanical composition, population and spatial structure in the study districts were compared. Tree DBH was measured by calipers while heights were measured using a laser hypsometer (Nikon Laser Hypsometer). To obtain an estimate of the shade provided by the shade trees to cacao and their contribution to canopy cover, crown area of each shade tree within the 1 ha sample plot of cacao farms was estimated by measuring the diameter of the crown in eight different directions, following the cardinal

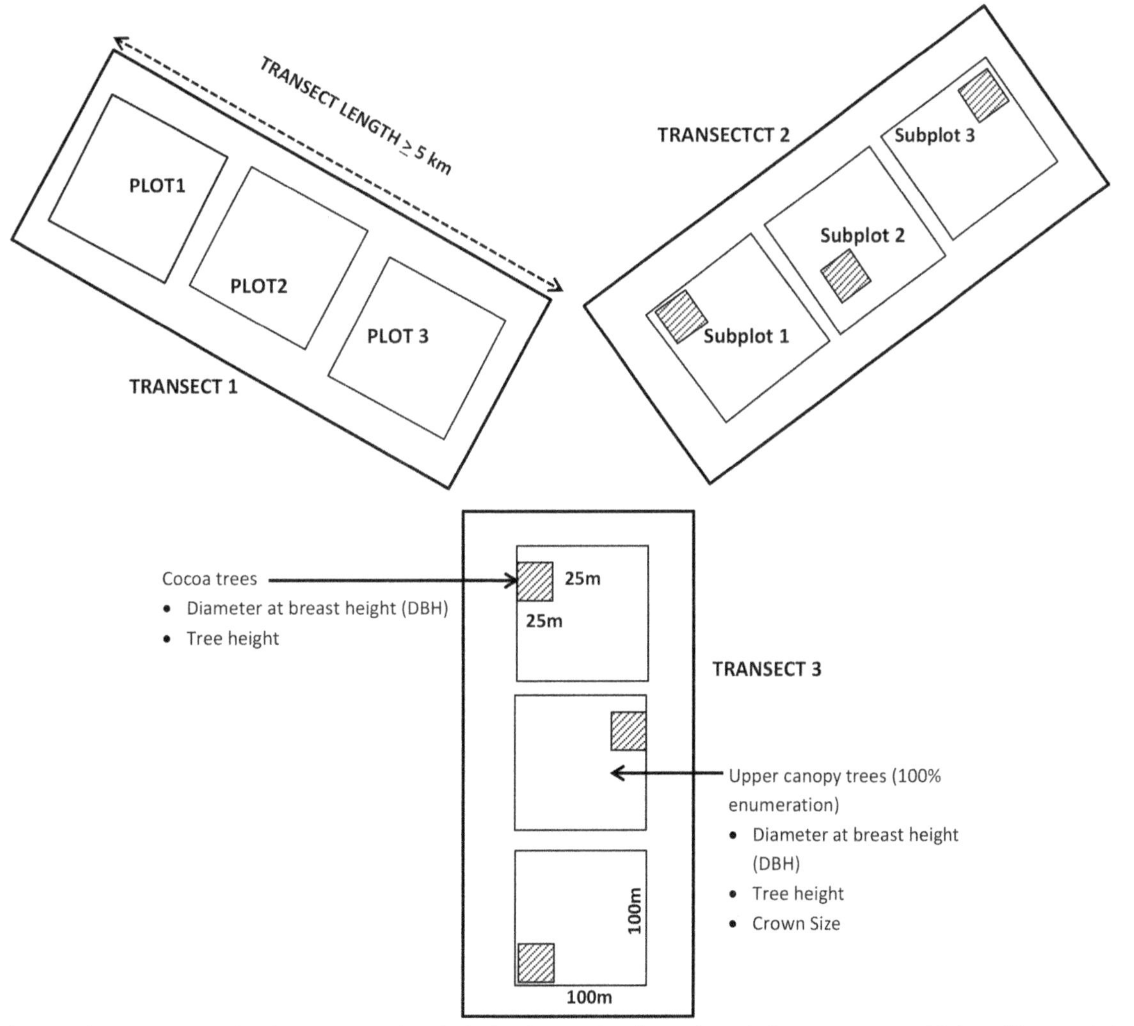


**Fig. 2** Layout of transects, main plots (100 m × 100 m) and subplots (25 m × 25 m) in each study district. *Lower arm* of *Y*-shape (Transect 3) shows dimensions of main and subplots and parameters measured (for both cocoa and shade trees) within each transect. Each transect is 5 km long and minimum distance between plots is 500 m

points and a subdivision within the cardinal points, i.e. North-South, East-West and then North-West, North-East, South-West and South-East. The total crown cover for all the upper canopy trees was expressed as a percentage of 1 ha to ensure easy comparison with the parameters of the national forest definition at the plot level. All data collection took place in April and May 2014.

### Tree diversity assessment

The Shannon-Weiner diversity indices (H', H max and J) were calculated using Biodiversity Professional Software (Version 2.) as follows:

$$H' = -\sum_{i=1}^{S}(p_i \times \ln p_i) = -\sum_{i=1}^{S}\left(\frac{n_i}{N} \times \ln\frac{n_i}{N}\right) \quad (1)$$

where $S$ is the total number of species in the landscape, $N$ is the total number of individuals, and $n_i$ is the number of individuals of the ith species, $\frac{n_i}{N}$ is equivalent to $p_i$, the probability of finding the i-th species. H max the maximum possible value of H', and J' the evenness (H'/H max) were also calculated. Spearman's rank correlation test of species similarity between the ten sites was also carried out using biodiversity professional software.

### Above ground biomass and C-stocks estimation

We estimated aboveground biomass (AGB) of cacao and upper canopy trees using the allometric equation developed by [19] for moist forests:

$$\text{Biomass (kg)} = \exp - 2.977 + \ln\left(X * Z^2 * W\right), \quad (2)$$

where X = species specific density; Z = diameter at breast height (DBH); and W = height. Species-specific wood densities were obtained from the World Agroforestry Centre's Wood Density Database [20]. If a species was not listed, we computed the average wood density of the plot and assigned it to the unknown species. To calculate carbon stocks in shade or cacao trees, we applied the general conversion factor of 0.5 to ABG tree biomass [21].

### Statistical analysis

Percentage crown cover and carbon stocks were compared using the pooled data at the transect level as replicates for each site. For each variable (crown cover, cacao carbon, shade tree carbon and stem diameter) normal distribution was tested using the Shapiro-Wilks W-test for homogeneity of variances. Variables that conformed to normal distribution were analyzed using one-way analysis of variance (ANOVA) while those that did not meet the assumptions for an ANOVA even when transformed were analyzed using Kruskal–Wallis parametric ANOVA on the mean ranks using the software package Statistix 8.0 for windows [22]. Separation of means was done at 5 % probability level using Tukey's HSD test. Regression and correlations analyses were also employed as tools for statistical tests and to establish trends and relationships between crown cover and carbon stocks, and shade tree diameter and carbon stocks and Shannon-Weiner Diversity Indices H' H max and J.

## Results

### Abundance and composition of shade species and families in the cacao landscape

A total of 1401 individual trees (shade) were recorded across all the 10 study sites (Table 1). These comprised 109 different species from 33 families on 90 ha (average 15 trees per hectare) surveyed. Out of the 109 species, the total number of species encountered in the districts ranged from 34 at Sewfi Wiawso to 49 at Goaso. *Newbouldia laevis* (family *Bignoniacea*) was the most frequently occurring species and constituted 43.2 % of all shade trees in the districts but was most dominant in four districts. The most predominant families across the landscape were *Sterculiaceae* and *Moraceae* (10 species each), followed by *Meliaceae* and *Mimosaceae* (eight species each) and *Caesalpiniacaea* (six species). *Anacardiaceae, Euphorbiaceae* and *Palpilioniaceae* were each represented by five species. Ten families were represented by two or three species. Table 2 shows the most dominant shade tree species in each district.

The average stand density of the shade trees varied and ranged from 9.33 ± 1.22 stems $ha^{-1}$ for Akontombra to 22.8 ± 1.71 stems $ha^{-1}$ for Offinso with an average of 15.6 ± 1.34 stems $ha^{-1}$ across districts. The total number of species was highest in Goaso, where the Shannon-Weiner indices (Shannon H' = 1.52, Shannon H max = 1.69 and Shannon J' = 0.90) were higher than all the other locations. Wiawso had the lowest indices (Shannon H' = 0.99, and Shannon J' = 0.65). The Spearman's rank correlation test of species similarity between the ten sites are shown in Table 3. The highest similarity index was 0.7 and the lowest 0.4.

### Relative abundance of timber and non-timber 1 species in the cacao landscape

All shade trees identified in the plots were grouped into timber and non-timber species. Generally, the relative abundance of non-timber species in all the districts was higher than timber species, except Asempaneye where

**Table 1 Overview of shade tree species diversity and selected dendrometric parameters in the different study districts**

| | Tree abundance[a] | Species richness[b] | Shade tree stem density ($ha^{-1}$) | Mean tree height (m) | Mean stem diameter (cm) | Shannon H' | Shannon H max | Shannon J' |
|---|---|---|---|---|---|---|---|---|
| Adabokrom | 97 | 35 | 10.8 ± 1.27 | 15.1 ± 0.74 | 51.6 ± 3.67 | 1.42 | 1.54 | 0.92 |
| Akontombra | 84 | 27 | 9.33 ± 1.22 | 11.9 ± 1.05 | 31.2 ± 4.71 | 1.22 | 1.43 | 0.85 |
| Asempaneye | 128 | 45 | 14.2 ± 1.11 | 19.1 ± 1.41 | 47.7 ± 3.91 | 1.51 | 1.65 | 0.92 |
| Asumura | 147 | 37 | 16.3 ± 2.58 | 13.8 ± 1.19 | 35.0 ± 3.97 | 1.04 | 1.57 | 0.66 |
| Goaso | 146 | 49 | 16.2 ± 3.00 | 15.3 ± 1.24 | 43.0 ± 3.55 | 1.52 | 1.69 | 0.90 |
| New Edubiase | 184 | 43 | 20.6 ± 3.25 | 18.2 ± 0.99 | 44.9 ± 2.85 | 1.34 | 1.63 | 0.82 |
| Nkawie | 136 | 40 | 15.1 ± 2.47 | 12.8 ± 0.64 | 36.6 ± 3.31 | 1.40 | 1.60 | 0.88 |
| Offinso | 204 | 46 | 22.8 ± 1.71 | 13.5 ± 0.82 | 40.2 ± 3.68 | 1.29 | 1.66 | 0.77 |
| Sankore | 108 | 35 | 12.0 ± 2.71 | 18.2 ± 1.58 | 49.6 ± 3.64 | 1.39 | 1.54 | 0.90 |
| Wiawso | 165 | 34 | 18.3 ± 3.57 | 13.7 ± 1.42 | 33.1 ± 3.72 | 0.99 | 1.53 | 0.65 |

[a] Total number of stems enumerated on a total plot size of 9 ha. These comprise all timber and non-timber tree species

[b] Number of different individual tree species recorded on a total plot size of 9 ha

**Table 2 Dominant shade tree species recorded in each study district (percentages are proportions of total individuals recorded in each district)**

| Community | Most dominant species | Tree abundance[a] | Proportion of total individual shade trees (%) |
|---|---|---|---|
| Adabokrom | *Ficus exasperata* | 97 | 10 |
| Akontombra | *Persea americana* | 84 | 16.7 |
| Asempaneye | *Triplochiton scleroxylon* | 128 | 10.2 |
| Asumura | *Newbouldia laevis* | 147 | 47.6 |
| Goaso | *Milicia excelsa* | 146 | 9.5 |
| New Edubiase | *Newbouldia laevis* | 185 | 24.3 |
| Nkawie | *Morinda lucida* | 136 | 13.2 |
| Offinso | *Citrus sinensis* | 205 | 24.0 |
| Sankore | *Newbouldia laevis* | 108 | 11.1 |
| Wiawso | *Newbouldia laevis* | 165 | 47.3 |

[a] Total number of stems enumerated on a total plot size of 9 ha

**Table 3 The Spearman's rank correlation test of species similarity between study districts in Ghana**

| Communities | Adabokrom | Akontombra | Asempaneye | Asumura | Goaso | New Edubiase | Nkawie | Offinso | Sankore | Wiawso |
|---|---|---|---|---|---|---|---|---|---|---|
| Adabokrom | 1 | * | * | * | * | * | | | | |
| Akontombra | 0.6 | 1 | * | * | * | * | * | * | * | * |
| Asempaneye | 0.6 | 0.4 | 1 | * | * | * | * | * | * | * |
| Asumura | 0.6 | 0.6 | 0.5 | 1 | * | * | * | * | * | * |
| Goaso | 0.7 | 0.5 | 0.5 | 0.7 | 1 | * | * | * | * | * |
| New Edubiase | 0.7 | 0.6 | 0.5 | 0.6 | 0.7 | 1 | * | * | * | * |
| Nkawie | 0.6 | 0.5 | 0.5 | 0.6 | 0.6 | 0.6 | 1 | * | * | * |
| Offinso | 0.5 | 0.5 | 0.4 | 0.5 | 0.5 | 0.5 | 0.5 | 1 | * | * |
| Sankore | 0.6 | 0.6 | 0.5 | 0.6 | 0.6 | 0.6 | 0.5 | 0.6 | 1 | * |
| Wiawso | 0.5 | 0.7 | 0.4 | 0.5 | 0.5 | 0.5 | 0.5 | 0.5 | 0.5 | 1 |

the abundance of timber and non-timber species were equal (Table 4). In addition to being important as shade trees for cacao, the most important uses of shade trees in cacao systems were for their productive values e.g. fruit, timber, firewood, fodder and medicines (Additional file 1: Appendix 1). Apart from the fruit trees, (*Persea americana, citrus* spp. and the oil palm *Elieas guineensis*) which were planted by farmers, most of the shade trees were selectively left on farmlands during land preparation and nurtured by farmers during cacao establishment. A list of current and potential uses of all identified non-timber trees is given in Additional file 1: Appendix 1.

### Carbon stocks in cacao and shade trees in the cacao landscape

Carbon stocks in shade trees ranged from $2.91 \pm 0.95$ Mg C $ha^{-1}$ at Akontombra to $15.6 \pm 2.89$ Mg C $ha^{-1}$ at New Edubiase and differed significantly ($F = 4.51$; $p = 0.0001$) between districts (Table 5). Carbon stocks in cacao trees on the other hand were similar ($F = 1.73$; $p = 0.095$) and averaged $7.45 \pm 0.41$ Mg C $ha^{-1}$ (range 5.84–10.2 Mg C $ha^{-1}$). Total carbon (shade + cacao trees) significantly differed ($F = 3.82$; $p = 0.0005$) between sites with New Edubiase having the highest ($23.4 \pm 3.23$ Mg C $ha^{-1}$) and Sefwi Wiawso the lowest ($10.9 \pm 1.56$ Mg C $ha^{-1}$), representing 85.9 and 40.0 Mg $CO_2$-eq. $ha^{-1}$ respectively. Carbon distribution between cacao and shade trees averaged of 48 and 52 % respectively and mean stocks across all study districts stood at $15.5 \pm 1.19$ Mg C $ha^{-1}$.

### Relationships between crown cover, diversity indices and carbon stocks

We analyzed the relationships and determined correlations between various shade tree parameters, diversity indices and total aboveground tree carbon to understand how these parameters affected carbon accumulation in shaded-cacao systems. Significant linear relationships were found to exist between crown cover % and total carbon C stocks ($R^2 = 0.7785$; $p = 0.0072$), and shade tree diameter and total carbon ($R^2 = 0.3574$, $p = 0.04912$)

**Table 4 Relative abundance of timber and non-timber species within smallholder cacao systems in ten cocoa growing districts in the high forest zone, Ghana**

| District | Timber species[a] | Relative abundance (%) | Non timber species[a] | Relative abundance (%) |
|---|---|---|---|---|
| Adabokrom | 45 | 45.0 | 55 | 55.0 |
| Akontombra | 19 | 22.6 | 65 | 77.4 |
| Asempanaye | 64 | 50.0 | 64 | 50.0 |
| Asumura | 39 | 26.6 | 108 | 73.5 |
| Goaso | 67 | 45.9 | 79 | 54.1 |
| New Edubiase | 85 | 45.9 | 100 | 54.1 |
| Nkawie | 48 | 35.3 | 88 | 64.7 |
| Offinso | 44 | 21.6 | 160 | 78.4 |
| Sankore | 31 | 28.7 | 77 | 71.3 |
| Sefwi Wiawso | 44 | 26.7 | 121 | 73.3 |

[a] Timber + Non-timber species = Tree abundance, i.e. the number of stems enumerated on a total plot size of 9 ha

(Fig. 2). The relationships between total carbon and Shannon-Weiner Diversity Indices H', H max and J' were not significant, p values being 0.2224, 0.1486 and 0.3539 respectively (Fig. 3). Apart from the equation relating crown cover % and total carbon stocks which had a moderately high predictive ability ($R^2 = 0.7785$), all the other regressions equations explained up to a maximum of 35 % of the variations in the relationships. Crown cover (%) was highly significantly ($p < 0.01$) correlated to shade tree C and total C (Table 6). Whereas Shannon H max correlated significantly ($p < 0.05$) positively with shade tree C and number of shade trees, it was non-significantly correlated with total C. Number of shade trees correlated positively but non-significantly ($p > 0.05$) with total C shade tree C and Crown %.

### Carbon stocks distribution and associated shade tree parameters

Shade tree characteristics (mean number of stems $ha^{-1}$, average height (m), average stem diameter (cm) and associated total C stocks (Mg C $ha^{-1}$) grouped according to districts with similar crown cover (%) are given in Table 7. Three groups of districts are discernible. Group 1 districts having crown cover <8 %, Group 2 districts 8.1–14.9 % and Group 3 districts have crown cover >15 %. Group 3 districts, i.e. New Edubiase and Asempanaye with the highest number of stems $ha^{-1}$ and stem diameter recorded the highest mean carbon stocks of 21.0 ± 2.45 Mg C $ha^{-1}$ because the biggest trees were found on these landscapes. Group 1 districts with fewer and smaller trees recorded the lowest total carbon storage of 12.1 ± 0.52 Mg C $ha^{-1}$. Progressively also, total tree carbon increased with increasing crown cover. Group 3 districts with crown cover equal to or more than the 15 % (the defined national threshold for forests in Ghana) compared to the other sites translated to high total tree carbon stocks.

## Discussions

### Abundance and composition of upper canopy tree species and families in the cacao landscape

Several workers (e.g., [24–26]) have emphasized that agrobiodiversity is an important feature of agro-ecological systems in terms of climate change adaptation. Shaded-cacao agroforestry may be considered promising in this context. Shade tree diversity in cacao farms offers farmers an array of agronomic, economic, cultural, and ecological benefits [9]. A diversified farm also enables farmers to exploit the different components as well as their interactions in the

**Table 5 Mean values ± SEM of shade tree carbon, cacao tree carbon, total carbon and their $CO_2$ equivalents in ten cacao growing districts in the Ashanti, Brong-Ahafo and Western regions of Ghana (numbers in parenthesis are standard errors of the means)**

| District | Shade trees C (Mg C $ha^{-1}$) | Cacao tree C (Mg C $ha^{-1}$) | Total C (Mg C $ha^{-1}$) | $CO_2$ equiv. of total C (Mg $CO_2$-eq. $ha^{-1}$) |
|---|---|---|---|---|
| Sefwi Wiawso | 5.05 ± 1.40[ab] | 5.84 ± 0.91[a] | 10.9 ± 1.56[b] | 40.0 |
| Asempanaye | 11.8 ± 1.99[a] | 6.77 ± 0.68[a] | 18.5 ± 2.12[ab] | 67.9 |
| Akontombra | 2.91 ± 0.95[b] | 10.2 ± 1.31[a] | 13.2 ± 2.07[ab] | 48.4 |
| Adabokrom | 7.70 ± 1.29[ab] | 8.02 ± 0.87[a] | 15.7 ± 0.87[ab] | 57.6 |
| Sankore | 8.60 ± 1.92[ab] | 7.00 ± 0.87[a] | 15.6 ± 2.40[ab] | 57.2 |
| Goaso | 9.17 ± 2.33[ab] | 7.79 ± 0.74[a] | 15.4 ± 2.58[ab] | 56.5 |
| Asumura | 5.04 ± 1.35[ab] | 7.58 ± 0.47[a] | 12.6 ± 1.68[ab] | 46.2 |
| New Edubiase | 15.6 ± 2.89[a] | 6.28 ± 0.82[a] | 23.4 ± 3.23[a] | 85.9 |
| Offinso | 9.29 ± 1.41[a] | 8.65 ± 0.70[a] | 17.9 ± 1.45[ab] | 65.7 |
| Nkawie | 8.02 ± 0.65[ab] | 6.40 ± 0.61[a] | 11.5 ± 1.19[b] | 42.2 |

Figures in the same column followed by similar alphabets do not differ significantly ($p \leq 0.05$) based on the Kruskal–Wallis parametric ANOVA applied to mean ranks. Total carbon stock was converted to tons of $CO_2$ equivalent by multiplying by 44/12 or 3.67 [23]

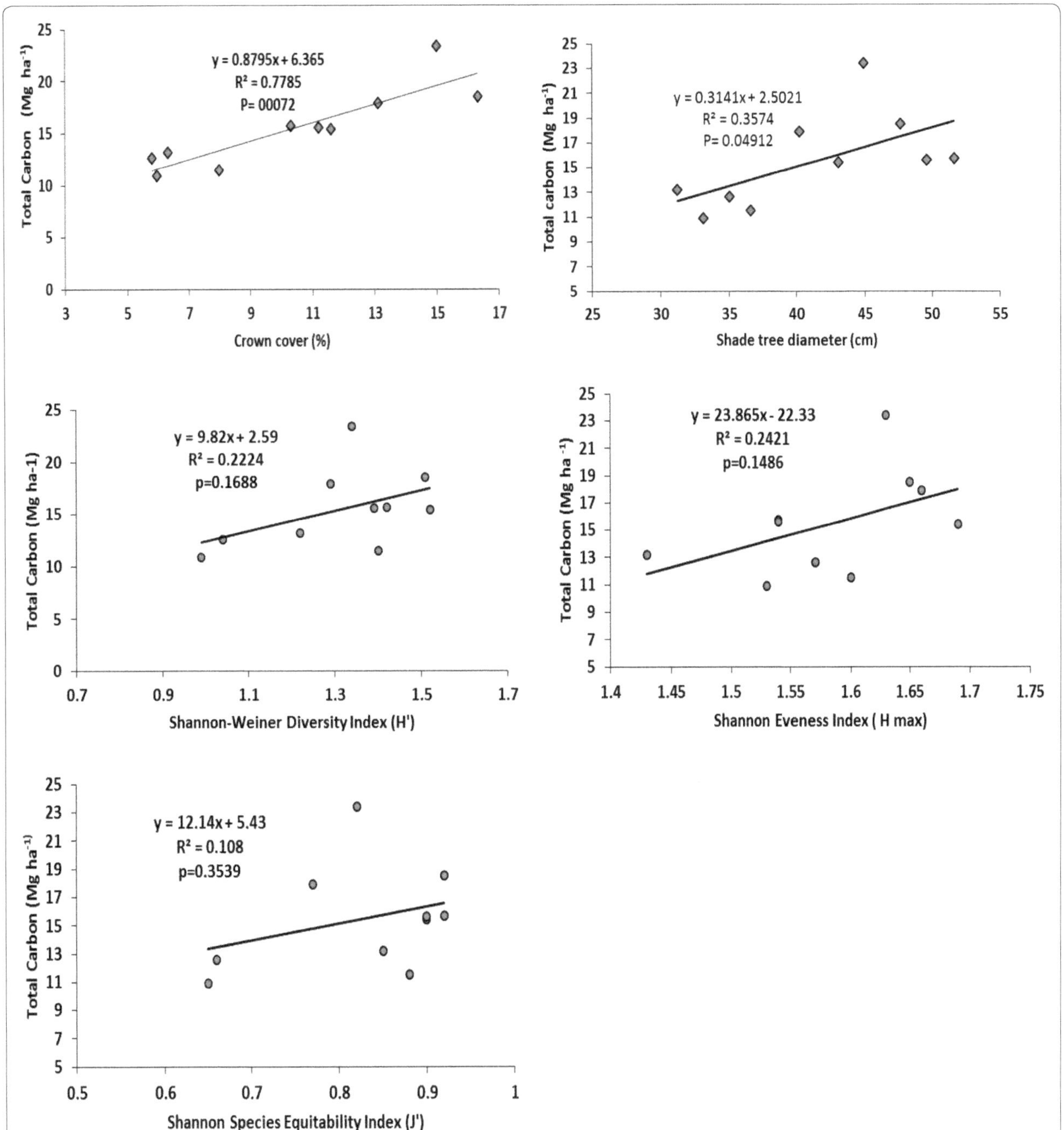


**Fig. 3** Relationships between selected shade tree dendrometric and diversity parameters and total aboveground (shade + cacao trees) carbon

system to meet subsistence needs, maximize incomes, and reduce risks against fluctuations in world market prices of cacao beans [27–29]. The 109 species (DBH ≥ 15 cm) with Shannon-Weiner diversity index H' ranging from 0.99 to 1.54 recorded from 90 ha of shaded-cacao systems in this study are relatively lower compared to values recorded in similar systems elsewhere. For instance, [30] found 293 tree species (DBH ≥10 cm) with Shannon-Weiner diversity indices ranging from 3.31 to 4.22 on 15 ha of five traditional cacao growing (*cabruca*) farms in Southern Bahia, Brazil, while in the dense and complex agroforestry systems of Southeast Cameroon. [31] sampled trees and pseudo-trees (e.g., banana) with DBH >2.5 cm and found 206 species in 60 cacao farms, with Shannon diversity indices ranging between 3.1 and 4.2 in each of the agroforestry systems studied. At two sites in the Atwima District of the Ashanti

**Table 6 Pearson correlation coefficients among diversity indices, number of shade trees and C content of trees in ten cacao growing districts in Ghana**

| | Crown % | Shannon H′ | Shannon H max | Shannon J | Shade trees C | Total C |
|---|---|---|---|---|---|---|
| Shannon H′ | 0.7107* | – | | | | |
| Shannon H max | 0.6788* | 0.4994 ns | – | | | |
| Shannon J | 0.5269 ns | 0.9333** | 0.1556 ns | – | | |
| Shade trees C | 0.8952** | 0.5791 ns | 0.7078* | 0.3692 ns | – | |
| Total C | 0.8823** | 0.4716 ns | 0.4920 ns | 0.3286 ns | 0.8807** | |
| No. of shade trees | 0.2988 ns | −0.2202 ns | 0.6467* | −0.5190 ns | 0.4560 ns | 0.3595 ns |

*ns* not significant

Asterisks denote statistical significance of correlations: * $p < 0.05$; ** $p < 0.01$

**Table 7 Shade tree dendrometric parameters and associated carbon stocks in the study districts with similar crown cover**

| Group | District | Crown cover % | Mean shade trees stem number ($ha^{-1}$) | Mean shade tree height (m) | Mean stem diameter (cm) | Shade and cacao trees C-stocks (Mg C $ha^{-1}$) |
|---|---|---|---|---|---|---|
| 1 | Asumura | 5.80 | 16.3 | 13.9 | 35.0 | 12.6 |
| | Sefwi Wiawso | 5.95 | 18.3 | 13.7 | 33.1 | 10.9 |
| | Akontombra | 6.31 | 9.3 | 11.9 | 31.2 | 13.2 |
| | Nkawie | 7.97 | 15.1 | 12.8 | 36.6 | 11.5 |
| | Mean ± SE | 6.51 ± 0.49 | 14.7 ± 1.93 | 13.1 ± 0.46 | 34.0 ± 1.17 | 12.1 ± 0.52 |
| | Range | 5.80–8.0 | 9.3–18.3 | 11.9–13.9 | 31.2–36.6 | 10.9–13.2 |
| 2 | Adabokrom | 10.3 | 10.8 | 15.1 | 51.6 | 15.7 |
| | Sankore | 11.2 | 12.0 | 18.2 | 49.6 | 15.6 |
| | Goaso | 11.6 | 16.2 | 15.3 | 43.0 | 15.4 |
| | Offinso | 13.1 | 22.8 | 13.5 | 40.2 | 17.9 |
| | Mean ± SE | 11.6 ± 0.58 | 15.4 ± 2.7 | 15.5 ± 0.98 | 46.1 ± 2.69 | 16.2 ± 0.587 |
| | Range | 8.1–14.9 | 10.8–22.8 | 13.5–18.2 | 40.2–51.6 | 15.4–17.9 |
| 3 | New Edubiase | 15.0 | 20.6 | 18.2 | 44.9 | 18.5 |
| | Asempanaye | 16.3 | 14.2 | 19.1 | 47.7 | 23.4 |
| | Mean ± SE | 15.5 | 17.4 ± 3.20 | 18.7 ± 0.45 | 46.3 ± 1.40 | 21.0 ± 2.45 |
| | Range | ≥15.0 | 14.2–20.6 | 18.2–19.1 | 44.9–47.7 | 18.5–23.5 |

Region of Ghana, [13] recorded a Shannon index of 4.69 for matured shaded-cacao farms. [13] study area falls within the same district as the study site for this study, which recorded a significantly lower tree diversity index of 0.88. It is likely that there has been an increased removal of shade trees in the district within the intervening 3-year period prior to this study as a result illegal chainsaw operations and the continuous giving out of these off-reserve areas to timber concessionaires [18]. Our species number is however comparable to that of [32] who recorded 105 species in *cabrucas* of the Espirito Santo state, Brazil, by sampling trees with DBH >10 cm in 4.8 ha of 20 farms studied.

### Relative abundance of timber and non-timber species in the cacao landscape

Most of the sites recorded wide disparities in the abundance of timber and non-timber tree species, a strong indication that farmers are shifting the composition of species in cacao agroforestry systems in favor of non-timber species. This disparity could also be a manifestation of logging impact or exploitation of timber species within the cacao landscape. It is obvious from the data that there is a shift in the species composition of the off-reserve landscape in the high forest zone, with a general decline in species diversity. The dominance of fruit trees (citrus and avocadoes) and *Newbouldia laevis* strongly indicates a deliberate transformation of the landscape by farmers from the naturally occurring pioneer species such as *Terminalia* spp. and other timber species that have been traditionally grown in tandem with cacao, and provide timber benefits, to non-timber species such as fruits, *Gliricidia sepium* and *N. laevis*. A prime factor that drove the transformation of the species composition of the landscape was logging [18]. However, general governance issues such as tree tenure and ownership and inconsistent understanding and weak implementation of forest policy and legal regime have also contributed to a strong negative perception about tree incorporation in the cacao systems.

Farmers consistently complained about the indiscriminate logging on their farms and continued award of concessions in the cacao landscape as well as the lack of proper compensation for cacao trees destroyed [18]. For a long time, the governance issues associated with off-reserve trees have bedeviled efforts at improving cacao agroforestry systems involving the incorporation of timber trees that could have multiple benefits for farmers in Ghana. The minimal occurrence and low dominance of the indigenous timber species such as *Terminalia* spp., *Melicia excelsa*, etc. within the cacao landscape have serious implications for the implementation of climate-smart cacao models and REDD+ strategies. Most of these models are posited on indigenous timber species that present better opportunities to contribute to climate change mitigation projects in cacao systems and also serve as preferred species in terms of shade and moderation of climatic parameters for optimal cacao productivity. In order to reverse the current trend in terms of dominant species within the landscape, it is important to overcome the governance issues associated with off-reserve timber trees exploitation at all levels, from the national to the community level. Once farmers' confidence has been gained, then mechanisms can be rolled out to introduce tree diversification strategies. It is also possible that the dominance of fruit trees is an indication that farmers want to diversify the sources of income as was indicated by respondents at Offinso, who incorporated citrus and pear to ensure that they could generate revenue in the off season period of cacao harvesting.

### Carbon stocks in cacao and shade trees

Agroforestry ecosystems are generally known to stock higher carbon than other cropping systems thereby contributing to climate change mitigation [33–35]. Closed forests in the study area are generally reported to hold about 155 Mg C ha$^{-1}$ with stocks in open forests ranging in the region between 85 and 96 Mg C ha$^{-1}$, while agricultural lands including no-shade cocoa systems stock on the average 15 Mg C ha$^{-1}$ [14, 36, 37]. Thus, the conversion of agricultural lands to cacao agroforests could be a management strategy for storing large quantities of carbon. As the largest component of total biomass, shade trees are crucial to carbon stocks in the cacao cultivation systems. Our findings of higher stocks in some districts compared to others reflects emerging or existing trends in the management of upper canopy trees and how this holds implications for landscape level carbon build-up or loss.

Our results for total tree carbon stocks in cocoa agroforests are similar to and fall within the range reported by [14] in the high forest cacao zone in the Western Region in Ghana, which is also similar to that of [38] in Central Sulawesi and [39] in Central Cameroon. [14] recorded aboveground carbon stocks of 16.8 and 15.9 Mg C ha$^{-1}$ respectively in 15 and 25-year-old shaded-cacao stands. Results of this study however fall in the lower range compared to studies by [40] in Bolivia, [7, 11, 39, 41] all in Cameroon, [36] in Ghana and [42] in Sulawesi, Indonesia. In these studies, stocks of tree carbon ranged from 34.4 to 135.5 Mg C ha$^{-1}$. Explanations for the lower carbon stocks found in this study include the relatively young age of most of the plots (40.7 % of all farms were 10 years or below) as well as the non-planting and removal of shade trees due to the perception that, shade trees are a major cause of pests, diseases and low yields. It also needs to be pointed out that, aboveground carbon depends on a number of factors including the canopy species, tree density, environment and the approach used in estimating total carbon. [7] have pointed out that using equations like [43] which do not incorporate wood density would overestimate the biomass and thus carbon stocks of a low wood density species such as cacao. Using allometric equations that integrate wood densities of separate species, as done in the current study, or developing location and species-specific allometric equations is likely to lead to a more accurate assessment of biomass and thus carbon stocks. This study used the equation developed by [19]. Hence, comparison of the carbon stocks figures generated in this study should also be placed in the context of the allometric equations used in the estimation of biomass. But most obviously, the fact that this study adopted a transect approach, also means that, all variations of cacao maturity stages and establishment were encountered. Given that biomass build-up directly relates to age, further comparisons of our results, should be placed within the context of random transect sampling.

### Relationships between shade tree dendrometric parameters and carbon stocks

Associated shade tree parameters (crown cover and shade tree stem diameter) significantly influenced aboveground C stocks in the cacao agroforestry systems. The predictive ability of the equation relating crown cover and c-stocks is moderately high while that between shade tree stem diameter and shade tree carbon and total tree carbon were generally low, the equations nevertheless demonstrate the relationship between the size of the shade tree and the amount of carbon stored. Progressively, total tree carbon increased with increasing crown cover. This is to be expected as increasing tree size (tree height and diameter) is generally associated with increasing crown cover. The bigger the stem diameter, the more the amount of carbon stored. It has been widely recognized that, shade trees in cacao systems account for the differences in carbon stocks in cacao systems. Though they are often assumed to negatively affect growth and yield of cacao plants through competitive water [44] and resource use, shade trees play a significant role on C sequestration. Empirical studies have also shown the positive effects of plant species-specific, complementary resource use in agroforestry systems [45]. Several authors (e.g. [6, 44]) have emphasized that the main drivers of C storage in cacao systems are shade trees, which usually has the highest amount of carbon. Studies by [6, 7] for instance all recorded shade trees storing up to 65 % of total tree carbon in cacao systems studied in Nigeria and Cameroon respectively.

The near-even distribution of carbon between cacao (48 %) and shade trees (52 %) in this study is possibly a reflection of the extent to which shade trees have been and continue to be progressively removed from farms. This could be the result of the widely held perception by farmers in some districts that presence of shade trees on farms is a major cause of disease incidence and low yields. There are also governance issues particularly that which relates tenure over naturally growing economic trees in cacao and other cash crop plantations. Existing tree tenure system where farmers have no ownership right over naturally growing economic trees in their cacao farms, has served as a disincentive for farmers to nurture and keep economic trees in cacao farms. When combined with a preference for full-sun cacao over shade production systems, as a result of relatively high yields obtained in the initial years of cultivation [10], there appears to be a strong incentive to remove shade trees from cacao farms in these and several other cacao growing districts [18]. However, since the government introduced the timber resources management (Amendment) Act of 2002, the right of tenure has now been assigned to the one who planted the tree, though this law appears to be more inclined towards plantation establishment and has largely not been tested in farming systems. This provides hope for REDD+ implementation in cacao farms, as this could motivate farmers to integrate trees into their farms through planting, if appropriate sensitization efforts are made, including enforcement of existing laws that that require that farmers consent is sought before trees are logged.

### Possibility of implementing REDD+ interventions in studied cacao landscapes

Results of this study also reveal that, contrary to existing assumptions, unshaded cacao systems meet the height requirements to qualify as forest per Ghana's forest definition for climate change mitigation projects. Coupled with the finding that only two of the studied districts have shade cover equal to or more than the 15 % threshold, significant opportunity exists to develop cacao-carbon and REDD+ projects in Ghana. This is certainly useful information and has diverse implications for the design of Ghana's REDD+ strategy. However, the extent to which cacao landscapes would contribute to the realization of REDD+ objectives hinges very much on Ghana's definition of forest. In line with requirements under the REDD+ readiness efforts, Ghana defined its forests as being a minimum of 1 ha, having at least 15 % canopy cover and containing trees that are 5 meters tall. The shade trees in the cacao system, however, could constitute a forest if they offer enough canopy cover beyond the 15 % and are taller than 5 m. Thus, the forest definition and type of cacao system (monoculture vs. shade) have serious implications for the type of REDD+ that is viable. Two out of the ten study districts, New Edubiase and Asempaneye, had crown cover of 15 and 16.3 % respectively and technically qualify to be defined as forests and inclusion for REDD+ intervention activities. By judiciously selecting shade trees and integrating them into cacao systems in a manner designed to offer optimal ecological, economic nutritional and financial benefits, cacao landscapes in the other study districts have the potential of contributing positively to the country's REDD+ efforts. Full-sun plantations with cacao trees up to the defined 5-m height would constitute a forest and therefore would qualify for REDD+ .

## Conclusions and recommendations

There were significant relationships between crown cover (%) and shade tree carbon stocks, shade tree stem diameter and shade tree carbon, and shade tree diameter (cm) and total aboveground carbon stocks. Progressively also, total tree carbon stocks increased with increasing mean stem numbers per hectare and stem diameter. Above ground carbon stocks in shaded-cacao systems in the studied landscape varied between 10.9 ± 1.56 and 23.4 ± 3.23 Mg C $ha^{-1}$ and is a reflection of the differences in tree density, tree diversity and management practices between farms. Additionally, only 20 % of cacao in the sampled districts is grown under shade regimes of 15 % crown cover and above, while 80 % of cacao is grown under shade regimes ranging from 5.8 to 14.9 %. In the light of these findings, significant opportunity exists to develop cacao carbon projects in Ghana. While different groups are making efforts to map out farms in the cacao landscape, comprehensive data and information such as those suggested above is needed to develop a well-planned national REDD+ strategy. By judiciously selecting preferred shade trees and integrating them into cacao systems in a manner designed to offer optimal ecological and socio-economic benefits, cacao landscapes in Ghana have the potential of contributing positively to the country's REDD+ efforts. There is also the need to educate farmers on the potential of monetary benefits from REDD+ activities. Broad stakeholder participation and access to credit and information are essential for the equitable and sustainable implementation of REDD+ policies. Increased participation is particularly important in terms of maximizing local knowledge and capacity to support the implementation of REDD+. However, the extent to which cacao landscapes would contribute to the realization of REDD+ objectives hinges very much on Ghana's definition of forest.

## Additional file

**Additional file 1: Appendix 1.** Shade tree species found in *Theobroma cacao* agroforestry systems in ten districts in the Ashanti, Brong-Ahafo and Western Regions of Ghana and reasons for their preference. (Uses: 1. Timber, 2. Food & beverages, 3. Fodder, 4. Fuelwood/Charcoal, 5. Soil fertility/Green manure, 6. Nitrogen-fixation, 7. Dyes/Colors, 8. Spices, 9. Honey production/Apiculture 10. Essential Oils, 11. Medicinal, and 12. Other uses).

### Authors' contributions

The study was carried out by ED, WA, EA and PB. All the authors participated in the design of the study, plot/field layouts and collection of field data. ED drafted the manuscript, performed the statistical analysis and its interpretation, drew all the graphs and constructed the tables. WA conceived the study and facilitated the acquisition of funds for its implementation, contributed to the interpretation of data, and made significant contribution in critically reviewing the manuscript. EA was team leader, coordinated all the study activities and read through the manuscript. PB contributed in reviewing the draft manuscript for its content. All authors read and approved the final manuscript.

### Author details

[1] Faculty of Renewable Natural Resources (FRNR), Kwame Nkrumah University of Science and Technology (KNUST), Kumasi, Ghana. [2] SNV Netherlands Development Organization, Accra, Ghana. [3] Forestry Research Institute of Ghana (FORIG), Council for Scientific and Industrial Research (CSIR), Kumasi, Ghana.

### Acknowledgements

This research was financed by SNV Netherlands Development Organisation, Ghana under its Cacao Eco Project. We are grateful to Messrs Seth Kankam, Elvis Nkrumah, Yaw Apau, Nana Prempeh Bandoh and Emmanuel Sey, who assisted in the fieldwork identifying tree species and compiling data. The ooperation of the survey participants and local community leaders in the research sites is also gratefully acknowledged. We acknowledge the contribution of Ernest Adzim, Arend, Richard McNally and Quirin Laumans for their insightful comments at various stages of the research. We would also like to thank the anonymous reviewers for their time and comments.

### Competing interests

The authors declare that they have no competing interests.

### References

1. Asante-Poku A., Angelucci F. Analysis of incentives and disincentives for cacao in Ghana. Technical notes series, Monitoring African Food and Agricultural Policies (MAFAP), FAO, Rome; 2013. 38 pp.
2. Ashitey E. Global agricultural information network (GAIN) Cacao Annual Report. No. 1202. Accra; 2012. p. 7.
3. Anthonio DC, Aikins ED. Reforming Ghana's cacao sector—an evaluation of private participation in marketing. Luleå University of Technology, Master Thesis, continuation courses marketing and e-commerce. Department of Business Administration and Social Sciences Division of Industrial marketing and e-commerce; 2009. pp. 114.
4. Anim-Kwapong G J, Frimpong EB. vulnerability and adaptation assessment under the Netherlands climate change studies assistance programme phase 2 (NCCSAP 2). Cocoa Research Institute of Ghana; 2005. pp. 44.
5. Sonwa DJ, Weise SF, Nkongmeneck BA, Tchatat M, Janssens MJJ. Carbon stock in smallholder chocolate forest in southern Cameroon and potential role in climate change mitigation. Earth Environ Sci 2009; 6. doi:10.1088/1755-1307/6/5/252008.
6. Oke D, Olatiilu A. Carbon storage in agroecosystems: a case study of the cocoa based agroforestry in Ogbese forest reserve, Ekiti State, Nigeria. J Environ Prot. 2011;2(8):1069–75.
7. Norgrove L, Hauser S. Carbon stocks in shaded *Theobroma cacao* farms and adjacent secondary forests of similar age in Cameroon. Trop Ecol. 2013;54(1):15–22.
8. Somarriba E, Cerda R, Orozco L, Cifuentes M, Dávila H, Espin T, Mavisoy H, Ávila G, Alvarado E, Poveda V, Astorga C, Say E, Deheuvels O. Carbon stocks and cacao yields in agroforestry systems of Central America. Agri Ecosyst Environ. 2013;173:46–57.
9. Gockowski J, Weise S, Sonwa D, Tchatat M, Ngobo M. Conservation because it pays: shaded cacao agroforests in West Africa. National Academy of Sciences in Washington DC on Feb 10 2004. (http://www.icraf.com/treesandmarkets/inaforesta/documents/shaded_cacao agroforests in_w_africa.pdf) Accessed 24 Aug 2014.
10. Ruf OF. The Myth of complex cacao agroforests: the case of Ghana'. Human Ecol. 2011;39:373–88.
11. Gockowski J, Sonwa D. Cacao intensification scenarios and their predicted impact on $CO_2$ emissions, biodiversity conservation, and rural livelihoods in the Guinea rain forest of West Africa. Environ Manag. 2011;48(2):307–21.
12. Sonwa DJ, Weise SF, Schroth G, Janssens MJJ, Shapiro HY. Plant diversity management in cacao agroforestry systems in west and central Africa – effects of markets and household needs. Agrofor Syst. 2014. doi:10.1007/s10457-014-9714-5.
13. Anglaaere LCN, Cobbina J, Sinclair FL, McDonald MA. The effect of land use systems on tree diversity: farmer preference and species composition of cacao-based agroecosystems in Ghana. Agrofor Syst. 2011;81:249–65.
14. Isaac ME, Gordon AM, Thevathasan N, Oppong SK, Quashie-Sam J. Temporal changes in soil carbon and nitrogen in West African multistrata agroforestry systems: a chronosequence of pools and fluxes. Agrofor Syst. 2005;65:23–31.
15. Taylor CJ. Synecology and silviculture in Ghana. Edinburgh: Thomas Nelson and Sons Ltd.; 1960. p. 418.
16. FAO/UNESCO soil map of the world. Revised Legend. FAO: Rome; 1990.
17. Ministry of Food and Agriculture, National Soil Fertility Management Action Plan, Directorate of crop services, MoFA, Accra, Ghana; 1998.
18. Acheampong E, Dawoe E, Bosu P, Asante W. Moving forward with REDD+ in Ghana: shade systems, crown cover, carbon stocks and socio-economic dynamics of smallholder Cacao, SNV Netherland Dev. Organization REAP Publication; 2014. pp. 81.
19. Chave J, Andalo C, Brown S, Cairns MA, Chambers JQ, Eamus D, Fölster H, Fromard F, Higuchi N, Kira T, Lescure JP, Nelson BW, Ogawa H, Puig H, Riera B, Yamakura T. Tree allometry and improved estimation of carbon stocks and balance in tropical forests. Oecologia. 2005;145:87–99.
20. Zanne AE, Lopez-Gonzależ G, Coomes DA, Ilic J, Jansen S, Lewis, SL, Miller RB, Swenson NG, Wiemann MC, Chave J. Global wood density database. 2009. Dryad. Identifier: http://hdl.handle.net/10255/dryad. 235. Assessed 15 June 2014.

21. IPCC. 2006 IPCC Guidelines for National Greenhouse Gas Inventories, Forestry. The Institute for Global Environmental Strategies (IGES), Japan. 2006.
22. Statistix 8.0 for Windows. Analytical software. Tallahassee; 2008.
23. Pearson T, Walker S, Brown S. Source book for land use change and forestry projects. Winrock international and biocarbon fund. 2007.
24. Ifejika CS. Resilient adaptation to climate change in African agriculture. DIE (Deutsches Institut für Entwicklungspolitik Studies 54) Bonn. 2010; 311.
25. Henry M, Tittonnell P, Manley RJ, Bermoux M, Albretch A, Vanlauwe W. Biodiversity, carbon stocks and sequestration potential in aboveground biomass in smallholder farming systems in western Kenya. Agric Ecosyst Environ. 2009;129:238–52.
26. Niggli U, Schmid H, Fliessbach A. Organic farming and climate change. Briefing prepared by Research Institute of Organic Agriculture FiBL, Frick, Switzerland for International Trade Centre ITC, Geneva; 2007. p. 30.
27. Rice RA, Greenberg R. Cacao cultivation and the conservation of biological diversity. Ambio. 2000;29:167–73.
28. Duguma B, Gockowski J, Bakala J. Smallholder cacao (*Theobroma cacao* Linn.) cultivation in agroforestry systems of West and Central Africa: challenges and opportunities. Agrofor Syst. 2001;51(3):177–88.
29. DiFalco S, Perrings C. Crop genetic diversity, productivity and stability of agroecosystems. A theoretical and empirical investigation. Scott J Polit Econ. 2003;50:207–16.
30. Sambuichi RHR, Widal DB, Piasentin FB, Jardim JG, Viana TG, Menezes AA, Mello DL, Ahnert D, Baligar VC. Caburca agroforests in southern Bahia, Brazil: tree component, management practices and tree species conservation. Biodivers Conserv. 2012;21(4):1055–77.
31. Sonwa DJ, Nkongmeneck BA, Weise SF, Tchatat M, Adesina A, Janssens MJJ. Diversity of plants in cacao agroforests in the humid forest zone of southern Cameroon. Biodivers Conserv. 2007;16(8):2385–400.
32. Rolim SG, Chiarello AG. Slow death of Atlantic forest trees in cacao agroforestry in southeastern Brazil. Biodivers Conserv. 2004;13:2679–94.
33. Paustian K, Six J, Elliot ET, Hunt HW. Management options for reducing CO2 emissions from agricultural soils. Biogeochemistry. 2000;48:147–63.
34. Albrecht A, Kandji ST. Carbon sequestration in tropical agroforestry systems. Agric Ecosyst Environ. 2003;99:15–27.
35. Montagnini F, Nair PKR. Carbon sequestration: an under-exploited environmental benefit of agroforestry systems. Agrofor Syst. 2004;61:281–95.
36. Dawoe E. Conversion of natural forest to cacao Agroforest in lowland humid Ghana: impact on plant biomass production, organic carbon and nutrient dynamics. Ph.D. thesis, Faculty of Renewable Natural Resources Kwame Nkrumah University of Science and Technology, Kumasi; 2011. pp. 260.
37. Asare RA, Kwakye Y, Quartey ET. Reducing deforestation in Ghana using a climate smart cocoa production strategy: the case for a cocoa forest REDD+ program. A paper presented at the 1st National Forestry Conference, Kumasi, Ghana, Sept. 16; 2014.
38. Leuschner C, Moser and G, Hertel D, Erasmi S, Leitner D, Culmsee H, Schuldt B, Schwendenmann L. Conversion of tropical moist forest into cacao agroforest: consequences for carbon pools and annual C sequestration. Agrofor Syst. 2013;87:1173–87.
39. Saj S, Jagoret P, Ngogue HT. Carbon storage and density dynamics of associated trees in contrasting *Theobroma cacao* agroforests of Central Cameroon. Agrofor Syst. 2013;87(6):1309–20.
40. Jacobi J, Andres C, Schneider M, Pillco M, Calizaya P, Rist S. Carbon stocks, tree diversity, and the role of organic certification in different cacao production systems in Alto Beni, Bolivia. Agrofor Syst. 2013. doi:10.1007/s10457-013-9643-8.
41. Kotto-Same J, Woomer PL, Appolinaire M, Louis Z. Carbon dynamics in slash-and-burn agriculture and land use alternatives of the humid forest zone in Cameroon. Agri Ecosyst Environ. 1997;65:245–56.
42. Smiley GL, Kroschel J. Temporal change in carbon stocks of cacao–gliricidia agroforests in Central Sulawesi. Indones Agrofor Syst. 2008;73(219):231.
43. Brown S, Gillespie AJR, Lugo AE. Biomass estimation methods for tropical forests with applications to forest inventory data. For Sci. 1989;35:881–902.
44. Tscharntke T, Clough Y, Bhagwat SA, Buchori D, Faust H, Hertel D, Hölscher D, Juhrbandt J, Kessler M, Perfecto I, Scherber C, Schroth G, Veldkamp E, Wanger TC. Multifunctional shade-tree management in tropical agroforestry landscapes—a review. J Appl Ecol. 2011;48(3):619–29.
45. Ong CK, Kho RM. Radersma S. In: van Noordwijk M, Cadisch G, Ong CK, editors. Ecological interactions in multispecies agroecosystems: concepts and rules. Belowground interactions in tropical agroecosystems. concepts and models with multiple plant components. Wallingford: CABI Publishing; 2004. p. 1–15.

10

# Future productivity and phenology changes in European grasslands for different warming levels: implications for grassland management and carbon balance

Jinfeng Chang[1,2*], Philippe Ciais[1], Nicolas Viovy[1], Jean-François Soussana[3], Katja Klumpp[4] and Benjamin Sultan[2]

**Abstract**

**Background:** Europe has warmed more than the global average (land and ocean) since pre-industrial times, and is also projected to continue to warm faster than the global average in the twenty-first century. According to the climate models ensemble projections for various climate scenarios, annual mean temperature of Europe for 2071–2100 is predicted to be 1–5.5 °C higher than that for 1971–2000. Climate change and elevated $CO_2$ concentration are anticipated to affect grassland management and livestock production in Europe. However, there has been little work done to quantify the European-wide response of grassland to future climate change. Here we applied ORCHIDEE-GM v2.2, a grid-based model for managed grassland, over European grassland to estimate the impacts of future global change.

**Results:** Increases in grassland productivity are simulated in response to future global change, which are mainly attributed to the simulated fertilization effect of rising $CO_2$. The results show significant phenology shifts, in particular an earlier winter-spring onset of grass growth over Europe. A longer growing season is projected over southern and southeastern Europe. In other regions, summer drought causes an earlier end to the growing season, overall reducing growing season length. Future global change allows an increase of management intensity with higher than current potential annual grass forage yield, grazing capacity and livestock density, and a shift in seasonal grazing capacity. We found a continual grassland soil carbon sink in Mediterranean, Alpine, North eastern, South eastern and Eastern regions under specific warming level (SWL) of 1.5 and 2 °C relative to pre-industrial climate. However, this carbon sink is found to saturate, and gradually turn to a carbon source at warming level reaching 3.5 °C.

**Conclusions:** This study provides a European-wide assessment of the future changes in productivity and phenology of grassland, and their consequences for the management intensity and the carbon balance. The simulated productivity increase in response to future global change enables an intensification of grassland management over Europe. However, the simulated increase in the interannual variability of grassland productivity over some regions may reduce the farmers' ability to take advantage of the increased long-term mean productivity in the face of more frequent, and more severe drops of productivity in the future.

**Keywords:** European grassland, Grassland management, Phenology, Warming levels, Climate change

*Correspondence: jinfeng.chang@locean-ipsl.upmc.fr
[1] Laboratoire des Sciences du Climat et de l'Environnement, UMR8212, CEA-CNRS-UVSQ, 91191 Gif-Sur-Yvette, France
Full list of author information is available at the end of the article

## Background

Global mean surface temperature (GMST) has increased since the late nineteenth century, and each of the past three decades has been warmer than all the previous decades [1]. If greenhouse gas (GHG) emissions continue unabated, GMST will continue to rise over the twenty first century [2]. Continual warming is predicted by the World Climate Research Program 'Climate Model Intercomparison Project' CMIP3 [3] and the more recent CMIP5 climate projections [4]; see [5] for the comparison between CMIP3 and CMIP5. For Europe, annual mean temperature for 2071–2100 is predicted to be 1–5.5 °C higher than that for 1971–2000. Specifically, a regional warming of 3–4.5 °C in the EU-FP6 ENSEMBLES multi-model ensemble for the scenario A1B of the Special Report on Emission Scenario (SRES) [6, 7], and of 1–4.5 and 2.5–5.5 °C in the new regional climate models ensemble EURO-CORDEX for the Representative Concentration Pathways RCP4.5 and RCP8.5 respectively [8]. Compared to pre-industrial climate, Europe has warmed more than the global average (land and ocean), and slightly more than global land temperature [9]. The decade from 2006 to 2015 was 1.5 °C warmer in Europe, against 0.83–0.89 °C at global scale. In the future, Europe is also projected to continue to warm faster than the global average (land and ocean; [9]). Spatially, greater warming in Southern Europe and towards the northeast is predicted in all climate scenarios. For annual precipitation, an increase in Central Europe and Northern Europe and a decrease in Southern Europe are predicted by regional climate change projections [8], i.e. a dryer Southern Europe and a wetter Northern Europe. In addition, more heavy precipitation events, extended dry spells and more heat waves are predicted in some regions of Europe in the future [8].

Grassland ecosystems cover 56.8 million ha (13.2%) of the land area in the EU-27 [10]. Most of these grasslands are used to feed animals, either directly by grazing or indirectly by grass harvest (mowing). Accounting for multiple drivers of climate change, rising $CO_2$, nitrogen addition, and land cover and management intensity changes, a recent study suggested that European grasslands acted as a C sink in the past five decades (15 ± 7 g C $m^{-2}$ $year^{-1}$ for 1961–2010; Chang et al. [11, 12]). In the future, emission scenarios and predicted climate change can profoundly impact the production (annual amount and seasonality), management (grazing timing and animal carrying capacity), and the carbon balance and non-$CO_2$ greenhouse gas emissions (i.e., $CH_4$ and $N_2O$) of grassland ecosystems.

The response of grasslands to climate change is complex as it implies interactions with water availability, nutrients, soil vegetation and management intensity [13]. Elevated $CO_2$ concentration has the dual effect of increasing leaf photosynthesis and leaf area index, and reducing stomatal conductance. These effects at ecosystem-scale result in an increase in above-ground dry matter (DM) production of grassland [14–16], increase water-use efficiency [17] and reduce the consumption of soil moisture by plant transpiration [18]. However, trends and variability in temperature and precipitation, as well as possible nitrogen limitations, all interact with the effects of elevated $CO_2$ to determine actual changes in grassland productivity [18–20]. In addition, there could be some risks for surviving/adapting of the grass species under rapid climate change, given the fact that the projected rate of climate change far outweighs the rate of niche change in grasses [21].

Climate change and elevated $CO_2$ concentration are anticipated to affect grassland management and livestock production in Europe [22, 23] with economic consequences that are yet to be sufficiently assessed [24]. The response of grazing systems may also vary markedly across European regions and with pasture type (e.g., [25, 26]). Under projected future conditions, Graux et al. [27] simulated an increased inter-annual and seasonal variation of grassland production at 12 contrasted French grassland sites using the Pasture Simulation model (PaSim). They also predicted to a significant increase in summer drought risk. Using a probabilistic risk analysis, Van Oijen et al. [28] estimated the drought vulnerability and risk of the carbon and water balance across Europe based on vegetation models, including PaSim for grassland ecosystems. Projections of climate change impacts on European grassland productivity are mostly based on local modeling studies that require many local variables, and this limits the up-scaling to regional or continental scale (e.g., [29]). Grid-based process-based vegetation models with equations representing biogeochemical and biophysical mechanisms have the advantage of being applicable from local to continental scale. This type of mechanistic models is increasingly used for global impact studies on agricultural productivity and terrestrial carbon fluxes (e.g. ISIMIP, http://www.isi-mip.org). However, most of the existing grid-based process-based ecosystem models currently simulate managed grassland either as natural grassland or as a sort of cropland with intensive harvest. Therefore, in order to make more realistic predictions of the impact of future climate change on European grassland, management processes must be included in grassland ecosystem models.

In the COP 21 UN Climate change conference, the Paris Agreement [30] recalls the article 2 of the United Nations Framework Convention on Climate Change [31], implementing the aim of *"holding the increase in the global average temperature to well below 2 °C above pre-industrial levels and to pursue efforts to limit the temperature increase to 1.5 °C above pre-industrial levels, recognizing that this would significantly reduce the risks and impacts of climate change".* In this context, the objective of this study is to predict changes in productivity, growing season, management and carbon fluxes of European grassland when global warming reaches these specific warming levels (SWL) of 1.5, 2 °C relative to pre-industrial climate. The impacts of high warming levels of 3 and 3.5 °C are also investigated to quantify the risks that could be avoided by the Paris agreement. Last, a sensitivity test is carried out to investigate the impact of fixed vs. adjustable livestock intensification on grassland carbon fluxes under climate change and the rising $CO_2$ concentration.

## Methods

### Model description

ORCHIDEE is a process-based ecosystem model built for simulating carbon cycling in ecosystems, and water and energy fluxes from site-level to global scale [32–34]. ORCHIDEE-GM is a version specifically developed to integrate the management of grassland with two options [35]. Either the model can be forced by observed animal density for grazing or by observed harvest time for forage removals, or it can calculate the *optimal* densities and practices that maximize the use of ecosystem productivity. The equations describing management in ORCHIDEE-GM are derived from PaSim [36]; Vuichard et al. [37–39]. ORCHIDEE-GM version 1 was evaluated and some of its parameters calibrated at 11 European grassland sites representative of a range of management practices, with eddy covariance net ecosystem exchange (NEE) and biomass measurements. The model showed capability to simulate net biome productivity (NBP; i.e. the C balance) of these managed grasslands [35] even though it does not include an explicit nitrogen cycle interacting with the carbon cycle. Chang et al. [40] further added a new parameterization to describe an adaptive management strategy of farmers who react to a climate driven change of previous-years' productivity. The positive effect of N addition on grass photosynthesis, and thus on subsequent ecosystem carbon balance, are parameterized with a simple empirical function calibrated from literature estimates (version 2.1; [40]. At continental scale, ORCHIDEE-GM v2.1 was applied over Europe to calculate the spatial pattern, recent trends and interannual variability of potential productivity (the productivity corresponding to an optimal management practice that maximizes livestock densities). This version was further used to simulate NBP and NBP trends over European grasslands during the last five decades at a spatial resolution of 25 km [11, 12]. In this study, ORCHIDEE-GM v2.2 is released with (1) an update of the general parameterizations from ORCHIDEE Trunk. rev3623 (https://forge.ipsl.jussieu.fr/orchidee/browser/trunk#ORCHIDEE), and (2) a new parameterization incorporated specific management strategies.

### Specific management strategies in ORCHIDEE-GM v2.2

Grazing when the air temperature is below freezing point 0 °C (frost-grazing) and grazing on grassland covered by snow (snow-grazing) increase farmers labor as additional energy is needed for maintenance and feed supplementation. Furthermore, as for wet soil described below, grazing of snow-covered ecosystems often leads to grassland degradation and trampling due to concentration of animals around feed points. Therefore, frost-grazing and snow-grazing are usually avoided by farmers given the fact that they are not economically efficient.

Moreover, livestock trampling over wet soil (wet-grazing) causes excessive soil compaction, represented by the increase in bulk density and soil strength, and the decrease in water infiltration rate [41]. Excessive soil compaction could result in soil degradation and further impact the sustainability of pasture productivity. Thus in practice, farmers tend to avoid grazing when the soil is too wet.

Here, the following new set of management rules has been incorporated into ORCHIDEE-GM v2.2 to represent the above limitations on grazing:

1. Grazing stops if daily mean air temperature drops below 0 °C.
2. Grazing stops when grassland is covered by snow, and for three consecutive days after snow melt. To simulate snow cover, the mechanistic intermediate-complexity snow scheme (ISBA-ES; Boone and Etchevers [42] has been implemented in ORCHIDEE-GM v2.2 (updated with ORCHIDEE Trunk.rev3623), which improves the representation of snow processes such as snowmelt timing [43].
3. Grazing stops when soil becomes too wet (i.e., soil moisture close to saturation), and can only be resumed after at least 10 consecutive days after the stop to avoid soil degradation due to trampling. Here, wet soil condition is defined in the model by daily mean soil moisture content (*mc*) of topsoil (0–9 cm in this study) being close to saturation:

$$mc \geq mc_{sat} - \Delta_{crit} \qquad (1)$$

where $mc_{sat}$ is the saturated soil moisture content derived from Carsel and Parrish [44]; the threshold $mc_{sat} - \Delta_{crit}$ represents an empirical value determining wet soil conditions; the value of $\Delta_{crit}$ is set to 0.05 $m^3\ m^{-3}$. To avoid grazing cessation during instantaneous soil saturation caused by discrete heavy precipitation, grazing only stops when wet soil conditions last for over 3 days during a running window of the previous 5 days. The 0–9 cm mean soil moisture is chosen here given the fact that (1) instead of the two-layer simple bucket used in previous ORCHIDEE-GM versions, ORCHIDEE-GM v2.2 here includes the more complex 11-layer soil–water diffusion scheme (implemented in ORCHIDEE Trunk.rev3623; De Rosnay et al. [45, 46], D'Orgeval [47, 48] to be able to simulate soil moisture in different soil layers from 1 mm depth for the first layer down to 2 m depth for the 11th layer (e.g., the top 5, 6, 7 layers of the soil represent 0–5, 0–9, and 0–19 cm soil depth respectively), (2) within the above vertical profile that model used, 9 cm is the most realistic depth impacted by livestock trampling (around 10 cm in practice). A sensitivity test using the top 6 (0–9 cm) or 7 layers (0–19 cm) of soil moisture as topsoil *mc* only shows marginal difference in both the *mc* value and the impact on grazing (data not shown).

#### Biological *potential* productivity, grazing capacity, and optimal livestock density

ORCHIDEE-GM simulates the two practices through which managed grassland provides grass biomass for livestock: mowing and grazing. Model rules automatically determine the frequency and magnitude of mowing (harvests) events in each grid cell [39], which simulates a potential (maximum) productivity. Here, we define grassland *potential* productivity as the annual maximum production of forage from mown grassland ($Y_{pot}$, kg of DM per hectare of mown grasslands). In terms of grazing, due to the impact of livestock on grass growth through trampling, defoliation (i.e., biomass intake) etc., and because grassland cannot be grazed continuously over the whole vegetation period, biomass production has to meet minimal conditions before animals can be applied (e.g., 18 kg DM intake day$^{-1}$ LU$^{-1}$; LU, livestock unit; [12]. Accordingly, thresholds of shoot biomass (Vuichard et al. [38]; updated in [37], snow, temperature and soil moisture conditions (see "Specific management strategies in ORCHIDEE-GM v2.2" for detail) are set for starting, stopping and resuming grazing. Consequently, the model estimates the number of grazing days ($N_{grazing}$) and barn days (i.e., off-site period; e.g., $N_{barn} = 365 - N_{grazing}$ for each year) for each grid and each time period (month or year). The grazing capacity for any time period ($C_{grazing,i}$; unit: LU days ha$^{-1}$) can be calculated as:

$$C_{grazing,i} = S_{opt,i} \times N_{grazing,i} \quad (2)$$

where $S_{opt,i}$ is the optimal instantaneous animal stocking rate at time period $i$ (LU per hectare of grazed grassland; $i$ can be 1 day, 1 month, one season or 1 year); $N_{grazing,i}$ is the number of days when animals are grazing the pasture.

Meanwhile a set of rules allowing the simulation of idealized self-sufficient herbage-based ruminant livestock farm [39] have been introduced into ORCHIDEE-GM v2.1 [40]. These rules are based upon two assumptions, (1) within each grid cell, livestock can only be fed by herbage (i.e. crop-fed animals are not considered), and (2) the use of grassland production is maximized in each grid cell, by calculating the mix of mowing and grazing that maximizes the number of animals in that grid cell. Under these assumptions, the optimal instantaneous animal stocking rate, $S_{opt}$, the optimal proportion of grazed vs. mown grasslands, $F_{opt}$ (within [0,1]), and thus an optimal livestock density, $D_{opt}$ (LU per hectare of total grassland) are calculated for each grid cell using the optimization algorithm of Vuichard et al. [39]. We further incorporated specific rules that were incorporated into ORCHIDEE-GM v2.1, in order to model adaptive management in response to climate variability. In other words, $S_{opt}$, $F_{opt}$, and $D_{opt}$ change in response to climate-driven changes in the grassland productivity of previous years [40].

#### Defining active growing season and grazing season

A positive NPP defines grass growth and the accumulation of biomass, which further impacts grazing practice. To detect a temporal shift (i.e., advance or delay of the active growing season and grazing season), we use four indicators in this study. The first one is the length of the active growing season in each grid cell ($L_{growing}$, unit: days per year) defined as the number of days per year when NPP is higher than 1 g C m$^{-2}$ day$^{-1}$. The second one is the beginning of the active growing season in each grid cell ($B_{growing}$, unit: day of year) being the first day of the year when (1) the grassland NPP is higher than 1 g C m$^{-2}$ day$^{-1}$, and (2) a positive NPP trend (i.e., growth rate is larger than 0.05 g C m$^{-2}$ day$^{-2}$) is found for the following 10 days. The third one is the length of grazing season in each grid cell ($L_{grazing}$, unit: days per year) defined as the number of days per year when grazing is possible with respect to above-ground biomass availability [40], air temperature, snow cover, and soil moisture conditions (see "Specific management strategies in ORCHIDEE-GM v2.2" for detail). The fourth indicator is the beginning of the grazing season ($B_{grazing}$, unit: day of year) defined in each grid-cell as the first day of the year when the following two criteria are met: (1) grazing is allowed over grassland by model, and (2) grazing will last more than 20 days in the following 30 days. The second criterion is to guarantee that the $B_{grazing}$ indicates the start of a continual grazing season rather than the sporadic grazing.

### Simulation set-up

ORCHIDEE-GM v2.2 is integrated over Europe using the harmonized climate dataset from the ERA-WATCH reanalysis at a spatial resolution of 0.25° by 0.25° for the period 1901–2010 [49]. A harmonized climate forcing and dynamical downscaling was used for future climate simulations based on the general circulation model (GCM) ECHAM5 (2010–2100) using 1970–2010 as the reference period to correct the GCM bias. Namely, results of the regional climate model (RCM) REMO driven by ECHAM5 GCM climate for the SRES A1B scenario provided by the ENSEMBLE project (http://www.ensembles-eu.org/) are bias-corrected [49] and used in this study. This forcing is described in Beer et al. [49], and hereafter referred to as bias-corrected REMO + ECHAM5 climate. The SRES A1B scenario has a rapid increase in fossil $CO_2$ emissions until 2050 and a decrease afterwards, and the projected atmospheric $CO_2$ concentration reaches 527.9 ppm in 2050 and 713.8 ppm in 2100. Climate change from ECHAM5 under this scenario leads to a GMST increase comparable to that of RCP6, or between RCP4.5 and RCP8.5 [50]. All climate data have the same spatial resolution of 0.25° by 0.25° and daily temporal resolution. This resolution (0.25° by 0.25°) is sufficient to represent regional meteorological regimes accurately in low-lying regions, but not in mountainous areas. Generally, the patterns of temperature and precipitation change in Europe projected by bias-corrected REMO + ECHAM5 climate (Additional file 1: Figures S1a, b and S2a, b) and non bias-corrected climate (Additional file 1: Figures S1c, d and S2c, d) are similar to those of the ensemble-mean from other RCMs and GCMs in the ENSEMBLES project for the same SRES A1B scenario (Additional file 1: Figures S1e, f and S2e, f; [8]. The changes in ensemble-mean temperature (from the ENSEMBLES project for the SRES A1B scenario) are comparable with those from the more recent EURO-CORDEX ensemble of RCMs for the Representative Concentration Pathways RCP4.5 and RCP8.5 [8]. Global atmospheric $CO_2$ concentration used to force ORCHIDEE-GM v2.2 is prescribed from the combination of ice core records and atmospheric observations for 1901–2010 [51] and update) and the SRES A1B scenario for 2011–2100.

The historical simulation for the period 1901–2010 (hereafter referred to as $E_{hist}$) is exactly the same than the one performed by Chang et al. [12]; see their Fig. 1. For the future period 2011–2100, three experiments are carried out to (1) separate the effect of rising $CO_2$ concentration and future climate change, and (2) to investigate the impact of fixed vs. adjustable management on the carbon balance of grasslands. Table 1 presents the drivers for different experiments.

The first experiment (hereafter referred to as $E_{control}$) is carried out with increasing atmospheric $CO_2$ concentration, and variable climate (bias-corrected REMO + ECHAM5 climate), considering no changes in the nitrogen status of plants (i.e., the same nitrogen effect on photosynthesis as that of 2010; see [12] for detail) and no land cover change (*HILDA* data set; [52]. In more details, a constant land cover map for the period 2011–2100 (Table 1) corresponded to (a) no land-cover change on forest, cropland and grassland area, (b) constant managed/unmanaged grassland fraction as that of 2010 (as calculated in [12], while (c) within managed grassland systems, the intensification/extensification of management (i.e., $D_{opt}$ increase/decrease), and the fraction between mown and grazed part is determined by the adaptive management change algorithm following the productivity change due to rising atmospheric $CO_2$ concentration and variable climate.

The second experiment has the same settings as $E_{control}$ but uses constant $CO_2$ concentration as in the year 2010 (hereafter referred to as $E_{noco2}$).

The third experiment is like $E_{control}$ but with constant livestock density over managed grassland as in the year 2010 (thus keeping the same livestock numbers over European grasslands; hereafter referred to as $E_{fixD}$).

The $E_{noco2}$ and $E_{fixD}$ simulations are used in "Discussion" for discussing the effect of rising $CO_2$ on productivity and the grassland carbon balance, and the impact of livestock numbers on grassland carbon balance respectively. The results presented in Sect. 3 are all derived from $E_{hist}$ for the reference period and $E_{control}$ for the future period.

### Represented time slices

Results for five time slices are presented in this study, namely the reference period (1981–2010) and different SWL of +1.5, +2, +3 and +3.5 °C relative to pre-industrial climate (hereafter referred to as SWL1.5, SWL2, SWL3, and SWL3.5). A SWL is defined as the 30-years period when the average global mean temperature reaches a given warming level compared to the pre-industrial period 1881–1910, following [53]. Due to the fast increase in the projected global mean temperature, there are overlaps across SWL time periods. Table 2 shows the central year and corresponding periods reaching the different SWLs in the ECHAM5 GCM simulation for SRES A1B scenario [54].

Our study domain covers 30 countries (EU-28 plus Norway and Switzerland), which were grouped into a number of major agricultural regions determined by environmental and socio-economic factors (Additional file 1: Table S1, detailed description in [55]. Regional mean biogenic potential productivity ($Y_{pot}$), grazing capacity ($C_{grazing}$), optimal livestock density ($D_{opt}$), and

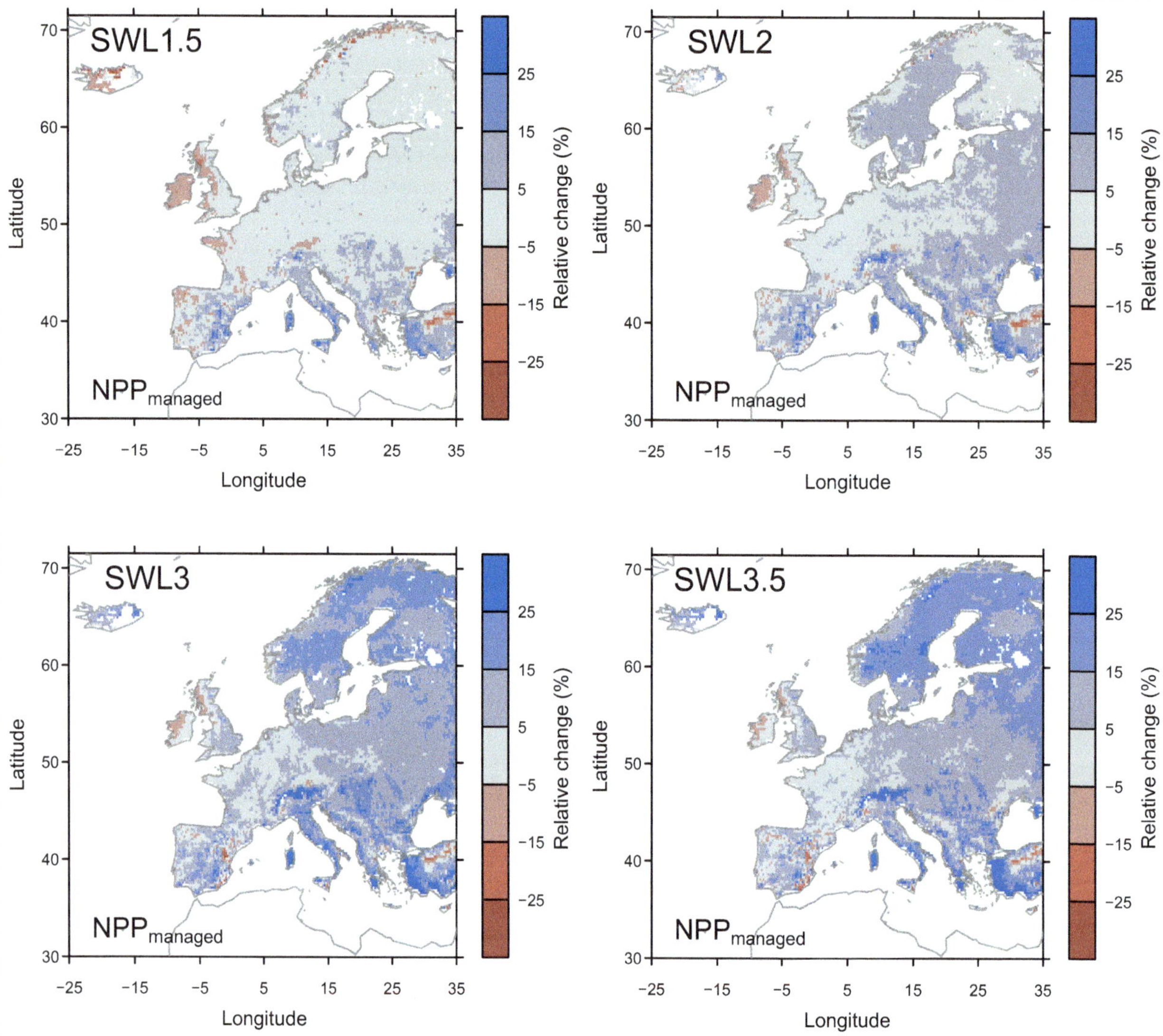


**Fig. 1** Simulated relative changes of $NPP_{managed}$ from experiment $E_{control}$. The change of NPP is defined as the difference between $NPP_{managed}$ during 30 years in the future corresponding to each SWL minus $NPP_{managed}$ in reference historical period (1981–2010)

NPP of managed grassland ($NPP_{managed}$) are averaged based on the managed grassland area in each grid cell. Regional mean net carbon fluxes such as NEE and NBP (calculated as in [12] are averaged based on the total grassland area in each grid cell (including both managed and unmanaged part).

#### Factorial attribution of the changes in the grazing season length ($L_{grazing}$)

In ORCHIDEE-GM v2.2, the grazing is determined by above-ground biomass availability [40] and by air temperature, snow cover, and soil moisture conditions ("Specific management strategies in ORCHIDEE-GM v2.2" for detail). Thus all these four factors could contribute to the simulated changes in $L_{grazing}$ (i.e., $\Delta L_{grazing}$ between the SWLs and the reference period). To determine the dominant factor affecting changes in grazing season length, $\Delta L_{grazing}$, we define $\Delta L_{grazing,\ biomass}$, $\Delta L_{grazing,\ frost}$, $\Delta L_{grazing,\ snow}$, and $\Delta L_{grazing,\ wet}$ as the individual contribution from changes on above-ground biomass availability, air temperature, snow cover, and soil moisture conditions respectively on the $\Delta L_{grazing}$. $\Delta L_{grazing,\ biomass}$ is calculated as:

$$\Delta L_{grazing,\ biomass} = L_{grazing,\ biomass,\ SWL} - L_{grazing,\ biomass,\ ref} \quad (3)$$

where $L_{grazing,\ biomass,\ SWL}$ and $L_{grazing,\ biomass,\ ref}$ are the mean number of days per year during when grazing is limited by above-ground biomass availability for a given SWL and for the reference period respectively. Here, positive $\Delta L_{grazing,\ biomass}$ indicates that the changes in above-ground

**Table 1 Drivers for different experiments in this study**

| Experiment | Climate[a] | Land cover[b] | Managed grassland area[c] | Atmospheric $CO_2$ concentration | Management strategy |
|---|---|---|---|---|---|
| $E^{d}_{hist}$ | Varied for 1901–2010 | Varied for 1901–2010 | Varied for 1901–2010 | Varied for 1901–2010 | Adaptive management change algorithm simulating potential livestock density |
| $E_{control}$ | Varied for 2011–2100 | Constant as in 2010 | Constant as in 2010 | Varied for 2011–2100 (A1B scenario) | Adaptive management change algorithm simulating potential livestock density |
| $E_{noco2}$ | Varied for 2011–2100 | Constant as in 2010 | Constant as in 2010 | Constant as in 2010 | Adaptive management change algorithm simulating potential livestock density |
| $E_{fixD}$ | Varied for 2011–2100 | Constant as in 2010 | Constant as in 2010 | Varied for 2011–2100 (A1B scenario) | Adaptive management change algorithm with constant livestock density |

[a] Climate change for 2011–2100 is predicted by REMO driven by ECHAM5 climate for A1B scenario [49] provided by the ENSEMBLE project (http://www.ensembles-eu.org/)

[b] Land cover is derived from HILDA dataset [52]

[c] Area of managed grassland is calculated using the same method as in previous study for historical period [40]

[d] See Chang et al. [40] for detail protocol of simulation

**Table 2 Time period and the corresponding annual mean temperature ($T_{Europe}$) and annual total precipitation ($P_{Europe}$) of Europe for which +1.5, +2, +3 and +3.5 °C global warming compared to pre-industrial times was reached in ECHAM5 GCM global climate for SRES A1B scenario provided by the ENSEMBLE project**

| | SWL[a] (°C) | $T_{Europe}$ (°C) | $P_{Europe}$ (mm year$^{-1}$) | Central year | Period |
|---|---|---|---|---|---|
| Reference | +0.68[b] | 8.6 | 818 | 1995 | 1981–2010 |
| SWL1.5 | +1.5 | 9.6 | 818 | 2035 | 2021–2050 |
| SWL2 | +2 | 10.2 | 815 | 2048 | 2034–2063 |
| SWL3 | +3 | 11.2 | 818 | 2072 | 2058–2087 |
| SWL3.5 | +3.5 | 11.7 | 821 | 2085 | 2071–2100 |

[a] *SWL* specific warming level relative to pre-industrial (1881–1910) climate. A SWL is defined as the 30-years period when the average global mean temperature reaches a given warming level compared to the pre-industrial period 1881–1910, following Vautard et al. [53]. Due to the fast increase in the projected global mean temperature, there are overlaps across SWL time periods

[b] The SWL in 'Reference' period (1981–2010) is the average SWL based on the three global observational datasets: NASA GISS (+0.70 °C; http://data.giss.nasa.gov/gistemp/), HadCRUT4 (+0.66 °C; http://www.metoffice.gov.uk/hadobs/hadcrut4/) and NOAA NCDC (+0.68 °C; http://www.ncdc.noaa.gov/cmb-faq/anomalies.html#anomalies). Data were Accessed in September, 2016

biomass availability tend to lengthen $L_{grazing}$ (i.e., leading to positive $\Delta L_{grazing}$) and vice versa. $\Delta L_{grazing,\,frost}$, $\Delta L_{grazing,\,snow}$, and $\Delta L_{grazing,\,wet}$ are calculated as:

$$\Delta L_{grazing,\,frost} = -(L_{grazing,\,frost,\,SWL} - L_{grazing,\,frost,\,ref}) \quad (4)$$

$$\Delta L_{grazing,\,snow} = -(L_{grazing,\,snow,\,SWL} - L_{grazing,\,snow,\,ref}) \quad (5)$$

$$\Delta L_{grazing,\,wet} = -(L_{grazing,\,wet,\,SWL} - L_{grazing,\,wet,\,ref}) \quad (6)$$

where $L_{grazing,\,frost,\,SWL}$, ($L_{grazing,\,snow,\,SWL}$, and $L_{grazing,\,wet,\,SWL}$) and $L_{grazing,\,frost,\,ref}$ ($L_{grazing,\,snow,\,ref}$ and $L_{grazing,\,wet,\,ref}$) are the mean number of days per year that grazing is allowed by above-ground biomass availability but prevented by the management strategy to avoid frost-grazing (snow-grazing, and wet-grazing; "Specific management strategies in ORCHIDEE-GM v2.2") during SWLs and during the reference period respectively. Here, a positive $\Delta L_{grazing,\,frost}$ value (or $\Delta L_{grazing,\,snow}$, and $\Delta L_{grazing,\,wet}$ value) indicates that the changes in the number of freezing days (snow-covering days, and wet soil days) tend to lengthen $L_{grazing}$ (i.e., leading to a positive $\Delta L_{grazing}$) and vice versa. We defined the *dominant* factor impacting $\Delta L_{grazing}$ as the factor (a) having the same sign of change than $\Delta L_{grazing}$, and (b) having the largest contribution (absolute value) among all the factors.

**Table 3 Simulated productivity of European managed grassland for the reference period (1981–2010), and their relative changes derived from experiment $E_{control}$**

| Variables | Period[a] | Average of Europe | Nordic | British Isles | Western | Mediterranean | Alpine | North eastern | South eastern | Eastern |
|---|---|---|---|---|---|---|---|---|---|---|
| $Y_{pot}$ (kg DM $m^{-2}$ $year^{-1}$) | Reference | 0.85 | 0.49 | 0.93 | 1.07 | 0.71 | 0.66 | 0.80 | 0.80 | 0.73 |
| | SWL1.5 | 6% | 6% | −3% | 5% | 10% | 10% | 9% | 13% | 11% |
| | SWL2 | 11% | 13% | 2% | 10% | 14% | 20% | 17% | 22% | 18% |
| | SWL3 | 18% | 28% | 8% | 16% | 18% | 37% | 26% | 31% | 28% |
| | SWL3.5 | 20% | 36% | 11% | 18% | 16% | 46% | 29% | 33% | 35% |
| $C_{grazing}$ (LU day $ha^{-1}$) | Reference | 208 | 61 | 241 | 286 | 175 | 128 | 187 | 196 | 123 |
| | SWL1.5 | 13% | 36% | 6% | 11% | 12% | 22% | 27% | 17% | 31% |
| | SWL2 | 20% | 56% | 10% | 15% | 15% | 36% | 42% | 29% | 51% |
| | SWL3 | 29% | 93% | 17% | 22% | 18% | 60% | 59% | 41% | 81% |
| | SWL3.5 | 32% | 119% | 21% | 23% | 16% | 73% | 66% | 43% | 98% |
| $D_{opt}$ (LU $ha^{-1}$) | Reference | 0.74 | 0.33 | 0.84 | 0.97 | 0.63 | 0.51 | 0.67 | 0.70 | 0.54 |
| | SWL1.5 | 10% | 17% | 4% | 9% | 11% | 18% | 18% | 13% | 20% |
| | SWL2 | 16% | 28% | 8% | 14% | 15% | 28% | 29% | 23% | 32% |
| | SWL3 | 24% | 47% | 14% | 20% | 19% | 49% | 44% | 35% | 51% |
| | SWL3.5 | 28% | 61% | 17% | 23% | 20% | 59% | 50% | 39% | 61% |
| $NPP_{managed}$ (g C $m^{-2}$ $year^{-1}$) | Reference | 861 | 597 | 937 | 1035 | 703 | 771 | 867 | 815 | 835 |
| | SWL1.5 | 0% | 1% | −7% | −1% | 7% | 3% | 1% | 6% | 1% |
| | SWL2 | 4% | 6% | −4% | 2% | 10% | 9% | 4% | 12% | 5% |
| | SWL3 | 8% | 14% | 0% | 5% | 13% | 17% | 8% | 16% | 12% |
| | SWL3.5 | 8% | 17% | 1% | 5% | 12% | 23% | 9% | 16% | 14% |

[a] The values for the reference period (1981–2010) are expressed with the original unit, and the relative changes under SWL of 1.5, 2, 3 and 3.5 °C are presented as percent (%) increase (+) or decrease (−) compared to the values for the reference period

## Results

### Changes in annual $NPP_{managed}$, $Y_{pot}$, $C_{grazing}$, and $D_{opt}$ from $E_{control}$ in response to global change

Compared to the reference period 1981–2010, $NPP_{managed}$ over European grasslands is projected to change by 0, +4 +8 and +8% under SWL of 1.5, 2, 3 and 3.5 °C respectively, but with large spatial variation (Table 3; Fig. 1). Under SWL of 1.5 °C, we found that (1) $NPP_{managed}$ decreases by more than 5% over Iceland, western British Isles, Brittany and some regions in south France, and the western Iberian Peninsula, (2) $NPP_{managed}$ increases by more than 5% over Mediterranean regions and Alps, and (3) little changes in other regions (Fig. 1), compared to the reference period. $NPP_{managed}$ is simulated to increase with further warming (i.e., under SWL of 2, 3 and 3.5 °C; Fig. 1) over most regions. One exception is the eastern Iberian Peninsula, where $NPP_{managed}$ decreases under SWL of 3 and 3.5 °C due to a strong decrease in precipitation (Additional file 1: Figure S2b). The spatial pattern of the relative changes in $Y_{pot}$ follows that of $NPP_{managed}$, while $Y_{pot}$ has larger relative increase (Additional file 1: Figure S3). The relative changes in $C_{grazing}$ show a little different spatial pattern from that of $Y_{pot}$. An annual $C_{grazing}$ increase is simulated over most regions of Europe under all SWLs (Additional file 1: Figure S4), including Iceland and the western part of British Isles. And the relative change in annual $C_{grazing}$ is stronger than that of $Y_{pot}$ (Table 3). It is noteworthy that consistent decreases in $NPP_{managed}$, $Y_{pot}$ and $C_{grazing}$ were simulated over eastern Iberian Peninsula under SWL of 3 and 3.5 °C. The relative change in $D_{opt}$ over managed grassland (Additional file 1: Figure S5) is in between changes in $Y_{pot}$ and in $C_{grazing}$ (Table 3).

We use the standard deviation (SD) to diagnose the magnitude of the interannual variability in modeled productivity during different time periods. Compared to the reference period, a large increase in SD of $NPP_{managed}$ is simulated over northern and eastern Europe, Iceland, Ireland, and the western Great Britain under all SWLs (Fig. 2), and over France and most part of the Iberian Peninsula under SWL of 3 and 3.5 °C. Model simulates decrease in SD of $NPP_{managed}$ over the eastern Great Britain, Denmark, Germany, Poland, northwestern Spain, and northern Greece and southern Sweden under all SWLs.

### Growing season and management shifts

ORCHIDEE-GM v2.2 simulates an earlier start of grassland growing season (i.e., negative $\Delta B_{growing}$ in Fig. 3a, b) over some regions of Europe under SWL of 1.5 °C. The earlier start under SWL of 3.5 °C becomes widespread all over the Europe except in Ireland, western France and eastern Portugal. However, this earlier start of the active growing season does not necessarily translate into a longer active growing season. Increasing $L_{growing}$ is simulated only over some Mediterranean regions, Alps, and mountains in Eastern Europe under all SWLs, and decreasing $L_{growing}$ over other regions (Fig. 3c, d).

An earlier start of the grazing season (i.e., negative $\Delta B_{grazing}$ in Fig. 4a, b) was simulated over some European grassland under SWL of 1.5 °C, and over most part of Europe under SWL of 3.5 °C. The largest advance was simulated over Alps and northern Europe under SWL of 3.5 °C. Longer grazing seasons (i.e., positive $\Delta L_{grazing}$ in Fig. 4c, d) were simulated over most parts of Europe, except in the Atlantic coast (western British Isles, western France, western Spain, and Portugal), southern Italy and Greece under all SWLs. Under SWL of 3.5 °C, shorter grazing seasons were simulated over France and the Iberian Peninsula, as well as in most parts of Italy and southeastern Europe. Attribution analysis of $\Delta L_{grazing}$ ("Factorial attribution of the changes in the grazing season length (*Lgrazing*)"; Fig. 4e, f) shows that (1) above-ground biomass availability is the dominant factor determining $\Delta L_{grazing}$ in most regions of Europe (Additional file 1: Figure S6), (2) snow cover is the dominant factor over some regions of British Isles, the coast of the North Sea and Baltic Sea, and eastern Europe (Additional file 1: Figure S7), (3) the impact of changes in frost days have similar spatial patterns but weaker than the impact of snow cover (Additional file 1: Figure S8), and (4) wet soils only play a dominant role in explaining $\Delta L_{grazing}$ over few regions (Additional file 1: Figure S9). It is noteworthy that snow-impact dominated regions do expand under SWL3.5 in mid-to-high latitude Europeans regions (Fig. 4f), indicating that changes in snow cover become an increasingly important factor impacting $L_{grazing}$ under higher warming level.

Following the simulated changes in grazing season and in $S_{opt}$, we found a $C_{grazing}$ increase for spring (March, April and May; Fig. 5a, b) under all SWLs and over most part of Europe except in high-latitude regions where grass leaf onset is usually later than May. European-wide $C_{grazing}$ increase was also simulated for summer (June, July and August; Fig. 5c) under SWL of 1.5 °C, while summer $C_{grazing}$ tended to decrease over the southern Iberian Peninsula, and Mediterranean coast of France under SWL of 3.5 °C. For autumn (September, October and November; Fig. 5e, f), a significant $C_{grazing}$ decrease was simulated over regions where grazing season become shorter in the future (Fig. 4c, d), due to an earlier end to the growing season.

### The carbon balance of European grassland in the future

Under SWL of 1.5 °C, ORCHIDEE-GM v2.2 simulates a carbon sink in grassland soils (i.e., positive NBP) over Southern and Eastern Europe, and a carbon source over the western British Isles, and some regions in Western and

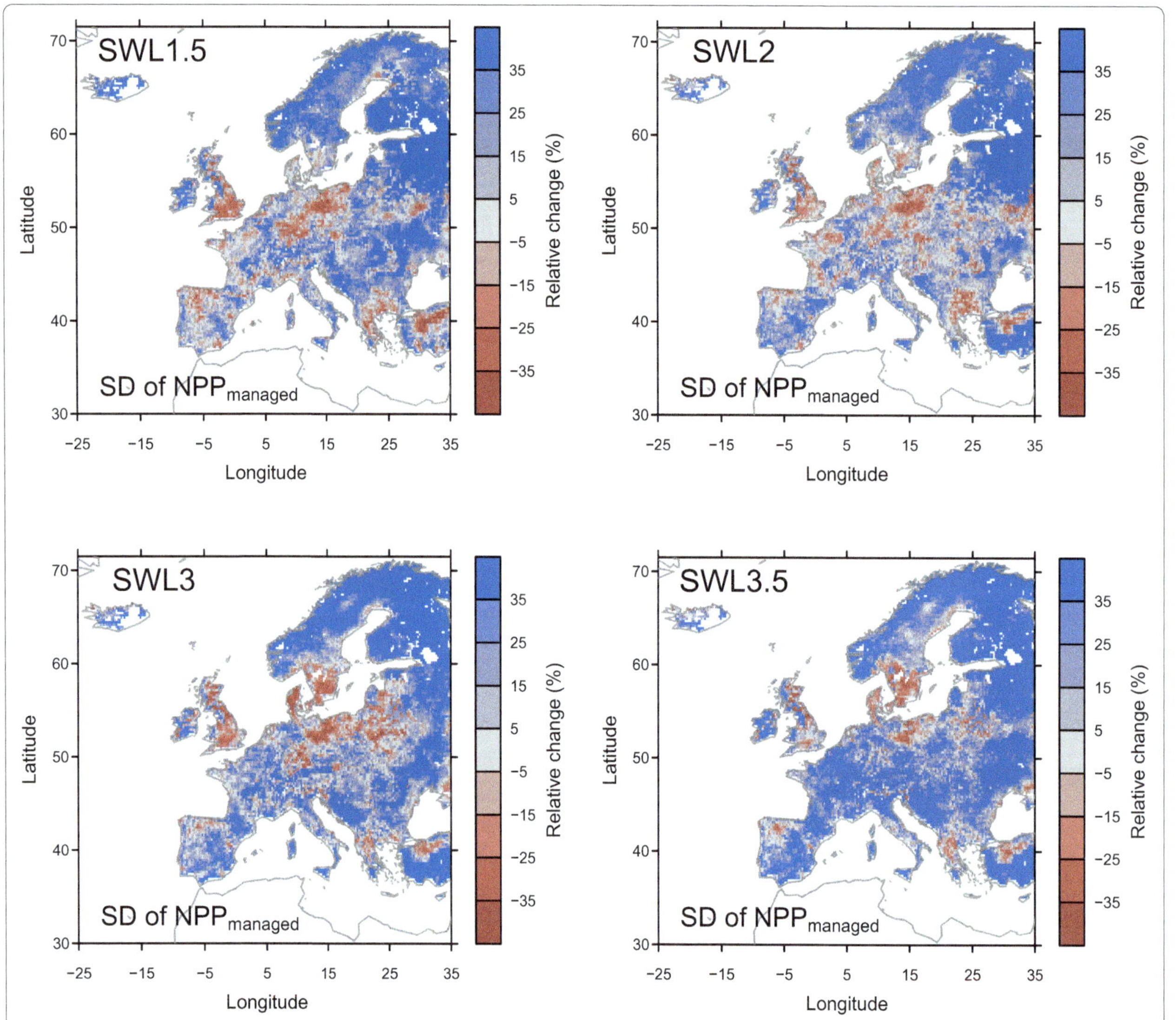


**Fig. 2** Simulated relative changes in the standard deviation (SD) of $NPP_{managed}$ from experiment $E_{control}$

Northern Europe (Fig. 6). Over most parts of the Europe, NBP was simulated to decrease along further warming (i.e., under SWL 2, 3 and 3.5 °C; Table 4). Under SWL of 3.5 °C, most part of the European grassland was simulated to be carbon neutral (e.g., western Iberian Peninsula, and some regions in northeastern Europe, and northern Europe; Fig. 6) or carbon source. One exception is the Alps, where carbon sink was simulated under all SWLs.

## Discussion

### Changes in productivity and phenology of European grassland under future global change: mechanisms and implications for grassland management

The increase in productivity over European grasslands under SRES A1B REMO + ECHAM5 bias-corrected climate and atmospheric $CO_2$ concentration prescribed to ORCHIDEE-GM v2.2 is mainly due to the effect of rising $CO_2$ concentration. In fact, the NPP increase is diminished (e.g., in some region of high-latitude Scandinavia, and in Alps) or NPP even decreases when excluding the effect of rising $CO_2$ ($E_{noco2}$; Fig. 7). Elevated $CO_2$ concentration has the dual effect of increasing leaf photosynthesis, leaf area index, and reducing stomatal conductance, thus increasing water-use efficiency [17] and indirectly reducing the consumption of soil moisture by transpiration [18]. An increase in productivity induced by elevated $CO_2$ has been documented in several grassland FACE experiments (e.g., [15, 56–58]), while actual productivity changes were impacted by interactions among the elevated $CO_2$ concentration, climate [59], soil conditions [60], nitrogen availability [61–64], and community composition [57, 65]. In this study, ORCHIDEE-GM v2.2

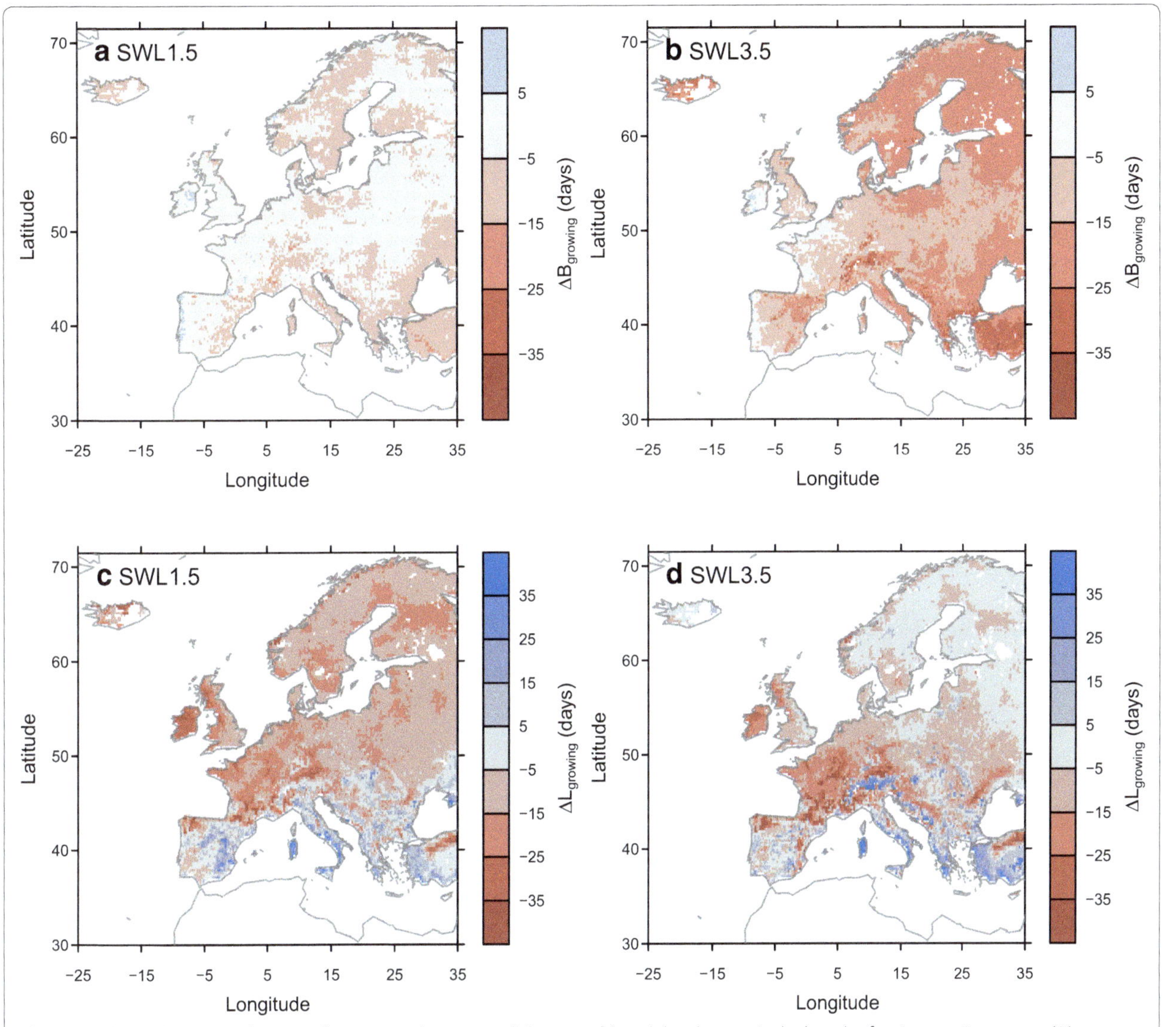


**Fig. 3** Simulated changes in the start of active growing season ($\Delta B_{growing}$; **a**, **b**), and the changes in the length of active growing season ($\Delta L_{growing}$; **c**, **d** from experiment $E_{control}$. Changes are defined as differences between future values and $B_{growing}$ (**a**, **b**) or $L_{growing}$ (**c**, **d**) during the reference period (1981–2010) respectively. Negative $\Delta B_{growing}$ indicated an advance in $B_{growing}$ (i.e., earlier start). Positive $\Delta L_{growing}$ indicated a longer $L_{growing}$ (i.e., extension of active growing season)

does not include nitrogen cycling, and we assumed a constant nitrogen status of grass in the future as that of 2010. Thus the projected productivity is only modulated by the effects of $CO_2$ and climate, but not by nitrogen availability. In the reality, changes in NPP will also be strongly modulated by the availability of soil nutrients [61] and their recycling by grazing animals. In particular, legumes could override $CO_2$ induced nitrogen limitation by increasing biological nitrogen fixation in grasslands [62, 66]. Besides affecting biomass production (yield), climate change can also impact the nutritional value of grass feed [67] through changing nutrition content of individual species and species composition [68], and altering the optimal timing and number of harvest (e.g., in northern Europe; [69]. The importance of nutrient effect on grassland productivity and forage quality represents a priority of including nutrient cycling (e.g., nitrogen and phosphorus) in ORCHIDEE-GM v2.2 in the future. Furthermore, grassland in different regions may have specific response to climate change, which brings challenges for grassland modeling [67]. For example, perennial grassland sward performance (e.g., productivity) may be affected by winter conditions (e.g., extreme cold events or snow cover depth; [70, 71], and Mediterranean grasslands with dominant annual species could have specific response to climate change different from perennial species [67].

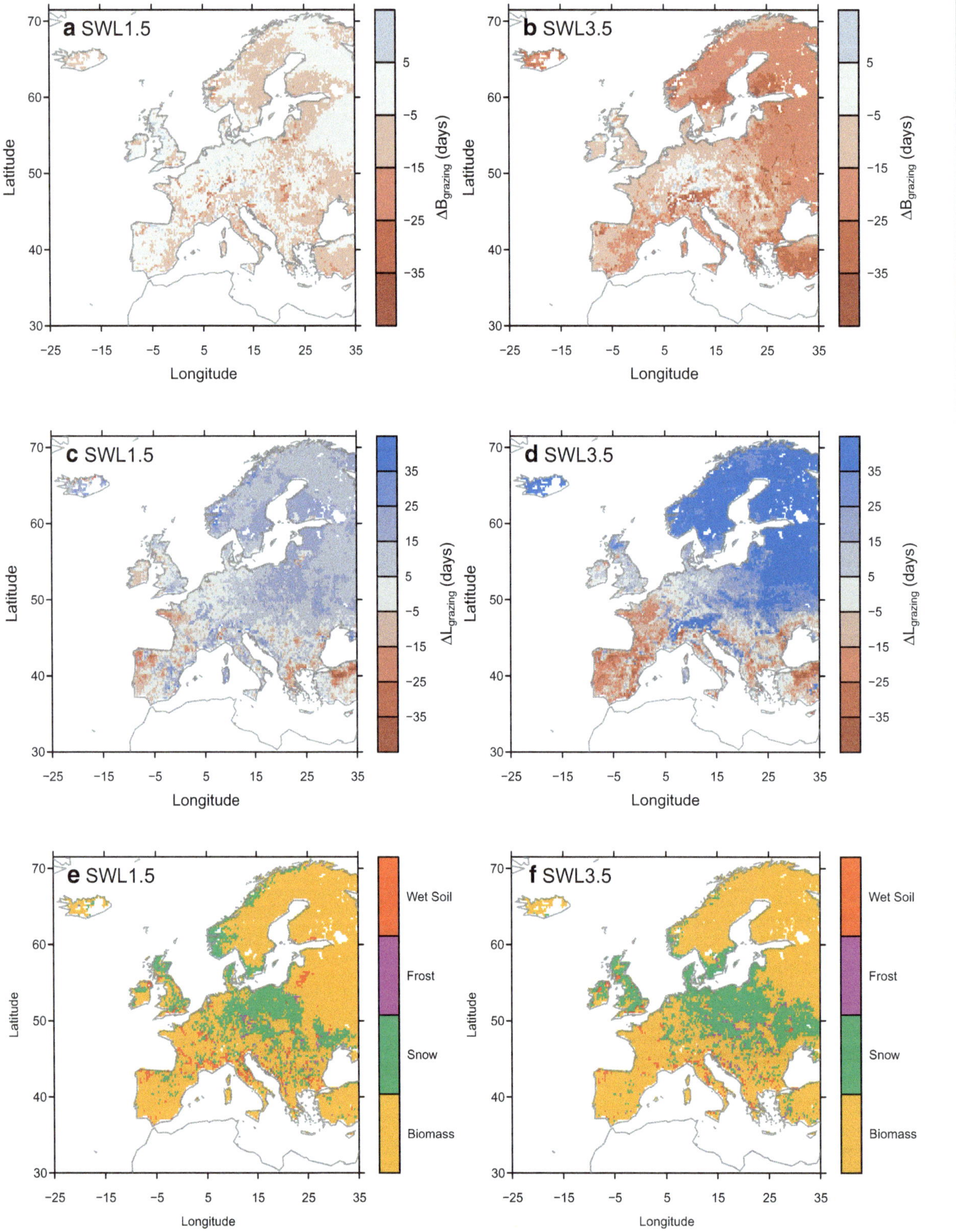


**Fig. 4** Simulated changes in the start of grazing season ($\Delta B_{grazing}$; **a**, **b**), the changes in the length of grazing season ($\Delta L_{grazing}$; **c**, **d**) from experiment $E_{control}$, and the dominant factor impacting the $\Delta L_{grazing}$ for SWL of **e** +1.5 °C and **f** +3.5 °C. Changes are defined as differences between future values and $B_{grazing}$ (**a**, **b**) or $L_{grazing}$ (**c** and **d**) during the reference period (1981–2010) respectively. Negative $\Delta B_{grazing}$ indicated an advance in $B_{grazing}$ (i.e., earlier start). Positive $\Delta L_{grazing}$ indicated a longer $L_{grazing}$ (i.e., extension of grazing season)

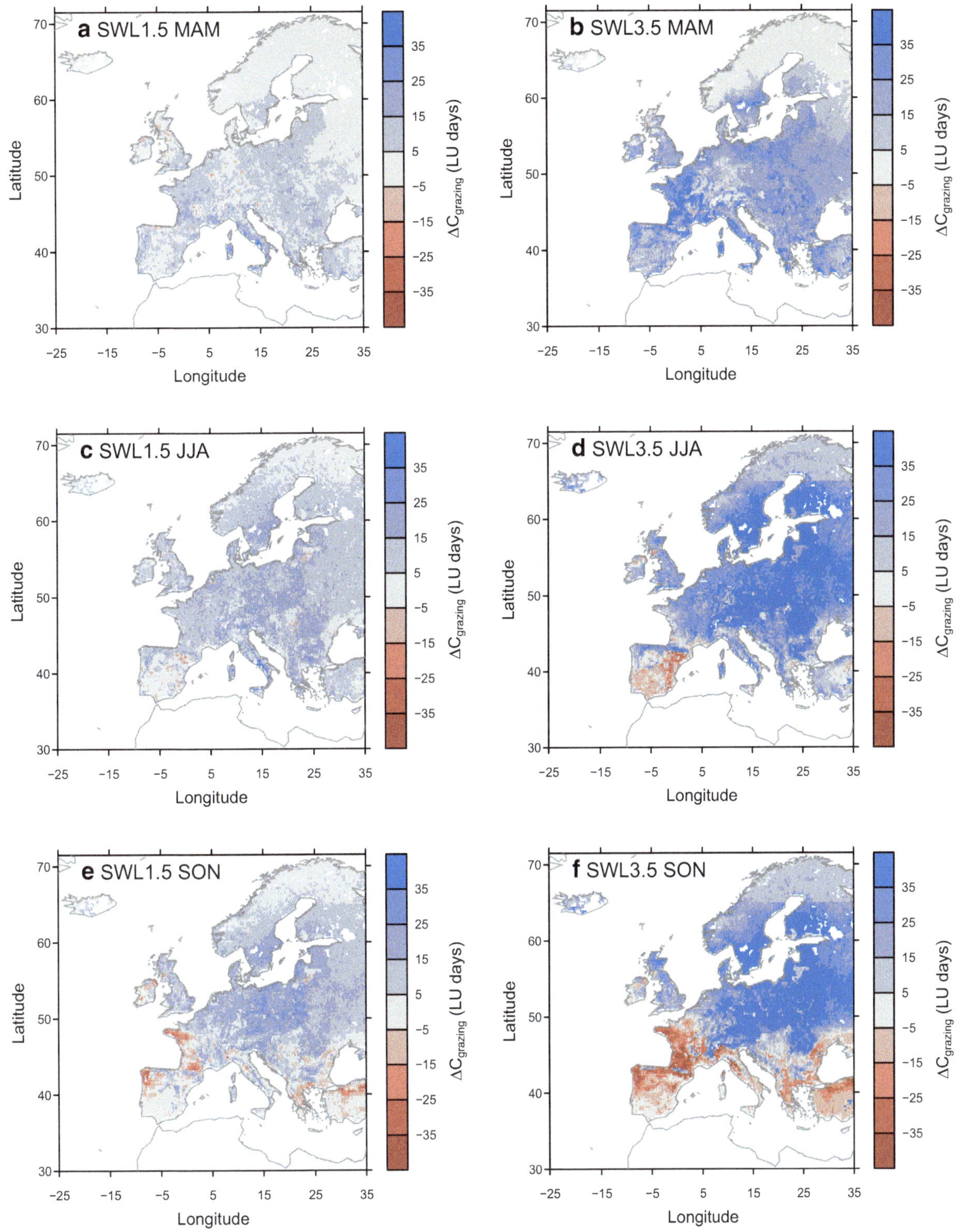


**Fig. 5** Projected seasonal changes of grazing capacity ($C_{grazing}$) from experiment $E_{control}$. Changes are defined as differences between future values and seasonal $C_{grazing}$ during the reference period (1981–2010) respectively. In this study, we defined seasons in Europe as: March, April and May (MAM) for spring; June, July and August (JJA) for summer; September, October and November (SON) for autumn; December, January and February (DJF) for winter. Due to the fact that few winter grazing could happened in Europe, we only showed the changes of grazing capacity for the other three seasons

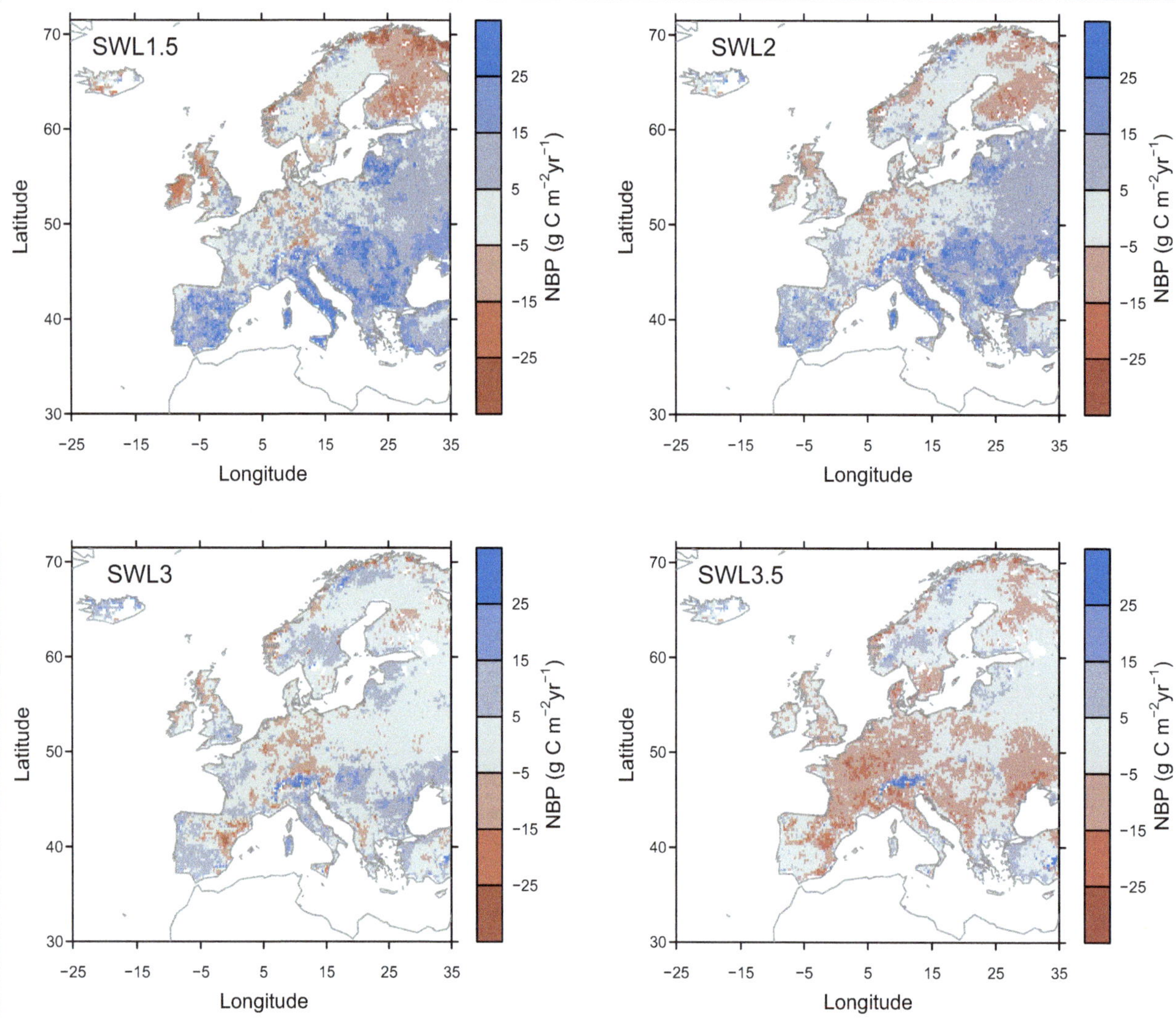


**Fig. 6** Simulated grassland NBP from experiment $E_{control}$. Positive NBP indicated a carbon sink

**Table 4 The NBP of European grassland simulated by ORCHIDEE-GM v2.2 from experiment $E_{control}$, $E_{fixD}$, and $E_{noco2}$**

| Experiment | Period | Europe | Nordic | British Isles | Western | Mediterranean | Alpine | North eastern | South eastern | Eastern |
|---|---|---|---|---|---|---|---|---|---|---|
| $E_{control}$ | SWL1.5 | 6 ± 17 | −3 ± 31 | −10 ± 23 | 1 ± 23 | 17 ± 46 | 10 ± 26 | 8 ± 23 | 19 ± 57 | 16 ± 35 |
| | SWL2 | 5 ± 17 | −1 ± 32 | −5 ± 20 | −0 ± 23 | 11 ± 61 | 11 ± 23 | 5 ± 24 | 17 ± 45 | 13 ± 36 |
| | SWL3 | 2 ± 24 | 2 ± 36 | −2 ± 23 | −1 ± 32 | 3 ± 51 | 9 ± 21 | 0 ± 21 | 7 ± 64 | 6 ± 26 |
| | SWL3.5 | −3 ± 32 | −0 ± 33 | −5 ± 25 | −9 ± 49 | −3 ± 59 | 10 ± 20 | −4 ± 26 | −4 ± 83 | 4 ± 28 |
| $E_{fixD}$ | SWL1.5 | 7 ± 17 | −2 ± 31 | −10 ± 23 | 1 ± 23 | 18 ± 46 | 11 ± 26 | 10 ± 23 | 19 ± 57 | 17 ± 35 |
| | SWL2 | 6 ± 17 | 0 ± 32 | −4 ± 20 | 0 ± 23 | 13 ± 61 | 14 ± 23 | 8 ± 24 | 19 ± 45 | 14 ± 36 |
| | SWL3 | 4 ± 25 | 4 ± 37 | −0 ± 23 | 0 ± 32 | 5 ± 52 | 14 ± 21 | 4 ± 20 | 10 ± 64 | 8 ± 26 |
| | SWL3.5 | −2 ± 32 | 2 ± 34 | −2 ± 24 | −8 ± 49 | −3 ± 60 | 15 ± 19 | −0 ± 26 | −2 ± 83 | 6 ± 29 |
| $E_{noco2}$ | SWL1.5 | −12 ± 16 | −14 ± 27 | −27 ± 20 | −20 ± 24 | −4 ± 42 | −5 ± 21 | −7 ± 23 | −2 ± 58 | −1 ± 31 |
| | SWL2 | −17 ± 16 | −15 ± 27 | −27 ± 22 | −25 ± 26 | −14 ± 54 | −8 ± 19 | −13 ± 22 | −7 ± 47 | −7 ± 30 |
| | SWL3 | −17 ± 22 | −13 ± 26 | −19 ± 23 | −22 ± 36 | −19 ± 42 | −9 ± 15 | −17 ± 25 | −15 ± 64 | −14 ± 22 |
| | SWL3.5 | −21 ± 30 | −14 ± 22 | −19 ± 25 | −28 ± 55 | −22 ± 46 | −9 ± 16 | −21 ± 34 | −24 ± 81 | −14 ± 19 |

The NBP is presented as 30-year mean value ± standard deviation of interannual variability. Positive value indicates a carbon sink, while negative value indicates a carbon source of grassland ecosystem

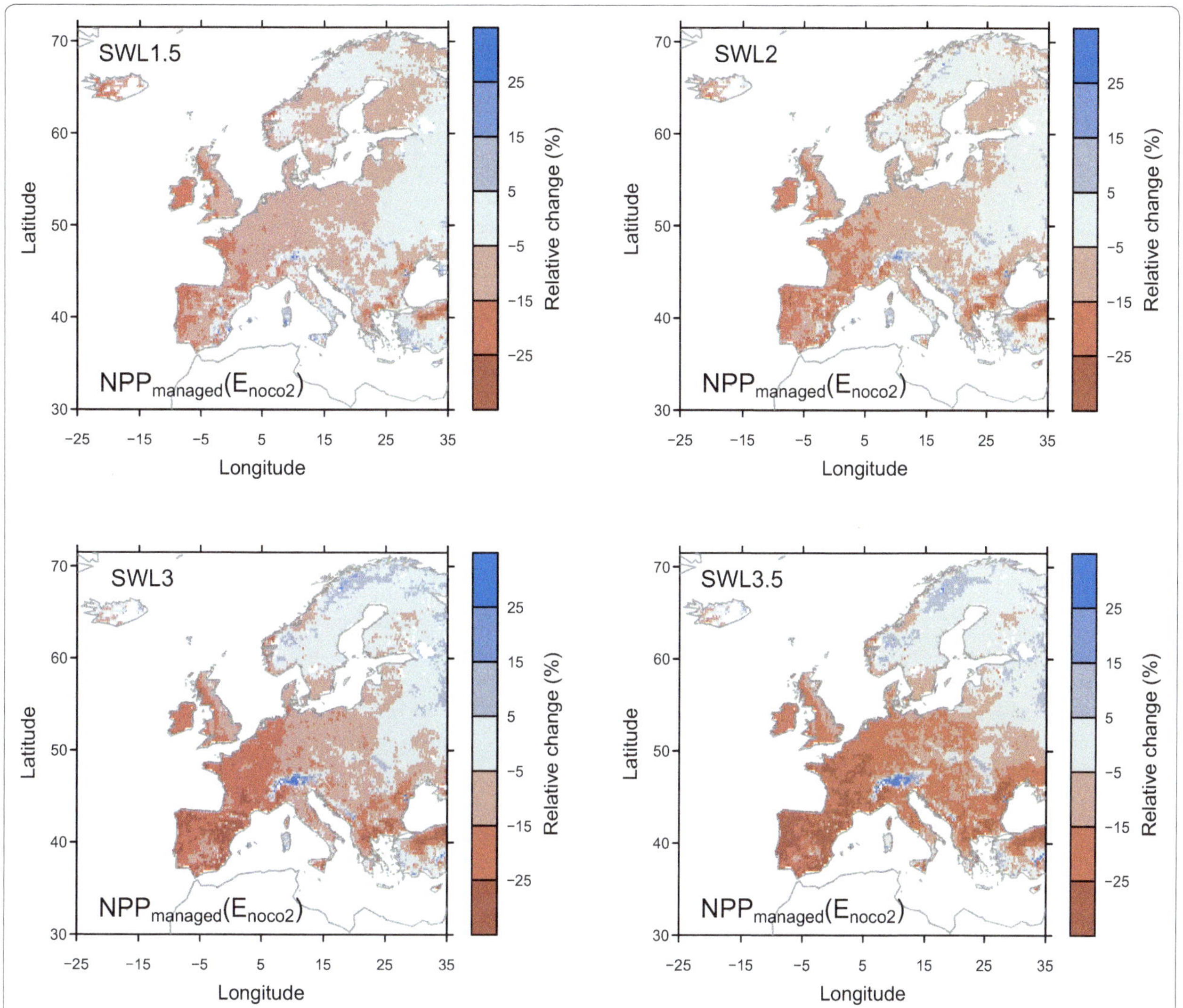


**Fig. 7** Simulated relative changes of $NPP_{managed}$ from experiment $E_{noco2}$

However, more experiment data is required for model development to account for these specific responses [67].

In addition to the changes in annual productivity, we also project a significant shift in grasslands phenology under future climate change, with an earlier spring onset of growth especially under high warming levels (Fig. 3a, b) and changes in growing season length (Fig. 3c, d). Trends towards earlier spring onset of growth were also documented by a number of studies (e.g., [72–75]) which could result from the advanced snowmelt date and warmer temperatures in subalpine and high latitude grassland ecosystems. However, the projected increase in winter precipitation across the mid- and high-latitude regions [2] makes the timing of modeled snowmelt, and hence future projections of phenological shifts, uncertain [76]. For European ecosystems, Menzel et al. [77] documented a general advance in spring leafing during the period 1971–2000. In Mediterranean shrub- and grasslands, advancement in leaf-out events was also observed during the recent decades since the 1970s [78, 79]. According to the projections by ORCHIDEE-GM v2.2, the earlier winter-spring onset will continue over most European grasslands (Fig. 3a, b).

One important result of this study is that the projected increase of summer drying and rising temperatures in Mediterranean regions (IPCC [2]) will probably result in a shift of the active season, with an earlier onset of winter-spring growth being (at least partially) offset by an earlier, and longer lasting, summer drought period. An earlier end to the growing season is predicted in this study over large part of Europe, especially in Western Europe and northern Spain (Fig. 3). This earlier end to the growing season and thus the lower productivity (either decrease or less increase in productivity in the future; Fig. 1) are related to the reduction of mid-summer soil moisture in the model, given the high positive correlation ($p < 0.01$) between detrended annual NPP variations and summer soil moisture variation over these regions (Additional file 1: Table S2). Our results are consistent with the findings from the simulation of 12 contrasted French grassland sites using PaSim, which show higher spring forage production, but a increasing risks of severe summer shortfalls in the future. The reduction in soil moisture (averaged over 0–9 cm in depth; Additional file 1: Figure S10) is not attributed to individual drivers in the model, but it reflects the combination of continued warming and the reduction in precipitation during summer and autumn from bias-corrected REMO + ECHAM5 climate, and possibly the fact that earlier leaf onset and warmer spring accelerates soil water depletion and thus increases water stress later during the season. Field experiments also suggest that warming and associated soil drying could reduce primary production in many temperate grasslands [80, 81].

With respect to grassland management in the future, the simulated general increase in annual grassland productivity (i.e., NPP) generally leads to an increase in the annual forage yield ($Y_{pot}$), annual grazing capacity ($C_{grazing}$), and thus of potential livestock density ($D_{opt}$), when management is allowed in the model to continually adjust to increasing NPP (i.e., $E_{control}$). However, there are some practical issues should be noted in this management intensification potential driven by the projected productivity increase. While the trampling effect on growth during grazing has been considered in our model [35]; derived from Vuichard et al. [38], the impact of intensive grazing on soil physics (e.g., soil compaction) and its consequence on soil hydrological and thermal properties [67] is not taken into account and should be addressed by models in the future. In addition, increasing livestock density might bring animal health problem in practice [67] due to various factors such as pathogen spread (e.g., [82]), parasite load, and exposure to environmental extremes.

Our results suggest that the phenological shifts in the future will enable an earlier start of grazing over European grasslands (Fig. 4a, b), which potentially should reduce costs of managing livestock indoors and reduce GHG emissions caused by manure management. We also project an extension of the grazing season over the Alps, and the north and east part of Europe (i.e., positive $\Delta L_{grazing}$, in Fig. 4c, d), even for the regions where the active growing season shortens in the future (Fig. 4c, d). The reason of this decoupling between the duration of growing vs. grazing season could possibly be attributed (Fig. 4e, f) to (1) the enhanced productivity (i.e., NPP) during growing season, which allows a longer grazing season, (2) the later snow cover (Additional file 1: Figure S7), (3) less days with daily mean air temperature drops below 0 °C (Additional file 1: Figure S8), and (4) reduction in waterlogging events (Additional file 1: Figure S9). In addition, the simulated shift in seasonal and annual grazing capacity ($C_{grazing}$; Fig. 5; Additional file 1: Figure S4) suggests that under future warming, 1) over most regions in Europe, more livestock could be put over pasture during growing season (Fig. 5a, b), (2) an earlier grazing is enabled by the advance of leaf onset (Fig. 4a, b), (3) for regions such as northern Spain and France, less livestock could be supported by grassland during summer and/or autumn (Fig. 5d–f), thus causing an increasing needs of conserved forage for summer and/or autumn utilization by ruminant livestock [27].

Given the coincident increase in the interannual variability of NPP and the mean value of NPP over some regions (e.g., northern Europe; Fig. 2), farmers may not be able to take a full advantage of the long-term mean productivity increase, or may even be negatively affected by larger productivity variations. In practice, farmers end to adopt a lower livestock density than the potential mean to avoid the risk of insufficient grass for livestock, since in the case of a bad year with low NPP, supplementary feeding is required (e.g., hay and silage) which increases the costs. To assess the possible impact of interannual variability in productivity on future management change, we assumed that farmers adapt their livestock density to the simulated minimum productivity in the past 10 years (i.e., 10-year minimum $NPP_{managed}$). In most regions of Europe, a weaker increase (or a stronger decrease) of 10-year minimum $NPP_{managed}$ (Fig. 8) than the increase (or the decrease) of mean $NPP_{managed}$ (Fig. 1) was found. However, stronger increase in 10-year minimum $NPP_{managed}$ than that in mean $NPP_{managed}$ was found over southeastern United Kingdom, Denmark, and some regions of Germany and Poland (Fig. 8), where strong decrease in SD of $NPP_{managed}$ was simulated (Fig. 2). Similar changes were found for simulated $Y_{pot}$ (Additional file 1: Figure S11) and $C_{grazing}$ (Additional file 1: Figure S12), if 10-year minimum $Y_{pot}$ and $C_{grazing}$ were adopted by farmers rather than mean value. The above findings suggest that if a conservative adaptation strategy is adopted (i.e., farmers adjust livestock density based on the lowest productivity in the past decade), farmers who confronted to the increased interannual variability in grassland production may, in the future, benefit less from the productivity increase or suffer more losses from the productivity decrease. Adopting a moderate increase in stocking density may reduce the vulnerability of livestock production systems to increased climate variability. Yet,

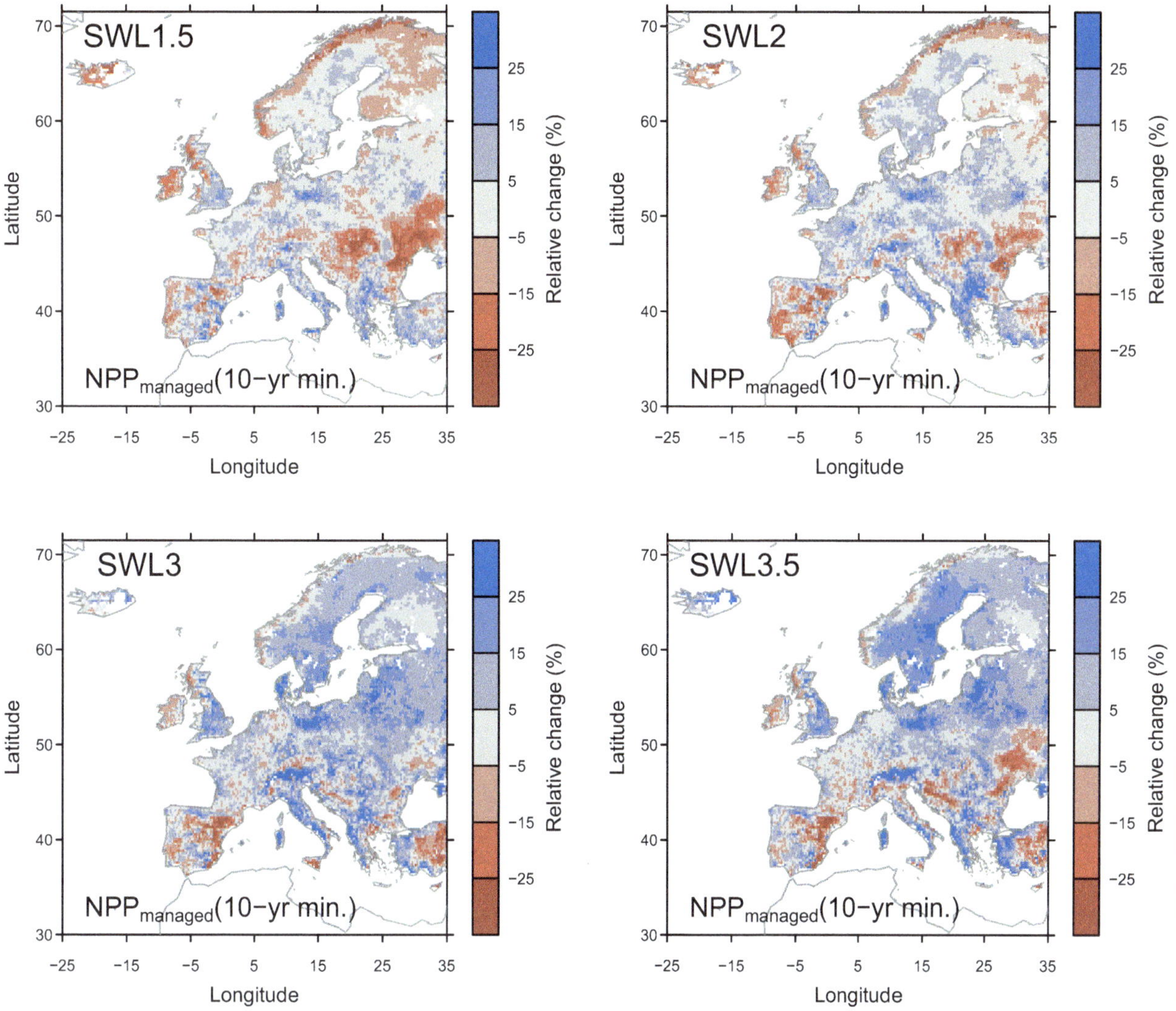


**Fig. 8** Simulated relative changes of 10-year minimum $NPP_{managed}$ from experiment $E_{control}$. The 10-year minimum $NPP_{managed}$ is defined as minimum $NPP_{managed}$ in the past 10 years. The change of 10-year minimum $NPP_{managed}$ is defined as the the difference between 10-year minimum $NPP_{managed}$ during 30 years in the future corresponding to each SWL minus 10-year minimum $NPP_{managed}$ in reference historical period (1981–2010)

an alternative adaptation strategy (not considered here) would also consist in increasing forage (hay, silage) stocks to reduce risks of low NPP available for grazing.

### Implications for grassland $CO_2$ fluxes

ORCHIDEE-GM v2.1 simulated a carbon sink (i.e., a positive NBP of 27 ± 8 g C m$^2$ year$^2$) over European grassland in the past decade (i.e., 2001–2010; [12]. In this study, we project with ORCHIDEE-GM v2.2 a persistence of the grassland carbon sink over Mediterranean, Alpine, North eastern, South eastern and Eastern regions under SWL of 1.5 and 2 °C (Table 4), associated with longer growing seasons (Fig. 3), and with higher productivity from elevated $CO_2$ (Fig. 1). In our simulation, elevated $CO_2$ concentration made the dominant contribution to the projected sustained carbon sink, given the fact that NBP becomes a carbon source when the effect of $CO_2$ is excluded (NBP from $E_{noco2}$ in Table 4), which is consistent with the findings from French grassland sites [27]. Both the amount [83] and interannual variability of NEE [84, 85] are sensitive to growing season length, with generally greater net C uptake at sites with a longer, rather than shorter growing season [86]. The extended growing season in Mediterranean, Alpine, and South-eastern regions (Fig. 3) likely drive the simulated carbon sink over those regions in our study. However, this carbon sink is modeled to be much weaker than that in the past, even with the beneficial effects of increasing $CO_2$, and the sinks gradually turns to carbon source with further climate change for SWL reaching above 3 °C (Fig. 6; Table 4). For British Isles and Western region, shorter future growing season (Fig. 3) could also contribute to the shift towards a carbon source. For France, slightly positive or neutral NBP was simulated under SWL of 1.5 °C to SWL of 3 °C, consistent with the findings from the detailed French sites simulations [27]. Increase in Rh under future warming (Additional file 1: Figure S13) largely offsets the C gain from the extension of growing season and elevated $CO_2$, resulting in most European grasslands becoming a carbon source under SWL of 3.5 °C (Fig. 6). In addition to changes in long-term grassland soil carbon status, there could be risks of grassland degradation and loss of carbon sink during climate extremes such as heat waves and drought events. For example, a probabilistic risk analysis based on simulations with the PaSim model showed increased drought risk in the future for grassland productivity in Southern Europe [28]. However, a series of experiments showed that annual C balance is preserved under elevated $CO_2$ despite the extreme [20] and that grasslands productivity decreased under drought event [87, 88] but showed some resilience under elevated $CO_2$ [87].

It should be noted that the future carbon balance of European grassland is projected in this study under the assumption of management practice is determined by our adaptive management change algorithm which gives a potential livestock density ($D_{opt}$) within the managed grassland ("Simulation set-up"). Under this assumption, we obtain an increase in livestock numbers sustained by European managed grassland by 9, 15, 24, and 29% under SWL of 1.5, 2, 3 and 3.5 °C respectively, which means an intensification of grassland management. To assess the

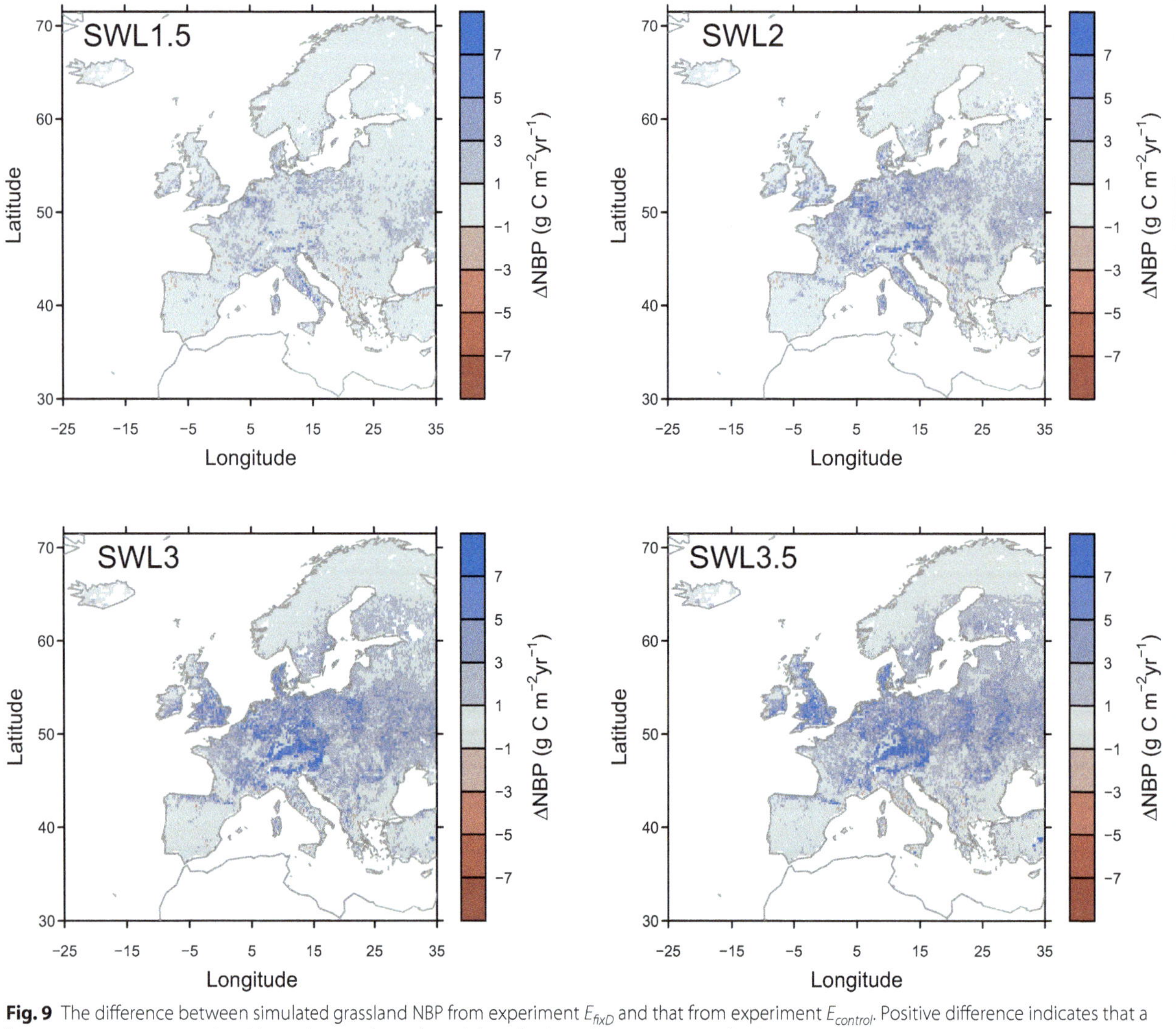


**Fig. 9** The difference between simulated grassland NBP from experiment $E_{fixD}$ and that from experiment $E_{control}$. Positive difference indicates that a lower management intensity either enhances the carbon sink or alleviates the appearance of carbon source

impact of adapting management intensity on the carbon balance of European grassland, we look at the results of the sensitivity experiment with constant livestock numbers at the value of the year 2010 (i.e., $E_{fixD}$). In $E_{fixD}$, the lower management intensity either enhances the carbon sink or alleviates the appearance of carbon source in the future (Fig. 9; Table 4) because of a lower carbon export from grassland ecosystem (i.e., a larger return of NPP to soils). The effect of the lower management intensity is mainly found over mid-latitude Europe (Fig. 9), and stronger effect can be found under higher SWL due to a larger difference in management intensity between $E_{control}$ and $E_{fixD}$. The simulated NBP change through lowering livestock density suggests that there is a possibility of managing (increasing) the carbon sink of European grassland in the future by maintaining/lowering current stocking densities, letting higher NPP under moderate climate change increase soil carbon storage.

An enhanced carbon sink might also be achieved by ameliorating the nutrient status of plants [61]. This could be implemented by additional mineral fertilization, more manure, and introducing/increasing legumes [62, 66]. However, additional fertilization (mineral or manure) may increase the $N_2O$ emission offsetting the carbon sink, as well as nitrogen leaching losses.

The modeled intensification in grassland management affects the carbon balance of European grassland, and should also impact $CH_4$ and $N_2O$ emissions. Enteric fermentation $CH_4$ emissions generally follow the livestock density (e.g., [12]. The simulation of the nitrogen related fluxes such as $N_2O$ emission requires an explicit nitrogen cycle interacting with the carbon cycle (e.g., [89], which is not included in the current version of ORCHIDEE-GM v2.2. Moreover, changes in livestock density could alter the associated nitrogen cycle during grazing, which potentially impact both the carbon balances and the $N_2O$ emissions.

## Conclusions

Using the ORCHIDEE-GM v2.2 managed grassland model, increase in grassland production was simulated in response to future global change over Europe, which is mainly due to the effect of rising $CO_2$ concentration. The model projects shifts in phenology characterized by an earlier winter-spring onset of grass growth over Europe caused by warming. A longer growing season is only projected over some regions in southern and southeastern Europe. In other regions, an earlier summer-autumn leaf fall caused an even shorter growing season, which could be related to the increased summer drying (i.e., lower soil moisture). The productivity increase and phenological shift enable an increase in potential management intensity with higher annual forage yield, grazing capacity and potential livestock density over European grassland, and a shift in seasonal grazing capacity. The model also projects a continual grassland carbon sink under SWL of 1.5 and 2 °C in most regions of Europe (Mediterranean, Alpine, North eastern, South eastern and Eastern regions). However, a reduced carbon sink or enhanced carbon source is projected over European grassland under continual warming. It should be kept in mind that the results obtained in this study are based on the climate projection from a single climate model. The changes in grassland productivity and phenology might depend on the choice of this specific climate model.

### Abbreviations

DM: dry matter; GCM: general circulation model; GHG: greenhouse gas; GMST: global mean surface temperature; LU: livestock unit; NBP: net biome productivity; NEE: net ecosystem exchange; NPP: net primary productivity; RCP: representative concentration pathways; Rh: heterotrophic respiration; SRES: special report on emission scenario; SWL: specific warming levels; UNFCCC: United Nations Framework Convention on Climate Change.

### Authors' contributions

Conceived and designed the experiments JC PC NV JFS. Performed the experiments JC. Analyzed the data JC. Contributed materials NV KK. Wrote the paper JC PC NV JFS KK BS. All authors read and approved the final manuscript.

### Author details

[1] Laboratoire des Sciences du Climat et de l'Environnement, UMR8212, CEA-CNRS-UVSQ, 91191 Gif-Sur-Yvette, France. [2] Sorbonne Universités (UPMC), CNRS-IRD-MNHN, LOCEAN/IPSL, 4 Place Jussieu, 75005 Paris, France. [3] INRA, UAR0233 CODIR Collège de Direction. Centre-Siège de l'INRA, Paris, France. [4] Grassland Ecosystem Research Unit, French National Institute for Agricultural Research (INRA), 63100, Clermont-Ferrand, France.

### Acknowledgements

Not applicable.

### Competing interests

The authors declare that they have no competing interests.

### Funding

We gratefully acknowledge funding from the European Union Seventh Framework Programme FP7/2007–2013 under Grant No. 266018 (Animal Change) and No. 603864 (HELIX). Philippe Ciais acknowledge support from the ERC Synergy grant ERC-2013-SyG-610028 IMBALANCE-P.

### References

1. Stocker T, Qin D, Plattner G-K, Alexander L, Allen S, Bindoff N, et al. Technical summary. Climate Change 2013: The Physical Science Basis Contribution of Working Group I to the Fifth Assessment Report of the Intergovernmental Panel on Climate Change. Cambridge: Cambridge University Press; 2013. p. 33–115.
2. IPCC (Intergovernmental Panel on Climate Change). Climate change 2013: the physical scientific basis (Contribution of Working Group I to the Fifth Assessment Report of the Intergovernmental Panel on Climate Change, 2013). Cambridge: Cambridge University Press; 2013. p. 2013.
3. Meehl GA, Covey C, Delworth T, Latif M, McAvaney B, Mitchell JFB, et al.

The WCRP CMIP3 multimodel dataset-a new era in climate change research. Bull Am Meteorol Soc. 2007;88(9):1383–94.
4. Taylor KE, Stouffer RJ, Meehl GA. An overview of Cmip5 and the experiment design. Bull Am Meteorol Soc. 2012;93(4):485–98.
5. Knutti R, Sedlacek J. Robustness and uncertainties in the new CMIP5 climate model projections. Nat Clim Change. 2013;3(4):369–73.
6. Hewitt CD, Griggs DJ. Ensembles-based predictions of climate changes and their impacts (ENSEMBLES). Eos. 2004;85(52):566.
7. Van der Linden P, Mitchell J, editors. ENSEMBLES: Climate Change and its Impacts: Summary of research and results from the ENSEMBLES project. Met Office Hadley Centre, Exeter. 2009.
8. Jacob D, Petersen J, Eggert B, Alias A, Christensen OB, Bouwer LM, et al. EURO-CORDEX: new high-resolution climate change projections for European impact research. Reg Environ Change. 2014;14(2):563–78.
9. EEA (European Environment Agency). Indicator assessment: Global and European temperatures; 2016. http://www.eea.europa.eu/data-and-maps/indicators/global-and-european-temperature-1/assessment.
10. Eurostat. 2010. http://ec.europa.eu/eurostat/web/agriculture/data/database. Accessed 10 May 2015.
11. Chang J, Ciais P, Viovy N, Vuichard N, Herrero M, Havlik P, et al. Effect of climate change, CO2 trends, nitrogen addition, and land-cover and management intensity changes on the carbon balance of European grasslands. Glob Change Biol. 2016;22(1):338–50.
12. Chang J, Ciais P, Viovy N, Vuichard N, Sultan B, Soussana JF. The greenhouse gas balance of European grasslands. Glob Change Biol. 2015;21(10):3748–61.
13. Soussana J-F, Barioni LG, Ari TB, Conant R, Gerber P, Havlik P, et al., editors. Managing grassland systems in a changing climate: the search for practical solutions. Proceedings of the 22nd International Grassland Congress, Sidney; 2013.
14. Ainsworth EA, Davey PA, Hymus GJ, Osborne CP, Rogers A, Blum H, et al. Is stimulation of leaf photosynthesis by elevated carbon dioxide concentration maintained in the long term? A test with *Lolium perenne* grown for 10 years at two nitrogen fertilization levels under free air CO(2) enrichment (FACE). Plant, Cell Environ. 2003;26(5):705–14.
15. Campbell BD, Smith DS, Pastures GC. M GPRN. A synthesis of recent global change research on pasture and rangeland production: reduced uncertainties and their management implications. Agric Ecosyst Environ. 2000;82(1–3):39–55.
16. Nowak RS, Ellsworth DS, Smith SD. Functional responses of plants to elevated atmospheric CO2-do photosynthetic and productivity data from FACE experiments support early predictions? New Phytol. 2004;162(2):253–80.
17. Rotter R, Van de Geijn SC. Climate change effects on plant growth, crop yield and livestock. Clim Change. 1999;43(4):651–81.
18. Soussana JF, Luescher A. Temperate grasslands and global atmospheric change: a review. Grass Forage Sci. 2007;62(2):127–34.
19. Jones MB, Donnelly A. Carbon sequestration in temperate grassland ecosystems and the influence of management, climate and elevated CO(2). New Phytol. 2004;164(3):423–39.
20. Roy J, Picon-Cochard C, Augusti A, Benot ML, Thiery L, Darsonville O, et al. Elevated CO2 maintains grassland net carbon uptake under a future heat and drought extreme. Proc Natl Acad Sci USA. 2016;113(22):6224–9.
21. Cang FA, Wilson AA, Wiens JJ. Climate change is projected to outpace rates of niche change in grasses. Biol Lett. 2016;12(9):20160368.
22. Holden NM, Brereton AJ. An assessment of the potential impact of climate change on grass yield in Ireland over the next 100 years. Ir J Agric Food Res. 2002;41(2):213–26.
23. Juin S, Brisson N, Clastre P, Grand P. Impact of global warming on the growing cycles of three forage systems in upland areas of southeastern France. Agronomie. 2004;24(6–7):327–37.
24. Parsons DJ, Armstrong AC, Turnpenny JR, Matthews AM, Cooper K, Clark JA. Integrated models of livestock systems for climate change studies. 1. Grazing systems. Glob Change Biol. 2001;7(1):93–112.
25. Bell MJ, Eckard RJ, Cullen BR. The effect of future climate scenarios on the balance between productivity and greenhouse gas emissions from sheep grazing systems. Livest Sci. 2012;147(1–3):126–38.
26. Cullen BR, Eckard RJ, Rawnsley RP. Resistance of pasture production to projected climate changes in south-eastern Australia. Crop Pasture Sci. 2012;63(1):77–86.
27. Graux A-I, Bellocchi G, Lardy R, Soussana J-F. Ensemble modelling of climate change risks and opportunities for managed grasslands in France. Agric For Meteorol. 2013;170:114–31.
28. Van Oijen M, Balkovi J, Beer C, Cameron DR, Ciais P, Cramer W, et al. Impact of droughts on the carbon cycle in European vegetation: a probabilistic risk analysis using six vegetation models. Biogeosciences. 2014;11(22):6357–75.
29. Soussana J-F, Graux A-I, Tubiello FN. Improving the use of modelling for projections of climate change impacts on crops and pastures. J Exp Bot. 2010;61(8):2217–28.
30. UNFCCC 2015 The Paris Agreements. United Nations Framework Convention on Climate Change http://unfccc.int/paris_agreement/items/9485.php. Accessed Sept 2016.
31. UNFCCC 1992 United Nations Framework Convention on Climate Change: full text of the convention. http://unfccc.int/essential_background/convention/items/6036.php. Accessed Sept 2016.
32. Ciais P, Reichstein M, Viovy N, Granier A, Ogee J, Allard V, et al. Europe-wide reduction in primary productivity caused by the heat and drought in 2003. Nature. 2005;437(7058):529–33.
33. Krinner G, Viovy N, de Noblet-Ducoudre N, Ogee J, Polcher J, Friedlingstein P, et al. A dynamic global vegetation model for studies of the coupled atmosphere-biosphere system. Glob Biogeochem Cycles. 2005;19(1).
34. Piao S, Friedlingstein P, Ciais P, de Noblet-Ducoudre N, Labat D, Zaehle S. Changes in climate and land use have a larger direct impact than rising CO2 on global river runoff trends. Proc Natl Acad Sci USA. 2007;104(39):15242–7.
35. Chang JF, Viovy N, Vuichard N, Ciais P, Wang T, Cozic A, et al. Incorporating grassland management in ORCHIDEE: model description and evaluation at 11 eddy-covariance sites in Europe. Geosci Model Dev. 2013;6(6):2165–81.
36. Riedo M, Grub A, Rosset M, Fuhrer J. A pasture simulation model for dry matter production, and fluxes of carbon, nitrogen, water and energy. Ecol Model. 1998;105(2–3):141–83.
37. Graux AI, Gaurut M, Agabriel J, Baumont R, Delagarde R, Delaby L, et al. Development of the Pasture Simulation Model for assessing livestock production under climate change. Agric Ecosyst Environ. 2011;144(1):69–91.
38. Vuichard N, Ciais P, Viovy N, Calanca P, Soussana J-F. Estimating the greenhouse gas fluxes of European grasslands with a process-based model: 2. Simulations at the continental level. Global Biogeochemical Cycles. 2007;21(1).
39. Vuichard N, Soussana J-F, Ciais P, Viovy N, Ammann C, Calanca P, et al. Estimating the greenhouse gas fluxes of European grasslands with a process-based model: 1. Model evaluation from in situ measurements. Global Biogeochem Cycles. 2007;21(1).
40. Chang JF, Viovy N, Vuichard N, Ciais P, Campioli M, Klumpp K, et al. Modeled changes in potential grassland productivity and in grass-fed ruminant livestock density in Europe over 1961–2010. PLoS ONE. 2015;10(5):e0127554.
41. Donkor NT, Gedir JV, Hudson RJ, Bork EW, Chanasyk DS, Naeth MA. Impacts of grazing systems on soil compaction and pasture production in Alberta. Can J Soil Sci. 2002;82(1):1–8.
42. Boone A, Etchevers P. An intercomparison of three snow schemes of varying complexity coupled to the same land surface model: local-scale evaluation at an Alpine site. J Hydrometeorol. 2001;2(4):374–94.
43. Wang T, Ottle C, Boone A, Ciais P, Brun E, Morin S, et al. Evaluation of an improved intermediate complexity snow scheme in the ORCHIDEE land surface model. J Geophys Res-Atmos. 2013;118(12):6064–79.
44. Carsel RF, Parrish RS. Developing joint probability distributions of soil water retention characteristics. Water Resour Res. 1988;24(5):755–69.
45. de Rosnay P, Bruen M, Polcher J. Sensitivity of surface fluxes to the number of layers in the soil model used in GCMs. Geophys Res Lett. 2000;27(20):3329–32.
46. de Rosnay P, Polcher J, Bruen M, Laval K. Impact of a physically based soil water flow and soil-plant interaction representation for modeling large-scale land surface processes. J Geophys Res-Atmos. 2002;107(D11).
47. d'Orgeval T, Polcher J, de Rosnay P. Sensitivity of the West African hydrological cycle in ORCHIDEE to infiltration processes. Hydroly Earth Syst Sci. 2008;12(6):1387–401.
48. d'Orgeval T. Impact du changement climatique sur le cycle de l'eau en

Afrique de l'Ouest: mod√©lisation et incertitudes. Paris: Universite Paris VI; 2006.
49. Beer C, Weber U, Tomelleri E, Carvalhais N, Mahecha M, Reichstein M. Harmonized European long-term climate data for assessing the effect of changing temporal variability on land-atmosphere CO2 fluxes. J Clim. 2014;27(13):4815–34.
50. Rogelj J, Meinshausen M, Knutti R. Global warming under old an new scenarios using IPCC climate sensitivity range estimates. Nat Clim Change. 2012;2(4):248–53.
51. Keeling CD, Whorf TP. Atmospheric CO2 records from sites in the SIO air sampling network. Trends: a compendium of data on global change. 2005:16–26.
52. Fuchs R, Herold M, Verburg PH, Clevers JGPW. A high-resolution and harmonized model approach for reconstructing and analysing historic land changes in Europe. Biogeosciences. 2013;10(3):1543–59.
53. Vautard R, Gobiet A, Sobolowski S, Kjellstr√∂m E, Stegehuis A, Watkiss, et al. The European climate under a 2 C global warming. Environ Res Lett. 2014;9(3):034006.
54. Roeckner E, Brokopf R, Esch M, Giorgetta M, Hagemann S, Kornblueh L, Manzini E, Schlese U, Schulzweida U. Sensitivity of simulated climate to horizontal and vertical resolution in the ECHAM5 atmosphere model. J Clim 2006;19(16):3771–91.
55. Olesen JE, Bindi M. Consequences of climate change for European agricultural productivity, land use and policy. Eur J Agron. 2002;16(4):239–62.
56. Edwards GR, Clark H, Newton PCD. The effects of elevated CO2 on seed production and seedling recruitment in a sheep-grazed pasture. Oecologia. 2001;127(3):383–94.
57. Fay PA, Newingham BA, Polley HW, Morgan JA, LeCain DR, Nowak RS, et al. Dominant plant taxa predict plant productivity responses to CO2 enrichment across precipitation and soil gradients. Aob Plants. 2015;7:plv027.
58. Lüscher A, Hendrey GR, Nosberger J. Long-term responsiveness to free air CO2 enrichment of functional types, species and genotypes of plants from fertile permanent grassland. Oecologia. 1998;113(1):37–45.
59. Hovenden MJ, Newton PC, Wills KE. Seasonal not annual rainfall determines grassland biomass response to carbon dioxide. Nature. 2014;511(7511):583–6.
60. Fay PA, Jin VL, Way DA, Potter KN, Gill RA, Jackson RB, et al. Soil-mediated effects of subambient to increased carbon dioxide on grassland productivity. Nat Clim Change. 2012;2(10):742–6.
61. Lüscher A, Daepp M, Blum H, Hartwig UA, N√∂sberger J. Fertile temperate grassland under elevated atmospheric CO 2, Äîrole of feedback mechanisms and availability of growth resources. Eur J Agron. 2004;21(3):379–98.
62. Lüscher A, Hartwig UA, Suter D, Nösberger J. Direct evidence that symbiotic N2 fixation in fertile grassland is an important trait for a strong response of plants to elevated atmospheric CO2. Glob Change Biol. 2000;6(6):655–62.
63. Reich PB, Hobbie SE, Lee TD. Plant growth enhancement by elevated CO2 eliminated by joint water and nitrogen limitation. Nat Geosci. 2014;7(12):920–4.
64. Richter M, Hartwig UA, Frossard E, Nosberger J, Cadisch G. Gross fluxes of nitrogen in grassland soil exposed to elevated atmospheric pCO(2) for seven years. Soil Biol Biochem. 2003;35(10):1325–35.
65. Morgan JA, Mosier AR, Milchunas DG, LeCain DR, Nelson JA, Parton WJ. CO2 enhances productivity, alters species composition, and reduces digestibility of shortgrass steppe vegetation. Ecol Appl. 2004;14(1):208–19.
66. Soussana J-F, Tallec T. Can we understand and predict the regulation of biological N-2 fixation in grassland ecosystems? Nutr Cycl Agroecosyst. 2010;88(2):197–213.
67. Kipling RP, Virkajaervi P, Breitsameter L, Curnel Y, De Swaef T, Gustavsson AM, et al. Key challenges and priorities for modelling European grasslands under climate change. Sci Total Environ. 2016;566:851–64.
68. Dumont B, Andueza D, Niderkorn V, Luscher A, Porqueddu C, Picon-Cochard C. A meta-analysis of climate change effects on forage quality in grasslands: specificities of mountain and Mediterranean areas. Grass Forage Sci. 2015;70:239–54.
69. Höglind M, Thorsen SM, Semenov MA. Assessing uncertainties in impact of climate change on grass production in Northern Europe using ensembles of global climate models. Agric For Meteorol. 2013;170:103–13.
70. Rapacz M, Ergon A, Hoglind M, Jorgensen M, Jurczyk B, Ostrem L, et al. Overwintering of herbaceous plants in a changing climate. Still more questions than answers. Plant Sci. 2014;225:34–44.
71. Thorsen SM, Höglind M. Modelling cold hardening and dehardening in timothy. Sensitivity analysis and Bayesian model comparison. Agric For Meteorol. 2010;150:1529–42.
72. Arft AM, Walker MD, Gurevitch J, Alatalo JM, Bret-Harte MS, Dale M, et al. Responses of tundra plants to experimental warming: meta-analysis of the international tundra experiment. Ecol Monogr. 1999;69(4):491–511.
73. Henry GHR, Molau U. Tundra plants and climate change: the International Tundra Experiment (ITEX). Glob Change Biol. 1997;3:1–9.
74. Post E, Forchhammer MC. Climate change reduces reproductive success of an Arctic herbivore through trophic mismatch. Philos Trans R Soc B-Biol Sci. 2008;363(1501):2369–75.
75. Post E, Pedersen C, Wilmers CC, Forchhammer MC. Warming, plant phenology and the spatial dimension of trophic mismatch for large herbivores. Proc R Socy B-Biol Sci. 2008;275(1646):2005–13.
76. Richardson AD, Keenan TF, Migliavacca M, Ryu Y, Sonnentag O, Toomey M. Climate change, phenology, and phenological control of vegetation feedbacks to the climate system. Agric For Meteorol. 2013;169:156–73.
77. Menzel A, Sparks TH, Estrella N, Koch E, Aasa A, Ahas R, et al. European phenological response to climate change matches the warming pattern. Glob Change Biol. 2006;12(10):1969–76.
78. Gordo O, Sanz JJ. Impact of climate change on plant phenology in Mediterranean ecosystems. Glob Change Biol. 2010;16(3):1082–106.
79. Gordo O, Sanz JJ. Long-term temporal changes of plant phenology in the Western Mediterranean. Glob Change Biol. 2009;15(8):1930–48.
80. De Boeck HJ, Lemmens CMHM, Gielen B, Bossuyt H, Malchair S, Carnol M, et al. Combined effects of climate warming and plant diversity loss on above- and below-ground grassland productivity. Environ Exp Bot. 2007;60(1):95–104.
81. De Boeck HJ, Lemmens CMHM, Zavalloni C, Gielen B, Malchair S, Carnol M, et al. Biomass production in experimental grasslands of different species richness during three years of climate warming. Biogeosciences. 2008;5(2):585–94.
82. Venglovsky J, Sasakova N, Placha I. Pathogens and antibiotic residues in animal manures and hygienic and ecological risks related to subsequent land application. Bioresour Technol. 2009;100:5386–91.
83. Churkina G, Schimel D, Braswell BH, Xiao XM. Spatial analysis of growing season length control over net ecosystem exchange. Glob Change Biol. 2005;11(10):1777–87.
84. Flanagan LB, Adkinson AC. Interacting controls on productivity in a northern Great Plains grassland and implications for response to ENSO events. Glob Change Biol. 2011;17(11):3293–311.
85. Ma SY, Baldocchi DD, Xu LK, Hehn T. Inter-annual variability in carbon dioxide exchange of an oak/grass savanna and open grassland in California. Agric For Meteorol. 2007;147(3–4):157–71.
86. Richardson AD, Black TA, Ciais P, Delbart N, Friedl MA, Gobron N, et al. Influence of spring and autumn phenological transitions on forest ecosystem productivity. Philos Trans R Soc B-Biol Sci. 2010;365(1555):3227–46.
87. Cantarel AAM, Bloor JMG, Soussana JF. Four years of simulated climate change reduces above-ground productivity and alters functional diversity in a grassland ecosystem. J Veg Sci. 2013;24(1):113–26.
88. Zwicke M, Alessio GA, Thiery L, Falcimagne R, Baumont R, Rossignol N, et al. Lasting effects of climate disturbance on perennial grassland above-ground biomass production under two cutting frequencies. Glob Change Biol. 2013;19(11):3435–48.
89. Graux A-I, Lardy R, Bellocchi G, Soussana J-F. Global warming potential of French grassland-based dairy livestock systems under climate change. Reg Environ Change. 2012;12(4):751–63.

11

# Combined effect of pulse density and grid cell size on predicting and mapping aboveground carbon in fast-growing Eucalyptus forest plantation using airborne LiDAR data

Carlos Alberto Silva[1,2*], Andrew Thomas Hudak[2], Carine Klauberg[2], Lee Alexandre Vierling[1], Carlos Gonzalez-Benecke[3], Samuel de Padua Chaves Carvalho[4], Luiz Carlos Estraviz Rodriguez[5] and Adrián Cardil[6]

**Abstract**

**Background:** LiDAR remote sensing is a rapidly evolving technology for quantifying a variety of forest attributes, including aboveground carbon (AGC). Pulse density influences the acquisition cost of LiDAR, and grid cell size influences AGC prediction using plot-based methods; however, little work has evaluated the effects of LiDAR pulse density and cell size for predicting and mapping AGC in fast-growing Eucalyptus forest plantations. The aim of this study was to evaluate the effect of LiDAR pulse density and grid cell size on AGC prediction accuracy at plot and stand-levels using airborne LiDAR and field data. We used the Random Forest (RF) machine learning algorithm to model AGC using LiDAR-derived metrics from LiDAR collections of 5 and 10 pulses $m^{-2}$ (RF5 and RF10) and grid cell sizes of 5, 10, 15 and 20 m.

**Results:** The results show that LiDAR pulse density of 5 pulses $m^{-2}$ provides metrics with similar prediction accuracy for AGC as when using a dataset with 10 pulses $m^{-2}$ in these fast-growing plantations. Relative root mean square errors (RMSEs) for the RF5 and RF10 were 6.14 and 6.01%, respectively. Equivalence tests showed that the predicted AGC from the training and validation models were equivalent to the observed AGC measurements. The grid cell sizes for mapping ranging from 5 to 20 also did not significantly affect the prediction accuracy of AGC at stand level in this system.

**Conclusion:** LiDAR measurements can be used to predict and map AGC across variable-age Eucalyptus plantations with adequate levels of precision and accuracy using 5 pulses $m^{-2}$ and a grid cell size of 5 m. The promising results for AGC modeling in this study will allow for greater confidence in comparing AGC estimates with varying LiDAR sampling densities for Eucalyptus plantations and assist in decision making towards more cost effective and efficient forest inventory.

**Keywords:** Carbon modeling, Remote sensing, Modeling, Forest inventory, Random forest

## Background

Atmospheric carbon dioxide concentration [$CO_2$] has increased by 40% since pre-industrial times, contributing greatly to climate change [1]. Managing the exchange of $CO_2$ and other greenhouse gases between the biosphere and the atmosphere is an important strategy for mitigating climate change [2]. Forest ecosystems play a key role in the global carbon cycle [3–6], since carbon is exchanged naturally between forests and the atmosphere through photosynthesis, respiration, decomposition and combustion [7]. Forest management can therefore alter

*Correspondence: carlos_engflorestal@outlook.com
[1] Department of Natural Resources and Society, College of Natural Resources, University of Idaho, (UI), 875 Perimeter Drive, Moscow, ID 83843, USA
Full list of author information is available at the end of the article

the amount and magnitude of $CO_2$ exchange between forests and the atmosphere, thus serving as an effective and economical means to help mitigate climate change [8].

Forest plantations cover approximately 1% of the tropics (40–50 million ha) and have great capacity to store carbon [9], both during growth and in the form of durable forest products after harvest. Forest plantations play a role in terms of carbon in reducing further degradation and deforestation of natural forests, as well as providing alternative to the fossil fuels. *Eucalyptus* spp. are fast-growing with desirable wood qualities and are, therefore, a preferred species in plantations; they are widely grown in the tropics and subtropics, especially in Brazil, India, and China [10]. At present, Eucalyptus is grown on more than 20 million ha of plantation land around the world [11]. Eucalyptus is the dominant hardwood plantation species in Brazil, where it has been planted on more than 3.1 million hectares, accounting for approximately 57% of the country's total reforested area [12]. Most Eucalyptus plantations are managed in short rotations (6–8 years) and the mean annual increment is approximately 40 $m^3$ $ha^{-1}$ $year^{-1}$ roundwood, ranging from 25 to 60 $m^3$ $ha^{-1}$ $year^{-1}$ depending on the level of environmental stress [13]. Quantifying the substantial roles of fast-growing Eucalyptus plantation on AGC stores, as sources of carbon emissions and as carbon sinks, has become key to understanding the global carbon cycle [6, 14].

Forest inventory in Eucalyptus plantations is usually conducted annually to monitor forest growth [13]. The aboveground carbon (AGC) production of *Eucalyptus* spp. is extremely high compared to natural forests especially when trees are grown for timber production and therefore contributes strongly to the reduction of atmospheric $CO_2$ [9, 14, 15]. Remote sensing of forest AGC has received increased attention during the last decade due to its relevance to global carbon cycle modelling and to international programs aimed at reducing greenhouse gas emissions [15, 16]. LiDAR (Light Detection and Ranging) is a powerful remote sensing technology for predicting forest attributes [17], since it enables precise mapping of the landscape distribution of forest attributes at high spatial resolution and in a relatively short time compared to conventional methods [15]. In particular, many studies showed good relationships between parameters derived from airborne LiDAR data and forest measures such as canopy height [18–20], basal area [21, 22], stem volume [23–26] and aboveground carbon [7, 15, 27–31].

While airborne LiDAR is increasingly used to map forest attributes at the landscape level, multitemporal LiDAR data acquisition over large areas is expensive [32]. Many factors influence the cost of LiDAR data, including normal cost variables, such as project size, location, and deliverables, as well as market variables, such as competition amongst LiDAR vendors. One of the most important variables affecting the cost of acquisition of LiDAR data is pulse density [33, 34]. Pulse density is defined by the number of pulses sent by the sensor per $m^2$ (pulses $m^{-2}$), and as pulse density increases the acquisition cost increases as well [32].

There has been a wide interest in understanding how reducing pulse density affects the accuracy of inventory information derived from LiDAR data [34–37]. For instance, a study conducted in a Douglas-fir plantation in South Island, New Zealand, found that the precision of models to predict forest attributes, such as mean height, volume and mean diameter at breast height (DBH), remained stable until densities of 10 pulses $m^{-2}$ were culled to 2–3 pulses $m^{-2}$ [38]. Moreover, they also found that for the scenario where the DTM created from high pulse density was used to height normalize a downscaled point clouds for corresponding LIDAR-derived metrics, little change in the precision of regression models was found until pulse densities of 0.2–0.04 pulses $m^{-2}$ were reached [38]. In mixed conifer forest in Washington state, USA, model precision was more affected by sample plot size than pulse density [39]. In mixed conifer-hardwood in Canada, pulse density could be reduced from 3.2 to 0.5 pulses $m^{-2}$ with little effect on the quality of inventory results [40].

Besides the effect of pulse density on forest attribute prediction, there is also interest in understanding how cell size combined with pulse density affects the prediction of AGC at the stand level. Most LiDAR studies have been mapping forest attributes at stand level with grid cell size similar to the size of the sample plots used for calibrating the models [19, 22, 37], which is the general recommendation [42]. However, the shapes of the inventory plots are usually different from the cell shapes utilized in the prediction maps [43] that are normally square. It seems that at high pulse density, it would be advantageous to map forest attributes with higher spatial precision; i.e., at a finer grid resolution than the plot size, especially in Eucalyptus plantation, where the trees are planted in a grid and in most cases are hybrid clones. However, it is not clear how different grid cell sizes could influence stand level forest attributes estimates due to the degree of variability within stands and how LiDAR predictions are derived within a cell [43]. For instance, a study evaluating the influence of six prediction cell sizes (2, 4, 10, 25, 50, and 100 m squares) and two prediction methods (parametric vs. non-parametric) on LiDAR-derived stand-level estimates of total volume found not trend of smaller cell sizes producing lower values of volume, while larger cell sizes tended to give higher estimates [43].

Understanding the combined influence of pulse density and grid cell sizes on AGC prediction and mapping allows us to understand how well AGC estimates can be compared over time, as LiDAR technology and flight characteristics vary. Although the effects of LiDAR pulse density on the prediction accuracy of forest attributes has been widely investigated in boreal and temperate forests [33–36, 38–40], no studies have been conducted in fast-growing plantations with the goal of assessing the combined effect of pulse density and grid cell size on predicting and mapping aboveground carbon. Therefore, the aim of this study was first to predict AGC in eight fast-growing Eucalyptus plantations in southeast Brazil using metrics derived from LiDAR with pulse densities of 5 and 10 pulses $m^{-2}$; and second, to assess the combined effect of these two pulse densities and grid cell sizes on AGC predictions at plot and stand levels. We hypothesized that due to the high homogeneity of this type of plantation, LiDAR data with a density of 5 pulses $m^{-2}$ would provide AGC prediction accuracies similar to those from LiDAR data with a density of 10 pulses $m^{-2}$, at grid cell sizes ranging from 5 to 20 m.

## Methods

### Study area description

The study area consisted of eight farms located within the Paraíba Valley in the state of São Paulo, Brazil (Fig. 1). The climate of the region is characterized as humid subtropical, with dry winters and hot summers [41]. Mean annual precipitation is ~1200 mm; mean temperature ranges from a minimum of 17.1 °C in the coldest month (July) to a maximum of 23.9 °C in the hottest month (February). The topography in the selected plantations is complex with high relief, ranging from 578 to 1310 m in elevation. The plantations contain hybrid clones of two Eucalyptus species, *Eucalyptus grandis* W. Hill ex Maid and *Eucalyptus urophylla* S.T. Blake. The plantations are managed by Fibria Celulose S/A, a pulp company. Stand age across the farms was variable and ranged from 3 to 8 years. All the trees were planted in 3 m between rows and 2 m tree within a row (1667 trees per ha).

### Field data collection

A total of 136 circular plots of 400 $m^2$ each were established across the eight farms (Site ID, Table 1) for stand measurement during the months of January and March of 2012. All plots were georeferenced with a geodetic GPS with differential correction capability (Trimble Pro-XR). Sample plot centers were geolocated with a horizontal error of up to 10 cm. In each plot, all trees were measured for diameter at breast height (DBH; cm) and a random subsample (15%) of trees for maximum height (Ht; m). For trees in the plot that were not directly

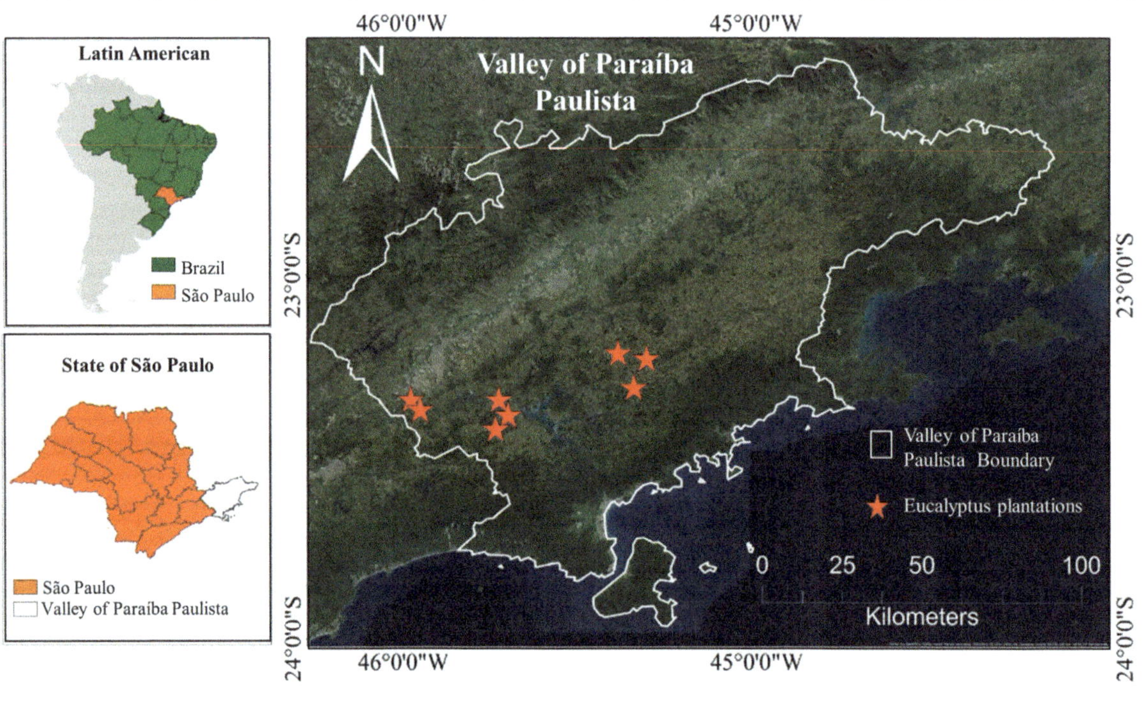


**Fig. 1** Location of the study area in the State of São Paulo, Brazil. The *stars* indicate the location of the Eucalyptus plantations

**Table 1 Descriptive values of the biometric parameters of the network of plots inventoried in the study area (January of 2012)**

| Sites ID | Area (ha) | DBH (cm) | | Ht (m) | | AGC (Mg.ha$^{-1}$) | | Age (years) | N plots |
|---|---|---|---|---|---|---|---|---|---|
| | | Mean | SD | Mean | SD | Mean | SD | | |
| F987 | 39.53 | 8.82 | 1.57 | 11.65 | 2.55 | 15.76 | 3.75 | 2.3 | 14 |
| F986 | 94.16 | 12.73 | 0.76 | 18.60 | 1.45 | 45.45 | 6.19 | 3.3 | 21 |
| F849 | 138.96 | 14.14 | 0.60 | 22.16 | 1.54 | 62.71 | 6.69 | 4.7 | 26 |
| F950 | 86.72 | 13.73 | 0.80 | 21.46 | 1.39 | 62.97 | 9.81 | 5.5 | 17 |
| F184 | 58.34 | 14.57 | 0.70 | 23.74 | 1.19 | 64.59 | 6.81 | 5.9 | 14 |
| F166 | 84.35 | 14.59 | 0.95 | 24.17 | 1.38 | 69.76 | 9.77 | 6.1 | 17 |
| F948 | 79.33 | 13.70 | 1.20 | 23.29 | 1.68 | 60.42 | 13.12 | 6.8 | 12 |
| F634 | 84.80 | 15.26 | 1.20 | 25.24 | 2.35 | 73.75 | 9.87 | 8.0 | 15 |

*SD* standard deviation

measured for Ht, the inventory team of Fibria Celulose S/A company predicted heights from hypsometric models, which use DBH as the predictor of Ht. For mapping validation purpose, trees were sub-sampled in square sub-plots of 25 m$^2$ (5 × 5 m), 100 m$^2$ (10 × 10 m) and 225 m$^2$ (15 × 15 m) located at the center of each plot. The total aboveground carbon—AGC (Mg.tree$^{-1}$) for the plots and sub-plots was obtained through the allometric model according to [15], employing as independent variables the logarithm of DBH and the Ht, and as dependent variables the AGC, as shown in the equation below:

$$\ln(AGC) = -2.87 + 1.95 \times \ln(DBH) + 0.44 \times \ln(Ht)$$

where: DBH is tree diameter over bark at breast height (1.37 m) in cm; Ht is tree height (m). The AGC model has adjusted coefficients of determination (Adj.R$^2$) of 0.97, absolute and relative root mean squared errors (RMSE) of 4.57 kg tree$^{-1}$ and 12.38%, respectively. The ln (AGC) was back transformed to natural scale, and multiplied by a correction factor of 1.03 [exp (0.5 × MSE)] to remove the bias added by the log transformation. The summed AGC stock of all trees within the plots and sub-plots was then divided by the plot and sub-plot area to calculate the AGC at plot-level in Mg.ha$^{-1}$. The summary statistics of AGC stocks (Mg.ha$^{-1}$) measured at plot-level in the farms evaluated are presented in Table 1.

### LiDAR data acquisition and data processing

LiDAR data were obtained with a Riegl LMS-Q680I sensor mounted on a Piper Seneca II aircraft during the months of January and March of 2012. Data were acquired subsequently with pulse densities of 5 and 10 pulses m$^{-2}$; the characteristics and precision of the LiDAR data are listed in Table 2. LiDAR data processing consisted of several steps that ingested the LiDAR point cloud data and provided two major outputs: the digital terrain model (DTM) and the LiDAR-derived canopy structure metrics.

LiDAR data processing for both pulse densities occurred with the following sequence of steps using FUSION/LDV toolkit software [44]. (I) Ground returns were classified using a filtering algorithm adapted from [45] based on linear prediction [46]. (II) Digital Terrain Models (DTMs) of 1 m spatial resolution were then developed using the classified ground returns. (III) Normalized LiDAR point clouds were obtained by subtracting the DTM elevation from each LiDAR return. (IV) Normalized point clouds were subset within each of the 136 sample plots, and then structure metrics were computed for each plot using only all returns above 2 m to remove the returns not belonging to tree crowns (e.g., hits on near-ground shrubs, forbs, grasses, etc.). (V) To evaluate the effect of grid cell size on the AGC prediction at stand level, we computed LiDAR metrics at grid cell sizes of 5, 10, 15, and 20 m. From the point cloud, it is possible to generate many LiDAR metrics, however, we generated only those metrics that have been frequently used as candidate predictors for forest attribute modeling in other studies [15, 18–22, 29, 47]. Therefore, a total of 27 LiDAR metrics calculated from all returns were considered for AGC modelling at the plot and stand levels (Table 3).

**Table 2 LiDAR flight characteristics**

| Parameter | Value |
|---|---|
| Average flight height | 422.94 m |
| Pulse density | 5 and 10 pulses m$^{-2}$ |
| Pulse frequency | 400 kHz |
| Scan angle | ±45° |
| Laser wavelength | 1055 nm |
| Average aircraft speed | 57 m/s (205.20 km/h) |
| Horizontal precision | 0.1–0.15 m (1.0 sigma) |

**Table 3 LiDAR-derived canopy height metrics considered as candidate variables for predictive AGC models [43]**

| Metric | Description |
|---|---|
| HMIN | Minimum canopy height |
| HMAX | Maximum canopy height |
| HMEAN | Mean canopy height |
| HMAD | Median canopy height |
| HSD | Standard deviation of canopy height |
| HSKEW | Skewness of canopy height |
| HKURT | Kurtosis of canopy height |
| HVAR | Variance of canopy height |
| HCV | Coefficient of variation canopy height |
| HMODE | Mode canopy height |
| H01TH | 1st percentile of canopy height |
| H05TH | 5th percentile of canopy height |
| H10TH | 10th percentile of canopy height |
| H20TH | 20th percentile of canopy height |
| H25TH | 25th percentile of canopy height |
| H30TH | 30th percentile of canopy height |
| H40TH | 40th percentile of canopy height |
| H50TH | 50th percentile of canopy height |
| H60TH | 60th percentile of canopy height |
| H70TH | 70th percentile of canopy height |
| H75TH | 75th percentile of canopy height |
| H80TH | 80th percentile of canopy height |
| H90TH | 90th percentile of canopy height |
| H95TH | 95th percentile of canopy height |
| H99TH | 99th percentile of canopy height |
| CR | Canopy relief ratio [(HMEAN-HMIN)/(HMAX-HMIN)] |
| COV | Canopy cover (percentage of first return above 1.30 m) |

## Random forest modeling

We used the random forest package [48, 49] in R [50] to create random forest (RF) models for predicting AGC at the plot and stand levels from the LiDAR dataset of 5 and 10 pulses $m^{-2}$ (RF5 and RF10). In the analysis, we defined the number of classification trees (ntree) as 1000, and for the number of variables randomly (mtry) sampled as candidates at each split, we used the default value, which for regression is defined as p/3, where p is the number of covariates. For the remaining parameters, we used the default values, which are all defined in [50] as well.

The selection of the best LiDAR metrics to be included in the models was performed in two steps. First, even though highly correlated variables won't cause multi-collinearity issues in RF, Pearson's correlation (r) was used to identify highly correlated predictor variables (r > 0.85) as presented in [29] and [15, 47]. If a given group (2 or more) of LiDAR metrics were highly correlated, we retained only one metric by excluding the others that were most highly correlated with the retained metrics. Second, we identified the most important metrics based on the Model Improvement Ratio (MIR), a standardized measure of variable importance [51, 52]. MIR scores are derived by dividing raw variable important scores (output from RF) by the maximum variable importance score, so that MIR values range from 0 to 1. MIR scores allow for variable importance comparisons among different RF models. For each of the two AGC models, we ran 1000 iterations of RF that included the <0.85 correlated preliminary set of LiDAR metrics to create distributions of MIR for each metric. Running 1000 iterations of RF produced consistent MIR distributions and avoided unnecessary processing time. To create parsimonious models, we reserved metrics for final RF models that were consistent and exhibited the highest mean MIR values.

The accuracy of estimates for each model was evaluated in terms of percentage of variation explained (Pseudo-$R^2$) in the RF models, Root Mean Square Error (RMSE), and Bias (both absolute and relative) computed by the linear relationship between observed and predicted AGC:

$$Pseudo - R^2 = \left[1 - \frac{\sum_{i=1}^{n}\left(y_i - \hat{y}_i\right)^2}{\sum_{i=1}^{n}\left(\frac{y}{n} - y_i\right)^2}\right] \tag{1}$$

$$RMSE = \sqrt{\frac{\sum_{i=1}^{n}\left(y_i - \hat{y}_i\right)^2}{n}} \tag{2}$$

$$Bias = \frac{1}{n}\sum_{i=1}^{n}\left(y_i - \hat{y}_i\right) \tag{3}$$

where n is the number of plots, $y_i$ is the observed value for plot i, and $\hat{y}_i$ is the predicted value for plot i. Moreover, relative RMSE and biases were calculated by dividing the absolute values (Eqs. 1, 2) by the mean of the observed AGC. We defined acceptable model precision and accuracy as a relative RMSE and Bias of $\leq$15% to have a model precision and accuracy higher than or equal to the conventional forest inventory standard in fast-growing Eucalyptus forest plantations in Brazil [14, 47].

The performance of the RF models to predict AGC was also evaluated by means of a leave-one-out cross-validation (LOOCV) strategy [e.g. 22, 49]. Statistical equivalence tests were employed [53, 54] to assess whether the AGC predictions were statistically similar at (i.e., equivalent; *p* value >0.05) to the field-based AGC stocks. Following [55], we employed a regression-based equivalence test for intercept equality to 0 (i.e., the mean of predicted AGC is equal to the mean of the field-based AGC) and slope equality to 1 (i.e., if the pairwise, predicted and observed AGC are equal, the regression will have a slope of 1). A description of equivalence tests can be also found in the 'equivalence' package in R [56], and examples of equivalence plots in LiDAR studies can be found in [15, 29, 45, 53, 55]. We also computed RMSE, Bias and the adjusted Coefficient of determination (Adj.$R^2$) from the observed and predicted AGC from LOOCV. Finally, the effects of pulse density of the AGC predictions at the plot-level were investigated by the comparison of RMSE and Bias statistics across pulse densities.

### Assessing effects of cell size on the AGC prediction and mapping at the stand level

As the study area is characterized by fast-growing plantations, with tree hybrid clones equally spaced in the ground, we assumed that AGC maps could be generated with high spatial resolution, even though the model is based on a sample unit of 400 $m^2$. Therefore, we applied the predictive model across the landscape to map AGC at spatial resolutions of 5, 10, 15, and 20 m, the last of which equates to the sample plot size of 400 $m^2$. To evaluate the effect of grid cell size on the AGC map prediction, we extracted the AGC for each study site, and boxplots were generated to compare variability of AGC at the stand level containing various grid cell sizes. Moreover, from the maps we extracted predicted AGC within the plot and sub-plot boundaries to validate the maps by the comparison with reference AGC through an equivalence test [53–57]. An overview of the methodology is outlined in Fig. 2.

## Results

### Canopy profile of Eucalyptus plantation across age and LiDAR pulse density

The canopy profiles of Eucalyptus plantations across age and pulse density are represented in Fig. 3.

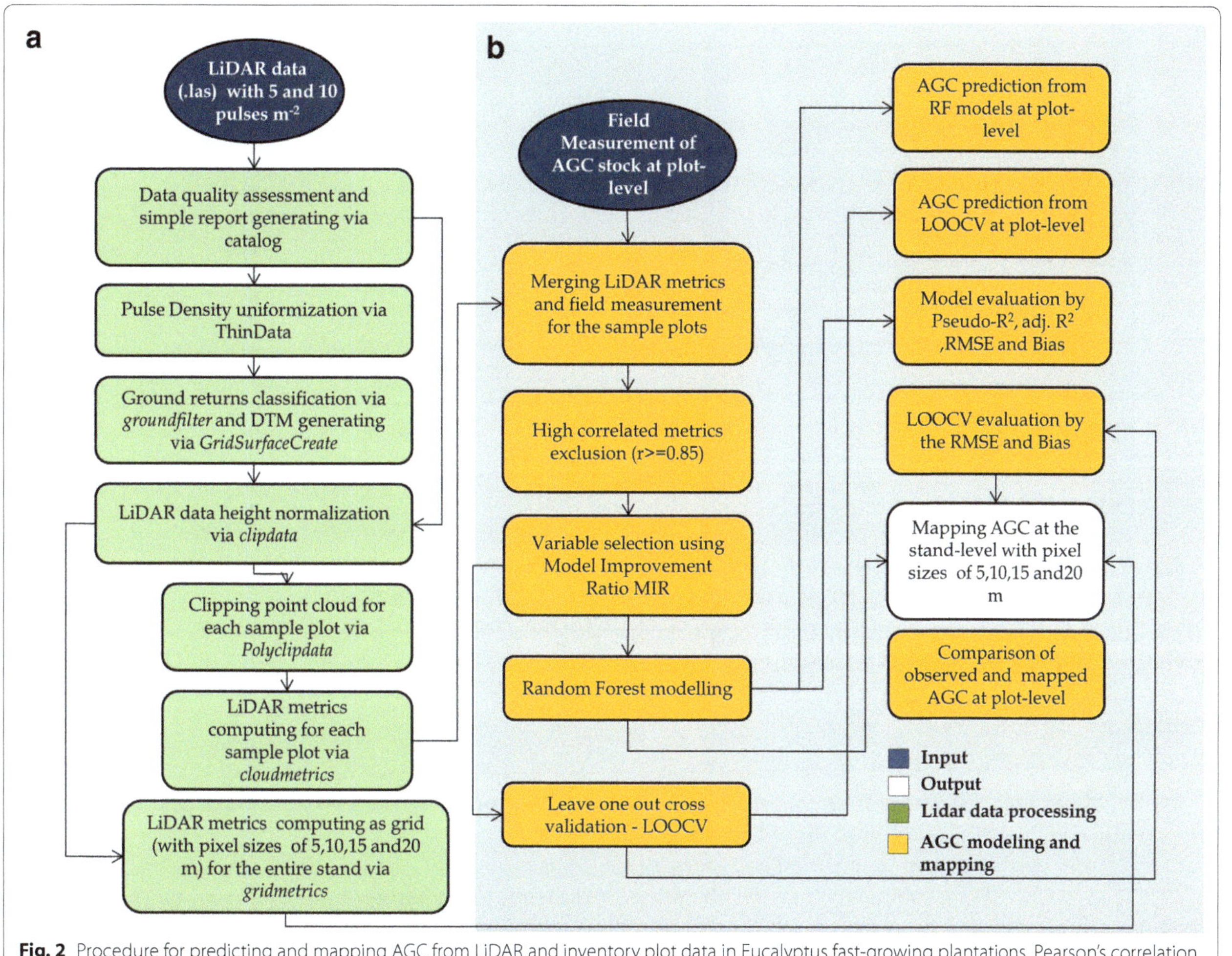


**Fig. 2** Procedure for predicting and mapping AGC from LiDAR and inventory plot data in Eucalyptus fast-growing plantations. Pearson's correlation (r); Model Improvement Ratio (MIR); Leave-one-out cross-validation (LOOCV); Adjusted coefficient of determination (adj.$R^2$); Percentage of variation explained (Pseudo-$R^2$); Root mean squared error (RMSE). **a** The light gray panel corresponds to the LiDAR data processing step and **b** the gray panel correspond to the AGC modeling step

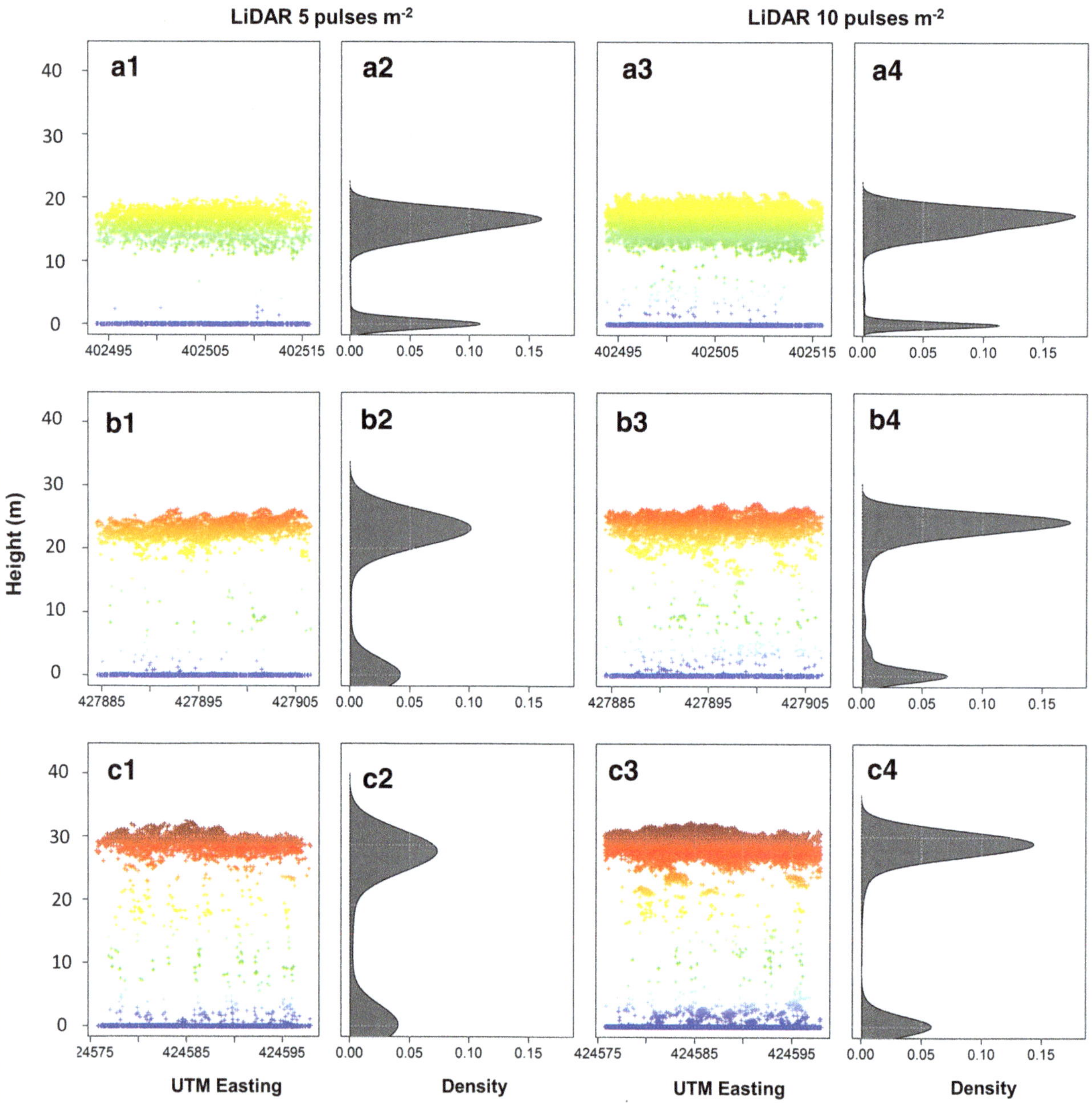


**Fig. 3** LiDAR profiles of selected sample plots (400 m$^2$) of Eucalyptus representative of early (i.e. 3.3 years) (**a**), intermediate (i.e. 5.5 years) (**b**) and advanced stages of development (i.e. 7.9 years) (**c**). (*1*) and (*3*) UTM easting profiles, (*2*) and (*4*) density plots. The *colors* represent height

LiDAR-derived height increased from early (Fig. 3a) to intermediate (Fig. 3b) and advanced ages (Fig. 3c). The variation in height was affected more by stand age than by pulse density. From a probabilistic perspective, the canopy height distribution remained the same using datasets with 5 and 10 pulses m$^{-2}$. The majority of LiDAR returns came from the upper canopy and the ground. However, as the stands approached harvest age (e.g. ~7–8 years), the number of LiDAR returns in the 0–5 m strata increased, indicating that the understory plant community increased in cover and height as the stand ages.

### LiDAR metrics selection

Pearson's correlation test (r) applied to the 27 candidate LiDAR metrics determined that 18 metrics were highly correlated (r > 0.85) for models using the 5 and 10 pulses m$^{-2}$. For each dataset, we excluded all but one of the highly correlated metrics (H99TH), and the remaining not highly correlated metrics were used to build prospective models after MIR analysis. The not highly correlated LiDAR metrics derived from the 5 and 10 pulses $m^{-2}$ dataset were HVAR, HCV, HSKE, HKUR, H01TH, H05TH, H10TH, H99TH and COV. In both correlation matrices for the LiDAR metrics derived from the 5 and 10 pulses $m^{-2}$, there were positively correlated metrics, such as H99TH and HVAR ($r = 0.65$; $r = 0.66$; p value $< 0.001$), and negatively correlated metrics, such as HCV and COV ($r = -0.34$; $r = -0.30$; p value $<0.001$), as shown in Table 4.

The remaining, not highly correlated LiDAR-derived metrics from 5 and 10 pulses $m^{-2}$ were correlated against each other and metrics such as H99TH and HVAR where highly stable across pulse density, while COV and H01TH showed to be less stable. Pearson's correlation between not highly correlated LiDAR-derived metrics from 5 and 10 pulses $m^{-2}$ are shown in the Table 5.

The most important LiDAR metrics for predicting AGC, both from the 5 and 10 pulses $m^{-2}$ datasets, were H99TH and HVAR. In addition to these two metrics, the H10TH and HSKE were also the most consistently important metrics with higher MIR (Table 6). The variation in metrics selection was less affected by pulse density, and the LiDAR metrics selected according to MIR were the same for both pulse densities.

### RF model fit and cross validation

Both RF models predicting AGC at plot level explained 83% of the variation in AGC. Greater pulse density did not improve model fit, and both RF5 and RF10 models resulted in a very low RMSE and Bias, indicating high model prediction accuracy. In addition to the RMSE and Bias, r and adj.$R^2$ did not significantly differ between the RF5 and RF10 models (Table 7).

The LOOCV analysis indicated that the RF5 and RF10 models were stable resulting RMSE of 13.61 and 13.38% and Bias of −0.09 and −0.27% for the models derived from the 5 and 10 pulses $m^{-2}$ datasets, respectively. Moreover, the equivalence plots of observed versus predicted AGC both from the two models showed that predicted and observed AGC were equivalent (p values >0.05 for intercepts and p values <0.05 for slopes (Fig. 4a, b).

**Table 4 Pearson's correlation among the derived metrics from LiDAR pulse density of 5 (white) and 10 (gray) pulses $m^{-2}$**

| r | HVAR | HCV | HKUR | H01TH | H05TH | H10TH | H99TH | COV |
|---|---|---|---|---|---|---|---|---|
| HVAR | – | *0.84*$^{***}$ | *−0.50*$^{***}$ | *−0.44*$^{***}$ | *−0.50*$^{***}$ | *−0.30*$^{***}$ | *0.66*$^{***}$ | *−0.21*$^{*}$ |
| HCV | 0.86$^{***}$ | – | *−0.73*$^{***}$ | *−0.58*$^{***}$ | *−0.79*$^{***}$ | *−0.66*$^{***}$ | *0.25*$^{**}$ | *−0.30*$^{***}$ |
| HKUR | −0.53$^{***}$ | −0.73$^{***}$ | – | *0.50*$^{***}$ | *0.83*$^{***}$ | *0.70*$^{***}$ | *0.06* | *0.10* |
| H01TH | −0.48$^{***}$ | −0.63$^{***}$ | 0.51$^{***}$ | – | *0.53*$^{***}$ | *0.26*$^{***}$ | *−0.13* | *−0.29*$^{***}$ |
| H05TH | −0.57$^{***}$ | −0.80$^{***}$ | 0.85$^{***}$ | 0.57$^{***}$ | – | *0.75*$^{***}$ | *0.17*$^{*}$ | *0.07*$^{**}$ |
| H10TH | −0.37$^{***}$ | −0.67$^{***}$ | 0.71$^{***}$ | 0.26$^{**}$ | 0.74$^{***}$ | – | *0.44*$^{***}$ | *0.40*$^{***}$ |
| H99TH | 0.65$^{***}$ | 0.29$^{***}$ | 0.02 | −0.18$^{*}$ | 0.10 | 0.39$^{***}$ | – | *0.49*$^{***}$ |
| COV | −0.30$^{***}$ | −0.34$^{***}$ | 0.12 | 0.08 | 0.25$^{**}$ | 0.18$^{*}$ | 0.45$^{***}$ | – |

The italic values indicates the values of r for metrics derived from 10 pulse $m^{-2}$

*** p value <0.001; ** p value <0.01; * p value <0.05; p value <0.1

**Table 5 Pearson's correlation between selected metrics from LiDAR pulse densities of 5 and 10 pulses $m^{-2}$**

| r | LiDAR-derived metrics (10 pulses $m^{-2}$) | | | | | | | |
|---|---|---|---|---|---|---|---|---|
| | HVAR | HCV | HKUR | H01TH | H05TH | H10TH | H99TH | COV |
| LiDAR-derived metrics (5 pulses $m^{-2}$) | | | | | | | | |
| HVAR | 0.93$^{***}$ | | | | | | | |
| HCV | 0.76$^{***}$ | 0.87$^{***}$ | | | | | | |
| HKUR | −0.38$^{***}$ | −0.55$^{***}$ | 0.69$^{***}$ | | | | | |
| H01TH | −0.38$^{***}$ | −0.46$^{***}$ | 0.40$^{***}$ | 0.60$^{***}$ | | | | |
| H05TH | −0.37$^{***}$ | −0.60$^{***}$ | 0.60$^{***}$ | 0.36$^{***}$ | 0.68$^{***}$ | | | |
| H10TH | −0.25$^{**}$ | −0.55$^{***}$ | 0.61$^{***}$ | 0.17 | 0.61$^{***}$ | 0.90$^{***}$ | | |
| H99TH | 0.64$^{***}$ | 0.28$^{***}$ | 0.03 | −0.17$^{*}$ | 0.11 | 0.40$^{***}$ | 1.00$^{***}$ | |
| COV | 0.14 | −0.09$^{***}$ | 0.13 | −0.19$^{*}$ | 0.18$^{*}$ | 0.45$^{***}$ | 0.50$^{***}$ | 0.13 |

*** p value <0.001; ** p value <0.01; * p value <0.05; p value <0.1

**Table 6 Mean of model improvement ratio (MIR) among derived metrics from LiDAR pulse density of 5 and 10 pulses m$^{-2}$ (N = 1000)**

| Pulse density (pulse m$^{-2}$) | LiDAR-derived metrics | | | | | | | |
|---|---|---|---|---|---|---|---|---|
| | **H01TH** | **H05TH** | **H10TH** | **H99TH** | **HVAR** | **HCV** | **HKUR** | **COV** |
| 5 | 0.04 | 0.13 | *0.47* | *1.00* | *0.68* | 0.21 | 0.17 | 0.02 |
| 10 | 0.03 | 0.11 | *0.33* | *1.00* | *0.52* | 0.13 | 0.07 | 0.07 |

MIR > 0.3 are highlighted in italics

**Table 7 Random Forest models fitted for predicting AGC from LiDAR pulse densities of 5 and 10 pulses m$^{-2}$**

| Pulse density | LiDAR derived metrics | Pseudo-R$^2$ | RMSE | | Bias | |
|---|---|---|---|---|---|---|
| | | | Mg ha$^{-1}$ | % | Mg ha$^{-1}$ | % |
| 5 | H99TH + HVAR + H10TH | 81.79 | 3.52 | 6.14 | −0.03 | −0.06 |
| 10 | | 82.17 | 3.44 | 6.01 | −0.01 | −0.01 |

Pearson's correlation coefficient (r); root mean square error (RMSE)

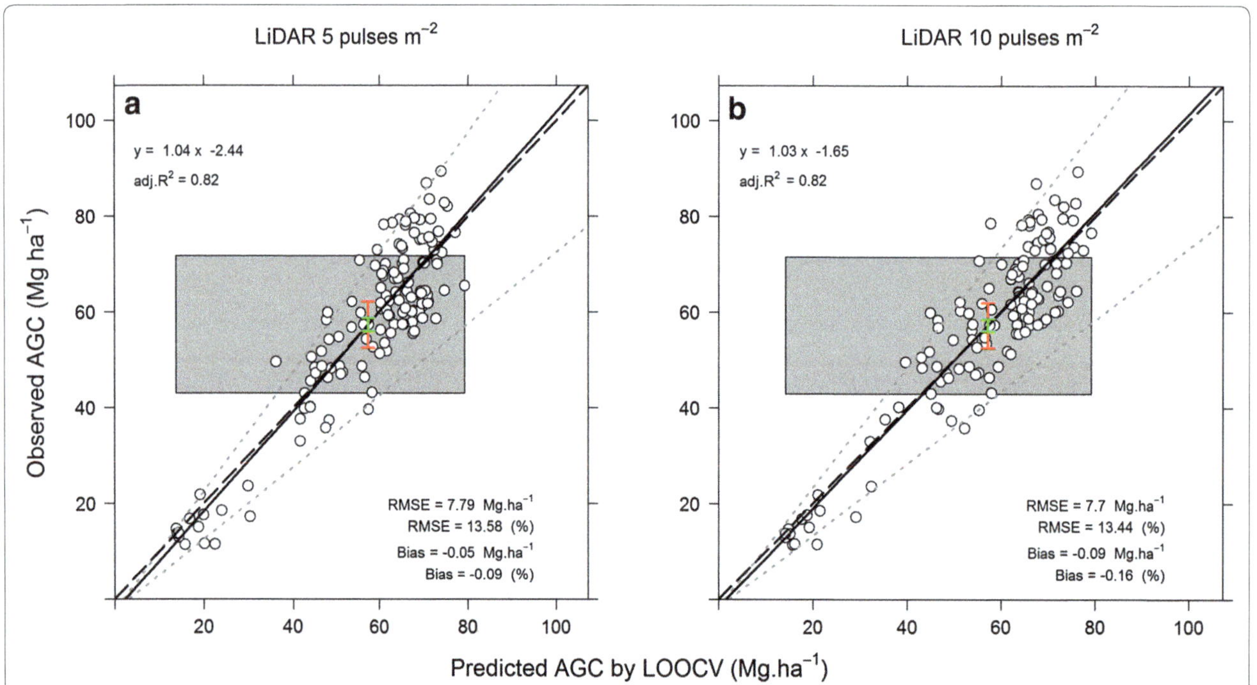


**Fig. 4** Equivalence plots of observed vs. predicted AGC by the LOOCV5 and LOOCV10 (**a**, **b**); (n = 136). The equivalence plot design presented herein is an adaptation of the original equivalence plots presented by [52]. The *grey polygon* represents the ±25% region of equivalence for the intercept, and the *red vertical bar* represents a 95% of confidence interval for the intercept. The predicted AGC from the LOOCV5 and LOOCV10 models are equivalents to the reference for the intercept if the *red bar* is completely within the *grey polygon*. If the *grey polygon* is lower than the *red vertical bar*, the predicted AGC is biased low; and if it is higher than the *red vertical bar*, the predicted AGC is biased high. Moreover, the *grey dashed line* represents the ±25% region of equivalence for the slope, and if the *green vertical bar* is contained completely within the *grey dashed line*, the pairwise measurements are equal. *Red and green bars* are wider than the region outlined by the *grey dashed lines* indicates highly variability predictions. The *white dots* are the pairwise measurements, and the *solid line* is a best-fit linear model for the pairwise measurements. The *light grey dashed line* represented the relationship 1:1

### AGC predictions at plot-level

There was no difference in the AGC prediction at plot level using LiDAR data of 5 or 10 pulses m$^{-2}$. The RF model AGC predictions for the 136 sample plots ranged from 12.99 to 82.37 for RF5, and 12.75 to 82.83 Mg. ha$^{-1}$ for RF10. In general, for both RF5 and RF10 the

AGC values were slightly overestimated during early and advanced stand ages, and slightly underestimated at intermediate ages. The Eucalyptus plantations containing younger stands showed the lowest AGC values (i.g. IDs F987 and F986), and advanced age stands contained the highest AGC stocks (i.g. IDs F166 and F634) (Fig. 5).

#### Effect of the pulse density and cell size on the AGC prediction and mapping at stand level

The variation of the predicted AGC at stand level, and across cell size and pulse density is presented in Fig. 6. Grid cell sizes higher than 5 m did not significantly affect AGC prediction at stand level. Even though the variability of AGC slightly decreased from 5 to 20 m cell size, we observed that in most of the sites, the AGC variability was similar in the maps with spatial resolution coarser than 5 m.

For each sample plot we extracted the predicted AGC from the maps. Equivalence plots indicated that the observed and predicted AGC were equivalent for the maps with all grid cell sizes (p values >0.05 for intercepts and p values <0.05 for slopes). RMSEs and Bias ranged from 11.01 to 12.3% and −0.37 to 0.87%, and showed lower values in the maps with cell size of 5 m (Fig. 7). Although the observed and predicted AGC values were equivalents at spatial resolution of 5 m for both 5 and 10 pulses $m^{-2}$, the maps are overestimating AGC in young stands (e.g. "F987") and underestimating AGC in older stands (e.g. "F634").

## Discussion

Forest managers are seeking new management strategies that integrate forest, industry and market goals that maximize financial return while ensuring sustainability in the forest production chain [14]. The carbon credit market has created optimistic prospects for expansion of the Brazilian forest sector. Although this market seems attractive, there are few studies available with carbon stock estimates for planted forests [15]. In this study, we evaluated the combined effect of pulse density and grid size on AGC prediction at plot and stand levels in a fast-growing Eucalyptus plantation. A significant amount of work has been done to evaluate the effect of LiDAR pulse density for forest inventory modeling. However, less attention has been given to the influence of the cell size used to generate stand-level maps of forest attributes [43].

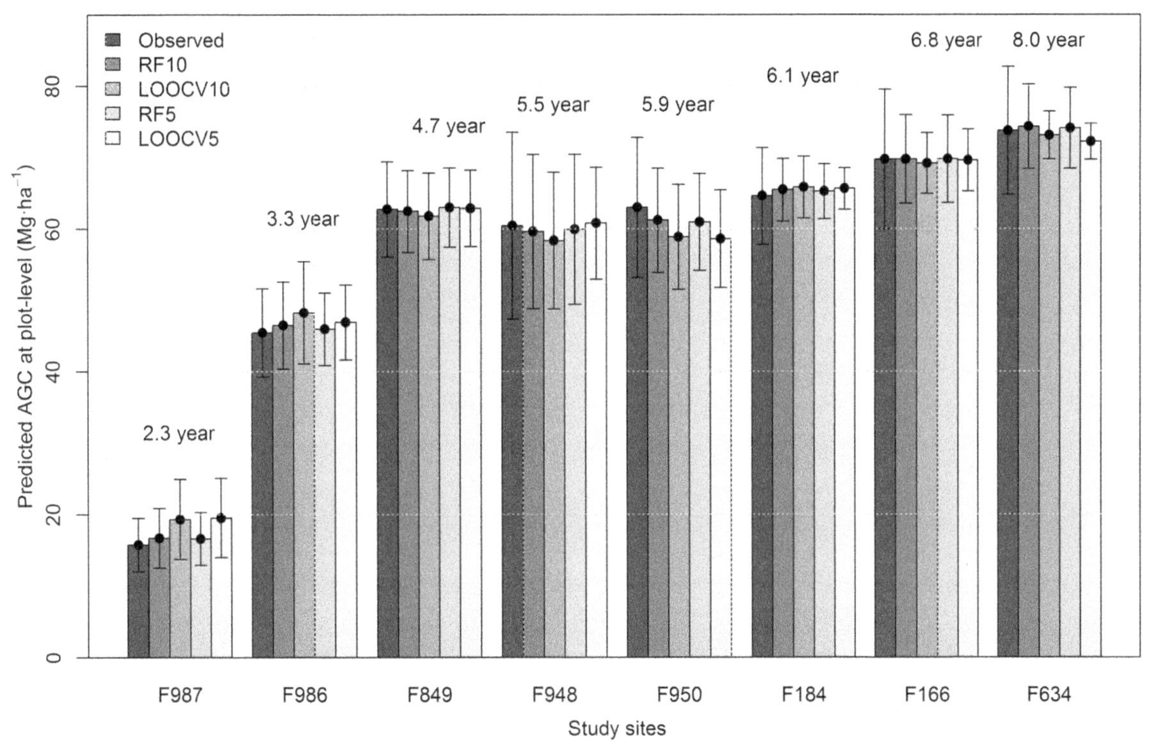


**Fig. 5** Observed versus predicted AGC by the RFs and LOOCVs for each plantation in the study area. *Black dots* represent averages and *vertical bars* represent standard deviations of AGC

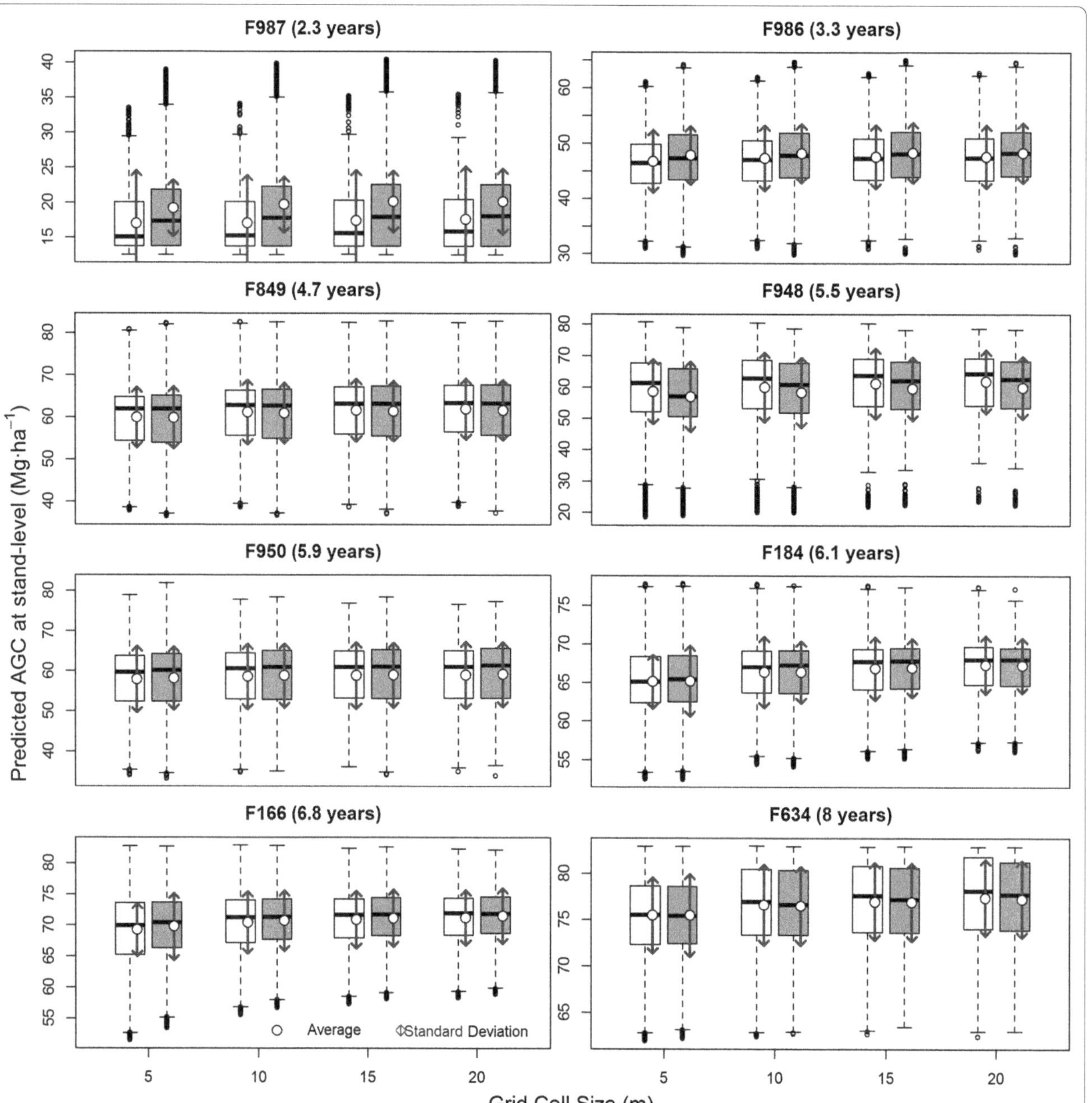


**Fig. 6** Predicted AGC at stand level. The *white and gray boxplots* represent the AGC predicted from 5 and 10 pulses m$^{-2}$, respectively

Accurate and timely measurements of AGC are critical for understanding the structure and function of terrestrial ecosystems, as well as sustainable forest resources management and carbon accounting [29]. Eucalyptus plantations are a major carbon sink compared to others types of plantations, efficiently transferring atmospheric carbon into forest biomass and soil [10, 11]. The need to increase the accuracy and the spatial coverage of carbon accounting in these types of ecosystems has economic, politic and scientific rationales [31]. Airborne LiDAR can acquire tree-dimensional information of all eucalyptus trees at a high spatial resolution in a short time [15], and whether there is benefit to using LiDAR for forest inventory depends in large part on characteristics specific to each forest [35]. The results presented in this current study provide a validation of plot and stand-based measurements of AGC. LiDAR has been shown to be a powerful technology for AGC prediction in Eucalyptus plantations, and our results have demonstrated that highly accurate estimates of AGC can be achieved using LiDAR data.

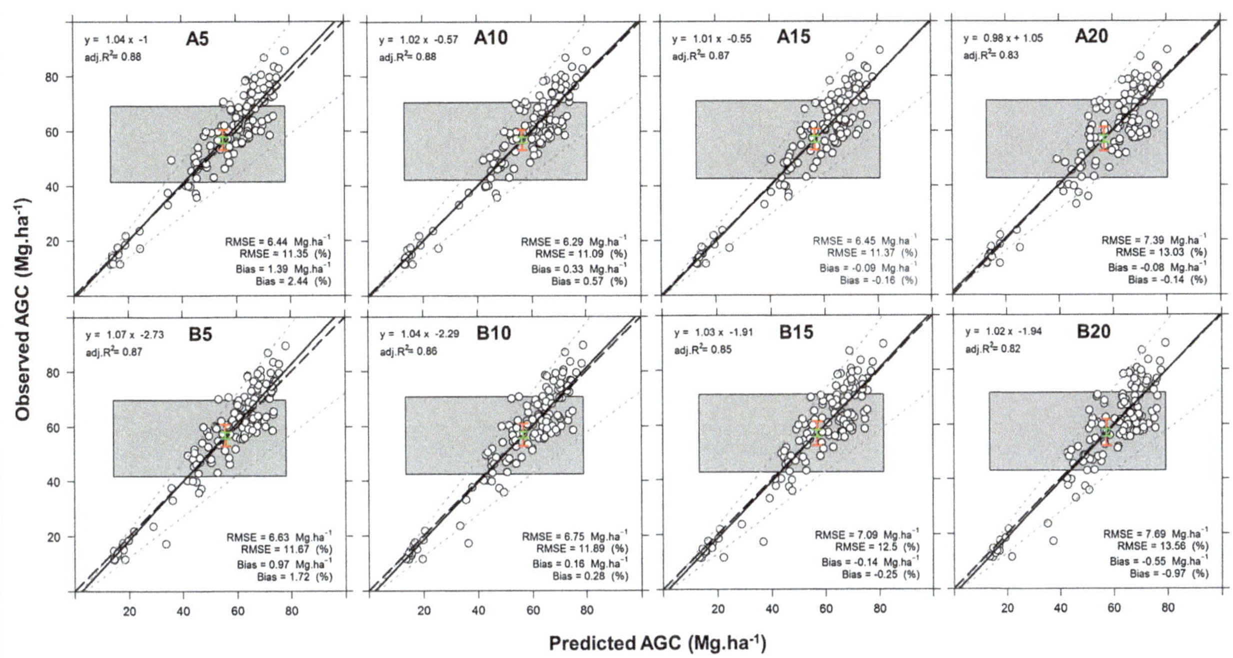


**Fig. 7** Equivalence plot of the observed and predicted AGC at stand level for the RF5 (**a**) and RF10 (**b**); (n = 136). The numbers *5*, *10*, *15*, and *20* m represent the grid cell sizes. The description of equivalence plot is presented in the caption of Fig. 4

Our results show that a LiDAR pulse density of 5 pulses $m^{-2}$ provides similar AGC prediction accuracy to that using a dataset with 10 pulses $m^{-2}$ in fast-growing eucalyptus plantations. Previous studies have demonstrated similar effect of pulse density in forest attribute estimation. For example, [21] found that pulse densities could be reduced from 1.13 to 0.25 pulses $m^{-2}$ with little effect on the quality of the forest inventory results in stands dominated by Norway spruce (*Picea abies* L. Karst.) and Scots pine (*Pinus sylvestris* L.). [35], whose objective was quantifying the effects of LiDAR pulse density and sample size on forest attributes prediction from LiDAR-derived metrics, found that model precision was more affected by sample plot sizes than by pulse density in mixed conifer forest in Washington state, USA. In a remnant forests in the rapidly urbanizing region of Charlotte, North Carolina, USA, LiDAR point density was responsible for a biomass prediction difference of 11.5% when the pulse density was downscaled from 100 to 1% [58]. The statistical parameters adj.$R^2$, RMSE and Bias presented in this study for the evaluation of AGC models are within the range reported by these other studies [33, 38, 47], though it is difficult to make a direct comparison due to differences in LiDAR sensors, pulse density testes, forest types and ground data collection.

LiDAR derived metrics, such as height percentiles, have been shown to be the most important LiDAR predictors of forest attributes in Eucalyptus plantations [15, 26, 47, 59]. In this study, the H99TH, H10TH, and HVAR were the most useful predictor variables for AGC prediction for the models derived from datasets of 5 and 10 pulses $m^{-2}$. These selected LiDAR metrics succinctly described the 3-dimensional canopy structure by capturing the majority of variation contained in the point cloud. In particular, H99TH captured canopy top height, and HVAR captured the canopy height variation of Eucalyptus stands. Also, the H99TH and HVAR were the most stable LiDAR-derived metrics when compared across pulse densities, accounting for a positive and strong linear correlation of r = 1 (H99TH5 × H99TH10) and r = 0.93 (HVAR5 × HVAR10).

The most accurate method of estimating AGC in Eucalyptus plantation is to physically sample it in the field. In a conventional inventory, one sample plot (300–500 $m^2$) per each 10 or 15 ha is normally measured with a goal of achieving a maximum acceptable RMSE of 10–15% [60]. However, this type of measurement over large areas is severely limited by cost and time. Approaches for deriving AGC information based on LiDAR data are of great utility and interest; in this study, we showed that airborne LiDAR with pulse densities of 5 pulses $m^{-2}$ can be used to predict AGC over large areas with RMSE and Bias equal to or less than in a conventional inventory. In this study, we achieved the RMSE goal of <15% established in the Methods section.

AGC maps are useful for forest management because they show the distribution of the AGC stock across the whole area. In theory, as spatial resolution increases, the higher is the probability to detect finer details and variability in forest stands. In this study, we found that AGC maps with spatial resolution ranging from 5 to 20 m show the same variability of AGC at stand level, indicating that in a fast-growing Eucalyptus plantation it is possible to map AGC in high detail even though the model was fitted using sample plots with larger sizes than the grid cell size used for mapping. However, LiDAR metrics computed in a grid cell of 5 m captured finer details of forest structure variability then those computed in a grid cell of 20 m; therefore, maps overestimated AGC in young stands and underestimated AGC in older stands, because the models calibrated from tree data collected within plots of 400 $m^2$ were not able to capture the forest variability at this fine resolution. The influence of the cell size used to generate forest attributes from LiDAR data, such as AGC, has not analyzed intensely at stand-level yet [43], and the range of examined cell sizes was limited [61, 62]. Whereas the differences among the cell sizes were almost negligible in [61], the effects of cell sizes on estimates was highlighted in [43], especially when comparing extreme cell sizes (2 and 100 m). These results concur with ours at resolutions between 5 and 20 m.

The maps with grid cell size of 5 m generated from 5 and 10 pulses $m^{-2}$ illustrated the broad similarities of AGC patterns across landscapes (Fig. 8). AGC stock was greatest in plantations with advanced ages (e.g. F634). AGC was more variable in the young plantations (such as F986 and F987). While the site F987 (Fig. 8a1, a2) showed relatively large variability in AGC, the map prediction was not affected by the LiDAR pulse density. Due to the large area and distance between the plantations, the LiDAR data at 5 and 10 pulses $m^{-2}$ at the same site were not collected on the same day. At the time of the first survey to collect LiDAR data with 5 pulses $m^{-2}$, a small stand, not studied herein, 8 years old at site F987 was not completely harvested, as indicated by the red spot in Fig. 8a1. On the other hand, the red spot was not detected in Fig. 8a2, because 1 week later, at the time of the second survey to collect the LiDAR of 10 pulse $m^{-2}$, the small stand was completely harvested.

Although the cost of LiDAR data acquisition was not a central objective that we evaluated in this study, it is nonetheless an important factor to consider. As already mentioned, pulse density has a strong influence on the acquisition cost of LiDAR data, and even though the cost for using LiDAR with high or low pulse density might be lower than the cost of a conventional inventory for AGC in fast-growing Eucalyptus plantations, as presented in [32], it will still be highly expensive for a large area. Although field-based carbon estimations remain necessary for these purposes, integrating LiDAR remote sensing into carbon inventory schemes allows recovery of spatially-explicit AGC estimates across landscapes, while reducing the total costs and need for extensive field-based sampling [31]. In a conventional inventory, normally the variability of the forest attributes at stand level is not always known and therefore less studied [63]. In this study we mapped AGC at the landscape level with a spatial resolution of 5 m, so that it is now possible to compute and capture the variability of AGC at the stand level as well. Moreover, the estimates of AGC in the stands are helpful to determine how much carbon will be stored as forest products made from the timber after harvest.

## Conclusion

The effect of LiDAR pulse density and grid cell size on prediction accuracy of AGC at plot and stand level has been evaluated and analyzed in this paper. First, we found that LiDAR measurements can be used to predict AGC across variable-age Eucalyptus plantations with adequate levels of precision and accuracy using both 5 and 10 pulses $m^{-2}$. Second, we found that H99TH, H10TH, and HVAR metrics were the most important LiDAR metrics for modeling AGC in this study. Third, we found that 5 pulses $m^{-2}$ and grid cell size of 5, 10, 15 and m did not affect the AGC predicted accuracy at plot and stand level. Finally, we demonstrated that the spatial distribution of AGC stocks can be precisely mapped at a grid cell size of 5 m using LiDAR pulse density of 5 pulses $m^{-2}$, which can be used to provide key information for carbon sequestration in this type of ecosystem. Moreover, the promising results for AGC modeling in this study will allow for greater confidence in comparing AGC estimates with varying LiDAR sampling densities for Eucalyptus plantations and assist in decision making towards more cost effective and efficient forest inventory modeling. Although this study presents results for assessing AGC in fast-growing eucalyptus plantations in Brazil, the framework presented herein can serve as a useful methodology, and we hope that the promising results for AGC modeling in this study will stimulate

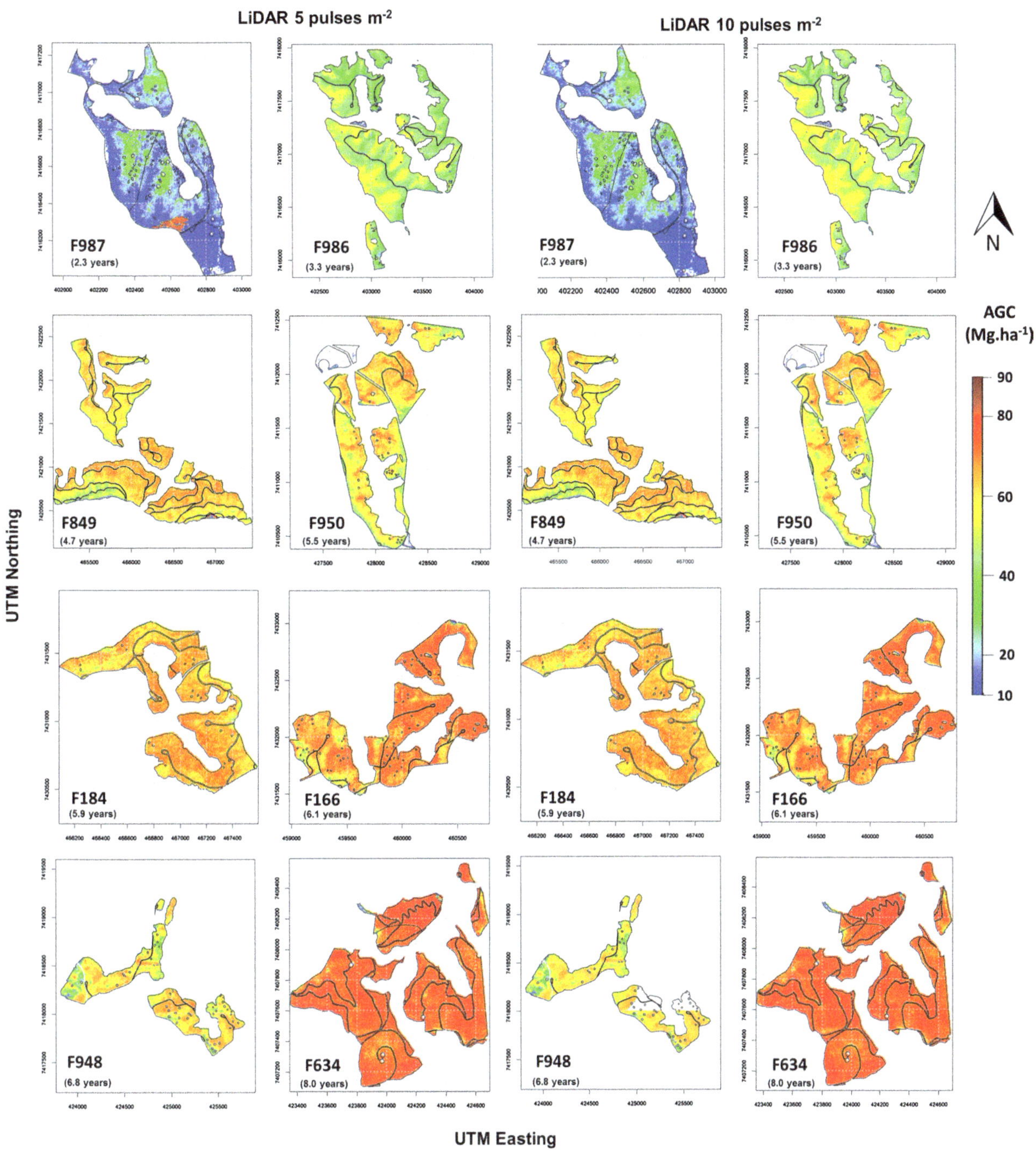


**Fig. 8** Predicted AGC of Eucalyptus at stand level with grid cell size of 5 m and pulse densities of 5 and 10 pulses m$^{-2}$ (see Table 1 for the sites codes)

further research and applications not just in Eucalyptus forest plantation in southeast Brazil, but elsewhere as well.

### Abbreviations

AGC: aboveground carbon; $CO_2$: carbon dioxide; CNPq: National Council of Technological and Scientific Development; DBH: diameter at breast height; DTM: digital terrain model; FAPESP: State of São Paulo Research Foundation; RF: random forest; RF5: random forest model of 5 pulses m$^{-2}$; RF10: random forest model of 10 pulses m$^{-2}$; GPS: global positioning system; Ht: tree height; LOOCV5: leave-one out cross validation 5 pulses m$^{-2}$; LOOCV10: leave-one out cross validation 5 pulses m$^{-2}$; PVE: percentage of the variance explained; RMSE: root mean squared errors; USFS: US Forest Service; 3D: tree-dimensional.

### Authors' contributions

All the authors have made substantial contribution towards the successful completion of this manuscript. They all have been involved in designing the study, drafting the manuscript and engaging in critical discussion. CAS, CK and ATH, contributed with the methodological framework, data processing analysis and write up. LAV, CG, LCER and AD contributed to the interpretation, quality control and revisions of the manuscript. SPC contributed with the field data survey. All authors read and approved the final manuscript.

**Author details**
[1] Department of Natural Resources and Society, College of Natural Resources, University of Idaho, (UI), 875 Perimeter Drive, Moscow, ID 83843, USA. [2] US Forest Service (USDA), Rocky Mountain Research Station, RMRS, 1221 South Main Street, Moscow, ID 83843, USA. [3] Department of Forest Engineering, Oregon State University, 269 Peavy Hall, Corvallis, OR 97331, USA. [4] College of Forestry, Federal University of Mato Grosso, Av. Fernando Correa da Costa, 2367, Boa Esperança, Cuiabá, MT 78060-900, Brazil. [5] Department of Forest Sciences, College of Agriculture Luiz de Queiroz (ESALQ), University of Sao Paulo (USP), Av. Pádua Dias, 11, Piracicaba, SP 13418-900, Brazil. [6] Tecnosylva, Parque Tecnológico de León, 24009 León, Spain.

**Acknowledgements**
The authors are very grateful for the LiDAR and field inventory data collections funded by Fibria S.A., a pulp and paper company. We thank Ben Bright and Jan Eitel for their helpful suggestions on an earlier draft of the manuscript.

**Competing interests**
The authors declare that they have no competing interests.

**Funding**
This research was funded through a Ph.D. scholarship from the National Council of Technological and Scientific Development - CNPq via the Science Without Borders Program (Process 249802/2013-9) and by the State of São Paulo Research Foundation-FAPESP (Process 12/03176-0).

**References**

1. IPCC. Climate change 2013: The physical science basis. contribution of working group I to the Fifth assessment report of the intergovernmental panel on climate change. Cambridge, UK, and New York, USA. http://www.ipcc.ch/report/ar5/wg1/. Accessed 10 Jan 2016.
2. IPCC. Climate change 2001: The scientific basis; Cambridge: Cambridge University Press; 2016. http://www.gridla.no/climate/ipcc_tar/wg1/pdf/wg1_tar-front.pdf. Accessed 10 Jan 2016.
3. Binkley CS, Apps MJ, Dixon RK, Kauppi P, Nilsson LO. Sequestering carbon in natural forests. Crit Rev Environ Sci Technol. 1998;27:23–45.
4. Fan S, Gloor M, Mahlman J, Pacala S, Sarmiento JL, Takahashi T, Tans P. The North American Sink. Science. 1815;1999:283. doi:10.1126/science.283.5409.1813q.
5. Houghton RA. Interannual variabiity in the global carbon cycle. J Geophys Res. 2000;105:20121–30.
6. Gálvez FB, Hudak AT, Byrne JC, Crookston NL, Keefe RF. Using climate-FVS to project landscape-level forest carbon stores for 100 years from field and LiDAR measures of initial conditions. Carbon Balance Manag. 2014;9:1–22. doi:10.1186/1750-0680-9-1.
7. García M, Riaño D, Chuvieco E, Danson FM. Estimating biomass carbon stocks for a Mediterranean forest in central Spain using LiDAR height and intensity data. Remote Sens Environ. 2010;114:816–30. doi:10.1016/j.rse.2009.11.021.
8. Bellassen V, Luyssaert S. Carbon sequestration: managing forests in uncertain times. Nature. 2014;506:153–5.
9. Stape JL, Binkley D, Ryan MG. Production and carbon allocation in a clonal Eucalyptus plantation with water and nutrient manipulations. For Ecol Manag. 2008;255:920–30. doi:10.1016/j.foreco.2007.09.085.
10. Du H, Zeng F, Peng W, Wang K, Zhang H, Liu L, Song T. Carbon Storage in a Eucalyptus Plantation chronosequence in Southern China. Forests. 2015;6:1763–78. doi:10.3390/f6061763.
11. Booth TH. Eucalypt plantations and climate change. For Ecol Manag. 2013;301:28–34. doi:10.1016/j.foreco.2012.04.004.
12. Abá. Brazilian tree industry. 2015. http://www.iba.org/images/shared/iba_2015.pdf.
13. Gonçalves JLDM, Alvares CA, Higa AR, Silva LD, Alfenas AC, Stahl J, Ferraz SFDB, Lima WDP, Brancalion PHS, Hubner A, Bouillet JPD, Laclau JP, Nouvellon Y, Epron D. Integrating genetic and silvicultural strategies to minimize abiotic and biotic constraints in Brazilian eucalypt plantations. For Ecol Manag. 2013;301:6–27. doi:10.1016/j.foreco.2012.12.030.
14. Silva CA, Klauberg C, Pádua SCD, Piccolo M, Rodriguez LCE. Estoque de carbono na biomassa aérea florestal em plantações comerciais de Eucalyptus spp. Sci For. 2015;43:135–46.
15. Silva CA, Klauberg C, Carvalho SPC, Hudak A, Rodriguez LCE. Mapping aboveground carbon stocks using LiDAR data in *Eucalyptus* spp. plantations in the state of São Paulo, Brazil. Sci For. 2014;42:591–604.
16. Laurin VG, Chen Q, Lindsell JA, Coomes DA, Del Frate F, Guerriero L, Pirotti F, Valentini R. Above ground biomass estimation in an African tropical forest with lidar and hyperspectral data. ISPRS J Photogramm Remote Sens. 2014;89:49–58. doi:10.1016/j.isprsjprs.2014.01.001.
17. Hudak AT, Evans JS, Smith AMS. LiDAR utility for natural resource managers. Remote Sens. 2009;1:934–51. doi:10.3390/rs1040934.
18. Næsset E. Determination of mean tree height of forest stands using airborne laser scanner data. ISPRS J Photogramm Remote Sens. 1997;52:49–56. doi:10.1016/S0924-2716(97)83000-6.
19. Næsset E, Bjerknes KO. Estimating tree heights and number of stems in young forest stands using airborne laser scanner data. Remote Sens Environ. 2001;78:328–40. doi:10.1016/S0034-4257(01)00228-0.
20. Næsset E, Økland T. Estimating tree height and tree crown properties using airborne scanning laser in a boreal nature reserve. Remote Sens Environ. 2002;79:105–15. doi:10.1016/S0034-4257(01)00243-7.
21. Gobakken T, Næsset E. Weibull and percentile models for lidar-based estimation of basal area distribution. Scand J For Res. 2005;20:490–502. doi:10.1080/02827580500373186.
22. Hudak AT, Crookston NL, Evans JS, Falkowski MJ, Smith AMS, Gessler P. Regression modeling and mapping of coniferous forest basal area and tree density from discrete- return LiDAR and multispectral satellite data. Can J Remote Sens. 2006;32:126–38. doi:10.5589/m06-007.
23. Nelson R, Krabill W, Tonelli J. Estimating forest biomass and volume using airborne laser data. Remote Sens Environ. 1988;24:247–67. doi:10.1016/0034-4257(88)90028-4.
24. Oderwald R, Popescu S. A simplified method of predicting percent volume in log portions. South J Appl. 2003;27:149–52.
25. Holmgren J. Prediction of tree height, basal area and stem volume in forest stands using airborne laser scanning. Scand J For Res. 2004;19:543–53. doi:10.1080/02827580410019472.
26. Tesfamichael SG, Van Aardt JAN, Ahmed F. Estimating plot-level tree height and volume of Eucalyptus grandis plantations using small-footprint, discrete return lidar data. Prog Phys Geogr. 2010;34:515–40. doi:10.1177/0309133310365596.
27. Næsset E. Estimation of above- and below-ground biomass in boreal forest ecosystems. Int Arc Photogramme Remote Sens Spat Inf Sci. 2004;36:145–8.
28. Næsset E, Gobakken T. Estimation of above- and below-ground biomass across regions of the boreal forest zone using airborne laser. Remote Sens Environ. 2008;112:3079–90.
29. Hudak AT, Strand EK, Vierling LA, Byrne JC, Eitel JUH, Martinuzzi S, Falkowski MJ. Quantifying aboveground forest carbon pools and fluxes from repeat LiDAR surveys. Remote Sens Environ. 2012;123:25–40. doi:10.1016/j.rse.2012.02.023.
30. Mascaro J, Detto M, Asner GP, Muller-Landau HC. Evaluating uncertainty in mapping forest carbon with airborne LiDAR. Remote Sens Environ. 2011;115:3770–4. doi:10.1016/j.rse.2011.07.019.
31. Patenaude G, Hill R, Milne R, Gaveau DLA, Briggs BBJ, Dawson TP. Quantifying forest above ground carbon content using LiDAR remote sensing. Remote Sens Environ. 2004;93:368–80. doi:10.1016/j.rse.2004.07.016.
32. Hummel S, Hudak AT, Uebler EH, Falkowski MJ, Megown KA. A Comparison of accuracy and cost of LiDAR versus stand exam data for landscape management on the Malheur National Forest. J For. 2011;109:267–73.
33. Watt MS, Adams T, Aracil SG, Marshall H, Watt P. The influence of

LiDAR pulse density and plot size on the accuracy of New Zealand plantation stand volume equations. N Z J For Sci. 2013;43:1–10. doi:10.1186/1179-5395-43-15.
34. Magnussen S, Næsset E, Gobakken T. Reliability of LiDAR derived predictors of forest inventory attributes: a case study with Norway spruce. Remote Sens Environ. 2010;114:700–12.
35. Magnussen S, Boudewyn P. Derivations of stand heights from airborne laser scanner data with canopy-based quantile estimators. Can J Remote Sens. 1998;28:1016–31.
36. Ruiz L, Hermosilla T, Mauro F, Godino M. Analysis of the influence of plot size and LiDAR density on forest structure attribute estimates. Forests. 2014;5:936–51.
37. Leitold V, Keller M, Morton DC, Cook BD, Shimabukuro YE. Airborne lidar-based estimates of tropical forest structure in complex terrain: opportunities and trade-offs for REDD+. Carbon Balance Manag. 2015;. doi:10.1186/s13021-015-0013-x.
38. Watt MS, Meredith A, Watt P, Gunn A. The influence of LiDAR pulse density on the precision of inventory metrics in young unthinned Douglas-fir stands during initial and subsequent LiDAR acquisitions. N Z J For Sci. 2014;44:18. doi:10.1186/s40490-014-0018-3.
39. Strunk J, Temesgen H, Andersen HE, Flewelling JP, Madsen L. Effects of LiDAR pulse density and sample size on a model-assisted approach to estimate forest inventory variables. Can J Remote Sens. 2012;38:644–54.
40. Treitz P, Lim K, Woods M, Pitt D, Nesbitt D, Etheridge D. LiDAR sampling density for forest resource inventories in Ontario. Can J Remote Sens. 2012;4:830–48. doi:10.3390/rs4040830.
41. Koppen W, Geiger R. Klimate der Erde. Gotha: Verlag Justus Perthes. Wall-map 150cm × 200cm. 1928.
42. White JC, Wulder MA, Varhola A, Vastaranta M, Coops N, Cook BD, Pitt D, Woods M. A best practices guide for generating forest inventory attributes from airborne laser scanning data using an area-based approach. Nat For Chron. 2013;89:5.
43. Hayashi R, Weiskittel A, Kershaw JA. Influence of prediction cell size on LiDAR-derived area-based estimates of total volume in mixed species and multi-cohort forests in north eastern North America. Can J Remote Sens. 2016;41:473–88.
44. Mcgaughey RJ. FUSION/LDV: Software for LiDAR data analysis and visualization. 3rd ed. USDA, Forest Service Pacific Northwest Research Station, Seattle; 2015.
45. Kraus K, Pfeifer N. Determination of terrain models in wooded areas with airborne laser scanner data. ISPRS J Photogramm Remote Sens. 1998;53:193–203.
46. Kraus K, Mikhail EM. Linear least squares interpolation. Photogramm Eng. 1972;38:1016–29.
47. Silva CA, Klauberg C, Hudak TA, Vierling LA, Carvalho SP, Rodriguez LC. A principal component approach for predicting the stem volume in Eucalyptus plantations in Brazil using airborne LiDAR data. Forestry. 2016;1:412. doi:10.1093/forestry/cpw016.
48. Breiman L. Random forests. Mach Learn. 2001;45(1):5–32.
49. Liaw A, Wiener M. Classification and regression by random-forest. R News. 2002;2:18–22.
50. R Development Core Team. R: A language and environment for statistical computing. Vienna: R Foundation for Statistical Computing Access: http://www.Rproject.org. 2015.
51. Evans JS, Cushman SA. Gradient modeling of conifer species using random forest. Landsc Ecol. 2009;24:673–83.
52. Evans JS, Murphy MA, Holden ZA, Cushman SA. Modeling species distribution and change using Random Forests. In: Drew CA, Huettmann F, Wiersma Y, editors. Predictive Modeling in Landscape Ecology. New York: Springer; 2011. p. 139–59.
53. Robinson AP, Froese RE. Model validation using equivalence tests. Ecol Mod. 2004;176:349–58.
54. Robinson AP, Duursma RA, Marshall JD. A regression-based equivalence test for model validation: shifting the burden of proof. Tree Physiol. 2005;25:903–13.
55. Smith AMS, Falkowski MJ, Hudak AT, Evans JS, Robinson AP, Steele CM. A cross-comparison of field, spectral, and lidar estimates of forest canopy cover. Can J Remote Sens. 2009;35(5):447–59.
56. Robinson, A. Equivalence: Provides Tests and Graphics for Assessing Tests of Equivalence, Version 0.7.2. https://cran.r-project.org/web/ packages/ equivalence/ (accessed on 20 January, 2016).
57. Silva CA, Hudak A, Vierling LA, Loudermilk L, O'brien JJ, Hiers J, Jack J, Gonzalez-Benecke CA, Lee H, alkowskie MJ, Khosravipour A. Imputation of individual longleaf pine ( Mill.) Tree attributes from field and LiDAR Data. Can J Remote Sens. 2016;42:554–73.
58. Singh KK, Chen G, McCarter JB, Meentemeyer RK. Effects of LiDAR point density and landscape context on estimates of urban forest biomass. ISPRS J Photogramm Remote Sens. 2015;101:310–22. doi:10.1016/j.isprsjprs.2014.12.021.
59. Packalen P, Maltamo M, Mehtatalo L. ALS-based estimation of plot volume and site index in a Eucalyptus plantation with a nonlinear mixed-effect model that accounts for the clone effect. Ann For Sci. 2011;2011(68):1085–92.
60. Batista JLF, Couto HTZ, Silva Filho DF. Quantificação de Recursos Florestais: árvores, Arvoredos e Florestas. 1st edn. Oficina de Textos; 2014.
61. Gonzalez P, Asner GP, Battles JJ, Lefsky MA, Waring KM, Palace M. Forest carbon densities and uncertainties from Lidar, QuickBird, and field measurements in California. Remote Sens Environ. 2010;114:1561–75.
62. Brubaker KM, Johnson SE, Brinks J, Leites LP. Estimating canopy height of deciduous forests at a regional scale with leaf-off, low point density LiDAR. Can J Remote Sens. 2014;40:123–34.
63. Wulder MA, White JC, Bater CW, Coops NC, Hokinson CH, Chen G. Lidar plots a new large-area data collection option: context, concepts, and case study. Can J Remote Sens. 2012;38:600–18.

# 12

# Historical effects of dissolved organic carbon export and land management decisions on the watershed-scale forest carbon budget of a coastal British Columbia Douglas-fir-dominated landscape

B. P. Smiley[1*] and J. A. Trofymow[1,2]

**Abstract**

**Background:** To address how natural disturbance, forest harvest, and deforestation from reservoir creation affect landscape-level carbon (C) budgets, a retrospective C budget for the 8500 ha Sooke Lake Watershed (SLW) from 1911 to 2012 was developed using historical spatial inventory and disturbance data. To simulate forest C dynamics, data was input into a spatially-explicit version of the Carbon Budget Model-Canadian Forest Sector (CBM-CFS3). Transfers of terrestrial C to inland aquatic environments need to be considered to better capture the watershed scale C balance. Using dissolved organic C (DOC) and stream flow measurements from three SLW catchments, DOC load into the reservoir was derived for a 17-year period. C stocks and stock changes between a baseline and two alternative management scenarios were compared to understand the relative impact of successive reservoir expansions and sustained harvest activity over the 100-year period.

**Results:** Dissolved organic C flux for the three catchments ranged from 0.017 to 0.057 Mg C ha$^{-1}$ year$^{-1}$. Constraining CBM-CFS3 to observed DOC loads required parameterization of humified soil C losses of 2.5, 5.5, and 6.5%. Scaled to the watershed and assuming none of the exported terrestrial DOC was respired to $CO_2$, we hypothesize that over 100 years up to 30,657 Mg C may have been available for sequestration in sediment. By 2012, deforestation due to reservoir creation/expansion resulted in the watershed forest lands sequestering 14 Mg C ha$^{-1}$ less than without reservoir expansion. Sustained harvest activity had a substantially greater impact, reducing forest C stores by 93 Mg C ha$^{-1}$ by 2012. However approximately half of the C exported as merchantable wood during logging (~176,000 Mg C) may remain in harvested wood products, reducing the cumulative impact of forestry activity from 93 to 71 Mg C ha$^{-1}$.

**Conclusions:** Dissolved organic C flux from temperate forest ecosystems is a small but persistent C flux which may have long term implications for C storage in inland aquatic systems. This is a first step integrating fluvial transport of C into a forest carbon model by parameterizing DOC flux from soil C pools. While deforestation related to successive reservoir expansions did impact the watershed-scale C budget, over multi-decadal time periods, sustained harvest activity was more influential.

**Keywords:** Carbon budgets, Deforestation, Dissolved organic carbon, Disturbance history, CBM-CFS3

*Correspondence: byron.smiley@canada.ca
[1] Natural Resources Canada, Canadian Forest Service, 506 West Burnside Road, Victoria, BC V8Z 1M5, Canada
Full list of author information is available at the end of the article

## Background

Climate change mitigation requires a global effort to reduce the amount of greenhouse gases (GHG) in the atmosphere. Strategies to decrease atmospheric concentrations of $CO_2$ require both reduction of anthropogenic emissions and improved means of C sequestration. The potential of forests in Canada to be net C sinks, while highly variable in space and time [1], can be considered to have a positive role in climate change mitigation. In temperate and boreal forests, while the natural disturbance regime is a primary driver of the ecosystem C balance, forest management activities also have an impact [2]. If forest management practices are amended to include C sequestration, management practices can be optimized to allow for the forested land base to sequester and store more C than it would have otherwise [3]. This can be accomplished through various management practices, including forest conservation in parks and protected areas [4], enhanced silviculture and harvest optimization [3] and longer-lived harvested wood products that displace more C intensive products [5].

The movement of C from the forested terrestrial system into the aquatic system is a subtle feature of the C cycle that has not been widely included in modelling efforts, dissolved organic C being a primary vector for C transport between these systems. Globally, human use of the terrestrial land base has increased the transfer of C to inland aquatic systems by as much as 1.0 pentagrams of C per year [6]. At the watershed-scale, accounting for the export of terrestrial C via fluvial systems is necessary when evaluating the C storage effect of different forest management practices. Anthropogenic disturbance can also have a considerable impact on the transport of suspended sediments, 90% of which do not make it to the ocean and deposit in lake and floodplain sediments [7]. Carbon burial in lake sediment is an important component of watershed-scale C budgets and has unique implications for areas managed for water supply [8, 9].

While the link between major hydrological events within a watershed and C being discharged in fluvial systems from that watershed are highly correlated, other watershed characteristics that may impact the concentration of C fluxes have not been well studied [10]. Dissolved organic matter, or dissolved organic carbon (DOC) as it commonly measured, is sourced from leached decaying plant material and mineral soil layers [11]. The fraction of lakes and wetlands within a catchment is known to be an important regulator of DOC export [12]. While the presence of bogs or wetlands within a catchment is a major source of DOC [11], natural or anthropogenic disturbance to forest cover and other land use classes [13] can also greatly influence the type and amount of C being exported from the terrestrial component of a watershed [10, 11, 14]. Forest cover disturbance affect both the short term discharge of DOC to the aquatic system due to factors such as amplified overland water flow [15] and rapid accumulation of organic matter [16], but also long term DOC discharge resulting from slow redevelopment of forest floor and soil C pools.

Research by Creed et al. [17] indicates that in North America, both environmental factors (summer precipitation, water residence time) and ecological factors (forest type and age) need to be considered when attempting to increase resilience of forested water supply watersheds against future climate warming. Considering, in the twentieth century, the area of inland river systems in the form of reservoirs increased by approximately 700% [18], the lateral transport of C from terrestrial systems to inland aquatic environments represents a significant C flux that may be altered by future climate change through increased sudden rainfall events and longer periods of summer drought [19]. Without understanding the existing magnitude of this C flux, the potential impact on watershed-scale C budgets is largely unknown.

The Carbon Budget Model of the Canadian Forest Sector 3 (CBM-CFS3) has been used in a C accounting and reporting capacity in numerous operational, regional and national scale analyses, both in Canada and internationally [20]. The model has also been used to evaluate the effectiveness of forest management strategies to mitigate climate change [3]. Recent work modelled carbon stocks and fluxes using spatially-explicit forest inventory and remotely-sensed disturbance datasets with a version of CBM-CFS3 that processes and outputs spatial layers [21, 22]. Within these analyses, the model assumes that any C transfers out of the forest system via dissolved C are included in decomposition releases to the atmosphere [20]. While small relative to the land–atmosphere exchange of C, the land-inland aquatic system exchange of C may account for a significant proportion of C that is generally assumed to be respired to the atmosphere in modeling efforts or remain within land ecosystems [23].

The main purpose of this study was to address a gap in current forest C budget research relating to the relative importance of including DOC as a dynamic C export mechanism from the terrestrial ecosystem. The specific objectives of this study were to: (1) parameterize the CBM-CFS3 modeled transfer of C from the terrestrial to the inland aquatic system using [DOC] and stream discharge data from 1996 to 2012 (Fig. 1), (2) apply the DOC parameterization to the Sooke Lake Watershed to estimate the impact on landscape C budgets over 100 years (1911–2012), and (3) compare the relative impacts of land management activities from reservoir creation and expansion and sustained harvest activity on the landscape C budget.

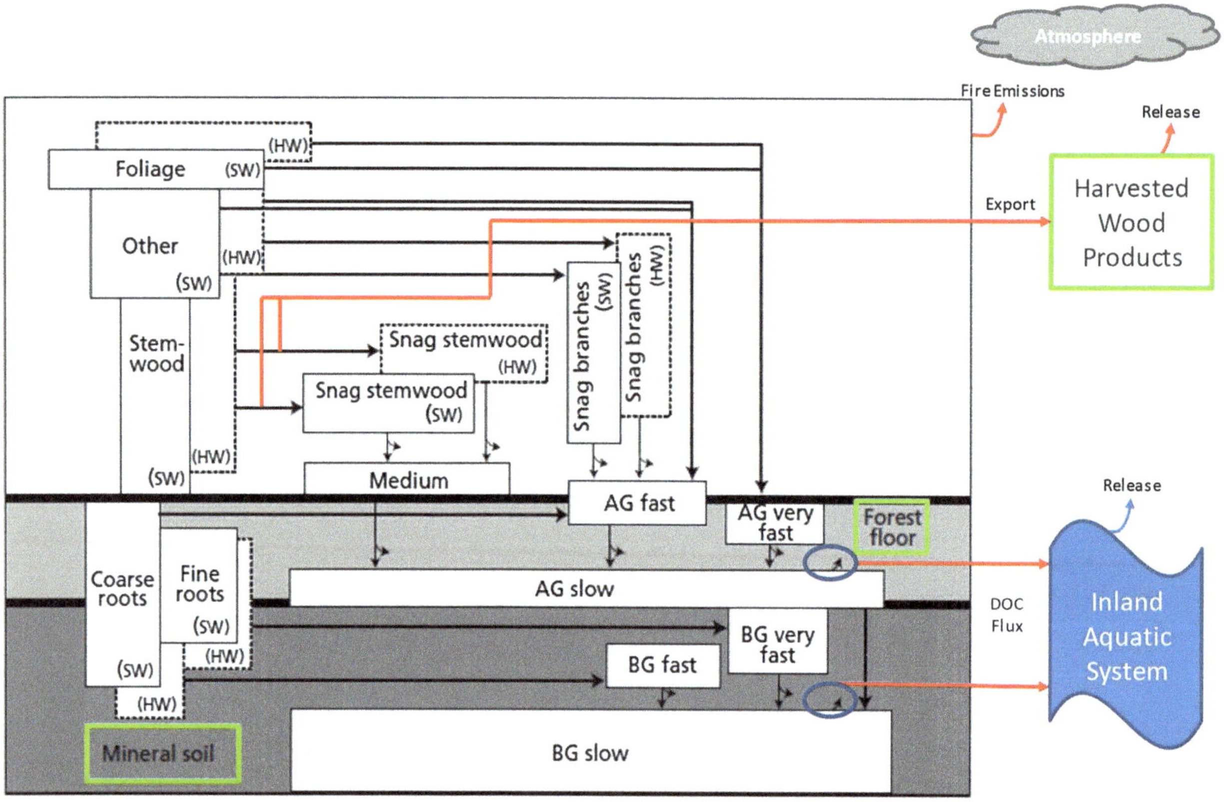


**Fig. 1** CBM-CFS3 carbon pool structure augmented to include transfers of C from the aboveground slow and belowground slow pools to the inland aquatic system via dissolved organic C (DOC) (Adapted from [36])

## Methods

### Study area

The Sooke Lake Watershed (SLW) Reservoir (48°31′30″N, 123°37′30″W) is located on southern Vancouver Island, British Columbia (BC), Canada (Fig. 2). The SLW, part of the Greater Victoria Water Supply Area, is approximately 40 km north of Victoria and is 8595 ha in size of which 810 ha is now reservoir. The Capital Regional District (CRD) ownership of the Sooke Lake water supply area constitutes over 90% of the area that drains into Sooke reservoir [24].

The SLW lies within the Nanaimo Lowlands Physiographic region and is dominated by the Coastal Western Hemlock, Very Dry Maritime biogeoclimatic zone [25]. It is a mild and moist climate with approximately 1640 mm of mean annual precipitation and warm dry summers with an average July air temperature of 16.4 °C. The wet season spans October to March and is characterized by a large hydrograph peak in the late fall followed by consistent rainfall for the remainder of the season until spring [26]. The winters are mild and typically free of extended sub-zero temperatures. During the winter some snowpack does exist in the watershed [27]. By April, precipitation begins to taper off; June has the least variable precipitation regime while July and August experience maximum temperatures and minimum precipitation [26].

Unlike the majority (95%) of forest land in BC which is in crown (public) possession [28], the SLW and adjacent areas became private land as part of the Esquimalt and Nanaimo Railway land grant in 1884. The majority (80%) of the SLW was bought and managed for Victoria's water supply in 1911 by the Greater Victoria Water District, now the CRD. Due to the potential negative implications of the remaining 20% of lands within the watershed being managed without consideration for water quality, the CRD, through a combination of land exchanges and purchases, eventually acquired much of the remaining lands within the SLW. Including the Council Lake drainage that is diverted into Sooke Reservoir, approximately 98% of the area that drains into Sooke Reservoir is now CRD-owned.

Data on forest disturbances in the SLW were consolidated into a geodatabase for the period 1911–2012 [21]. Sooke Lake was dammed for Greater Victoria's water supply (1915) and the reservoir system was expanded three times (1970, 1980, and 2002). The SLW experienced three

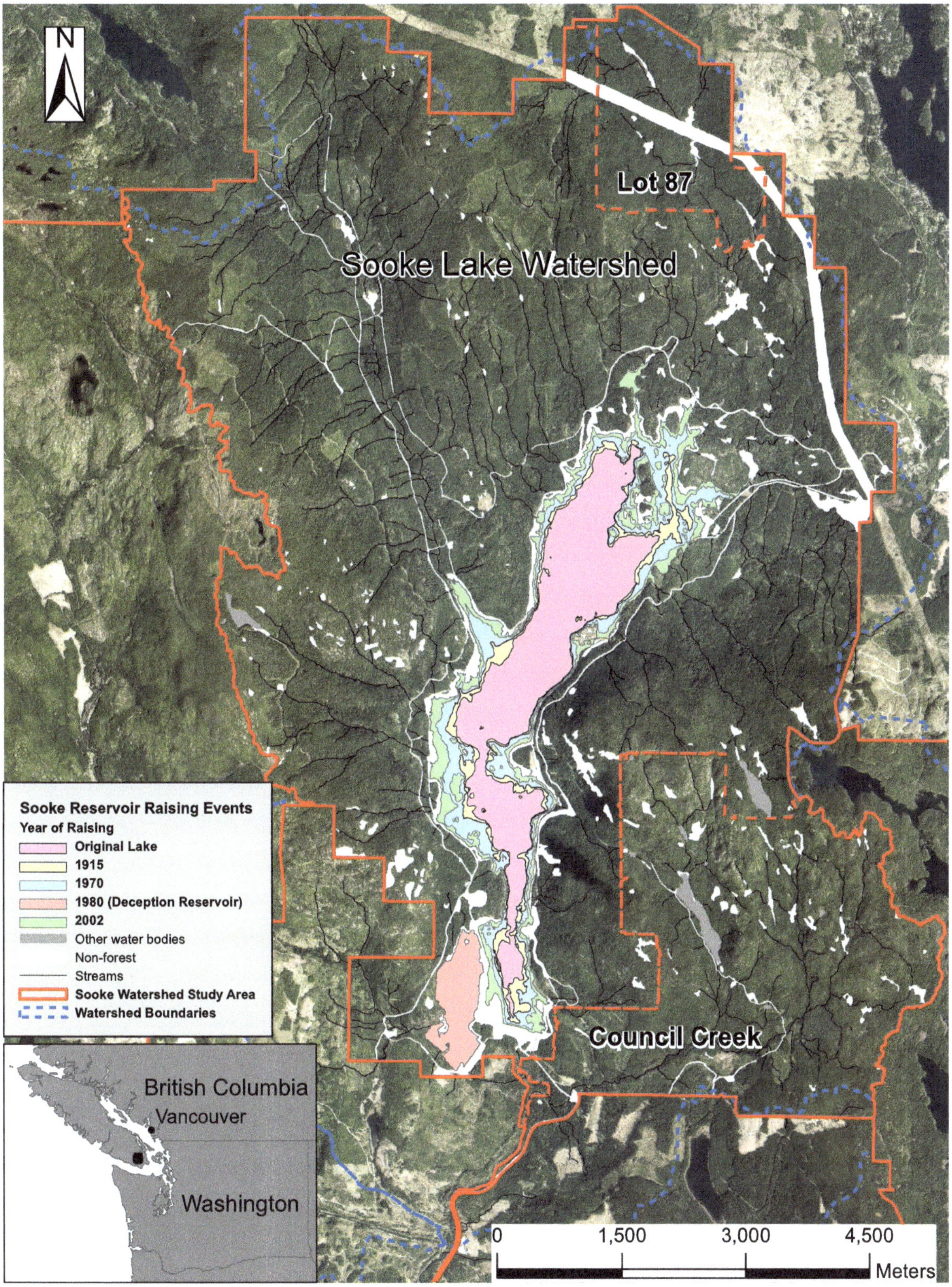

**Fig. 2** Sooke Lake Watershed study area

distinct management periods. Until the mid-1950s, very few disturbances occurred in areas owned and managed for water supply. Conversely, Council Creek catchment and Lot 87 (Fig. 2) were owned by logging companies until the 1990s and were intensively harvested during the 1920s and 1930s. Beginning in the 1950s, a period of sustained harvest activity began within areas owned for water supply and lasted until the mid-1990s. Harvesting then ceased and the SLW in its entirely, owned and managed fully for water supply, experience no further stand-replacing disturbances, other than those associated with the reservoir expansion of 2002.

### Gauged catchments

The catchments of the three gauged creeks of Rithet, Judge, and Council within the SLW constitute 44% of the total watershed area (Table 1). Rithet is the largest and only catchment with perennial stream flow and consequently is the largest contributor of water to the reservoir. On average, Rithet catchment is the steepest at 17°, and has the largest range of elevation, from 188 m at lakeside to 840 m (average elevation is 450 m). Sustained yield forestry occurred in the Rithet valley between 1954 and 1996, harvesting high quality old growth (>250 years) Coastal Douglas-fir stands. Yet, of the three gauged catchments, Rithet has the highest proportion of forest considered to be mature forest (≥80 years) at 67% (Table 2) and has the least extensive disturbances over the last 100 years. Due to the low proportion of both lakes and wetlands, Rithet catchment has limited capability to buffer stream discharge or alter constituent loading once the runoff enters Rithet Creek.

In contrast to Rithet, the Council catchment has had an intense and distributed disturbance history, spanning from the 1930s through the 1990s and has the highest proportion of juvenile and immature forest (<80 years) at 79% (Table 2). Council has roughly the same mean slope (16.5°) and elevation (450 m) as Rithet, yet Council has a much lower peak elevation (630 m). Council catchment contains a 14 ha lake into which the majority of the catchment drains before exiting into Council Creek. This hydrologic feature has important implications for constituent flux from the terrestrial land base of Council to Sooke Reservoir.

Judge Creek is the most northern and smallest of the three catchments (Table 1). The disturbance history of Judge is characterized by a short period of intense clear-cut logging and broadcast burning during the late 1920s. Other areas of Judge were harvested from the early 1950s until the mid-1980s and by 2012 56% of the catchment

**Table 1 Individual catchment (Rithet, Council, Judge) and landscape units (Rithet + Rithet-like, Council + Council-like, Judge + Judge-like) sharing similar physiographic and hydrologic characteristics [29] for scaling up to SLW level of analysis**

| Catchment | Area (ha) | % of SLW |
|---|---|---|
| Rithet | 1824.9 | 21.2 |
| Council | 1189.4 | 13.8 |
| Judge | 765.1 | 8.9 |
| Rithet + Rithet-like | 3926.4 | 45.7 |
| Council + Council-like | 1473.2 | 17.1 |
| Judge + Judge-like | 2822.5 | 32.8 |
| Not modelled (non-forest) | 373.1 | 4.3 |
| SLW total | 8595.1 | |

**Table 2 Individual catchment (Rithet, Council, Judge) and landscape unit (Rithet + Rithet-Like, Council + Council-Like, Judge + Judge-Like) characteristics in 2012 including areas of forest seral stage, wetlands and lakes (total area and percent of catchment)**

| Catchment | Rithet | Council | Judge | Rithet + Rithet-like | Council + Council-like | Judge + Judge-like |
|---|---|---|---|---|---|---|
| Immature forest[a] | | | | | | |
| Area (ha) | 601.8 | 899.2 | 322.6 | 1353.8 | 1007.2 | 1150.3 |
| % of catchment | 33.2 | 78.7 | 43.5 | 36.6 | 71.4 | 46.5 |
| Mature forest[a] | | | | | | |
| Area (ha) | 1210.8 | 243.0 | 418.6 | 2347.2 | 403.3 | 1325.4 |
| % of catchment | 66.8 | 21.3 | 56.5 | 63.4 | 28.6 | 53.5 |
| Wetlands | | | | | | |
| Area (ha) | 7.8 | 15.3 | 23.5 | 22.5 | 16.3 | 59.8 |
| % of catchment | 0.4 | 1.3 | 3.1 | 0.6 | 1.1 | 2.1 |
| Lakes | | | | | | |
| Area (ha) | 0.8 | 16.1 | 0.0 | 14.6 | 16.1 | 1.0 |
| % of catchment | 0.0 | 1.4 | 0.0 | 0.4 | 1.1 | 0.0 |

[a] Immature forests are considered to be stands less than 80 years while mature forests are equal to or older than 80 years

was considered to be mature and 44% was immature forest (Table 2). The most pronounced physiographic and hydrologic difference between Judge and the other catchments is the prevalence of relatively large wetland areas. Judge catchment has the lowest proportion of area covered by lakes/ponds, yet over 3% of Judge land cover is considered wetlands compared to Rithet's 0.5% and Council's 1.3%. These wetlands are contiguous with the drainage in Judge Creek and thus have a significant impact on the load of dissolved stream constituents into Sooke Reservoir.

## Catchment scale analyses

### *Hydrological data*

Quarter-hourly stream discharge measurements from January 1st 1996 to December 31st 2012 for the, Rithet, Judge, and Council catchments were supplied by the CRD (J. Blaney, personal. communication). The diversion from the Council catchment makes up 90% of the combined Council-Trestle discharge where stream flow is measured using a mechanical totalizer. Both Rithet and Judge catchments use a concrete weir and water level recording device to determine stream discharge [29] (F. Hall, personal. communication).

DOC concentration (mg/L) was taken intermittently between 1997 and 2008 at the Rithet, Judge and Council outflow points into Sooke reservoir. 50 ml water samples were collected using either a Sutek sampler or sampling rod close to the water surface and transported to the CRD lab in a cooler (J. Blaney, personal. communication) and a Shimadzu TOC analyzer used to determine DOC (<0.45 μm) in the sample.

### *Software*

The 'R' environment [30] and related time series package (i.e. zoo package) were used to merge stream flow and DOC measurement data files into an acceptable format to process into daily values for further analysis. The R package rLOADEST [31] derived from the FORTRAN Load Estimator (LOADEST) program was used to estimate annual DOC loads from concentration and stream flow measurements [32]. While three statistical estimation methods are available in LOADEST, for the purposes of this study, adjusted maximum likelihood estimation (AMLE) was used (Eq. 1). Instantaneous load estimates are derived from all observation in the estimation dataset using:

$$\hat{L}_{AMLE} = exp\left(a_0 + \sum_{j=1}^{M} a_j X_j\right) H(a, b, s^2, \alpha, \kappa) \quad (1)$$

where $\hat{L}_{AMLE}$ is the AMLE instantaneous load estimate, $a$ and $b$ are explanatory variable functions, $\alpha$ and $\kappa$ are gamma distribution parameters and $s^2$ is the residual variance [32–34]. AMLE allows for a "nearly unbiased" estimation of instantaneous dissolved stream constituent load [33]. Of the other two methods, maximum likelihood estimation (MLE) is more commonly used when the observation data set is uncensored (no observation concentration less than the laboratory detection limit) and least absolute deviation (LAD) when model residuals are not normally distributed, both of which were not the case with this study [32].

rLOADEST provides both a collection of predefined models that can be selected based on the 'best fit' with the data, and the ability for the user to define a unique model form. In this case, 'best fit' is defined as the lowest Akaike information criterion (AIC). AICc (c for correction) is an extension of AIC that corrects for small sample size by including an 'effective sample size' variable (n). Model coefficients are developed using ordinary least squares (OLS) regression. This regression equation (Eq. 2) is then used to calculate estimates of log- load for each observation in the time series. The full form of AICc is:

$$\text{AICc} = -2(\log - \text{likelihood}) + 2\text{K} + \frac{2K(K+1)}{(n-K-1)} \quad (2)$$

where K is the number of estimated parameters included in the model and $n$ is the effective sample size.

### *Empirical DOC load estimation and application*

Annual DOC load reconstructions for Rithet, Judge and Council were determined for the period 1996–2012 (calibration period). Chemical concentration data are often scarce compared to measurements of stream flow. However, through development of a regression relationship (for model form see Eq. 3), missing concentration data were interpolated using available site-specific concentration and stream discharge measurements.

The model form used was:

$$\begin{aligned} InstantaneousLoad &= a_0 + a_1 \ln Qa_2 \sin(2\pi dtime) \\ &+ a_3 \cos(2\pi dtime) \end{aligned} \quad (3)$$

where lnQ = ln(stream flow) − center of ln(stream flow); *dtime* = decimal time − center of decimal time and $a_0$ to $a_3$ are model coefficients.

Once the regression model form was defined, various temporal scales of DOC load and concentration were predicted to interrogate the output data. Daily DOC concentration and flow values for the three catchments were examined in relation to measured DOC concentrations in order to gauge the model's ability to interpolate concentration at the daily temporal scale. For CBM-CFS3 parameterization, annual DOC load values were required; therefore calendar year DOC load in Mg C per day were exported from rLOADEST. These are considered to be the 'observed' values (see Additional file 1: Figure S1 for daily measured and empirically fit DOC values). As allochthonous carbon (i.e., the terrestrial environment) is the primary source of carbon for most small streams [35], the in-stream DOC load was used as a surrogate for DOC flux from the CBM-CFS3 soil C pools. Therefore the DOC load values were annualized to Mg C per year and converted to a unit area value given the area of each catchment of interest for use in parameterizing CBM-CFS3 DOC fluxes.

## Watershed scale DOC fluxes and baseline C budget using CBM-CFS3

### *CBM-CFS3*

Smiley et al. [21] describes the use of CBM-CFS3 in a fully spatial mode and the development of the retrospective C budget for the SLW; a brief description of the model function follows. CBM-CFS3 runs on annual time-steps and uses growth and yield curves and forest cover inventory attributes to estimate stand- and landscape-level biomass C dynamics [36]. The model estimates annual biomass turnover (e.g., litter fall) which then flows to detrital C pools each of which have varying temperature-dependant decay rates based on the type of plant material represented. Carbon from decaying plant material is either lost as $CO_2$ to the atmosphere or is transferred to humified soil C pools with slow decay rates released from those pools, by default, as $CO_2$. Using this pool structure, the model accounts for C stocks and stock changes in tree biomass and dead organic matter [36]. The model can assess past changes in C stocks by using management and disturbance information as well as evaluate future changes that might result from modified management schemes or altered disturbances patterns [35].

As a forest-sector model, CBM-CFS3 only simulates C pools for the forested areas of a landscape. Gaseous C fluxes occur between the terrestrial system and the atmosphere while forest harvesting results in a C export from the ecosystem as round wood. Integration of terrestrial-to- aquatic C fluxes in CBM-CFS3 occur through the fraction of the slow aboveground and slow belowground dead organic matter (DOM) C pools that respire to the atmosphere. The aboveground slow DOM pool includes the F, H, and O soil horizons and corresponds to the 'Litter' pool in the IPCC good practice guidance (GPG) [37]. These horizons include humified organic matter that develop from the decomposition of litter and woody material [38]. The belowground slow DOM corresponds to a segment of the "soil organic matter" GPG pool and specifically includes humified organic matter in the mineral soil layer [20]. Heterotrophic respiration from these slow pools is dependent on the annual base decay rate, 1.5% per year for the aboveground slow DOM pool and 0.33% per year for the belowground slow DOM pool [20]. The path by which decaying slow C exits the terrestrial system is determined by the "fraction to atmosphere" parameters which have a default value of 1, (i.e., 100% of respired C goes to the atmosphere). Adjusting the default value to less than 1 result in a fraction of the C exported from the forest system as DOC.

### *Watershed-scale DOC fluxes*

The per hectare observed DOC load values were used to calibrate CBM-CFS3 model runs for the Rithet, Council and Judge catchments to partition the decay losses from the slow aboveground and slow belowground DOM C pools to either DOC flux or $CO_2$ to the atmosphere. This DOC fraction parameter was calibrated so that modelled annual $ha^{-1}$ DOC flux matched the observed DOC loads for each catchment separately. High variance in observed annual DOC loads was most likely a result of higher or lower stream flow years; the mean DOC load for each catchment for the 1996–2012 period were used to adjust the DOC fraction parameters. Through multiple iterations of CBM-CFS3 runs, the aboveground and belowground DOC fraction parameter values were calibrated by first adjusting the parameter values prior to a model run, comparing the annual modelled DOC fluxes for the 1996–2012 period to the observed mean DOC load values, then repeating until modelled DOC flux $ha^{-1}$ for each catchment was within 0.001 Mg C $ha^{-1}$ of the observed DOC load. The remaining SLW catchments were then assigned DOC fraction parameter values based on the physiographic and hydrologic similarities of the 35 ungauged to the three gauged catchments as defined by Werner [29]. A model run was then conducted on the entire SLW using the same disturbance history as the existing retrospective C budget [22] using the calibrated DOC fraction parameters for all catchments.

## Land management scenarios

Alternative historic land management scenarios were conducted to allow for the quantification and direct comparison of the effects that different land management decisions had on the C budget of the land base over an extended timeframe. Alternative management scenarios were only applied to areas that were owned and operated by the CRD for the entire study period (80%). The conditions for the baseline and alternative management scenarios include:

- *Baseline*—disturbance and management history as occurred from 1911 to 2012 and as described in the "Study area" section

- *Scenario #1 (SC1)—water supply without deforestation or forest management*—no forest harvesting or reservoir raising (flooding) between 1911 and 2012 within the original ownership boundary (disturbances in Lot 87 and Kapoor land maintained)
- *Scenario #2 (SC2)—water supply without forest management*—reservoirs are created and raised as in baseline model runs, however, no forest harvesting occurs between 1911 and 2012 within the original ownership boundary (disturbances in Lot 87 and Kapoor land maintained).

The SC1 scenario represents a situation where the SLW was left in its original state and supposes that the Greater Victoria demand for water could be entirely supplied by the original Sooke Lake. No reservoir expansion or logging occurs within the original CRD tenure.

In the baseline and SC2 scenario, observed reservoir expansion required to meet water demand were maintained. However, the capital projects necessary for these events, including land clearing, engineering and dam construction, are assumed to be financed by means other than logging revenue and therefore these disturbances do not take place. This management regime mimics that which has been in place since the mid-1990s whereby population increases in Greater Victoria have occurred (and are incorporated into future plans) but forestry activity within the water supply area has ceased.

For both SC1 and SC2, disturbances related to natural events, adjacent lands or transportation access were preserved within the CRD tenure. These included wildfire and insect outbreaks, transportation corridor clearing for railway, access road and transmission line right-of-ways and escaped fires from adjacent lands.

Table 3 shows the breakdown of forest and non-forest areas between the Baseline, SC1 and SC2 scenarios. By comparing the difference in cumulative net biome production (ΣNBP) for SC1 vs SC2 scenarios (for which neither experience forest management on CRD ownership lands but the latter includes deforestation due to reservoir creation) the C budget consequences of just the deforestation events can be investigated. Comparing the difference in ΣNBP for Baseline vs SC2 scenarios (for which both experience reservoir expansion but the latter includes no forest management) examines the consequences of forestry activities alone.

#### Fate of exported round wood

Under IPCC guidelines in 2012, the existing assumption within CBM-CFS3 is that round wood C exported to wood products is immediately released to the atmosphere as $CO_2$ at the time of harvest [37]. More recent guidelines allow for tracking the fate of round wood in harvest wood products (HWP) using separate models [39]. An alternative British Columbia-specific framework has been developed to run in conjunction with CBM-CFS3 C pool and flow capabilities (CBMF-HWP) [5]. The HWP model tracks the C storage of wood products post-harvest by simulating primary milling, construction and secondary manufacturing, retirement

**Table 3 Area of analysis units and non-forest in 1911 and baseline, scenario 1 and scenario 2 in 2012**

| Analysis unit | Description | Area in 1911 (ha) | Area in 2012 (ha) | | |
|---|---|---|---|---|---|
| | | | Baseline | SC1 | SC2 |
| Productive forested land | | | | | |
| 1 | Fir | 5371 | 4326 | 5209 | 4830 |
| 2 | Fir–cedar | 1048 | 760 | 938 | 964 |
| 3 | Fir–hemlock/Grand fir/Sitka Spruce | 1097 | 1479 | 1182 | 1075 |
| 4 | Fir–alder/maple/poplar/arbutus | 9 | 357 | 75 | 57 |
| 5 | Cedar leading with conifer mix | 35 | 48 | 61 | 55 |
| 6 | Hemlock | 5 | 6 | 11 | 6 |
| 7 | Hemlock–Fir | 217 | 247 | 276 | 258 |
| 8 | Hemlock–cedar | 33 | 34 | 34 | 43 |
| 9 | Broadleaf greater than 75% composition | 8 | 20 | 33 | 34 |
| 10 | Alder–conifer mix | 18 | 13 | 14 | 14 |
| Total | | 7841 | 7290 | 7833 | 7336 |
| Non-forest land | | | | | |
| Sooke Lake/reservoir | | 373 (49%) | 813 (62%) | 373 (49%) | 813 (65%) |
| Other un-established/non-forest land | | 382 (51%) | 492 (38%) | 389 (51%) | 446 (35%) |
| Total | | 754 | 1305[a] | 762 | 1259 |

[a] 80% of the change in non-forest land is due to reservoir creation, the remainder is from road/railway creation, etc.

from material in-use, and disposal and decay of forest products. Once considered "in-use", the C is stored between 2 and 90 years, depending on the half-live of the pool (single family homes being the longest and shipping products being the shortest) [5].

While our analysis does not explicitly track disposal and decay of HWP over time, we investigated the implications of harvested round wood between the alternative management scenarios. The simple emissions factor for whitewood from Dymond [5] (0.52) was adjusted (0.58) to include emissions associated with outside-bark dimensions (i.e. round wood) exported from CBM-CFS3.

## Results

### DOC fluxes

#### *Catchment scale*

Over the course of the calibration period (1996–2012) mean DOC load from the Rithet catchment was approximately 72.5 and 29.1 Mg C $year^{-1}$ for judge. Council exported the lowest on average of 18.3 Mg C $year^{-1}$ to DOC (Fig. 3a). As Council is the second largest catchment by area, on a per hectare basis, mean annual DOC load was significantly lower (0.0154 Mg C $ha^{-1}$ $year^{-1}$) than the other two catchments (0.0397 Mg C $ha^{-1}$ $year^{-1}$ for Rithet vs 0.0381 Mg C $ha^{-1}$ $year^{-1}$ for judge) (Fig. 3b). In all three catchments the annual variability in DOC load is closely tied to the annual stream flow. This is most likely due to the higher positive correlation between DOC load and stream flow as opposed to DOC concentration. Over the course of the study period Council was the only catchment that showed an upward trend in DOC flux, while Judge and Rithet trended downward slightly; however, the trends for all three catchments were not significant (Fig. 3). As the observed DOC load magnitudes varied spatially among the three catchments, the CBM-CFS3-model parameters of DOC flux from the slow aboveground and belowground DOM pools required different DOC fractions.

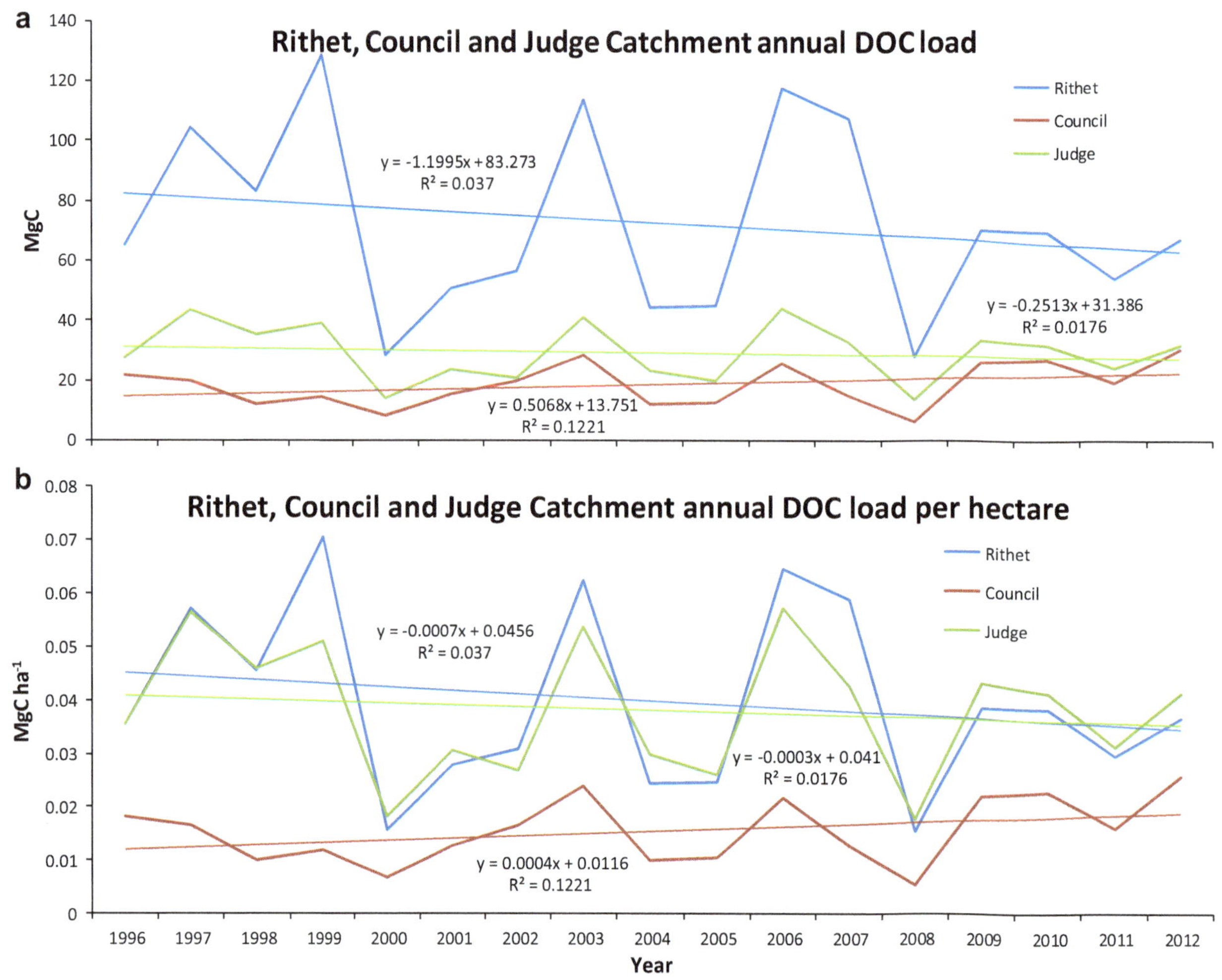


**Fig. 3** Dissolved organic carbon (DOC) load and trendlines from 1996–2012 on a total DOC flux $yr^{-1}$ (**a**) and DOC flux $ha^{-1}$ $yr^{-1}$ (**b**) basis

After model calibration, the three unique fraction-to-atmosphere parameters derived for Rithet, Council and Judge catchments were 0.945, 0.975, and 0.935, respectively (Table 4). The fraction-to-atmosphere parameter from the slow belowground pool remained constant at 0.99 for all three catchments. Carbon in the slow aboveground DOM pool is more labile relative to belowground DOM owing to the constituents that are represented within it [16]. Carbon from the aboveground DOM pool was considered to have a greater fraction of labile C exported as DOC. This supposition was based on a higher probability of aboveground DOM in overland flow, and it reaching a watercourse. Some of this C is transferred to the slow belowground pool, and is represented in the model as a 0.6% annual transfer. In contrast, DOC from the relatively large belowground slow DOM component was considered less mobile and that microbes consumed more than 90% before it could enter a watercourse [40]. As a result, a DOC fraction value of 1% was considered appropriate for the belowground slow DOM pool.

Judge catchment had high per hectare DOC flux relative to the size of the slow aboveground DOM pool, consequently resulting in the largest DOC fraction parameter at 6.5%. The DOC fraction parameter was 5.5% for Rithet and 2.5% for Council. The relative size of the modelled slow aboveground DOM pools over the course of the study period [Rithet (63 Mg C $ha^{-1}$ $\pm 1$) > Judge (49 Mg C $ha^{-1}$ $\pm 1$) > Council (41 Mg C $ha^{-1}$ $\pm 2$) ranked similarly to the observed DOC load for all three catchments.

#### *Watershed scale*

The DOC fraction/fraction to atmosphere parameter values for the gauged catchments were applied to the ungauged catchments based on physiographic and hydrologic similarities [29] and model runs conducted for the entire SLW. Combined Rithet and Rithet-like catchments made up the largest proportion of the modelled area (Table 1); while, the area of the combined judge and judge-like catchments was greater than the Council and council-like catchments. Stands with high DOC fluxes in 2012 (Fig. 4) had higher soil C stocks and tended to be older. Areas west and south of Sooke Lake typically had lower DOC fluxes compared to forests east and northeast of the lake. The non-gauged catchments have differing amounts of C in the slow above and belowground DOM pools compared to the gauged catchments. As the ungauged catchments were assigned DOC parameters based on their hydrologic and physiographic characteristics and not on similar DOM pool sizes, the $ha^{-1}$ DOC flux values differed slightly from those of the gauged catchments (Table 5). Significantly higher DOC fluxes were observed from polygons that recently had forests greater than 300 years on highly productive sites.

For the calibration period, the average DOC flux from the terrestrial area of the SLW was 93.0 Mg C $year^{-1}$ (0.0308 Mg C $ha^{-1}$ $year^{-1}$) with 81% of that coming from the slow aboveground DOM pool. Total DOC export for this period was 4740 Mg C. Over the 100-years historic period, 30,657 Mg C was exported from the terrestrial system via DOC, representing the upper bounds for what could be sequestered in lake sediment in this watershed.

### Land management scenarios carbon budgets

Three distinct periods of management are apparent in the baseline scenario. Until the mid-1950s, only a few large disturbances occurred. As these disturbances were

**Table 4 Calibrated CBM-CS3 parameters partitioning C losses from decaying slow aboveground (AG) and belowground (BG) DOM pools to the atmosphere (fraction to atmosphere) or as dissolved organic carbon (DOC) (fraction to DOC)—modelled and observed values of mean and mean $ha^{-1}$ Mg of carbon 1996–2012 for Rithet, Judge and Council catchments used to derive parameter values**

| Catchment | Slow DOM pool | Fraction to atmosphere | Fraction to DOC | Modelled value | | Observed value | |
|---|---|---|---|---|---|---|---|
| | | | | 17 year mean (Mg C $year^{-1}$) | 17 year mean (Mg C $ha^{-1}$ $year^{-1}$) | 17 year mean (Mg C $year^{-1}$) | 17 year mean (Mg C $ha^{-1}$ $year^{-1}$) |
| Rithet | AG | 0.945 | 0.055 | 60.4 | 0.0331 | | |
| | BG | 0.99 | 0.01 | 11.9 | 0.0065 | | |
| | Total | | | 72.4 | 0.0397 | 72.5 | 0.0397 |
| Council | AG | 0.975 | 0.025 | 11.4 | 0.0096 | | |
| | BG | 0.99 | 0.01 | 7.4 | 0.0062 | | |
| | Total | | | 18.7 | 0.0157 | 18.3 | 0.0154 |
| Judge | AG | 0.935 | 0.065 | 23.4 | 0.0306 | | |
| | BG | 0.99 | 0.01 | 5.6 | 0.0073 | | |
| | Total | | | 29.0 | 0.0379 | 29.1 | 0.0381 |

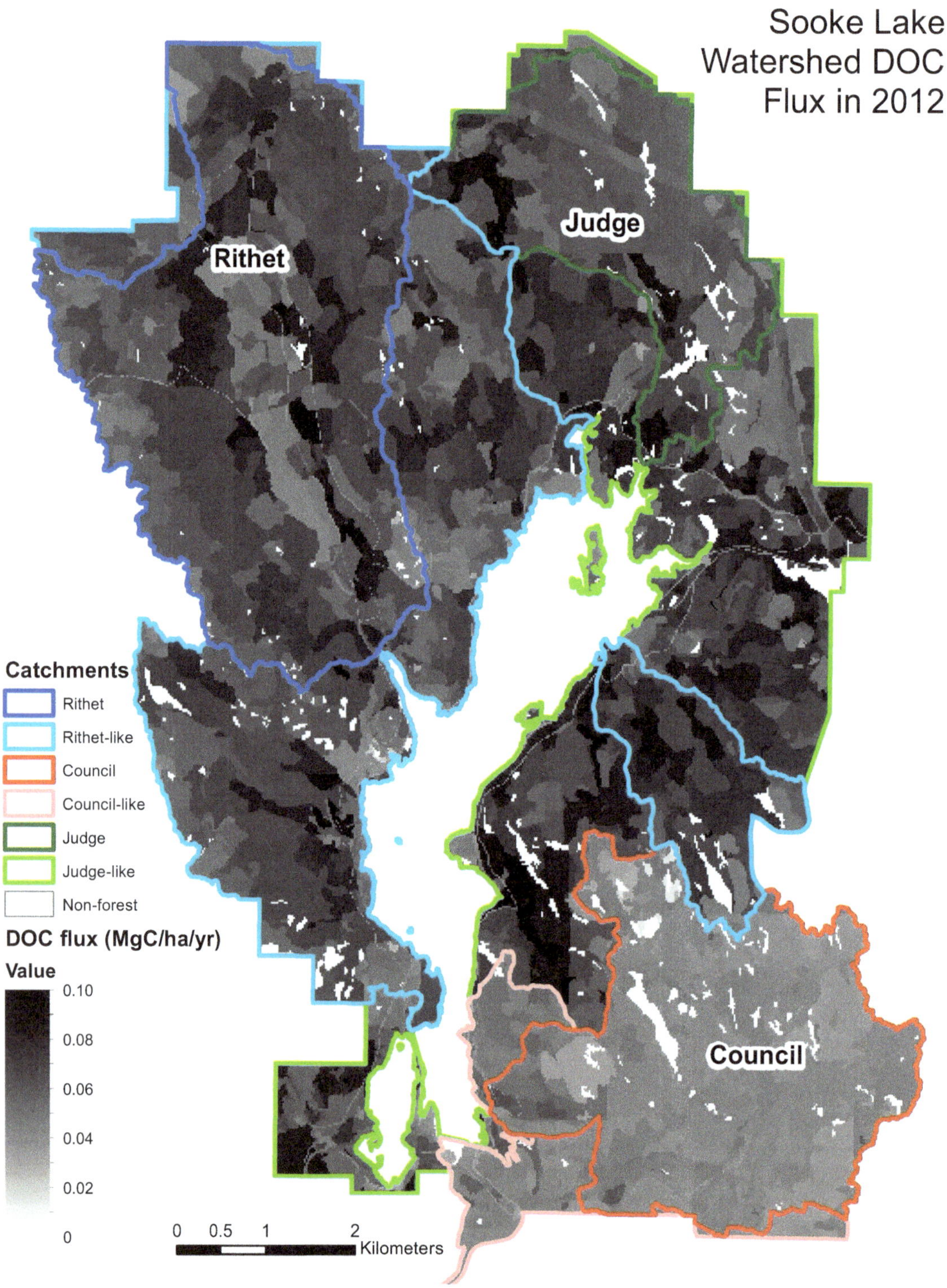


**Fig. 4** Sooke Lake Watershed DOC flux in Mg C $ha^{-1}$ $year^{-1}$ in 2012 and catchment delineation

**Table 5 CBM-CFS3 dissolved organic carbon (DOC) flux from slow aboveground (AG) and belowground (BG) DOM pools from 1996 to 2012 for -gauged and ungauged catchments and Sooke Lake Watershed totals**

| Landscape unit | Value | Mean | Max | Min | Total |
|---|---|---|---|---|---|
| Rithet + Rithet-like | AG | 123.2 | 124.6 | 122.0 | 2094.3 |
| | BG | 24.7 | 24.9 | 24.5 | 419.5 |
| | Total | 147.9 | 149.6 | 146.5 | *2513.8* |
| | $ha^{-1}$ | 0.0377 | 0.0381 | 0.0373 | 0.6402 |
| Council + Council-like | AG | 14.5 | 14.8 | 14.2 | 246.2 |
| | BG | 9.3 | 9.4 | 9.2 | 158.2 |
| | Total | 23.8 | 24.0 | 23.6 | *404.4* |
| | $ha^{-1}$ | 0.0168 | 0.0169 | 0.0166 | 0.2745 |
| Judge + Judge-like | AG | 89.1 | 90.6 | 88.0 | 1515.4 |
| | BG | 18.1 | 18.4 | 17.8 | 307.2 |
| | Total | 107.2 | 109.0 | 105.8 | *1822.6* |
| | $ha^{-1}$ | 0.0380 | 0.0386 | 0.0375 | 0.6457 |
| Watershed total | AG | 75.6 | 124.6 | 14.2 | 3855.8 |
| | BG | 17.4 | 24.9 | 9.2 | 884.9 |
| | Total | 93.0 | 149.6 | 23.6 | *4740.7* |
| | $ha^{-1}$ | 0.0308 | 0.0386 | 0.0166 | 0.5766 |

All totals in Mg C or Mg C $ha^{-1}$

mostly outside of the original CRD ownership the forest disturbances for this period are mirrored in the other scenarios, with the exception of the absent flooding event in SC1 (Fig. 5). The period of sustained yield forestry began after 1955 when clearcut, and residue burning and thinning events start to vary among the three scenarios and influenced forest age class structure in 2012. Both SC1 (4360 ha) and SC2 (3947 ha) had considerably more forest greater than 200 years than the Baseline scenario (2057 ha) (Fig. 6). In 2012, over 3500 ha were less than 80 years in the Baseline vs only 1306 ha and 1472 ha for SC1 and SC2, respectively. The third management period is denoted by the cessation of logging activity in the Baseline scenario in the mid-1990s and the resulting recovering in C stocks.

Comparisons among scenarios in live biomass C (above- and below-ground), detritus (litter and deadwood) and soil C stocks over the historical period are shown in Fig. 7. Because of the inherent stability of the soil C pools, differences due to management scenario were minimal over the study period), ranging between 2.8 and 3.1 Mg C $ha^{-1}$ by 2012 (Table 6). Detritus stocks exhibited more differences, with SC2 and SC1 25.1 and 26.0 Mg C $ha^{-1}$ greater than the Baseline scenario by 2012 (Table 6). Post-1960, detritus C stocks stabilize in SC1 and SC2 while in the Baseline they continue to decline until the end of the study period (Fig. 7). Live biomass stocks in all three scenarios began to recover after 1940 from a low between 231.0 Mg C $ha^{-1}$ (baseline) and 240.0 Mg C $ha^{-1}$ (SC1) (Fig. 7). However, by the mid-1950s, the recovery in stocks began to diverge, with SC1 and SC2 continuing to accumulate biomass whereas the Baseline scenario declined until the early-1990s. In 1991, the differences in biomass C for Baseline vs SC1 and SC2 scenarios reached a high of 93.5 and 83.5 Mg C $ha^{-1}$, respectively, and then narrowed by 2012 (Fig. 7; Table 6). NBP describes the overall ecosystem C exchange of a landscape over multi-decadal time spans [41], and includes the removal of C due to disturbances [20]. Figure 8 shows the cumulative NBP (ΣNBP) for the three management regimes and the influence that DOC export has on the C budget. In the first 15 years ΣNBP remained approximately C neutral for all three scenarios. In areas outside CRD tenure, the large removals of live biomass C through HWP export and release to the atmosphere from slash burning resulted in a watershed-wide decline to −98.7 (baseline), −83.2 (SC1), and −85.8 (SC2) Mg C $ha^{-1}$ in 1955 when including DOC export ($\Sigma NBP^{DOC}$). All scenarios were approximately 2.0 Mg C $ha^{-1}$ lower without DOC export. $\Sigma NBP^{DOC}$ of the SC1 and SC2 scenarios began to recover after 1955, whereas the Baseline scenario continued to decrease. $\Sigma NBP^{DOC}$ of SC1 and SC2 remained within ~10.0 Mg C $ha^{-1}$ of one another until the mid-1960s when deforestation for reservoir expansion in 1970, 1980 and 2002 in SC2 increased HWP exports, and no biomass regrew on deforested lands. $\Sigma NBP^{DOC}$ for the Baseline scenario began to recover in 1994 from a low of −167.4 Mg C $ha^{-1}$ (−170.7 Mg C $ha^{-1}$ ΣNBP) to its current (2012) level of −142.4 Mg C $ha^{-1}$ (−146.2 Mg C $ha^{-1}$ ΣNBP). In contrast, SC1 did not decline below −85.0 Mg C $ha^{-1}$ (1956) and recovered to −35.4 Mg C $ha^{-1}$ by 2012. The $\Sigma NBP^{DOC}$ for SC2 was also at its lowest point in 1956 (−88.5 Mg C $ha^{-1}$). While deforestation events in SC2 did dampen the ability to recuperate C losses from earlier in the century, $\Sigma NBP^{DOC}$ had recovered to −49.4 Mg C $ha^{-1}$ by 2012. Not unexpectedly, total HWP export for SC2 and SC1 were 46 and 60% lower, respectively, as compared to the Baseline of 882,746.2 Mg C (Table 6). More HWP was exported from SC2 than SC1 due to activities related to reservoir expansion and road access.

Cumulative DOC (ΣDOC) export was greatest in SC1 (4.1 Mg C $ha^{-1}$) (Table 6) as the DOM stocks which feed DOC export in the model increased with the higher proportion of mature forest. As well, the lack of deforestation meant DOM stocks that were removed from the land base in the Baseline and SC2 management regimes were maintained in SC1 and continued to decay and release DOC. The SC1 and SC2 management scenarios could have potentially sequestered slightly higher amounts (2162 and 1390 Mg C, respectively). The impacts of

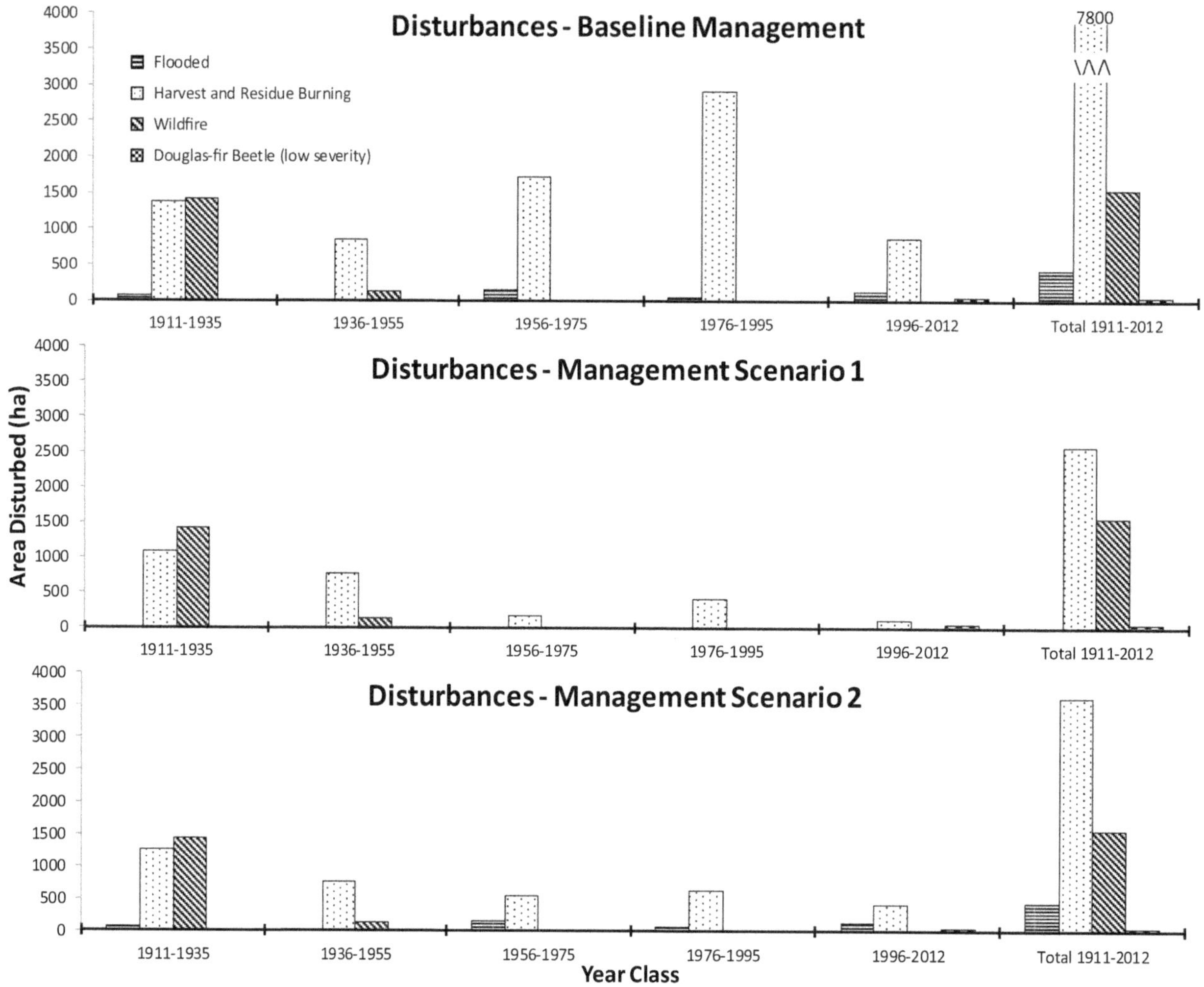


**Fig. 5** Disturbances by period for baseline, scenario 1 and scenario 2 management regimes

forestry (Baseline vs SC2) and deforestation of reservoir expansion (SC1 vs SC2) on the land base (excluding Lot 87 and the Council Creek catchment) are further illustrated by the differences among scenarios in the spatial distribution of forest ecosystem C stocks in 2012 values (see Additional file 2: Figure S2 for total forest ecosystem C stocks in 2012 for Baseline, SC1 and SC2 across the SLW).

## Discussion

### Watershed scale DOC fluxes

Rithet summer stream flow is thought to be sourced from a small bedrock aquifer [29, 41], since there are no snowpacks, glaciers or significant lakes to contribute to summer discharge. Groundwater can also be a source of high DOC [42]. While Kenny [43] investigated aquifer extent across the CRD, little is known about the geological formations and their porosity and permeability within the SLW. Therefore, groundwater DOC input into the reservoir was not considered in the watershed DOC fluxes.

Modelling the C exported from the terrestrial to the inland aquatic system on a watershed scale suggests allochthonous C storage in lake sediment may be a significant C sink. Physiographic differences, specifically percent area of wetlands and lakes, forest cover age structure (Table 2), and size of slow above and belowground DOM pools were the primary terrestrial forces driving long term DOC export to fluvial systems. The inundation of littoral wetlands areas due to reservoir raising events can also have a significant impact on the nutrient loading within a lake, generally [44, 45] and in Sooke reservoir in particular [46]. However, the impact of reservoir expansion on terrestrial-to-aquatic DOC transfers was not

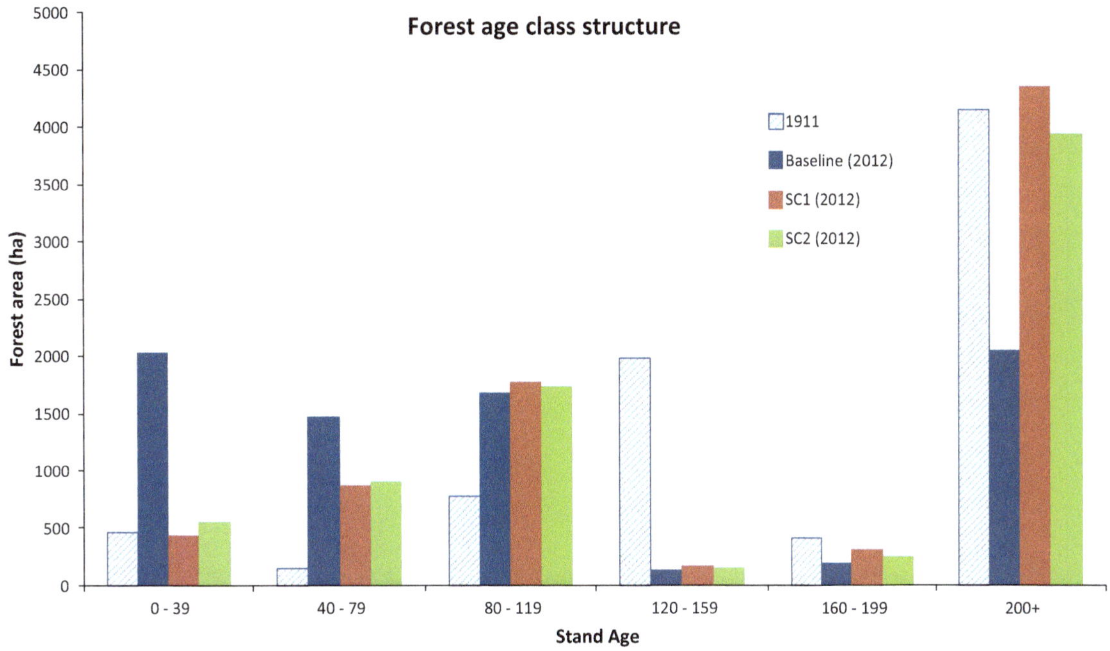


**Fig. 6** Forest age class structures in 1911 and 2012 for baseline, scenario 1 and scenario 2 management regimes

included in this study, and therefore reported DOC flux values to the reservoir may be an underestimation in this respect. The slow DOM pools and selected DOC fraction parameters capture well the trend and magnitude of long term DOC loads observed in the gauged catchments. Long term trends in DOC load increases have been observed in areas of western and northern Europe, most likely due to acid deposition histories resulting from industrial development [47, 48]. The current configuration of the CBM-CFS3 does not include a mechanism to model the short term (1–5 years) event-driven spikes in DOC load due to effects of disturbance on stream DOC concentrations. On some forested landscapes hydrologic events (i.e., storms and snowmelt) can be the source of approximately 86% of terrestrially-derived DOC to the aquatic environment [10]. If more mobile sources of DOM (i.e., litter) are available due to disturbances such as forest harvesting or wildfire then this terrestrially-sourced DOC will be magnified initially and then be depleted. The introduction of a DOC fraction parameter to another, more mobile C pool (i.e., the aboveground very fast DOM) or a transfer function built into the disturbance matrices might improve the ability of the model to simulate the short term DOC export that would occur after disturbance.

DOC fraction parameters must be calibrated based on the physiographic and hydrological characteristics of the study area in question. Differences in DOC transfer rates are highly variable spatially and sensitive to temperature and resulting decomposition rates. Study area-specific mean annual temperature could increase the accuracy of the soil decomposition rates compared to the ecozone normals used in this study. The impact of precipitation on DOC fluxes is considerable as well. In similar sized ocean-draining watersheds on the central coast of British Columbia, where annual rainfall is double and forest soils are thicker organic layers and have higher soil C contents than that observed in the SLW [49], DOC fluxes (0.377 Mg C $ha^{-1}$ $year^{-1}$) are almost 10 times those estimated in this study [50]. The annualized DOC flux parameters selected for the three catchment types only represent a small fraction of the slow above and below-ground DOM pools; however, accumulation over many years could impact the C sequestration expectations, and therefore the watershed-scale C budget [10].

While the question of DOC fate was beyond the scope of this study, the final destination of terrestrially-sourced C is an important component of coupled terrestrial–inland aquatic modelling efforts. Dean and Gorham [51] estimated that average long-term C burial rates of lakes of 14 g C $m^{-2}$ $year^{-1}$, with reservoirs sequestering on average a much higher amount (400 g C $m^{-2}$ $year^{-1}$). The upper bounds of annual carbon burial for the SLW may be up to 37 g C $m^{-2}$ $year^{-1}$; integrating $CO_2$ respired from the reservoir will adjust this figure downward. Average DOC concentrations from the Sooke reservoir spillway were lower than those recorded for streams draining the three gauged catchments (Sooke Reservoir: 2.43 mg C/l;

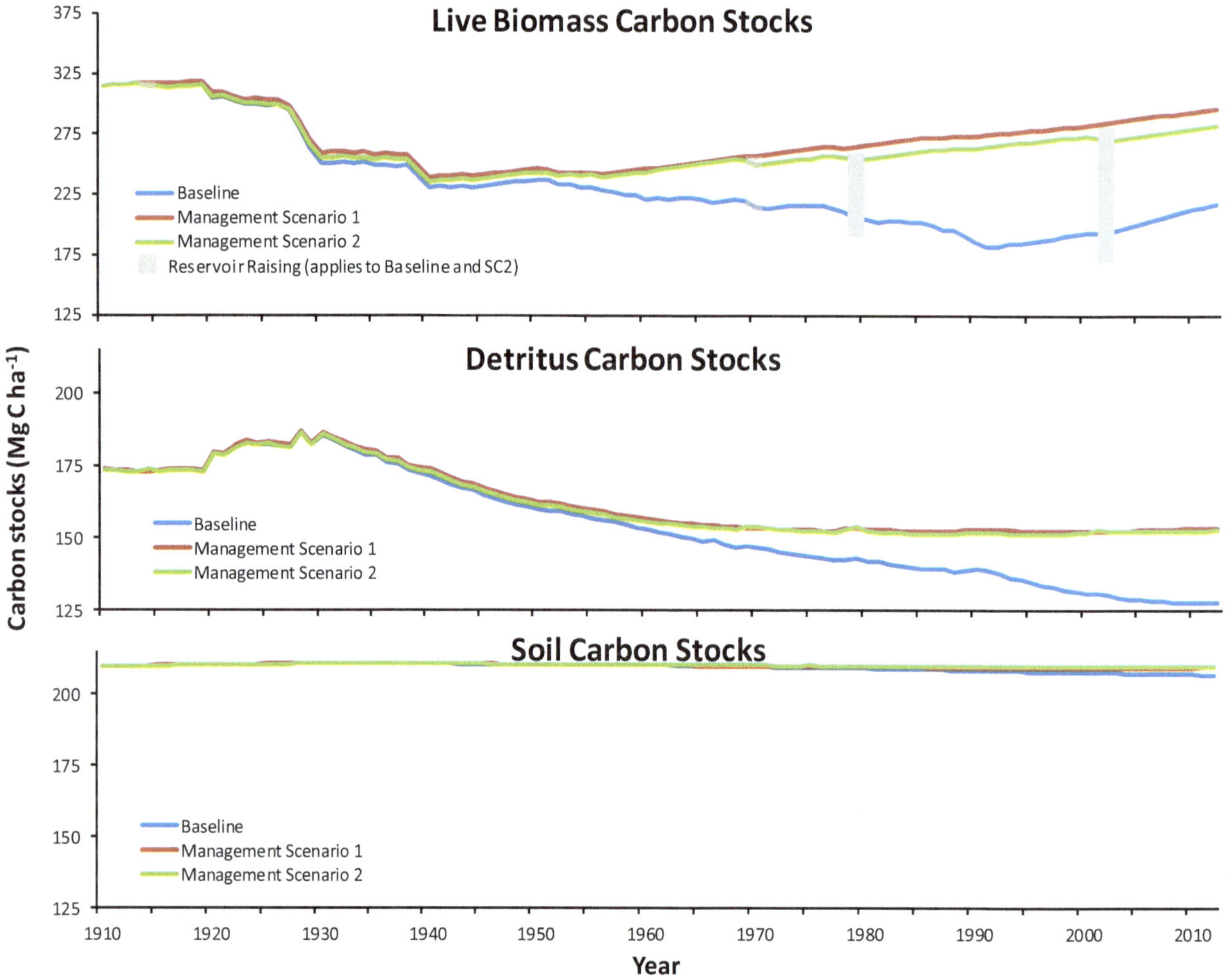


**Fig. 7** Baseline and alternative management scenario live biomass, detritus and soil C stocks 1911–2012

Judge: 5.67 mg C/l; Rithet: 3.47 mg C/l; Council: 3.43 mg C/l). While within-lake C fixation through aquatic gross primary production is considered to be a net source of C, the addition of terrestrially-sourced C into the system, which can be equal to or greater than autochthonous C [52], can potentially accumulate and result in long term C storage in lake sediments [53]. Thus, the increased reservoir area and sediment deposition resulting from reservoir creation could, over time, potentially offset the sudden release of C that occurs during deforestation from reservoir expansion.

Potential increases in the frequency and magnitude of rainfall events with a changing climate may result in increased DOC export to the Sooke reservoir and this reinforces the need for more consistent DOC monitoring in order to inform adaptation strategies. Dore et al. [54] reported that precipitation patterns have changed since monitoring began in the SLW in 1914. The IPCC predicts that in the Pacific Northwest and Western Canada, the variance in seasonal precipitation will increase and temperatures will rise steadily over the next century [55]. Drier summer soils, changes in decomposition rates and more rapid, intense flushes of DOC through higher intensity rainfall events could have water quality implications.

#### Possible reservoir expansion effects on methane fluxes

An important consideration in both terrestrial and aquatic C cycling is the significance of methane ($CH_4$) because of its role as a potent GHG which can affect the intensity of global climate change. As a GHG, $CH_4$ is 28 times more potent than $CO_2$ [56]; this fact coupled with the speed at which it is accumulating in the atmosphere relative to $CO_2$, averaging 1% per year over the last few decades [57], makes it an important component to study in terrestrial-inland aquatic ecosystems. The major natural source of $CH_4$ stems from methanogenesis which mainly occurs in wetlands and wet lowland areas where C is released from wetland and lakebed sediments [57]. In

**Table 6 Baseline, scenario 1 and scenario 2 carbon stocks and fluxes as of 2012**

| Flux/pool | Management scenarios | | |
|---|---|---|---|
| | Baseline | Scenario 1 | Scenario 2 |
| Cumulative NBP (Mg C $ha^{-1}$) | | | |
| No DOC export | −146.2 | −39.4 | −53.4 |
| DOC export | −142.4 | −35.2 | −49.4 |
| Carbon stocks (Mg C $ha^{-1}$) | | | |
| Live biomass | 217.4 | 296.1 | 282.0 |
| Detritus | 127.8 | 153.9 | 152.9 |
| Soil C | 207.1 | 209.9 | 210.2 |
| Cumulative DOC export (Mg C $ha^{-1}$) | | | |
| Aboveground slow | 3.2 | 3.4 | 3.3 |
| Belowground slow | 0.7 | 0.7 | 0.7 |
| Total | 3.9 | 4.1 | 4.0 |
| Total DOC export (Mg C) (1911–2012) | 30,657.2 | 32,819.5 | 32,047.2 |
| Total round wood export (Mg C) (1911–2012) | 882,746.2 | 354,247.0 | 475,183.9 |

upland regions, a small amount of $CH_4$ is absorbed into the soil by methanotrophic bacteria, although this is only a fraction of what is released from lowland areas [57]. $CH_4$ cycling in forest ecosystems can also be impacted by many forestry practices such as land clearing (for quarries, roads, etc.) and nitrogen fertilization which have been found to produce nitrite that persistently inhibit methanotrophic bacteria [57].

**Relative impacts of deforestation and forest management**

The multiple reservoir raisings had a stepped effect on ΣNBP over the study period (Fig. 8). At the watershed scale, the impact of deforestation (SC1 vs SC2) resulted in a cumulative decrease of approximately 14.0 Mg C $ha^{-1}$ by 2012 equivalent to 110,991 Mg C less being sequestered. In contrast, sustained yield forestry activity within the CRD's tenure (Baseline vs SC2) accounts for a 93.0 Mg C $ha^{-1}$ difference in ΣNBP by 2012, equivalent to 738,809 Mg C less being sequestered. This shows that while deforestation due to reservoir creation removes biomass stocks and ends forest C sequestration on those lands, over 100 years, the recurring removal of C in the form of harvested round wood (Fig. 9a) had a substantially greater impact on the landscape C budget than did reservoir creation. That said, the removal of C from the SLW during forestry operations is partially offset by renewed sequestration after stands establish and tree growth resumes.

For different ecosystems, and different scales of analysis a mix of forest management techniques is more likely to optimize forest C sequestration [3]. Man et al. [58] explored two general forest management methods for increasing C sequestration and found that strategies that reduced harvest levels had greater C sequestration benefits than strategies that increased growth. In the SLW, the harvest reduction strategy exhibited in SC2 whereby the CRD-owned land becomes a reserve shows a stark increase in C stored in biomass pools in comparison with the Baseline.

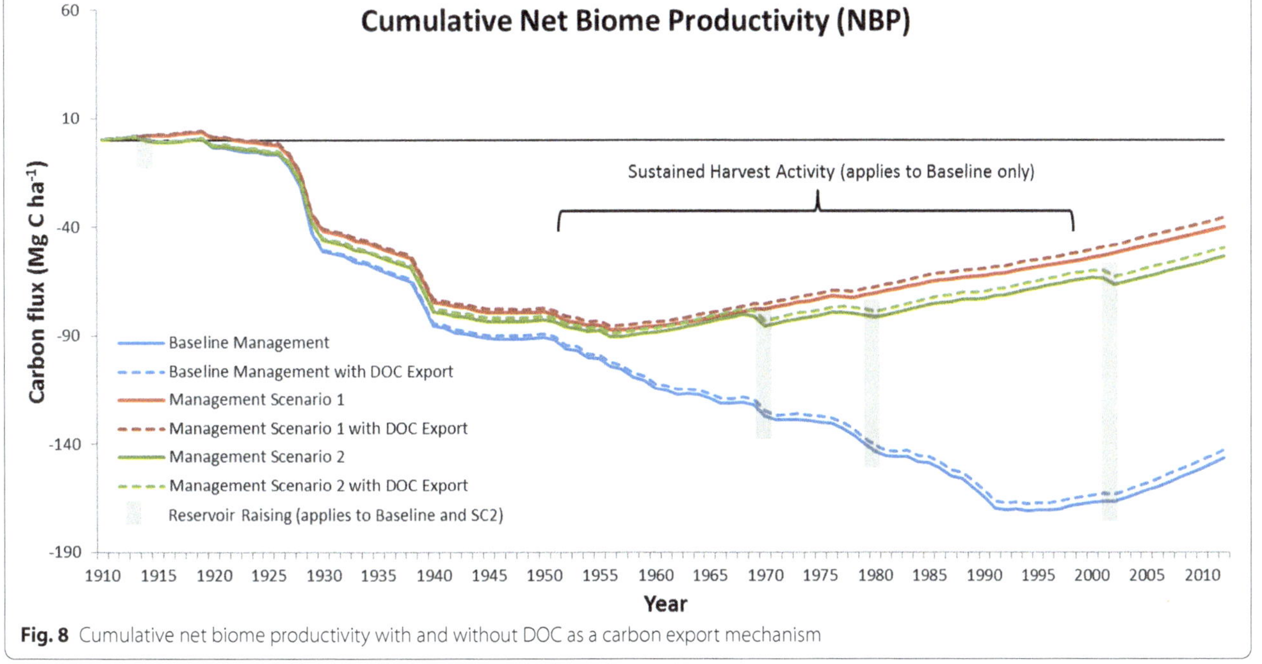


**Fig. 8** Cumulative net biome productivity with and without DOC as a carbon export mechanism

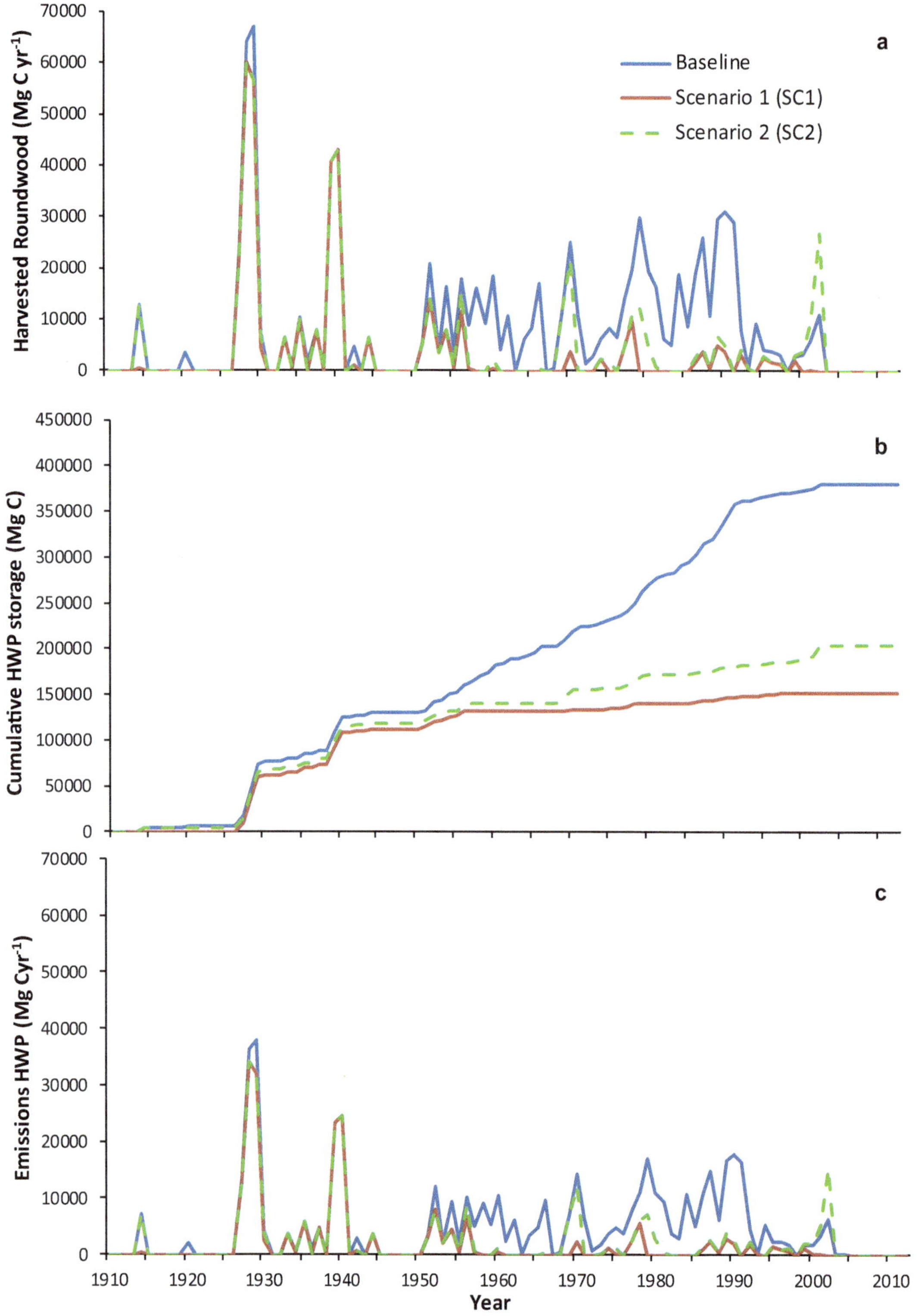


**Fig. 9** Baseline, scenario 1 and scenario 2 harvested round wood (**a**), cumulative storage in harvested wood products (**b**) and emissions from processing (**c**) from 1911–2012

#### Accounting for exported round wood in harvested wood products

The preceding analysis assumes all C in exported round wood (IPCC rules until 2012) is emitted to the atmosphere and forgoes accounting for C stored in HWP, the cumulative difference of which is 176,222 Mg C (Fig. 9b). Emissions associated with HWP (Fig. 9c) are 2.5 and 1.8 times greater in the Baseline scenario than in SC1 and SC2, respectively, but accounting for the fate of HWP enables a commensurate fraction of C to be stored in products. Considering current harvest rotation ages of less than 50 years in some managed forests [59], the residency time of C in manufactured products could in fact be longer than that sequestered in managed forests.

Managing forests for conservation purposes often increases the C stocks on the land base; however, the risk of natural disturbance (e.g., wildfire, drought, insects or disease) means the ecosystem C storage can be at risk [4]. In their case study Man et al. [58] found that greater than 25% stand mortality can nullify the C storage gains from the reserved forest in the short-term, while 50% stand mortality has a permanent negative effect. Using a forest reserve strategy whereby an area is removed from the harvesting land base might have detrimental impacts on ecosystem C storage due to unforeseen natural disturbances or climate change impacts on decay rates. Projected future changes of natural disturbance patterns call into question the effectiveness of existing forest management mechanisms to achieve C sequestration objectives [20].

Different forest management regimes can have a considerable impact on forest C biomass and DOM stocks, especially when these management decisions are compared over decadal and longer time scales. In BC, current C credit legislation dictates that C credits may not be granted unless the atmospheric effect of the C removals endures for a minimum of 100 years [60]. This requires that the effects of management decision must be considered, at minimum, on a multi-decadal scale. The comparison of SC1 and SC2 with the 100-year Baseline C budget of the SLW enables the C budget effect of the specific management decisions that led to deforestation for reservoir creation as well as sustained forest harvest to be quantified. Also, the Baseline C budget allows for future extrapolation of C stocks and C fluxes. While CBM-CFS3 implicitly includes environmental differences through temperature input, and growth curve manipulation, it does not explicitly integrate the potential effects of climate change into growth, decay or decomposition rates. Work is progressing to investigate environmental change effects on forest ecosystem carbon stocks [61, 62]. Changing growth and decomposition dynamics observed in the Pacific Northwest over the twentieth century [63] need to be integrated to examine CBM-CFS3's ability to model the effects of climate change on future forest ecosystem C budgets.

## Conclusions

DOC flux from temperate forest ecosystems is spatially complex and a small but persistent C flux which may have long term implications for C storage in inland aquatic systems. CBM-CFS3 parameterization of DOC flux from the SLW forest ecosystem used [DOC] and stream flow measurements (1996–2012) from three catchments. Model calibration yielded three distinct DOC transfer fractions from the aboveground slow pool resulting in DOC fluxes between 0.0154 and 0.0381 Mg C $ha^{-1}$ $year^{-1}$.

When applied to the entirety of the SLW, the modelled accumulation of DOC from uplands sources totalled 30,657 Mg C for the 100 year period. While we do not assert all fluvial transported C remains within the inland aquatic system in the long term, our estimate represents an upper bound for what could be sequestered through burial in reservoir/lake sediment for this watershed. This is a first step to integrating fluvial transport of C into a forest carbon model by parameterizing DOC flux from the detrital and soil C pools.

Employing alternative management scenarios is an effective means of understanding how past management decisions influence current and future C stocks and fluxes. By 2012, deforestation due to reservoir creation and expansion resulted in the watershed sequestering 14 Mg C $ha^{-1}$ less than it otherwise would have with no deforestation. Sustained harvest activity had a substantially greater impact with sequestration reduced by an additional 93 Mg C $ha^{-1}$. However as approximately half of the round wood C removed during logging ends up in wood products, over 176,000 Mg C could have remained in storage and out of the atmosphere reducing the cumulative impact of forestry activity from 93 to 71 Mg C $ha^{-1}$.

While successive deforestation related to reservoir expansion does influence watershed-scale C budgets, over multi-decadal time periods, sustained harvest activity was more impactful in the SLW. Understanding the role forest ecosystems play in the global C cycle and, more specifically, integrating the aquatic components of those landscapes into modelling efforts will enable a more accurate determination of anthropogenic impacts on the C cycle.

## Additional files

**Additional file 1: Figure S1.** Daily stream flow and dissolved organic carbon (DOC) concentration, measured and simulated, for Rithet, Judge and Council catchments 1996–2012.

**Additional file 2: Figure S2.** Total forest ecosystem C stocks in 2012 for Baseline Scenario 1 and Scenario 2 across the Sooke Lake watershed.

**Abbreviations**

C: carbon; GHG: greenhouse gas; BC: British Columbia; CBM-CFS3: Carbon Budget Model of the Canadian Forest Service; CO2: carbon dioxide; CH4: methane; IPCC: Intergovernmental Panel on Climate Change; SLW: Sooke Lake Watershed; DOC: dissolved organic carbon; ha: hectares; yr: years; CRD: Capital Regional District; AMLE: study adjusted maximum likelihood estimation; AIC: Akaike information criterion; OLS: ordinary least squares; DOM: dead organic matter; GPG: good practice guidance; HWP: harvested wood products; NBP: net biome production; SC1: alternative management scenario #1; SC2: alternative management scenario #2.

**Authors' contributions**

BPS developed database, contributed intellectual content, conducted R script and CBM-CFS3 model runs, and wrote paper. JAT wrote proposal and advised on study, contributed intellectual content, and wrote paper. Both authors read and approved the final manuscript.

**Author details**

[1] Natural Resources Canada, Canadian Forest Service, 506 West Burnside Road, Victoria, BC V8Z 1M5, Canada. [2] Biology Department, University of Victoria, Victoria, BC V8W 3R4, Canada.

**Acknowledgements**

Many thanks to Dr. Daniel Peters for his assistance and direction during DOC model parameterization. From the CRD, thanks to Joel Ussery for his continuous support of the Sooke Lake Watershed carbon project and Jennifer Blaney for information on lab analysis procedures and providing dissolved organic carbon measurement data. The assistance of CRD Engineering Department employees Fraser Hall, Sigi Gudavicius and Adrian Betanzo for their insights into CRD hydrological measurement installations and for supplying stream flow data was greatly appreciated. As well, many thanks to supporting members of the Canadian Forest Service Carbon Accounting Team Max Fellows, Scott Morken and Gary Zhang. Basil Veerman of the Pacific Climate Impacts Consortium provided useful guidance during the initial stages of hydrologic data manipulation in R. We also thank the two anonymous reviewers for providing thoughtful and constructive comments on the manuscript.

**Competing interests**

The authors declare that they have no competing interests.

**Funding**

Funding for this research was provided by the Capital Regional District Integrated Water Services Division and the Canadian Forest Service.

**References**

1. Stinson G, Kurz WA, Smyth CE, Neilson ET, Dymond CC, Metsaranta JM, et al. An inventory-based analysis of Canada's managed forest carbon dynamics, 1990 to 2008. Global Chang Biol. 2011;17(6):2227–44.
2. Pan Y, Birdsey RA, Fang J, Houghton RA, Pekka K, Kurz WA, et al. A large and persistent carbon sink in the world's forests. Science. 2011;333:988–93.
3. Smyth CE, Stinson G, Neilson E, Lemprière TC, Hafer M, Rampley GJ, et al. Quantifying the biophysical climate change mitigation potential of Canada's forest sector. Biogeosciences. 2014;11(13):3515–29.
4. Sharma T, Kurz WA, Stinson G, Pellatt MG, Li Q. A 100-year conservation experiment: impacts on forest carbon stocks and fluxes. For Ecol Manag. 2013;310:242–55.
5. Dymond CC. Forest carbon in North America: annual storage and emissions from British Columbia's harvest, 1965–2065. Carbon Balance Manage. 2012;7(1):1–20.
6. Regnier P, Friedlingstein P, Ciais P, Mackenzie FT, Gruber N, Janssens IA, et al. Anthropogenic perturbation of the carbon fluxes from land to ocean. Nat Geosci. 2013;6(8):597–607.
7. Lal R. Global soil erosion by water and carbon dynamics. Advances in soil Science: Soils and global change, 1st edn. Boca Raton, Florida: CRC Lewis Publishers; 1995. p. 131–142.
8. Mulholland PJ, Elwood JW. The role of lake and reservoir sediments as sinks in the perturbed global carbon cycle. Tellus. 1982;34(5):490–9.
9. Marcé R, Moreno-Ostos E, López P, Armengol J. The role of allochthonous inputs of dissolved organic carbon on the hypolimnetic oxygen content of reservoirs. Ecosystems. 2008;11(7):1035–53.
10. Raymond PA, Saiers JE. Event controlled DOC export from forested watersheds. Biogeochemistry. 2010;100(1–3):197–209.
11. Hillman GR, Feng JC, Feng CC, Wang Y. Effects of catchment characteristics and disturbances on storage and export of dissolved organic carbon in a boreal headwater stream. Can J Fish Aquat Sci. 2004;61(8):1447–60.
12. Alvarez-Cobelas M, Angeler DG, Sánchez-Carrillo S, Almendros G. A worldwide view of organic carbon export from catchments. Biogeochemistry. 2012;107(1):275–93.
13. Oni SK, Futter MN, Dillon PJ. Landscape-scale control of carbon budget of Lake Simcoe: a process-based modelling approach. J Gt Lakes Res. 2011;37:160–5.
14. Hope D, Billett MF, Cresser MS. A review of the export of carbon in river water: fluxes and processes. Environ Pollut. 1994;84(3):301–24.
15. Hornberger GM, Bencala KE, McKnight DM. Hydrological controls on dissolved organic carbon during snowmelt in the Snake River near Montezuma, Colorado. Biogeochemistry. 1994;25(3):147–65.
16. Schlesinger WH, Bernhardt ES. Chapter 6—the biosphere: biogeochemical cycling on land. Biogeochemistry: an analysis of global change, 3rd edn. Boston: Academic Press; 2013. p. 173–231.
17. Creed IF, Spargo AT, Jones JA, Buttle JM, Adams MB, Beall FD, et al. Changing forest water yields in response to climate warming: results from long-term experimental watershed sites across North America. Glob Chang Biol. 2014;20(10):3191–208.
18. Vörösmarty CJ, Sharma KP, Fekete BM, Copeland AH, Holden J, John M, et al. The storage and aging of continental runoff in large reservoir systems of the world. Ambio. 1997;26(4):210–9.
19. Romero-Lankao P, Smith JB, Davidson DJ, Diffenbaugh NS, Kinney PL, Kirshen P, et al. North America. Climate Change 2014: impacts, adaptation, and vulnerability part B: regional aspects contribution of working group II to the Fifth Assessment Report of the Intergovernmental Panel of Climate Change. Cambridge: Cambridge University Press; 2014. p. 1439–98.
20. Kurz WA, Dymond CC, White TM, Stinson G, Shaw CH, Rampley GJ, et al. CBM-CFS3: a model of carbon-dynamics in forestry and land-use change implementing IPCC standards. Ecol Model. 2009;220(4):480–504.
21. Smiley BP, Trofymow JA, Niemann KO. Spatially-explicit reconstruction of 100 years of forest land use and disturbance on a coastal British Columbia Douglas-fir-dominated landscape: implications for future watershed-scale carbon stock recovery. Appl Geogr. 2016;74:109–22.
22. Boisvenue C, Smiley BP, White JC, Kurz WA, Wulder MA. Improving carbon monitoring and reporting in forests using spatially-explicit information. Carbon Balance Manag. 2016;11(1):23.
23. Billett MF, Palmer SM, Hope D, Deacon C, Storeton-West R, Hargreaves KJ, et al. Linking land-atmosphere-stream carbon fluxes in a lowland peatland system. Global Biogeochem Cycles. 2004;18(1):1–12.
24. Capital Regional District. Facts and figures for the greater Victoria water supply area. Victoria: Capital Regional District; 2014. http://www.crd.bc.ca/education/in-your-community/public-tours/watershed-tours/facts-figures.
25. Pojar J, Klinka K, Demarchi DA. Chapter 6: coastal Western Hemlock Zone. In: Meidinger D, Pojar J, editors. Ecosystems of British Columbia. Victoria, BC, Canada: Research Branch BC Ministry of Forests; 1991.
26. Werner A, Prowse T, Bonsal B. Characterizing the water balance of the Sooke reservoir, British Columbia over the last century. Climate. 2015;3(1):241–63.
27. Zhu JZ, Mazumder A. Estimating nitrogen exports in response to forest vegetation, age and soil types in two coastal-forested watersheds in British Columbia. For Ecol Manag. 2008;255(5–6):1945–59.
28. Province of British Columbia. Forest Governance in the Province of British Columbia. Victoria, BC: Forest, Lands and Natural Resource Operations; 2016.
29. Werner A. Seasonality of the water balance of the Sooke reservoir, BC, Canada [thesis]. Victoria, BC: University of Victoria; 2007.
30. R Development Core Team. A language and environment for statistical

computing. Vienna: R Foundation for Statistical Computing; 2014. http://www.R-project.org.
31. Lorenz D, Runkel R, De Cicco L. rloadest: river load estimation. R package version 0.2. 2013.
32. Runkel RL, Crawford CG, Cohn TA. Load Estimator (LOADEST): A FORTRAN program for estimating constituent loads in streams and rivers. Reston, Virgina: United States Geological Survey; 2004.
33. Cohn TA. Adjusted maximum likelihood estimation of the moments of lognormal population from type I censored samples. Open-File Report 88-350. Reston, Virgina: U.S. Geological Survey; 1988.
34. Cohn TA, Gilroy EJ, Baier WG, editors. Estimating fluvial transport of trace constituents using a regression model with data subject to censoring. Boston: Proceedings of the section on statistics and the environment of the American Statistical Association; 1992.
35. Schlesinger WH, Bernhardt ES. Chapter 8—Inland waters. Biogeochemistry: an analysis of global change, 3rd edn. Boston: Academic Press; 2013. p. 275–340.
36. Kull SJ, Rampley GJ, Morken S, Metsaranta J, Neilson ET, Kurz WA. Operational-scale carbon budget model of the Canadian forest sector (CBM-CFS3) version 1.2: user's guide. Edmonton, Ab.: Canadian Forest Service; 2011.
37. IPCC. Good practice guidance for land use, land-use change and forestry. Hayama, Japan: Institute for Global Environmental Strategies; 2003.
38. Soil Classification Working Group. The Canadian system of soil classification, 3rd edn. Ottawa: Research Branch Agriculture and Agri-Food Canada; 1998. Report No. 0-660-17404-9, Contract No. 1646.
39. IPCC. 2013 Revised supplementary methods and good practice guidance arising from the Kyoto Protocol. Switzerland: IPCC; 2014.
40. Cleveland CC, Neff JC, Townsend AR, Hood E. Composition, dynamics, and fate of leached dissolved organic matter in terrestrial ecosystems: results from a decomposition experiment. Ecosystems. 2004;7(3):175–85.
41. Chapin FS, Woodwell GM, Randerson JT, Rastetter EB, Lovett GM, Baldocchi DD, et al. Reconciling carbon-cycle concepts, terminology, and methods. Ecosystems. 2006;9:1041–50.
42. Striegl RG, Dornblaser MM, Aiken GR, Wickland KP, Raymond PA. Carbon export and cycling by the Yukon, Tanana, and Porcupine rivers, Alaska, 2001–2005. Water Resour Res. 2007;43(2). doi:10.1029/2006WR005201.
43. Kenny S. Aquifers of the Captial Regional District. Victoria, BC: BC Ministry of Water, Land and Air Protection; 2004.
44. Glazebrook HS, Robertson AI. The effect of flooding and flood timing on leaf litter breakdown rates and nutrient dynamics in a river red gum forest. Aust J Ecol. 1999;24(6):625–35.
45. Langhans AD, Tockner K. The role of timing, duration and frequency of inundation in controlling leaf litter decompostion in a river-floodplain ecosystem (Tagliamento, northeastern Italy. Oecologia. 2005;174:501–9.
46. Das B, Narwani A, Mathews B, Nordin R, Maxumder A. Anthropogenic disturbance history influences the temporal coherence of paleoproductivity in two lakes. J Paleolimnol. 2009;42(2):167–81.
47. Evans CD, Monteith DT, Cooper DM. Long-term increases in surface water dissolved organic carbon: observations, possible causes and environmental impacts. Environ Pollut. 2005;137(1):55–71.
48. de Wit HA, Mulder J, Hindar A, Hole L. Long-term increases in dissolved organic carbon in streamwaters in Norway in response to reduced acid deposition. Environ Sci Technol. 2007;41:7706–13.
49. Trofymow JA, Blackwell BA. Changes in ecosystem mass and carbon distributions in coastal forest chronosequences. Northwest Sci. 1998;72(2):40.
50. Oliver AA, Tank SE, Giesbrecht I, Korver MC, Floyd WC, Sanborn P, et al. Globally significant yields of dissolved organic carbon from small watersheds of the Pacific coastal temperate rainforest. Biogeosciences Discuss. 2017;1–44. doi:10.5194/bg-2017-5
51. Dean WE, Gorham E. Magnitude and significance of carbon burial in lakes, reservoirs, and peatlands. Geology. 1998;26(6):535–8.
52. Polis A, Power ME, Huxel GR. Food webs at the landscape level. Chicago: University of Chicago Press; 2004.
53. Cole JJ, Prairie YT, Caraco NF, McDowell WH, Tranvik LJ, Striegel RG, et al. Plumbing the global carbon cycle: integrating inland waters into the terrestrial carbon budget. Ecosystems. 2007;10(1):171–84.
54. Dore MHI, Matilla-García M, Marín MR. Changing patterns of precipitation at the Sooke reservoir in British Columbia. Atl Econ J. 2012;41(2):97–113.
55. Climate Change 2014. Impacts, adaptation, and vulnerability. part B: regional aspects. Contribution of Working Group II to the Fifth Assessment Report of the Intergovernmental Panel on Climate Change. Cambridge: Cambridge University Press; 2014. p. 14–18.
56. IPCC. Climate change 2014: synthesis report. Contribution of working groups I, II and III to the Fifth Assessment Report of the Intergovernmental Panel on Climate Change. Geneva: IPCC; 2014.
57. Schlesinger WH, Bernhardt ES. Chapter 11—The global carbon cycle. Biogeochemistry: an analysis of global change. 3rd edn. Boston: Academic Press; 2013. p. 419–444.
58. Man CD, Lyons KC, Nelson JD, Bull GQ. Potential of alternate forest management practices to sequester and store Carbon in two forest estates in British Columbia, Canada. For Ecol Manag. 2013;305:239–47.
59. Trofymow J, Stinson G, Kurz W. Derivation of a spatially explicit 86-year retrospective carbon budget for a landscape undergoing conversion from old-growth to managed forests on Vancouver Island, BC. For Ecol Manag. 2008;256(10):1677–91.
60. Protocol for the creation of forest carbon offsets in British Columbia Version 1.0. http://www.env.gov.bc.ca/cas/mitigation/fcop.html. Accessed 12 Aug 2011.
61. Metsaranta JM, Dymond CC, Kurz WA, Spittlehouse DL. Uncertainty of 21st century growing stocks and GHG balance of forests in British Columbia, Canada resulting from potential climate change impacts on ecosystem processes. For Ecol Manag. 2011;262(5):827–37.
62. Hember RA, Kurz WA, Coops NC. Increasing net ecosystem biomass production of Canada's boreal and temperate forests despite decline in dry climates. Global Biogeochem Cycles. 2017;31(1):134–58.
63. Boisvenue C, Running SW. Impacts of climate change on natural forest productivity—evidence since the middle of the 20th century. Global Chang Biol. 2006;12(5):862–882.

# Revised methane emissions factors and spatially distributed annual carbon fluxes for global livestock

Julie Wolf[1*], Ghassem R. Asrar[2] and Tristram O. West[2,3]

**Abstract**

**Background:** Livestock play an important role in carbon cycling through consumption of biomass and emissions of methane. Recent research suggests that existing bottom-up inventories of livestock methane emissions in the US, such as those made using 2006 IPCC Tier 1 livestock emissions factors, are too low. This may be due to outdated information used to develop these emissions factors. In this study, we update information for cattle and swine by region, based on reported recent changes in animal body mass, feed quality and quantity, milk productivity, and management of animals and manure. We then use this updated information to calculate new livestock methane emissions factors for enteric fermentation in cattle, and for manure management in cattle and swine.

**Results:** Using the new emissions factors, we estimate global livestock emissions of $119.1 \pm 18.2$ Tg methane in 2011; this quantity is 11% greater than that obtained using the IPCC 2006 emissions factors, encompassing an 8.4% increase in enteric fermentation methane, a 36.7% increase in manure management methane, and notable variability among regions and sources. For example, revised manure management methane emissions for 2011 in the US increased by 71.8%. For years through 2013, we present (a) annual livestock methane emissions, (b) complete annual livestock carbon budgets, including carbon dioxide emissions, and (c) spatial distributions of livestock methane and other carbon fluxes, downscaled to $0.05 \times 0.05$ degree resolution.

**Conclusions:** Our revised bottom-up estimates of global livestock methane emissions are comparable to recently reported top-down global estimates for recent years, and account for a significant part of the increase in annual methane emissions since 2007. Our results suggest that livestock methane emissions, while not the dominant overall source of global methane emissions, may be a major contributor to the observed annual emissions increases over the 2000s to 2010s. Differences at regional and local scales may help distinguish livestock methane emissions from those of other sectors in future top-down studies. The revised estimates allow improved reconciliation of top-down and bottom-up estimates of methane emissions, will facilitate the development and evaluation of Earth system models, and provide consistent regional and global Tier 1 estimates for environmental assessments.

**Keywords:** Methane emissions, Carbon monitoring system, Livestock, Enteric fermentation, Manure management, Greenhouse gas, Carbon dioxide, IPCC

## Background

Livestock play an important role in agricultural carbon (C) cycling and are associated with large annual greenhouse gas emissions [1, 2]. The IPCC [3, 4] provides guidelines for bottom-up estimation of livestock emissions based on inventory, which have been employed at the global [5, 6] and national levels (e.g. annual reports to the United Nations Framework Convention on Climate Change). In inventory-based estimation of national livestock methane ($CH_4$) emissions, annual standing populations of each animal type are multiplied by species- and region-specific emissions factors to obtain annual

*Correspondence: julie.wolf2@ars.usda.gov; thejuliewolf@gmail.com
[1] USDA-ARS, Adaptive Cropping Systems Laboratory, 10300 Baltimore Ave., Building 001, Room. 342, BARC-WEST, Beltsville, MD 20705, USA
Full list of author information is available at the end of the article

emissions quantities. The emissions factors are derived using sets of mathematical formulae with inputs that vary depending on regional livestock qualities and management (e.g. feed intake quantity and quality; milk production quantity; amount of energy used for growth, draft work, foraging, and pregnancy; and utilization of various manure management systems) [4].

The input information in the IPCC 2006 guidelines is based on literature reflecting earlier decades; e.g. sources listed for tables in Annexes 10.A1 and 10A.2 in [4] were published between 1976 and 2004, with most from the 1980s and 1990s. In at least some regions, this information no longer reflects the state of livestock. For example, in many industrialized or industrializing nations, management of manure in pits or lagoons, instead of on pasture or cropland, has become more prevalent [7, 8] and animals perform less draft work [9] than in earlier decades. For example, IPCC 2006 guidelines and recent publications based on them [10, 11] consider 12% of US dairy cattle manure to be managed in anaerobic lagoons, while more recent data from the US EPA [12] suggest that anaerobic lagoons are now much more widely used. Because $CH_4$ emissions from anaerobic lagoons are calculated to be nearly twice the magnitude of those from aerobic systems per unit of manure input, these changes must be taken into account in new bottom-up inventories.

The IPCC 2006 default information is used to calculate bottom-up $CH_4$ emissions in important global earth system simulation studies and environmental assessments [13, 14]. For example, in addition to reports from the IPCC [15], the US Environmental Protection Agency's report on global emissions [16], IIASA's greenhouse gas and air pollution interactions and synergies (GAINS) model [17], and the emissions database for global atmospheric research (EDGAR) [18] use IPCC 2006 default information, although the latter modifies cattle inputs based on carcass weight or milk productivity. IIASA's RAINS model, an earlier source of global $CH_4$ emissions used in a recent longer-term study along with EDGAR and EPA data [13], is based on IPCC 1996 [3] emissions factors [19]. Recent top-down estimates for the US, however, suggest that even revised methods based on IPCC guidelines underestimate livestock $CH_4$ emissions in recent years at the national or state level [20–23]. Additionally, since 2007, global atmospheric concentrations of $CH_4$ began increasing again after several stable years, and the $^{13}C$ isotopic ratio of atmospheric $CH_4$ concurrently become more negative; these changes may indicate increasing $CH_4$ emissions from biogenic sources such as wetlands, rice paddies, and/or livestock in various global regions [24–28]. These changes and discrepancies illustrate the need for updated livestock $CH_4$ emissions coefficients for bottom-up inventories.

Many factors are likely to impact recent livestock $CH_4$ emissions quantities, such as the proportion of animals in large animal feeding operations that use various manure management systems; animal traits, such as body mass or productivity, which have changed with animal breeding and increased use of improved breeds; and animal feed quality and quantity, which may change over sub-annual and longer time periods. Here, we re-evaluated inputs used to calculate IPCC tier 1 $CH_4$ emission factors for (1) enteric fermentation emissions in dairy cows and in meat/other cattle and (2) manure management emissions in dairy cows, meat/other cattle, and swine.

## Methods

### Revision of annual, per-animal $CH_4$ emissions factors and other livestock C fluxes

The 2006 IPCC $CH_4$ emissions factors were revised by (1) collecting updated regional input information (Tables 1, 2) and (2) following the Tier 2 equations for enteric fermentation and manure management $CH_4$ emissions [4] with the updated inputs. This resulted in new emissions factors suitable for Tier 1 bottom-up inventory based estimates. To revise enteric fermentation emissions *factors for lactating dairy cows, for example*, Equations 10.2, 10.3, 10.4, 10.6, 10.8, 10.11, 10.13, 10.14, 10.16, 10.18b, and 10.21 were used with input from Tables 10.2, 10.4, 10.5, 10.8, 10.12, 10.A.1 [4] (Table 1). To revise manure $CH_4$ emissions factors for dairy cows, meat/other cattle, and swine, Equations 10.23 and 10.24 were used with input from Tables 10.17, 10A-4, 10A-5, 10A-7, and 10A-8 [4] (Table 2). Some information on total dry matter intake and/or gross energy intake and manure production are also provided by IPCC; these quantities were also updated and used to create complete livestock C budgets (see below). Manure production for cattle was estimated from updated regional animal body weights, assuming that dairy cattle produce 2205 kg manure dry matter per animal unit per year, and meat/other cattle produce 1510 kg manure dry matter per animal unit per year [29]. Manure production for swine was estimated using IPCC 1996 regional swine body weight and manure production information [3] along with revised (recent) regional body weights, based on the approximation that intake scales with a three-fourths fractional exponent of body mass [30]:

$$\text{manure-production}_{\text{revised}} = \text{manure-production}_{\text{1996IPCC}} \times \left[\text{weight}_{\text{revised}}/\text{weight}_{\text{1996IPCC}}\right]^{0.75} \quad (1)$$

To evaluate our bottom-up approach to estimating C stocks and fluxes, the equations and default inputs were first used to recalculate the IPCC 2006 $CH_4$ emissions factors. Literature search results were then used to revise inputs and recalculate these equations.

**Table 1 Dairy cow enteric fermentation emissions factor inputs**

| Region | Based on | Average animal body mass (kg, lactating cows only) | | % Stall fed animals | | Average milk production (kg/head/year) | | Gross energy intake | | Digestibility of feed (%) | | $Y_m$ (%) | |
|---|---|---|---|---|---|---|---|---|---|---|---|---|---|
| | | IPCC 2006[a] | This study | IPCC 2006[a] | This study | IPCC 2006[a] | This study | IPCC 2006[a] | This study[a,b] | IPCC 2006[a] | This study | IPCC 2006[a] | This study |
| US and Canada | US | 600 | 680 [12] | 100 | 93 [12, 39] | 8400 | 9732 [12, 39] | 301.7 | 403.2 | 75 | 66.7 [12] | 6.5 | 6 [70] |
| W. Europe | EU-15 | 600 | 624 [71] | 100 | 77 [70] | 6000 | 6935 [71] | 275.4 | 304.9 | 70 | 70.2 [70] | 6.5 | 6.53 [70] |
| E. Eur.–Ctrl. Asia | Russian Fed. | 550 | 572 [4, 71] | 100 | 80 [70] | 2550 | 3898 [70] | 232.5 | 233.9 | 60 | 68.6 [70] | 6.5 | 6 [70] |
| Oceania | Australia | 500 | 565 [70] | 0 | 7 [70] | 2190 | 5789 [70] | 234.6 | 253 | 60 | 76.1 [70] | 6.5 | 7.24 [70] |
| Latin America | Brazil | 400 | 458 [72] | 0 | 2 [73] | 800 | 1208 [74] | 179.3 | 207.1 | 60 | 60[c] | 6.5 | 8 [74] |
| E.-S.E. Asia | China | 350 | 500 [59, 75] | 100 | 100 [58, 59, 76] | 1650 | 3011 [36] | 160.9 | 333.7 | 60 | 60[c] | 6.5 | 7 [76] |
| Africa | Ethiopia | 275 | 354.75[d] | 100 | 50 [77] | 475 | 687.2 [36, 77] | 107.3 | 147.2 | 60 | 60[c] | 6.5 | 8[e] |
| S. Asia | India | 275 | 281 [78, 79] | 100 | 70 [77] | 900 | 1355.2 [36, 77] | 135.4 | 160.2 | 55 | 55[c] | 6.5 | 8[e] |

Only lactating dairy cows are included in the dairy cow category. Reported values that included dairy calves, heifers, dry cows, and/or replacements were not used or were adjusted to reflect producing dairy cow populations

[a] Calculated using information in supplemental tables in Chapter 10 of IPCC 2006 [4]

[b] Net energy for growth and wool production are assumed to be zero for mature dairy cows

[c] Unchanged from information in supplemental tables in Chapter 10 of IPCC 2006 [4] due to lack of newer information

[d] Based on average increases found for Latin America and E.-S.E. Asia, due to lack of information

[e] As reported for Latin America, due to lack of information

**Table 2 Cattle and swine manure management emissions input factors used for this study, as compared with those used/published by IPCC**

| Region[e] | Average animal body mass (kg) | | Volatile solids (kg $year^{-1}$) | | $B_o$ ($m^3$ $CH_4$/kg VS) | | MCF for dry systems[f] | | MCF liquid | |
|---|---|---|---|---|---|---|---|---|---|---|
| | IPCC 2006[d] | This study | IPCC 2006[d] | This study | IPCC 2006[d] | This study | IPCC 2006[d] | This study | IPCC 2006[d] | This study |
| Dairy cattle | | | | | | | | | | |
| US and Canada | 600 | 680[e] | 1971 | 2703 [12, 39] | 0.24 | 0.24 [12] | 1.2 | 1 [12, 70] | 20.0 | 29.5 [12] |
| W. Europe | 600 | 624[e] | 1862 | 1632 [70] | 0.24 | 0.23 [70] | 1.5 | 3.1 [70] | 20.0 | 26 [70] |
| E. Eur.–Ctrl. Asia | 550 | 572[e] | 1643 | 1705 [70] | 0.24 | 0.24 [70] | 1.7 | 1.8 [70] | 17.0 | NA |
| Oceania | 500 | 565[e] | 1278 | 1033 [70] | 0.24 | 0.24 [70] | 1.4 | 1.4 [70] | 50.0 | 35.0 [70] |
| Latin America | 400 | 458[e] | 1059 | 1507[h] | 0.13 | 0.13[g] | 0.9 | 0.9[g] | 65.0 | 65.0[g] |
| E.–S.E. Asia | 350 | 500[e] | 1022 | 1602[h] | 0.13 | 0.13[g] | 1.7 | 2.0 [9] | 19.0 | 20.0 [9] |
| Africa | 275 | 355[e] | 694 | 1071[h] | 0.13 | 0.13[g] | 1.9 | 1.9[g] | 46.0 | NA |
| S. Asia | 275 | 281[e] | 949 | 975[h] | 0.13 | 0.13[g] | 5.6 | 5.6[g] | 55.0 | 55.0[g] |
| Meat/other cattle | | | | | | | | | | |
| US and Canada | 389 | 420 [70] | 876 | 1059 [70] | 0.19 | 0.20 [12, 39][a] | 1.0 | 1.4 [12, 39] | 20.0 | 34.0 [12, 39] |
| W. Europe | 420 | 401 [70] | 949 | 726 [70] | 0.18 | 0.19 [70] | 1.5 | 4.5 [70] | 20.0 | 29.0 [70] |
| E. Eur.–Ctrl. Asia | 391 | 351 [70] | 986 | 923 [70] | 0.17 | 0.17 [70] | 1.6 | 1.7 [70] | 17.0 | 20.0 [70] |
| Oceania | 330 | 376 [70] | 1095 | 965 [70] | 0.17 | 0.17 [70] | 1.5 | 1.5 [70] | NA | NA |
| Latin America | 305 | 323 [80, 81] | 913 | 779[h] | 0.10 | 0.10[g] | 1.5 | 1.5[g] | NA | NA |
| E.–S.E. Asia | 319 | 259 [59] | 840 | 546[h] | 0.10 | 0.10[g] | 1.2 | 1.2[g] | NA | NA |
| Africa | 173 | 173[g] | 548 | 548[g] | 0.10 | 0.10[g] | 1.7 | 1.7[g] | NA | NA |
| S. Asia | 110 | 208 [78] | 511 | 676[h] | 0.10 | 0.10[g] | 5.9 | 5.9[g] | 55.0 | 55.0[g] |
| Swine | | | | | | | | | | |
| US and Canada | 61 | 60 [70] | 99 | 106 [70] | 0.48 | 0.48 [70] | 1.5 | 1.5 [12, 70] | 20.0 | 30.7 [12, 70] |
| W. Europe | 50 | 76 [70] | 110 | 113 [70] | 0.45 | 0.41 [70] | 1.8 | 4.4 [70] | 20.0 | 26.5 [70] |
| E. Eur.–Ctrl. Asia | 50 | 55 [70] | 110 | 168 [70] | 0.45 | 0.45 [70] | 2.3 | 2.0 [70] | 17.0 | 20.0 [70] |
| Oceania | 45 | 61 [70] | 102 | 113 [70] | 0.45 | 0.45 [70] | 1.4 | 1.0 [70] | 65.0 | 65.0 [70] |
| Latin America | 28 | 60 [82] | 110 | 106 [82] | 0.29 | 0.48 [82] | 1.5 | 1.5 [82] | 60.0 | 60.0 [82] |
| E.–S.E. Asia | 28 | 60 [62, 83] | 110 | 110[g] | 0.29 | 0.29[g] | 1.0 | 1.0[g] | 19.0 | 19.0[g] |
| Africa | 28 | 28[g] | 110 | 110[g] | 0.29 | 0.29[g] | 1.7 | 1.7[g] | 60.0 | 60.0[g] |
| S. Asia | 28 | 28[g] | 110 | 110[g] | 0.29 | 0.29[g] | 3.0 | 3.0[g] | 55.0 | 55.0[g] |

| Region | MCF lagoon | | MCF digester/other | % of manure managed in dry systems a | | % of manure managed in liquid and deep pit systems | | % of manure managed in lagoon systems | | % of manure managed in digesters/other | |
|---|---|---|---|---|---|---|---|---|---|---|---|
| | IPCC 2006[d] | This study | This study | IPCC 2006[d] | This study | IPCC 2006[d] | This study | IPCC 2006[d] | This study | IPCC 2006[d] | This study |
| Dairy cattle | | | | | | | | | | | |
| US and Canada | 70.0 | 71 [12, 39] | NA | 58 | 42 [12, 39] | 30 | 24.3 [12, 39] | 15 | 33.6 [12, 39] | 0 | 0 [12] |
| W. Europe | 70.0 | 42 [70] | 18.0 [70] | 64 | 54 [70] | 36 | 41 [70] | 0 | 0 [70] | 0 | 5 [70] |
| E. Eur.–Ctrl. Asia | 66.0 | NA | NA | 83 | 100 [70] | 18 | 0 [70] | 0 | 0 [70] | 0 | 0 [70] |

**Table 2 continued**

| Region | MCF lagoon | | MCF digester/other | % of manure managed in dry systems a | | % of manure managed in liquid and deep pit systems | | % of manure managed in lagoon systems | | % of manure managed in digesters/other | |
|---|---|---|---|---|---|---|---|---|---|---|---|
| | IPCC 2006[d] | This study | This study | IPCC 2006[d] | This study | IPCC 2006[d] | This study | IPCC 2006[d] | This study | IPCC 2006[d] | This study |
| Oceania | 78.0 | 90.0 [70] | NA | 84 | 95 [70] | 1 | 0 [70] | 16 | 5 [70] | 0 | 0 [70] |
| Latin America | 78.0 | NA | NA | 99 | 99 [73] | 1 | 1 [73] | 0 | 0 [73] | 0 | 0 [73] |
| E.–S.E. Asia | 68.0 | NA | 10.0[d] | 56 | 66 [59] | 38 | 25 [9] | 4 | 0 [59] | 2 | 9 [59] |
| Africa | 79.0 | NA | NA | 100 | 100[g] | 0 | 0[g] | 0 | 0[g] | 0 | 0[g] |
| S. Asia | 79.0 | NA | 10.0[d] | 99 | 80 [84] | 1 | 0 [84] | 0 | 0 [84] | 1 | 20 [84][b] |
| Meat/other cattle | | | | | | | | | | | |
| US and Canada | NA | NA | NA | 100 | 100 [12, 39] | 0 | 1 [12, 39][c] | 0 | 0 [12, 39] | 0 | 0 [12, 39] |
| W. Europe | NA | NA | NA | 75 | 72 [70] | 25 | 26 [70] | 0 | 0 [70] | 0 | 3 [70] |
| E. Eur.–Ctrl. Asia | NA | NA | NA | 78 | 86 [70] | 23 | 14 [70] | 0 | 0 [70] | 0 | 0 [70] |
| Oceania | NA | NA | NA | 100 | 100 [70] | 0 | 0 [70] | 0 | 0 [70] | 0 | 0 [70] |
| Latin America | NA | NA | NA | 100 | 100[g] | 0 | 0[g] | 0 | 0[g] | 0 | 0[g] |
| E.–S.E. Asia | NA | NA | NA | 100 | 98 [59] | 0 | 0 [59] | 0 | 0 [59] | 0 | 2 [59] |
| Africa | NA | NA | NA | 100 | 100[g] | 0 | 0[g] | 0 | 0[g] | 0 | 0[g] |
| S. Asia | NA | NA | NA | 99 | 99[g] | 1 | 1[g] | 0 | 0[g] | 0 | 0[g] |
| Swine | | | | | | | | | | | |
| US and Canada | 70.0 | 72.2 [12, 39] | NA | 8 | 6 [12, 39] | 59 | 58 [12, 39] | 33 | 36 [12, 39] | 0 | 0 [12] |
| W. Europe | 70.0 | 41.7 [70] | 16.5 | 22 | 13 [70] | 70 | 58 [70] | 9 | 3 [70] | 0 | 26 [70] |
| E. Eur.–Ctrl. Asia | 66.0 | NA | NA | 72 | 37 [70] | 25 | 63 [70] | 3 | 0 [70] | 0 | 0 [70] |
| Oceania | 78.0 | 90.0 [70] | 10.0 | 46 | 23 [70] | 0 | 0 [70] | 54 | 76 [70] | 0 | 1 [70] |
| Latin America | 79.0 | 79.0 [82] | NA | 93 | 30 [82] | 8 | 0 [82] | 0 | 70 [82] | 0 | 0 [82] |
| E.–S.E. Asia | 68.0 | 68.0[g] | 10.0 | 54 | 28 [63, 85–87] | 40 | 55 [63, 85–87] | 0 | 0[g] | 7 | 18 [63, 85–87] |
| Africa | 79.0 | NA | NA | 94 | 94[g] | 6 | 6[g] | 0 | 0[g] | 0 | 0[g] |
| S. Asia | 79.0 | NA | NA | 69 | 69[g] | 22 | 22[g] | 9 | 9[g] | 0 | 0[g] |

*NA* not applicable

[a] $B_o$ for US meat/other cattle was calculated as the weighted average of on-feed and not-on-feed live-animal biomass. We estimated that 17.2% of US meat/other cattle live animal mass is on-feed in 2012, based on reported body masses and populations [39]. Values of $B_o$ for these populations was taken from [12]

[b] $CH_4$ emissions from the estimated 20% of manure that is burned for fuel in this region were considered similar to emissions from manure treated in anaerobic digestors for lack of specific information; $CH_4$ and/or volatile solids assumed to be oxidized to $CO_2$ when burned

[c] Table A-206 in [12] states "because manure from beef feedlots…may be managed for long periods of time in multiple systems (i.e. both drylot and runoff collection pond), the percent of manure that generates emissions is greater than 100%."

[d] Calculated using information in supplemental tables in Chapter 10 of IPCC 2006 [4]

[e] See Table 1 for the country/countries on which regional values are based, and for dairy cattle weight sources

[f] Dry systems include dry lot, pasture/range, solid, daily spread, burned, and pit storage <1 month

[g] Unchanged from information in supplemental tables in Chapter 10 of IPCC 2006 [4] due to lack of newer information

[h] Calculated using the updated information presented here in Eq. 10.24 in Chapter 10 of IPCC 2006 [4]

For dairy cow enteric fermentation $CH_4$ emissions factors, revisions focused on changes in mature animal weight, percent of animals that are stall fed as opposed to grazing/ranging for feed, annual milk productivity, changes in total feed intake, and on reported values of $Y_m$ (the $CH_4$ conversion factors for feed energy intake during enteric fermentation). For these calculations, we assumed that mature lactating dairy cows do not gain or lose weight, so that net energy for growth takes a value of zero. For enteric fermentation $CH_4$ emissions from meat/other cattle, we use recently reported emissions factors from national UNFSCCC reports where available, and where such information was not available, we calculated revised factors based on changes in animal body weight only. This approach was taken due to the complexity and variability in important management factors for meat cattle, particularly in industrialized systems (e.g. type of diet provided, timing of placement from pasture to feedlot, slaughter age and weight).

For manure management $CH_4$ emissions factors, revisions focused on changes in animal weight at slaughter, changes in total feed intake and feed digestibility, and changes in the percentage of manure managed in various manure management systems (e.g. deposited on pasture, drylot storage, short-term pit storage, long-term anaerobic lagoon treatment), and MCFs (methane conversion factors, the $CH_4$ conversion factors for manure volatile solids during manure storage and/or treatment) for different manure management systems at various temperatures. Because of the difficulty in obtaining recent information for all regions of the world, we did not revise $B_o$ (the amount of $CH_4$ produced per quantity of manure volatile solids). Manure management $CH_4$ emissions factors were revised for (1) lactating dairy cattle; (2) meat/other cattle (encompassing meat and dairy calves and heifers and all other cohorts of non-lactating cattle grown for slaughter, replacement, breeding, or other purposes, weighted using mean weights and reported population cohorts), and (3) swine (encompassing farrowing sows, nursing piglets, and feeders, weighted using mean weights and reported population cohorts). For meat/other cattle in the US, where in recent years animals weighed 27–45 kg at birth [31], were weaned at ~260 kg [31], were placed on feedlots at ~317 kg [32], and were slaughtered at ~610 kg [33], the amounts of manure managed on pasture and on feedlot were weighted by average cohort masses accordingly.

### Uncertainty analysis

We employed IPCC 2006 Uncertainty Approach I: Propagation of Error [34] to arithmetically combine the uncertainties associated with livestock carbon fluxes of interest:

> *Where uncertain quantities are to be combined by multiplication, the standard deviation of the sum will be the square root of the sum of the squares of the standard deviations of the quantities that are added, with the standard deviations all expressed as coefficients of variation, which are the ratios of the standard deviations to the appropriate mean values...Where uncertain quantities are to be combined by addition or subtraction, the standard deviation of the sum will be the square root of the sum of the squares of the standard deviations of the quantities that are added with the standard deviations all expressed in absolute terms ... [34]*

When the uncertainties being combined can be considered independent, their standard deviations or coefficients of variation are added in quadrature (i.e. the square root of the sum of the squares of each standard deviation or coefficient of variation) [35]. This has the effect of reducing overall propagated uncertainty. We added in quadrature when propagating uncertainties within a livestock type, because we independently assembled separate estimates of the various carbon fluxes and their uncertainties (e.g. intake, manure production, milk production, $CH_4$ emissions) except for $CO_2$, which is calculated by subtraction. We then used these uncertainties to calculate fractional standard deviations (equal to the coefficient of variation, the standard deviation divided by the mean value) for each per-animal carbon flux quantity in each global region. However, when combining uncertainties across livestock types within a nation or from multiple nations to the regional or global level, the uncertainties were simply added (not in quadrature), because these estimates are not independent [35]—i.e. the livestock in all nations within a region share the same carbon flux estimates, emissions coefficients, and uncertainties, and all livestock within a nation share many regional attributes. Using the arithmetic sum, as opposed to adding in quadrature, results in larger uncertainties, which may be considered more conservative.

Uncertainty on all non-$CH_4$ quantities is derived from the coefficients of variation (the standard deviation/mean value of the quantity) that we calculated for these quantities in previous work [2]. Uncertainty on IPCC livestock $CH_4$ emissions factors is given as ±30% [4], and is defined as representing ±1.96 times the standard deviation of the mean [34]. In order to be combined mathematically [34, 35] with our estimates of uncertainty on other C fluxes, we used 15.3% (30% divided by 1.96) as the uncertainty for all calculated $CH_4$ quantities.

#### Derivation of annual livestock C fluxes, including emissions of $CO_2$ and $CH_4$

We assumed a linear transition from IPCC 2006 emissions to revised emissions factors during the years 1990–2012:

$$f_{yeari} = f_{IPCC} + (f_{revised} - f_{IPCC}) \cdot (Y/22) \quad (2)$$

where $f_{yeari}$ is the flux of $CH_4$, feed, or other C containing quantity per animal in the year of interest; $f_{IPCC}$ is the flux of $CH_4$, feed, or other C quantity per animal given or calculated from data provided by 2006 IPCC guidelines [4]; $f_{revised}$ is the revised flux of $CH_4$, feed, or other C quantity per animal (resulting from this work); and Y is equal to 0 for years before 1990, to (year—1990) for 1990–2012; and to 22 for years after 2012.

Livestock carbon dioxide ($CO_2$) emissions associated with respiration were estimated as the deficit between the C contained in annual livestock feed intake and the sum enteric fermentation $CH_4$ emissions, production of milk or eggs, and manure production. Similarly, $CO_2$ emissions associated with manure management were estimated as the difference between total manure C production and manure management $CH_4$ emissions, assuming that all manure C is emitted as either $CH_4$ or $CO_2$ within one year of production.

#### Livestock populations

Annual national livestock populations of meat and milk-producing cattle, meat and milk-producing buffaloes, meat and egg-laying chickens, swine, sheep, turkeys, ducks, geese and guinea fowl, goats, horses, mules, asses, camels, and other camelids (i.e. llamas and alpacas) were compiled for years 1961–2013 from FAOSTAT [36]. Annual producing populations of egg-laying chickens and milk-producing cattle and buffalo were subtracted from conspecific total populations to estimate populations raised for meat production. For all calculations made here, the dairy cattle livestock populations include only milk-producing mature dairy cows; calves, heifers, breeding steers, and any other dairy cattle 'replacements' are categorized with meat/other cattle. For nine large countries (Argentina, Brazil, Canada, Chile, China, India, Kazakhstan, Mexico, and the Russian Federation), state- or province-level livestock population data were compiled for available years between 2000 and 2011 [37, 38], and used to improve the spatial distribution of inventory data. For the United States, livestock populations were refined to the county level using National Agricultural Statistical Service Census and Survey data [39]. Livestock in all other nations of the world are constrained at the national level only.

#### Livestock C fluxes and $CH_4$ emissions

Accounting of livestock C fluxes was conducted as described in Wolf et al. [2]. Annual per-animal dry weight feed intake, dry weight manure production, manure C content, milk and egg production C, and manure management and enteric fermentation $CH_4$ emissions are from IPCC [4] or were estimated from existing literature. Livestock dry matter intakes were assumed to be 44% C by weight. The difference between total livestock feed intake C and total C produced or emitted by live animals (i.e. the sum of C contained in manure, enteric fermentation $CH_4$, and milk and eggs) approximates the amount of C respired in the form of $CO_2$ over a given year, excluding C stored in livestock biomass. Although herd sizes do change over time, C stored in livestock biomass is assumed constant in this effort. Similarly, the difference between total manure C content and manure management $CH_4$ provides an estimate of $CO_2$ released by livestock manure management, all of which is assumed to be emitted in the same year of manure production.

#### Estimating livestock consumption of fodder and forage

For purposes of tracking the use of all harvested crop C and estimating amounts of livestock forage, total livestock feed was disaggregated into fodder (i.e. biomass harvested by humans from croplands) and forage (i.e. biomass grazed or scavenged by livestock from non-cropland sources) [2]. Fodder was further subdivided into (a) market feed items derived from primary harvests (e.g., grains, brans, crop by-product feeds), derived from FAO [36] (food balance: commodity balances, crops primary equivalent, feed category), (b) hay and fodder crops (e.g., harvested quantities of alfalfa, clovers, grasses, corn and sorghum silage) derived from FAO [36] (production: crops, crops primary list), including maize, alfalfa, and other grains, grasses, legumes, roots, and vegetables denoted as produced for forage and/or silage; category no longer available), and (c) crop residue feed, consisting of crop residue collected from the field for livestock feed, estimated from annual production of several utilized crops [2]. Annual national quantities of all market feed items and hay crops available were converted into units of C using fractional item-specific dry weights and C contents [2]. The crop residue feed quantities were estimated by applying crop-specific regional percentages of residues collected for feed [40] to the crop- and country-specific estimates of annual residue production. Total annual available fodder per nation is the sum of market feeds, hay and fodder crop production, and crop residues collected for feed. At the national level, annual available fodder was subtracted from total livestock feed intake requirement (calculated from national annual populations and per-animal feed intake values) to approximate national livestock forage intake, including grazing and scavenging. Because national quantities of market feeds and hay crops were not available for years after 2011 at the time of download, fodder and forage intake for 2012 and 2013 were estimated using average available quantities for each country over 2005–2011.

### Downscaling and spatial distribution of C fluxes

Livestock C fluxes were downscaled and spatially distributed to 0.05 × 0.05 degree resolution using the MODIS Land Cover Type 5 data product for year 2005, following methods documented by West et al. [41] and Wolf et al. [2]. Downscaling started with the reconciling of land class areas between satellite-based land cover in 2005 and crop harvest area inventory data in each year from 2000 to 2011. Cropland area in 2005, based on MODIS, was compared to the sum of area inventoried for harvest per geopolitical region. The MODIS cropland areas were then adjusted to equal the sum of harvested areas for respective geopolitical regions and years. Cropland area was expanded or contracted as necessary, using a global kernel density representing the combined density of cropland and distance of each grid-cell to the nearest cropland region. Based on reconciled land cover information within each nation, state or province, or county, a separate amount of area was allocated to livestock. The livestock area requirement per nation, state/province, or county was derived from the livestock population therein, along with estimated area per animal required for each livestock type, for housed and free-ranging animals, and regional estimates of the proportion of animals that are free-ranging. Livestock were spatially distributed to grasslands, based on the livestock area requirement, per nation, state/province, or county. If there was insufficient grassland area, livestock were then distributed to shrubland areas. If grassland and shrubland areas together were smaller than the estimated required livestock area, the livestock area requirement was reduced to a smaller housed-animal area requirement value, thereby increasing livestock density. Respective carbon fluxes were subsequently applied to spatial livestock distributions.

## Results

### Revised livestock emissions factors

The revised emissions factors calculated here are greater than those given by IPCC 2006 for many, but not all, livestock types and regions (Table 3). The information we assembled to revise emissions factors highlights important recent changes in regional livestock systems. Mature dairy cattle body mass and milk productivity were greater in all global regions than IPCC 2006 default values, although the magnitude of increase varied (Table 1). Revised enteric fermentation emissions factors for dairy cows range from 7% smaller (E. Europe and W. and Central Asia) to 125% larger (E. and S.E. Asia) than IPCC 2006 emissions factors (Table 3). Dairy manure management strategies changed along with increasing dairy cow body mass and productivity (Table 2). This resulted in more variable changes in manure management emissions factors among global regions than enteric fermentation emissions factors for dairy cows. Changes in dairy cow manure management emissions factors ranged from a 68% decrease in Oceania to a 158% increase in the US and Canada region (Table 3).

In contrast to the increases in mature dairy cow body mass, we found that body mass at time of slaughter for meat/other cattle decreased in several regions (Table 2). The mature weights of producing dairy cows are determined by breed/genetics and nutritional status of the animals. While this is also true for meat/other cattle, slaughter weights for meat animals are also determined by management decisions, and as such may vary with changing economic or environmental factors (e.g. weather extremes, feed costs, or meat prices and demand). For meat cattle in many regions, external factors also influence the weight at which grazing animals are placed on feedlots to be grain-finished—with very large differences in manure management $CH_4$ emissions between these situations (Table 2). Changes in emissions factors for enteric fermentation in meat/other cattle ranged from an 18% decrease (W. Europe) to a 54% increase (E. and S.E. Asia). Manure management $CH_4$ emissions factors for meat/other cattle are overall much smaller than those for dairy cows, and the IPCC 2006 default factors are rounded to the nearest integer value (e.g. "1"). Therefore, some of the changes reported here result merely from inclusion of additional significant digits. Given the large global populations of meat/other cattle, these small changes are nevertheless important. Revision of manure management emissions factors for meat/other cattle resulted in variable changes among regions, ranging from a 60% decrease (E. and S.E. Asia) to a 140% increase (US and Canada).

Changes in swine manure management emissions factors, relative to IPCC 2006 reported values, range from −4% (W. Europe) to +1800% in Latin America. The latter large increase is due to modernization of swine production in that region, including use of improved breeds with larger potential body mass, changing animal diet, and in particular a shift from dry manure management systems to anaerobic lagoons.

### Revised global livestock C fluxes

Fluctuations in annual livestock populations [FAO, 36] play a large role in the magnitude of C fluxes associated

**Table 3 Emissions factors (kg $CH_4$-animal$^{-1}$ year$^{-1}$) as given by IPCC 2006 [4] and as revised resulting from this study**

| Region | Dairy cow enteric fermentation | | Meat and other cattle enteric fermentation | | Dairy cow manure management | | Meat and other cattle manure management | | Swine manure management | |
|---|---|---|---|---|---|---|---|---|---|---|
| | IPCC 2006[a] | This study[b] | IPCC 2006[a] | This study[d] | IPCC 2006[a] | This study[b] | IPCC 2006[a] | This study[b] | IPCC 2006[a] | This study[b] |
| US–Canada | 128 | 158.7 | 53 | 58.8 | 53 | 137.0 | 1 | 2.4 | 12 | 15 |
| W. Europe | 117 | 130.6 | 57 | 46.5 | 25 | 31.0 | 7 | 9.9 | 7 | 6.7 |
| E. Europe | 99 | 92.1 | 58 | 56.1 | 12 | 4.9 | 6 | 4.5 | 3 | 6.8 |
| Oceania | 90 | 120.1 | 60 | 71.9 | 29 | 9.4 | 2 | 1.6 | 13 | 23.6 |
| Latin America | 72 | 108.6 | 56 | 57.9 | 1 | 2.0 | 1 | 0.8 | 1 | 19 |
| E.–S.E. Asia | 68 | 153.2 | 47 | 42.4 | 10 | 10.1 | 1 | 0.4 | 2 | 2.7 |
| Africa | 46 | 77.2 | 31 | 31.0 | 1 | 1.8 | 1 | 1[c] | 1 | 1[c] |
| S. Asia | 58 | 62.4 | 27 | 41.6 | 5 | 5.5 | 2 | 3 | 5 | 5[c] |

[a] IPCC 2006 [4], Chapter 10: Agriculture, supplemental tables

[b] Calculated using the updated information presented here in the equations in IPCC 2006 [4], Chapter 10

[c] Emissions factors not modified from IPCC 2006 [4] due to sparse information

[d] 2014 National Inventory Submissions to the UNFCCC, CRF, Table 4. A reported for all non-dairy cattle in year 2012 (US, EU15, Russian Fed., and Australia); or calculated using IPCC 2006 [4] defaults except for revised body weights listed in Table 2

with livestock, including $CH_4$ emissions. Global populations of most livestock species did not change greatly over the years between 1990 and 2013, except for goats and chickens (Fig. 1). However, when separated by region, changes in the distribution of global cattle and swine populations are apparent (Fig. 2). For dairy cows, meat/other cattle, and swine, populations in W. Europe and US and Canada regions remained steady or declined slightly over the years in this study. In contrast, meat/other cattle populations increased dramatically in Latin America during the early 2000s, and the already large swine population in E. and S.E. Asia has continued to increase in recent decades.

Total livestock $CH_4$ emissions account for ca. 3% of total livestock C fluxes (Fig. 3; Table 4). Nevertheless, estimating livestock $CH_4$ emissions with our revised

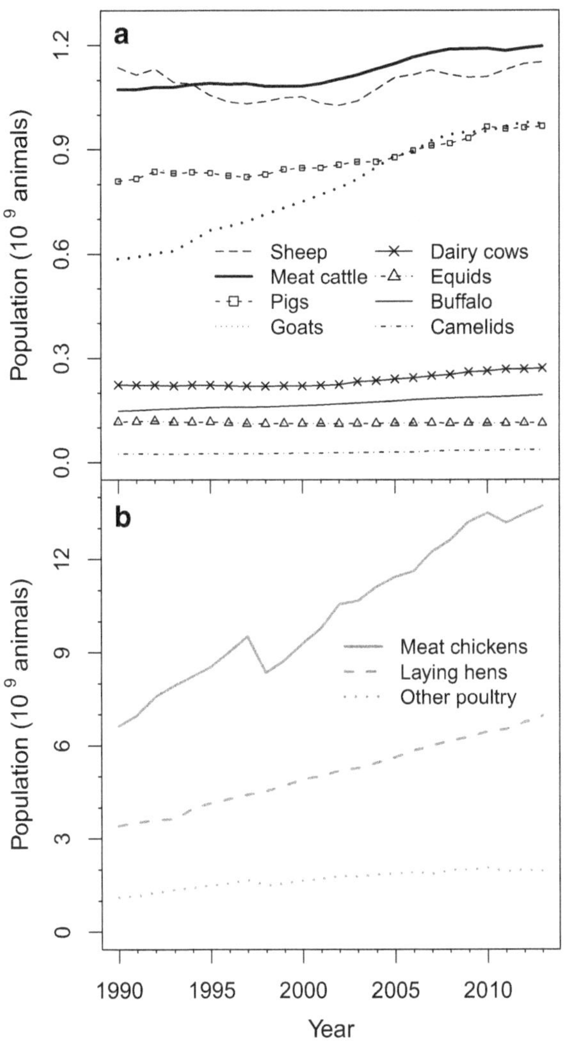


**Fig. 1** Annual global populations of mammalian livestock (**a**) and poultry (**b**), in billions, for 1990–2013 period

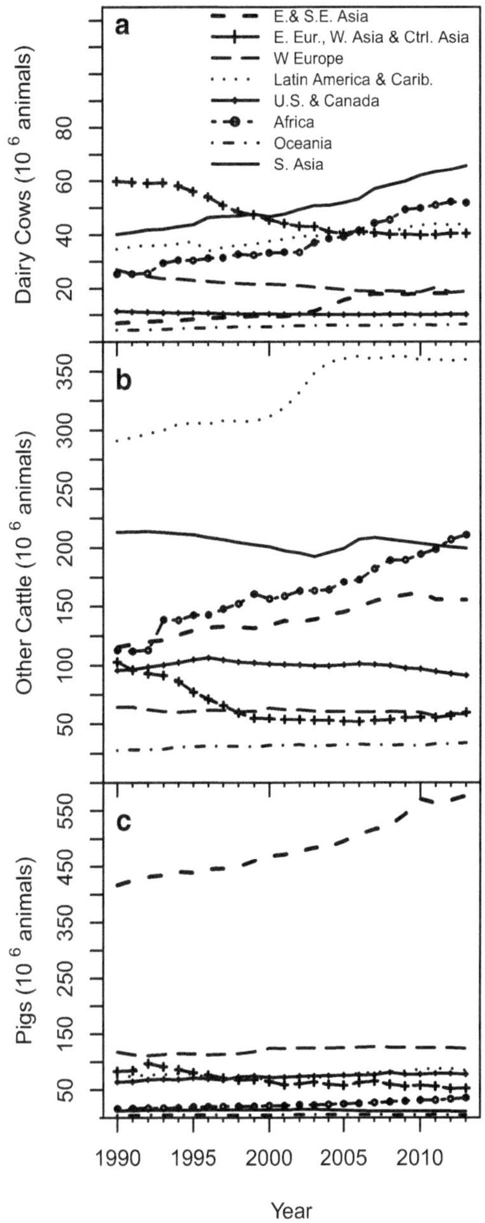


**Fig. 2** Annual regional populations of dairy cattle, meat/other cattle, and swine

emissions factors results in discernably larger emissions relative to calculations made using IPCC 2006 emissions factors. Revised global total $CH_4$ C emission quantities for 2011 are 89.4 ± 13.7 Tg C (119.1 ± 18.2 Tg $CH_4$), an increase of 11% over estimates made using IPCC 2006 emissions factors. This change encompasses an 8.4% increase in enteric fermentation $CH_4$ C and a 36.7% increase in manure management $CH_4$ C (Fig. 4a). In certain regions, these changes are more pronounced, such as in the US and Canada (Fig. 4b), where 2011 total

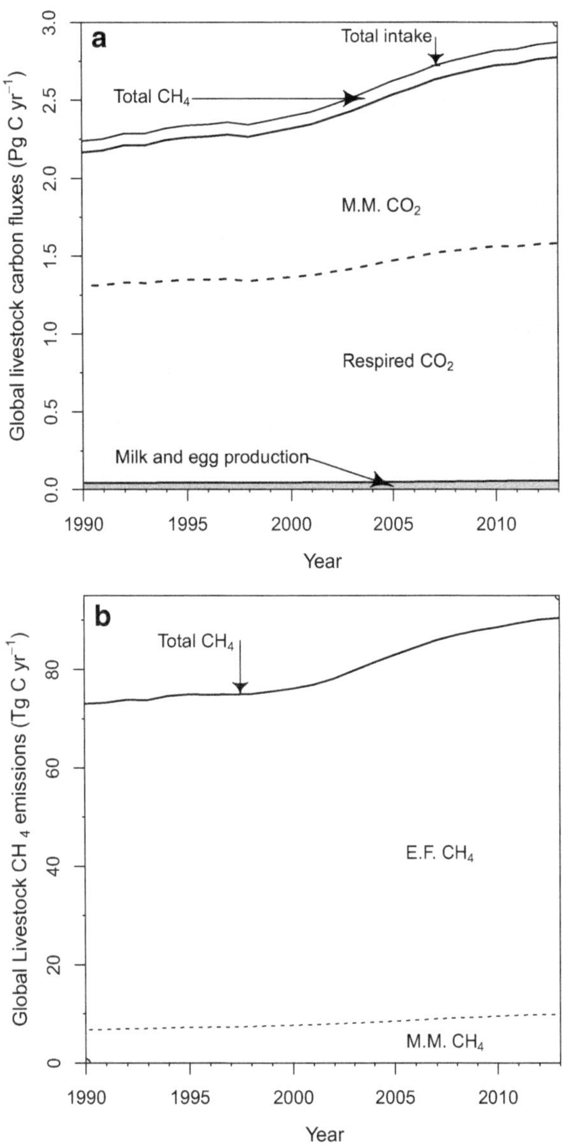


**Fig. 3** Revised annual global livestock carbon fluxes, 1990–2013: carbon contained in all fluxes associated with livestock (**a**), and carbon contained in methane emissions associated with manure management (M.M.) and enteric fermentation (E.F.). Note different units

livestock $CH_4$ emissions were 24.2% greater than when calculated with IPCC 2006 emissions factors, including a 12.3% increase in enteric fermentation $CH_4$ C and a 71.8% increase in manure management $CH_4$ C (Fig. 4b).

Over the 1990–2013 period, total livestock $CH_4$ C emissions exhibit contrasting dynamics among global regions (Fig. 5a) due to trends in livestock populations (Fig. 2) as well as to revision of emission factors (Table 3). The changes in total livestock emissions relative to IPCC 2006 calculations vary by region (Fig. 5b). The largest changes are seen in the US and Canada region, despite declines in dairy (−7.7%) and meat/other cattle (−5.8%) populations there (Figs. 2, 5b).

Livestock C fluxes, including solids (i.e. feed intake and manure production) and gases (i.e. respiration and manure management $CO_2$ C, and enteric fermentation and manure management $CH_4$ C) are downscaled and mapped at 0.05 × 0.05° resolution, in both g C per $m^2$ and Mg C per 0.05° gridcell formats for years 2000–2013. The maps show the interplay between regional livestock characteristics and emissions factors, national, state, or county level cohorts of various livestock species and types, and local densities of livestock. For livestock $CH_4$ C fluxes in 2011 (Fig. 6), the percent change from calculations made using IPCC 2006 emissions factors are also downscaled and mapped (Fig. 7).

#### Revised livestock forage intake and global livestock C budget

We show the revised global livestock C budget for 2011 in Fig. 8, using boxes with areas proportional to the magnitudes of the C flux represented. Our revised data are available through 2013, but because livestock fodder items were not available beyond the year 2011 at the time of data download [36], we estimated 2012 and 2013 fodder quantities based on 2005–2011 average availability. Crop NPP, primary (main crop) harvest C, and residue collected for feed were calculated as in Wolf et al. [2]; reported meat production is converted to Tg C from FAO reports [36] of total global meat production in 2011 (292 Tg of meat entering food supply) multiplied by conversion factors to estimate C content [2]. Estimated milk and egg production are the result of our calculations, based on estimated per-animal production by region. Our global value of 55.7 Tg C is similar to the value of 55.2 Tg C obtained by multiplying FAO reported global production [36] (743 Tg of milk, 71 Tg of eggs), and conversion factors for milk and egg C content in 2011 [2]. In comparison to primary crop harvest, crop residue harvest, and the quantity of livestock-based food produced, the magnitude of livestock fodder and forage consumption is apparent. Emissions of $CO_2$ associated with livestock respiration and manure management are also shown, which are calculated by subtraction of all other fluxes from total intake or total manure production at the per-animal level, assuming static standing live populations with no net change in biomass across years.

Note that the 'market feed' category includes primary crop products as well as crop by-products that are unsuited or undesirable for human consumption, such as distillers grains (a by-product of bioethanol production) and various oil-crop extraction by-products (e.g. oil seed meal or cake). The C contained in and used for production of biofuels is included in the harvest/other uses box, but

**Table 4 Livestock C fluxes by region for year 2011**

| 2011 livestock C fluxes ± 1 SE | Region | | | | | | | | |
|---|---|---|---|---|---|---|---|---|---|
| | Africa | E.-S.E. Asia | E. Europe, W. Asia, and Central Asia | Latin America | Oceania | S. Asia | US and Canada | W. Europe | Globe |
| Intake C (Tg C) | 415 ± 52.2 | 610.7 ± 67.8 | 261 ± 31.8 | 559.5 ± 80.4 | 76.6 ± 10.5 | 451.2 ± 66.7 | 256.1 ± 34 | 195.7 ± 26.3 | 2825.7 ± 369.8 |
| Manure production C (Tg C) | 176.4 ± 30.8 | 219.8 ± 35.4 | 111.4 ± 19.2 | 259.1 ± 49.5 | 34.7 ± 6.2 | 197.4 ± 37.8 | 99.8 ± 18.4 | 82.5 ± 15 | 1181.1 ± 212.4 |
| Enteric Fermentation $CH_4$ C (Tg C) | 12.69 ± 1.94 | 10.5 ± 1.61 | 6.16 ± 0.94 | 20.26 ± 3.1 | 2.97 ± 0.45 | 16.89 ± 2.58 | 5.64 ± 0.86 | 4.56 ± 0.7 | 79.67 ± 12.19 |
| Manure Management $CH_4$ C (Tg C) | 0.45 ± 0.07 | 1.53 ± 0.23 | 0.73 ± 0.11 | 1.63 ± 0.25 | 0.21 ± 0.03 | 1.39 ± 0.21 | 2.16 ± 0.33 | 1.58 ± 0.24 | 9.68 ± 1.48 |
| Total $CH_4$ C | 13.14 ± 2.01 | 12.03 ± 1.84 | 6.89 ± 1.05 | 21.89 ± 3.35 | 3.18 ± 0.49 | 18.29 ± 2.8 | 7.79 ± 1.19 | 6.14 ± 0.94 | 89.35 ± 13.67 |
| Milk and egg production C (Tg C) | 3.41 ± 0.84 | 8.34 ± 1.95 | 9.52 ± 2.35 | 4.75 ± 1.15 | 2.32 ± 0.58 | 9.9 ± 2.46 | 7.27 ± 1.79 | 10.21 ± 2.53 | 55.73 ± 13.65 |
| Respiration $CO_2$ C (Tg C) | 222.5 ± 87.7 | 372 ± 108.3 | 133.9 ± 55.4 | 275.4 ± 137.1 | 36.6 ± 18.4 | 227 ± 112 | 143.3 ± 56.3 | 98.5 ± 45.4 | 1509.2 ± 620.6 |
| Manure management $CO_2$ C (Tg C) | 176 ± 31 | 218.3 ± 35.8 | 110.7 ± 19.5 | 257.4 ± 50 | 34.5 ± 6.3 | 196 ± 38.2 | 97.7 ± 19 | 80.9 ± 15.5 | 1171.4 ± 215.3 |
| Available fodder (Tg C) | 83 | 260.5 | 212.7 | 157.3 | 8.6 | 231.6 | 136.6 | 156.9 | 1247.2 |
| Unused/waste fodder[a] | 4.9 | 8.6 | 38.4 | 0 | 0 | 0.5 | 0 | 19.4 | 71.8 |
| Fodder intake[a] (Tg C) | 78.1 | 251.9 | 174.2 | 157.3 | 8.6 | 231.1 | 136.6 | 137.6 | 1175.4 |
| Forage intake[a] (Tg C) | 336.9 | 358.7 | 86.8 | 402.1 | 68 | 220.1 | 119.5 | 58.2 | 1650.3 |
| % of intake from forage | 81.2 | 58.7 | 33.2 | 71.9 | 88.8 | 48.8 | 46.7 | 29.7 | 58.4 |

[a] Unused/waste fodder occurs when the amount of available fodder C is greater than livestock feed requirements per nation in a given year. In 2011, this occurred in: the Czech Republic, Denmark, Egypt, Germany, Hungary, Japan, Kazakhstan, Malaysia, the Russian Federation, Slovakia, Sri Lanka, and the Ukraine, and could be due to waste, stockpiling, misreporting, or other errors. Actual fodder losses per nation were not estimated

C contained in biofuel by-products sold and consumed as feed are pushed back into the primary harvest for feed/market feeds box; this results in a smaller total amount of C devoted to biofuels than in calculations that do not account for use of by-products in livestock feeds [2].

For most livestock types and regions, default livestock body weights and total feed requirements increased in our revision (Tables 1, 2). Reported amounts of annual available fodder, however, did not change [36]. In our accounting, the gap between total feed requirements and available fodder in each nation, if any, is filled by forage intake (i.e. grazing). Therefore, our revision of total livestock feed requirements also necessitated revision of livestock forage intake and the percentage of total livestock C intake supplied by forage. The revised percent of global livestock intake supplied by forage was 58.4% in 2011, reflecting 1.65 Pg C of forage intake from global rangelands (Table 4; Fig. 9b). These percentages are similar to estimates reported by other researchers; Bouwman et al. [42] estimated 59.2% of total livestock intake from forage in 1990 (our value is 56.6% for that year), and Krausman et al. [40] estimated 54.5% in 2000 (our value is 58.6%). When feed intake requirements were calculated using IPCC 2006 or IPCC 1996 [2, 3] livestock total intake values, the estimates for 2011 were 55.2 and 52.4%, respectively.

Although the percentage of forage intake increased based on our revision, the global average percentage did not change greatly over the 1990–2011 time period, ranging between 56.6 and 60.7% (Fig. 9b). This suggests that, at the global level, amounts of forage and fodder intake

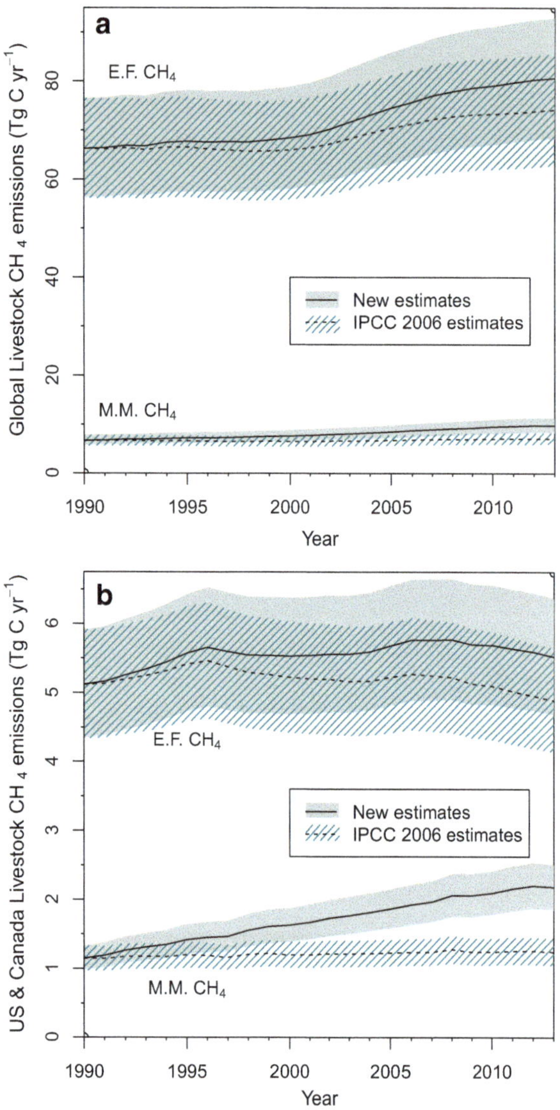


**Fig. 4** Comparison of enteric fermentation (E.F.) and manure management (M.M.) methane emissions estimated using IPCC 2006 and revised emissions factors resulting from this study, for the globe (**a**), and the US and Canada region (**b**)

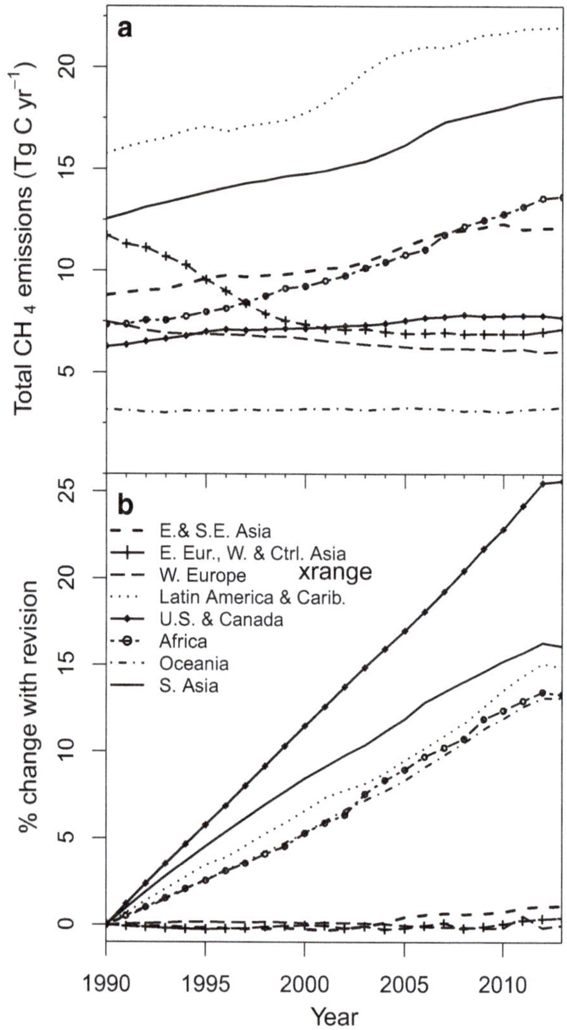


**Fig. 5** Revised total livestock methane emissions by region (**a**) and percent change in annual emissions relative to calculations made based on IPCC 2006 emissions factors

C have increased apace over this time period to meet increasing total livestock intake requirements. This is also true in most regions during this time period, except for E. Europe and West and Central Asia after the breakup of the Soviet Union. However, in the US and Canada, the percentage of intake from forage increased sharply in 1995 and again after 2009, with the 2011 value (45.1%) doubled from 2005 (22.4%). Additional data from USDA (annual grain quantities fed, hay harvests, and by-product feed quantities excluding distiller's grains) [43] and from the Renewable Fuels Association (annual quantities of distiller's grains by-products from bioethanol production in the US, decreased by the estimated one-third that is exported annually) [44] were converted to units of C [2] to provide approximate annual amounts of available fodder in the US (Fig. 10). These data support the observed increases in percent livestock intake from forage in those years. The spike in and after 1995 can be attributed to drought in the Midwest US and other factors [45]. Uncertain harvests in the US and in E. Europe, along with the increasing use of corn for bioethanol production, may be the causes of the sharp increase after 2009 [46]. Corn prices, which averaged $2.75 per bushel in the 2000s, jumped to an average of $6.10 in 2010–2013 [43]. The jump in corn prices could have driven farmers to delay moving cattle from pasture to feedlots, without deterring the subsidized and mandated production of bioethanol

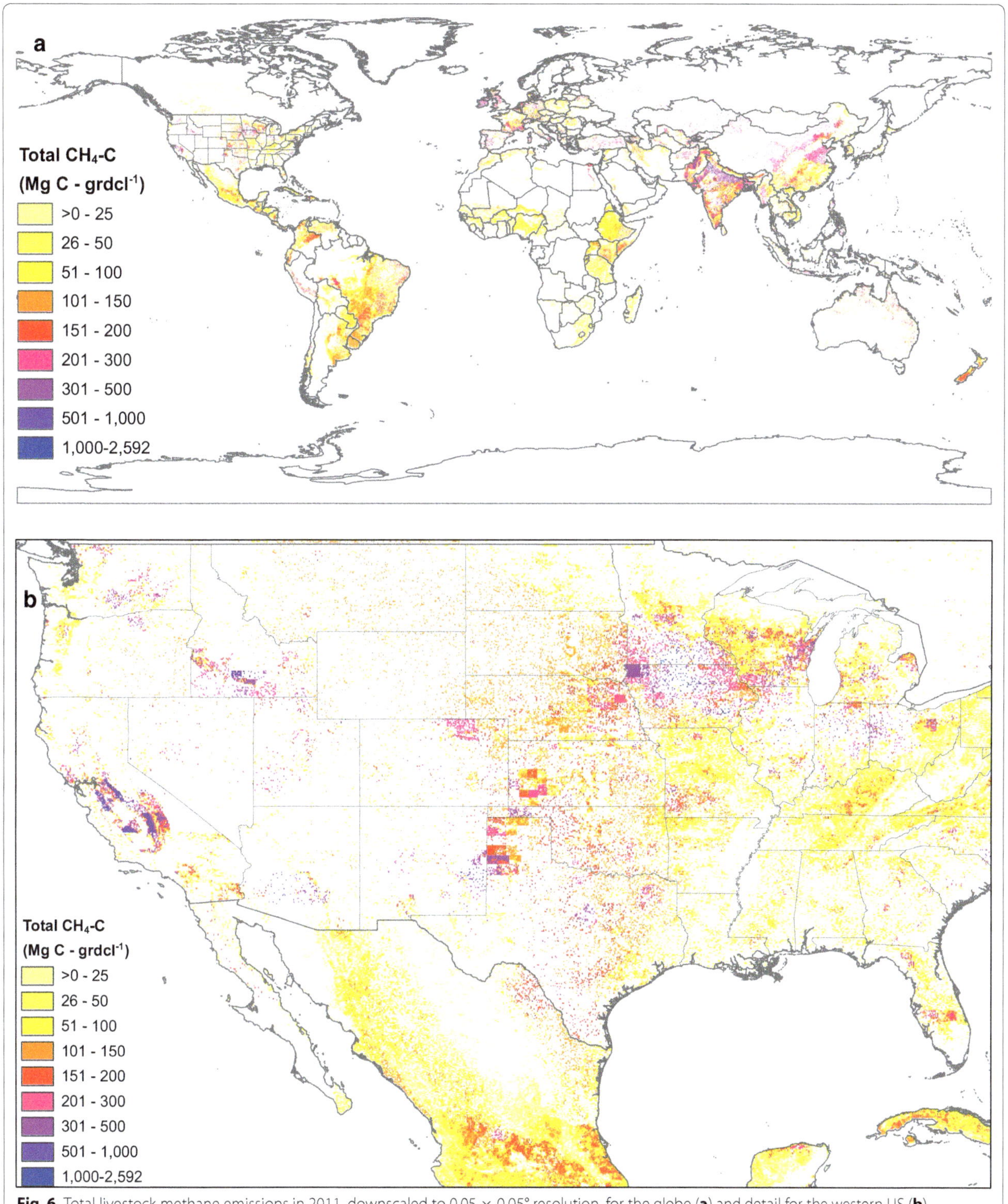


**Fig. 6** Total livestock methane emissions in 2011, downscaled to 0.05 × 0.05° resolution, for the globe (**a**) and detail for the western US (**b**)

in the US. In addition to the impacts of corn prices, the by-products of corn bioethanol production (i.e. distiller's grains) are used as a high energy, high protein livestock feed supplement, which affects other components of livestock feed and forage intake [47] and potentially $CH_4$ emissions from livestock consuming them [48].

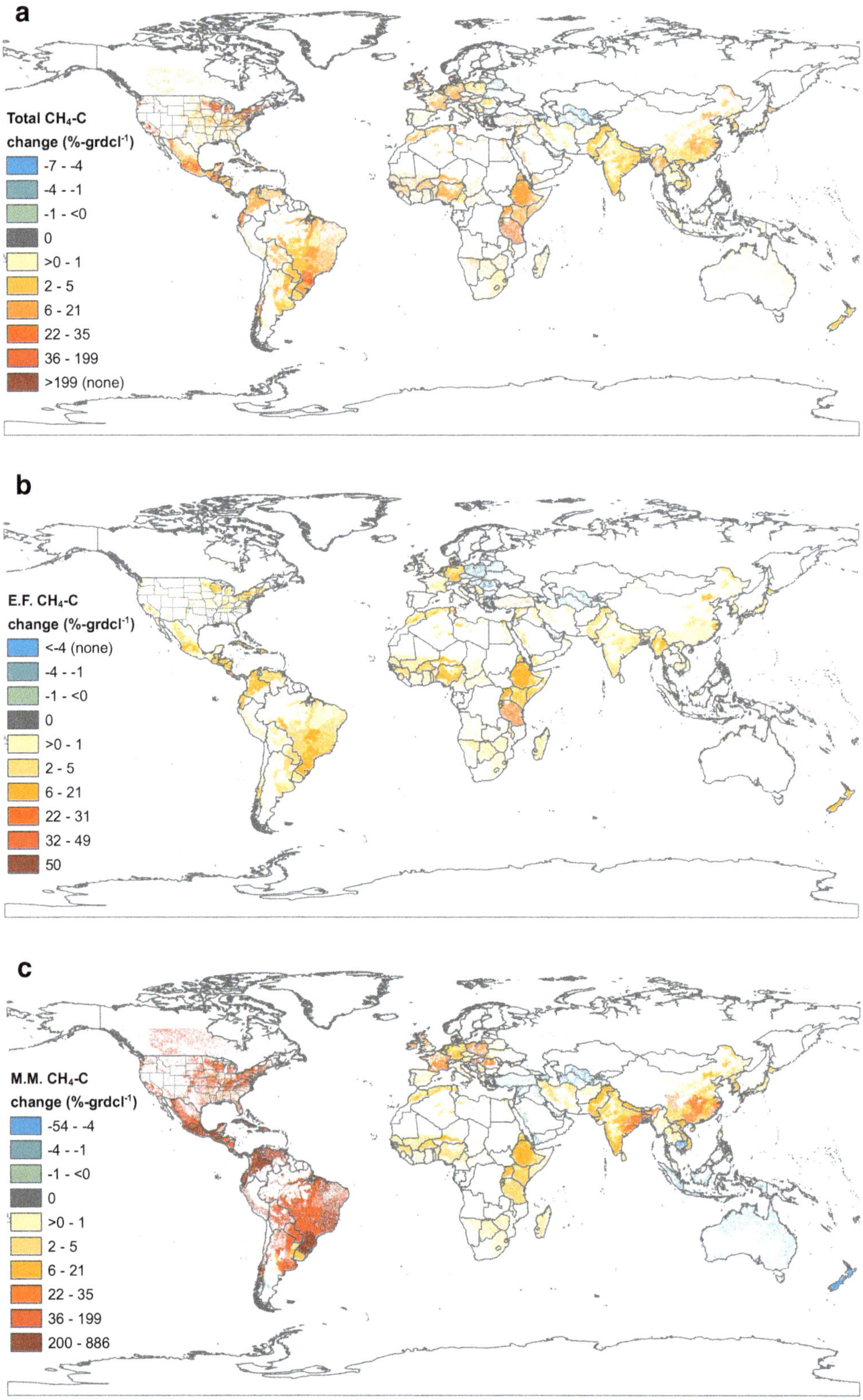


**Fig. 7** Percent change in global livestock methane emissions with revision, downscaled to 0.05 × 0.05° resolution, for the total (**a**), enteric fermentation (E.F.) (**b**), and manure management methane (M.M.) emissions (**c**), in 2011

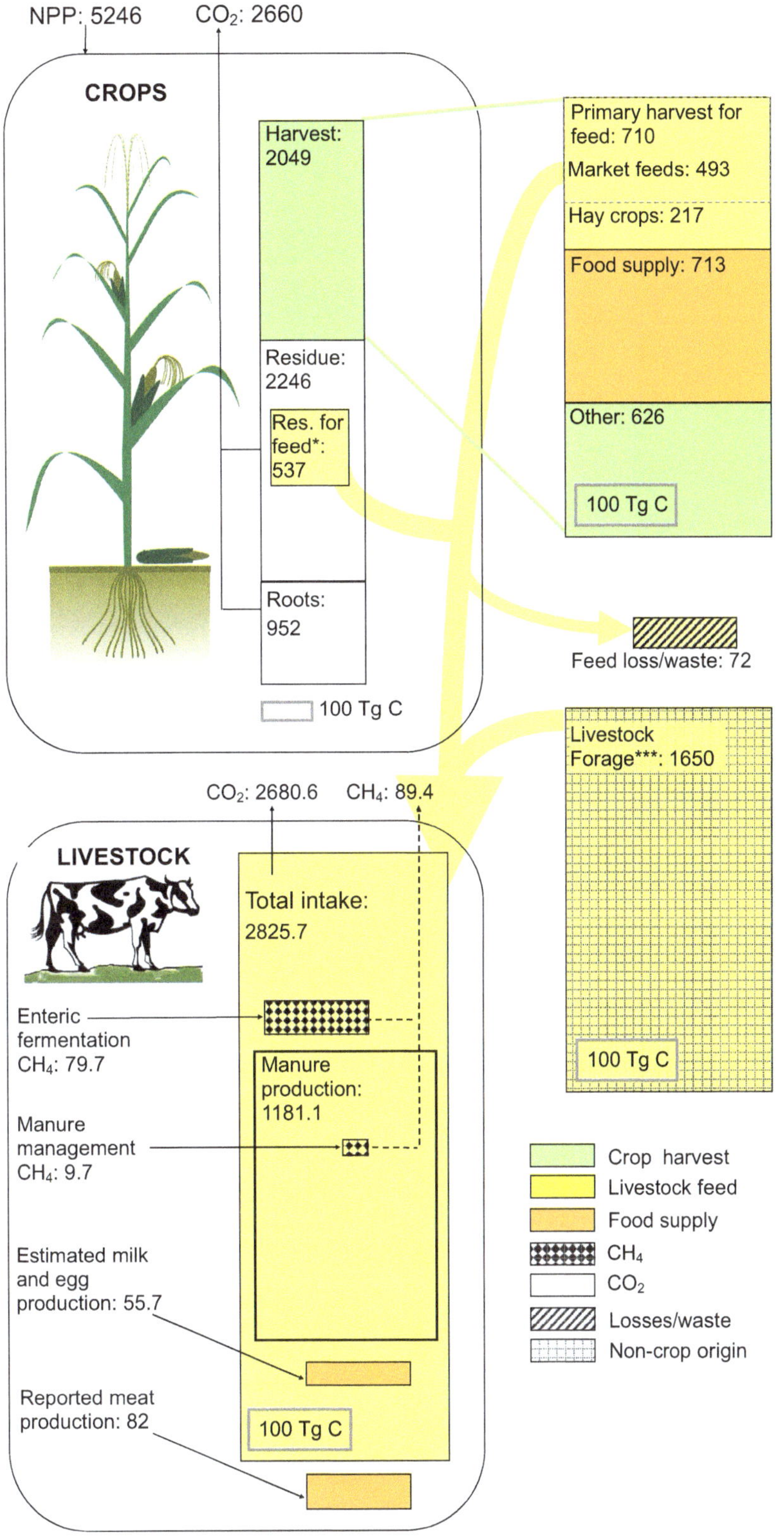


**Fig. 8** Revised livestock C budget for 2011. All non-harvested crop biomass C, and all manure C not emitted as $CH_4$, are assumed to be decomposed and respired as $CO_2$ by decomposing organisms within the same year as production

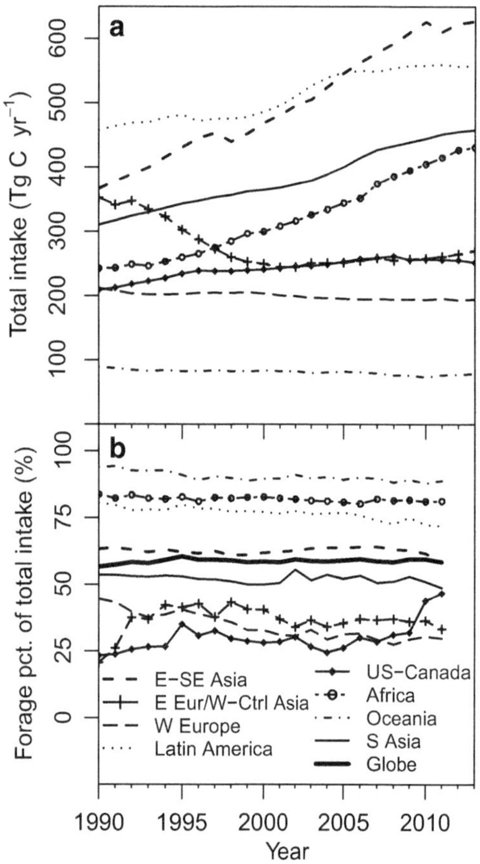


**Fig. 9** Revised total livestock feed intake carbon (**a**) and percent supplied by forage (**b**) by region

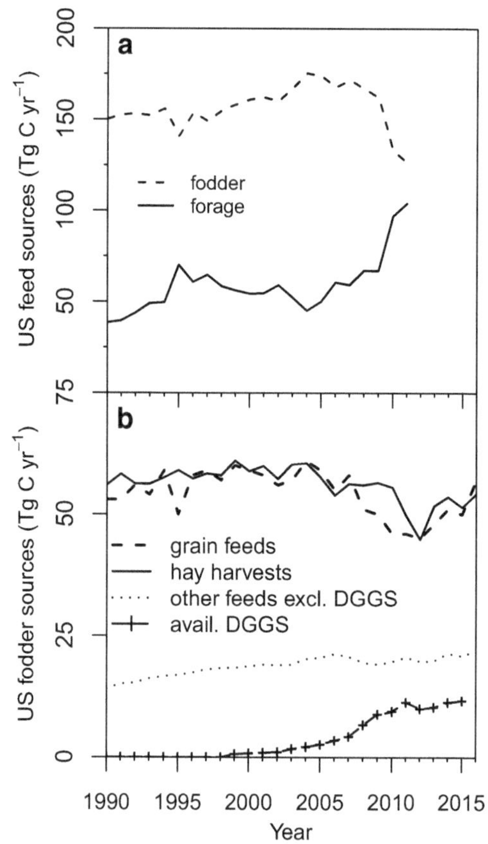


**Fig. 10** Detail of US livestock intake of fodder and forage (**a**) and US fodder sources (**b**)

## Discussion

### Evaluation of revised livestock CH4 emissions estimates

Compared to other bottom-up estimates for recent years (Table 5), our revised emissions factors yield annual $CH_4$ C emission estimates that are: 11% larger than global estimates made using IPCC 2006 emissions factors; 15% larger than EPA global estimates but similar or slightly smaller than EPA US estimates; and 4% larger than EDGAR global estimates, 3% larger than EDGAR US estimates, but 54% larger than EDGAR estimates for the state of California. Our global estimates are slightly larger than those published for the 2000s by Tian et al. [49] based on a suite of bottom-up estimates, but have larger uncertainties. EDGAR uses IPCC 2006 Tier 2 calculations but modifies cattle emissions factors based on body weight or milk productivity; such modifications would not capture the effects of recent changes in manure management systems and other factors. EPA, in contrast, uses models with annually modified inputs for the US [50], but uses 2006 IPCC coefficients for its global estimates [16]. Our US emissions estimates are not significantly different from those made by EPA. This is not unexpected, as we use similar estimates for enteric fermentation emissions in US meat/other cattle, and rely on information from EPA to derive the new emissions factors for other livestock categories.

Our estimate of global livestock $CH_4$ C emissions is similar to top-down estimates made using atmospheric inversion methods [20] (Table 5). Our estimates for the US, however, are smaller than recent top-down estimates by 21–51% [20], 46% [23], or 30% [21]. For the state of California only, our total-livestock estimate is 17% smaller than top-down [22] for 2010; for 2013, our estimate for non-dairy livestock was smaller but comparable, while our dairy cattle estimate was 37–64% smaller, than top-down [51] (Table 5). The differences over the entire US may be due in part to the difficulty in separating livestock $CH_4$ emissions from other sources for the entire country in top-down studies [20]. US emissions could indeed be larger than our estimates, as suggested by these top-down studies; however, further investigation of this possibility will require more quantitative research on recent per animal emissions, particularly from

**Table 5 Comparison of livestock $CH_4$ C emissions reported in literature to revised values obtained in this study**

| Area | Years | Methane quantity | Source/emissions factors used | Method | Value (Tg C $year^{-1}$) | Revised value (Tg C $year^{-1}$) |
|---|---|---|---|---|---|---|
| Globe | 2013 | Total lvstk. $CH_4$ C | IPCC 2006[a] | Bottom-up | 81.3 ± 12.4 | 90.4 ± 13.8 |
| | | | IPCC 2006 [36] | | 81.0 | |
| | | E.F. $CH_4$ C | IPCC 2006[a] | Bottom-up | 74.2 ± 11.4 | 80.6 ± 12.3 |
| | | | IPCC 2006 [36] | | 73.8 | |
| | | M.M. $CH_4$ C | IPCC 2006[a] | Bottom-up | 7.14 ± 1.09 | 9.79 ± 1.5 |
| | | | IPCC 2006 [36] | | 7.24 | |
| Globe | 2012 | Total lvstk. $CH_4$ C | IPCC 2006[a] | Bottom-up | 80.9 ± 12.4 | 90.08 ± 13.78 |
| | | | IPCC 2006 [36] | | 80.6 | |
| | | E.F. $CH_4$ C | IPCC 2006[a] | Bottom-up | 73.8 ± 11.3 | 80.3 ± 12.3 |
| | | | IPCC 2006 [36] | | 73.4 | |
| | | M.M. $CH_4$ C | IPCC 2006[a] | Bottom-up | 7.10 ± 1.1 | 9.81 ± 1.5 |
| | | | IPCC 2006 [36] | | 7.19 | |
| Globe | 2009–2011 | Total lvstk. $CH_4$ C | [20] | Top-down | 88.94 | 87.88 ± 13.44 (2009)<br>89.35 ± 13.67 (2011) |
| Globe | 2010 | E.F. $CH_4$ C | [88] | Bottom-up | 69.0 | 79.0 ± 12.1 |
| | | M.M. $CH_4$ C | | | 8.19 | 9.51 ± 1.5 |
| Globe | 2000s (average) | E.F. $CH_4$ C | [49] | Bottom-up | 70.0 ± 3.3 | 72.0 ± 11.0 |
| | | M.M. $CH_4$ C | | | 8.0 ± 0.3 | 8.4 ± 1.29 |
| US | 2012 | E.F. $CH_4$ C | IPCC 2006[a] | Bottom-up | 4.38 ± 0.67 | 4.95 ± 0.76 |
| | | | IPCC 2006 [36] | | 4.4 | |
| | | | [89] | | 5.0 | |
| | | M.M. $CH_4$ C | IPCC 2006[a] | Bottom-up | 1.09 ± 0.17 | 1.93 ± 0.3 |
| | | | IPCC 2006 [36] | | 1.02 | |
| | | | [89] | | 1.91 | |
| US | 2009–2011 | Total lvstk. $CH_4$ C | [90] | Top-down | 8.75–14.09 | 6.83 ± 1.05 (2009)<br>6.90 ± 1.06 (2011) |
| | | | [91] | Bottom-up | 6.82 (2010); 6.77 (2011) | |
| | | | [12] | | 7.02 (2010); 6.97 (2011) | |
| | | | [89] | | 6.97 (2010); 6.91 (2011) | |
| California | 2013–2014 | Dairy cattle $CH_4$ C | IPCC 2006[a] | Bottom-up | 0.230 ± 0.035 | 0.382 ± 0.058 (2013) |
| | | | [51] | Top-down | 0.603–1.06 | |
| | | All non-dairy lvstk. $CH_4$ C | IPCC 2006[a] | Bottom-up | 0.15 ± .022 | 0.168 ± 0.026 (2013) |
| | | | [51] | Top-down | 0.149–0.259 | |
| California | 2010 | Total lvstk. $CH_4$ C | IPCC 2006[a] | Bottom-up | 0.38 | 0.54 ± 0.08 |
| | | | [5, as analyzed in 22] | | 0.35 | |
| | | | [22] | Top-down | 0.65 | |
| US | 2008 | Total lvstk. $CH_4$ C | [5, as analyzed in 23] | Bottom-up | 6.7 | 6.88 ± 1.05 |
| | | | [12] | | 7.09 | |
| | | | [23] | Top-down | 12.7 ± 0.5 | |
| Globe | 2008 | E.F. $CH_4$ C | [Sum over all gridcells from 5] | Bottom-up | 75.12 | 77.82 ± 11.91 |
| | | M.M. $CH_4$ C | | | 8.65 | 9.14 ± 1.4 |
| US | 2004 | Total lvstk. $CH_4$ C | IPCC 2006[a] | Bottom-up | 5.5 | 6.42 ± 0.98 |
| | | | [92] | | 5.8 | |
| | | | [21] | Top-down | 9.15 ± 0.98 | |

*E.F.* enteric fermentation, *M.M.* manure management

[a] Our calculations, using IPCC 2006 [4] Tier 1 regional emissions factors

increasingly used anaerobic manure treatment lagoons, such as recent studies on dairy cattle emissions by Owen and Silver [52, 53].

Our estimates for the state of California result in livestock emissions of 0.540 g $CH_4$ C in 2010 and 0.550 Tg $CH_4$ C in 2013, of which 0.165 and 0.177 Tg $CH_4$ C are due to dairy cattle manure management, respectively. California, however, utilizes anaerobic lagoon manure management systems at a higher rate than the US national average (59% of manure is managed in anaerobic lagoons in California, compared to 34% for the US, based on state population-weighted values [50]). If we calculate emissions using the California manure management utilization rates in place of national average rates, California dairy cattle manure management emissions in 2010 and 2013 increase to 0.263 and 0.274 Tg $CH_4$ C, bringing total California livestock emissions up to 0.638 and 0.647 Tg $CH_4$ C for 2010 and 2013, respectively. These totals approach Wecht et al.'s value of 0.65 Tg $CH_4$ C for 2010 [22], but are well below the range of 0.752–1.32 Tg $CH_4$ C presented by Jeong et al. for 2013–2014 [51]. If we also employ MCF values from Owen and Silver's recent field observations of anaerobic lagoon manure management systems [52], California dairy cattle manure management emissions in 2010 and 2013 increase to 0.306 and 0.318 Tg $CH_4$ C, respectively, bringing total emissions increase to 0.681 and 0.691 Tg $CH_4$ C in those years. These results show that our emissions estimates, if modified to reflect local conditions, are similar to or smaller than recent top-down estimates in California, where livestock and fossil fuel-sector $CH_4$ emissions are spatially well separated. The discrepancies between top-down and bottom-up estimates may arise from factors influencing either or both of the methods. Our estimates could be too low for several reasons, including underreported usage of anaerobic manure treatment lagoons, recent increases in local temperatures impacting emissions, and/or MCF values that are too low. Because our emissions factors were calculated at the regional level, it will remain important to modify them when characterizing localized emissions; this can be done by using the equations published by the IPCC [4] with the inputs provided here in Tables 1 and 2, modified by relevant localized information such as manure management system utilization rates.

#### Role of livestock $CH_4$ in global atmospheric $CH_4$ dynamics

In the early 2000s, annual increases in atmospheric $CH_4$ concentrations temporarily flattened [24, 54]. After 2006, however, atmospheric $CH_4$ concentration abruptly began to rise each year, and at the same time, its $^{13}C$ isotopic signature began to grow more negative [26, 27]. Several possible explanations are offered for the causes and geographical distribution of this renewal in growth. Bergamaschi et al. [55] find that annual $CH_4$ emissions (from all sources) in 2007–2010 were 16–20 Tg larger than emissions in 2003–2005 period, with the increase mostly in the Northern and Southern tropics and Northern mid-latitudes, and Nisbet et al. [25] indicate that global $CH_4$ emissions (from all sources) were 15–22 Tg larger in 2010 than in 2005. Schaefer et al. [27] suggest that increases after 2007 are most likely from agricultural sources in the Northern hemisphere tropics and subtropics. In contrast, Nisbet et al. [26] suggest that these increases originate in the Southern hemisphere and Northern and Southern tropics, and are more likely due to wetland responses to meteorological conditions than agriculture, because of the abrupt step-change after 2006.

Our global estimates for annual livestock $CH_4$ emissions are 118.0 Tg $CH_4$ (88.5 Tg $CH_4$ C) in 2010, 11.7 Tg $CH_4$ greater than 2003 emissions of 106.3 Tg $CH_4$ (79.7 Tg $CH_4$ C). These quantities represent ca. one-fifth of total global methane emissions of 540–568 Tg $CH_4$ year$^{-1}$ estimated for this time period by a suite of top-down inversions [14]. The 11.7 Tg $CH_4$ year$^{-1}$ increase in annual livestock emissions reported here accounts for ca. one half to three-fourths of the increases over this time period reported by Bergamaschi et al. [55] and Nisbet et al. [25]. These proportions support the idea that livestock $CH_4$ emissions, while not the dominant overall source of global $CH_4$ emissions, may be a major contributor to the observed increases in emissions in the 2000s to 2010s. As suggested by Saunois et al. [28], the importance of agricultural emissions in the global $CH_4$ budget is highlighted by our results, which provide quantitative estimates with associated uncertainties. It is important to note, however, that our results cannot reveal any sharp changes from year-to-year, because we have imposed a linear transition from IPCC-based to revised coefficients over the years from 1990 to 2012; therefore, a larger magnitude of change over this time period is possible. In summing the changes in annual livestock $CH_4$ emissions over time by latitude (Fig. 11; Table 6), we find that the largest increases are between 30N and the equator (Northern tropics), potentially lending support to the conclusions of Schaefer et al. [27]. In the northern and southern tropics (30N to equator and equator to 30S), our results are comparable to the results of multiple models reported by Bergamaschi et al. [Table 3 in 55]; whereas in higher latitudinal zones, the changes over time that we document are the same in sign but smaller in magnitude than the output of most of the inversions reported by those authors. These longitudinal patterns may improve future discernment of $CH_4$ sources and dynamics over time.

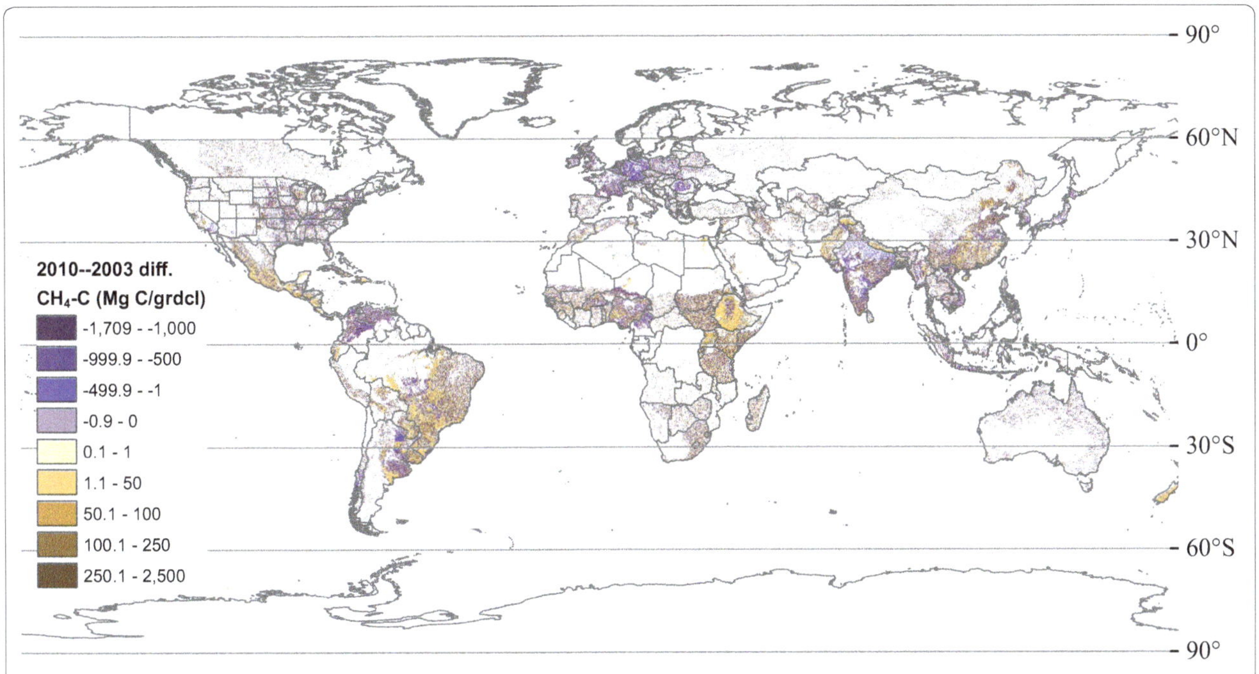


**Fig. 11** Change in total livestock $CH_4$ emissions between 2003 and 2010 (2010–2003 emissions), per 0.05 × 0.05° gridcell

**Table 6 Temporal changes in annual livestock $CH_4$ emissions by latitudinal zone**

| Time period | 90N–60N | 60N–30N | 30N-equator | Equator-30S | 30S–60S | 60S–90S |
|---|---|---|---|---|---|---|
| Tg $CH_4$ C year$^{-1}$ | | | | | | |
| 2003–2010 | −0.08 ± 0.15 | 1.7 ± 8.96 | 5.18 ± 9.87 | 2.01 ± 5.26 | −0.16 ± 1.37 | 0 |
| 2000–2013 | −0.12 ± 0.16 | 2.1 ± 8.97 | 5.59 ± 9.39 | 4.23 ± 5.19 | 0.28 ± 1.34 | 0 |
| Tg $CH_4$ year$^{-1}$ | | | | | | |
| 2003–2010 | −0.11 ± 0.21 | 2.27 ± 11.95 | 6.91 ± 13.17 | 2.69 ± 7.02 | −0.21 ± 1.83 | 0 |
| 2000–2013 | −0.16 ± 0.21 | 2.8 ± 11.97 | 7.45 ± 12.51 | 5.64 ± 6.92 | 0.38 ± 1.78 | 0 |

Tg $CH_4$ C and Tg $CH_4$ both shown to facilitate comparisons to other studies

### Limitations of revised emissions factors

The revised per-animal emissions factors and/or total $CH_4$ emissions reported here may differ from recent national self-reported emissions factors. This can be due to several factors, including (1) the inclusion of dairy calves and heifers with mature dairy cow populations, despite large differences in emissions between those groups, which can lead to low emissions factors; (2) interannual and sub-regional variation in diet and other factors. The revised emissions factors were developed for global analyses based on recent information, and the switch from IPCC 2006 was made linearly over a long time period (1990–2012) because information about their temporal dynamics was lacking. Therefore, variability at subregional and interannual scales are embedded in our estimates, and the revised emissions factors may not provide the best representation of emissions at local scales and/or for earlier years during the transition.

Emissions factors for poultry manure management $CH_4$ were not revised in this study, but they should be reevaluated in future work. In the IPCC 2006 guidelines, poultry emissions factors for manure management are small (i.e. <0.10 kg $CH_4$ per bird per year, except for laying hens with manure managed in lagoons). However, global poultry populations are large and continue to increase (Fig. 1b). Poultry manure management $CH_4$ emissions are likely to be larger than estimated by IPCC 2006 guidelines for similar reasons as in cattle and swine: breeding has increased body sizes and growth rates [56] and utilization of liquid manure management systems is increasing [57].

The region of E. and SE Asia has proportionally less cattle and more swine than other regions of the world, making manure management of $CH_4$ emissions much more prominent there (because swine do not have appreciable enteric fermentation emissions). The quantities of $CH_4$ emitted from swine manure in this region, however, do have greater uncertainty. While the bulk of cattle manure in this region is collected and applied to cropland, particularly where cash crops are grown [9, 58–60], some of the manure produced by China's large swine population (ca. half of the world's total) is discharged to surface waters [61–64], and $CH_4$ emissions have not been characterized for this situation. In this study, we considered emissions from manure discharged to surface waters to be similar to emissions from 'liquid and deep pit' treatments. If emission quantities from manure discharged to surface waters are more similar to emissions in anaerobic lagoons (i.e. liquid storage systems to combine waste stabilization and storage, in which solids are not removed more frequently than 15–20 years [65]), then the swine manure emissions factor may be as much as three times higher than the value used here.

In the Latin America region, meat cattle populations and their management practices are changing rapidly. In this region, our estimates of cattle on feedlots and other rapidly changing attributes may already be outdated [66], and frequent reassessment of this region will be warranted.

### US livestock $CH_4$ emissions in recent decades

Total $CH_4$ emissions for the US and Canada show a slight but steady increase over recent decades despite decreasing populations of dairy cows and other cattle. This contrasts with W. Europe, where total emissions trajectories decline slightly in parallel with declines in livestock populations. The means exist to further reduce $CH_4$ emissions, and they are available in the US and Canada (e.g. covered manure storage). The centralization of manure management in this region increases profitability but is also associated with increasing $CH_4$ emissions, decreased potential for cropland application of manure, and other threats to common resources and public health [7, 61, 67–69]. Studies examining the overall tradeoffs associated with increasing centralization vs. decentralization (e.g. potential C sequestration from manure application to cropland soils; NOx and $CH_4$ emissions; costs of transportation for livestock, milk and meat, and manure; air and water quality; and impacts on rural communities such as odors and health risks) are needed. For example, if the efficiency of centralization outweighs other negative impacts, then capping lagoons to capture $CH_4$ should be considered.

## Conclusions

In this study, we found the revised bottom up estimates of C fluxes and stocks from agricultural systems to be higher than those based on IPCC 2006 guidelines. The estimated global livestock $CH_4$ emissions were $119.1 \pm 18.2$ Tg $CH_4$ in year 2011; this quantity is 11% greater than that obtained using the IPCC 2006 emissions factors, encompassing an 8.4% increase in enteric fermentation $CH_4$ and a 36.7% increase in manure management $CH_4$, with notable variability among regions and sources. Likewise, the revised manure management $CH_4$ emissions for year 2011 in the US were 71.8% higher than IPCC-based estimates, consistent with recently reported top-down estimates. Summing changes in annual livestock $CH_4$ emissions geographically, by latitude and over time, we found that the largest increases over time were between 30N and the equator (i.e. Northern tropics). Our results suggest that livestock $CH_4$ emissions, while not the dominant overall source of global $CH_4$ emissions, may be a major contributor to the recent increases in global $CH_4$ emissions. The new regional and global C fluxes and stocks estimates improve the ability to reconcile top-down and bottom-up estimates of $CH_4$ production, and provide consistent estimates of $CH_4$ emissions at the national, regional and global level, for use in development and evaluation of Earth system models and environmental assessments. The results reported here are useful to scientists and policy decision makers, given the importance of agricultural systems for food, fiber and bioenergy production, and their contributions to global methane emissions.

#### Authors' contributions

The study was conceived by TOW. The study was designed by all three authors. Data acquisition, analysis, interpretation, and drafting of the initial manuscript were conducted by JW, with critical revision of the manuscript, expert interpretation, and intellectual content provided by GRA and TOW. All authors read and approved the final manuscript.

#### Author details

[1] USDA-ARS, Adaptive Cropping Systems Laboratory, 10300 Baltimore Ave., Building 001, Room. 342, BARC-WEST, Beltsville, MD 20705, USA. [2] Joint Global Change Research Institute, 5825 University Research Court, Suite 3500, College Park, MD 20740, USA. [3] US Department of Energy, SC-23, 1000 Independence Ave., Washington, DC 20585, USA.

#### Competing interests

The authors declare that they have no competing interests.

#### Funding

This research was conducted with support from the National Aeronautics and Space Administration (NASA), Carbon Monitoring System (CMS) program under Project #NNH13AW58I.

**References**


1. Herrero M, Havlík P, Valin H, Notenbaert A, Rufino MC, Thornton PK, et al. Biomass use, production, feed efficiencies, and greenhouse gas emissions from global livestock systems. Proc Natl Acad Sci USA. 2013;110:20888–93.
2. Wolf J, West TO, Le Page YL, Kyle GP, Zhang X, Collatz GJ, et al. Biogenic carbon fluxes from global agricultural production and consumption. Glob Biogeochem Cycles. 2015;29:1617–39.
3. Revised IPCC. IPCC guidelines for national greenhouse gas inventories, reference manual, vol. 3. Bracknell: Hadley Centre; 1996. p. 1996.
4. IPCC. 2006 IPCC guidelines for national greenhouse gas inventories, volume 4: agriculture, forestry and other land use. Kanagawa: Institute for Global Environmental Strategies; 2006. http://www.ipcc-nggip.iges.or.jp/public/2006gl/vol4.html.
5. European Comission. Emission database for global atmospheric research (EDGAR), release version 4.2; 2010. http://edgar.jrc.ec.europa.eu.
6. Tubiello FN, Salvatore M, Rossi S, Ferrara A, Fitton N, Smith P. The FAOSTAT database of greenhouse gas emissions from agriculture. Environ Res Lett. 2013;8:015009.
7. MacDonald J, McBride W. The transformation of U.S. livestock agriculture: scale, efficiency, and risks. USDA Economic Research Service-EIB43. http://www.ers.usda.gov/publications/eib-economic-information-bulletin/eib43.aspx. Accessed 29 Jan 2014.
8. MacDonald J, Ribaudo M, Livingston M, Beckman J, Huang W-Y. Manure use for fertilizer and for energy: report to congress USDA ERS-AP-037. USDA Economic Research Service; 2009. http://ers.usda.gov/publications/ap-administrative-publication/ap-037.aspx. Accessed 11 June 2015.
9. Xue B, Wang LZ, Yan T. Methane emission inventories for enteric fermentation and manure management of yak, buffalo and dairy and beef cattle in China from 1988 to 2009. Agric Ecosyst Environ. 2014;195:202–10.
10. Gerber P, Vellinga T, Opio C, Henderson B, Steinfeld H. Greenhouse gas emissions from the dairy sector: a life cycle assessment. Rome: Food and Agriculture Organization Animal Production and Health Division; 2010.
11. Petersen SO, Blanchard M, Chadwick D, Del Prado A, Edouard N, Mosquera J, et al. Manure management for greenhouse gas mitigation animal. 2013;7:266–82.
12. EPA. Inventory of U.S. Greenhouse Gas Emissions and Sinks: 1990–2012. Washington: US Environmental Protection Agency, Climate Change Division Report No.: EPA 430-R-14-003. http://www.epa.gov/climatechange/ghgemissions/usinventoryreport.html. Accessed Apr 2014.
13. Kirschke S, Bousquet P, Ciais P, Saunois M, Canadell JG, Dlugokencky EJ, et al. Three decades of global methane sources and sinks. Nat Geosci. 2013;6:813–23.
14. Saunois M, Bousquet P, Poulter B, Peregon A, Ciais P, Canadell JG, et al. The global methane budget: 2000–2012. Earth Syst Sci Data Discuss. 2016;1–79.
15. Smith P, Bustamante M, Ahammad H, Clark H, Dong E, Elsiddig EA, et al. Agriculture, forestry and other land use (AFOLU). In: Edenhofer O, Pichs-Madruga R, Sokona Y, Farahani E, editors. Climate Change 2014 Mitigation Climate Change Contribution Work. Group III Fifth Assessment Report Intergovernmental Panel Climate Change. Cambridge: Cambridge University Press; 2014.
16. EPA. Global anthropogenic non-$CO_2$ greenhouse gas emissions: 1990–2030. Washington: US EPA, Office of Atmospheric Programs, Climate Change Division; 2012. https://www3.epa.gov/climatechange/Downloads/EPAactivities/EPA_Global_NonCO2_Projections_Dec2012.pdf.
17. Höglund-Isaksson L. Global anthropogenic methane emissions 2005–2030: technical mitigation potentials and costs. Atmos Chem Phys. 2012;12:9079–96.
18. European Comission. EDGAR: the emissions database for global atmospheric research, part III: greenhouse gas emissions. 2012. http://edgar.jrc.ec.europa.eu/docs/IEA_PARTIII.pdf. Accessed 3 Nov 2016.
19. Dentener F, Stevenson D, Cofala J, Mechler R, Amann M, Bergamaschi P, et al. The impact of air pollutant and methane emission controls on tropospheric ozone and radiative forcing: CTM calculations for the period 1990–2030. Atmos Chem Phys. 2005;5:1731–55.
20. Turner AJ, Jacob DJ, Wecht KJ, Maasakkers JD, Lundgren E, Andrews AE, et al. Estimating global and North American methane emissions with high spatial resolution using GOSAT satellite data. Atmos Chem Phys. 2015;15:7049–69.
21. Wecht KJ, Jacob DJ, Frankenberg C, Jiang Z, Blake DR. Mapping of North American methane emissions with high spatial resolution by inversion of SCIAMACHY satellite data. J Geophys Res Atmos. 2014;119:7741–56.
22. Wecht KJ, Jacob DJ, Sulprizio MP, Santoni GW, Wofsy SC, Parker R, et al. Spatially resolving methane emissions in California: constraints from the CalNex aircraft campaign and from present (GOSAT, TES) and future (TROPOMI, geostationary) satellite observations. Atmos Chem Phys. 2014;14:8173–84.
23. Miller SM, Wofsy SC, Michalak AM, Kort EA, Andrews AE, Biraud SC, et al. Anthropogenic emissions of methane in the United States. Proc Natl Acad Sci. 2013;110:20018–22.
24. Kai FM, Tyler SC, Randerson JT, Blake DR. Reduced methane growth rate explained by decreased northern hemisphere microbial sources. Nature. 2011;476:194–7.
25. Nisbet EG, Dlugokencky EJ, Bousquet P. Methane on the rise—again. Science. 2014;343:493–5.
26. Nisbet EG, Dlugokencky EJ, Manning MR, Lowry D, Fisher RE, France JL, et al. Rising atmospheric methane: 2007–2014 growth and isotopic shift. Glob Biogeochem Cycles. 2016;30:1356–70.
27. Schaefer H, Fletcher SEM, Veidt C, Lassey KR, Brailsford GW, Bromley TM, et al. A 21st-century shift from fossil-fuel to biogenic methane emissions indicated by 13CH4. Science. 2016;352:80–4.
28. Saunois M, Jackson RB, Bousquet P, Poulter B, Canadell JG. The growing role of methane in anthropogenic climate change. Environ Res Lett. 2016;11:120207.
29. Manure management plan nutrient balance worksheet user guide. Penn State Extension; 2013. http://extension.psu.edu/plants/nutrient-management/manure/plan-writing/manure-management-plan-nutrient-balance-worksheet-user-guide. Accessed Apr 2017.
30. Müller DWH, Codron D, Meloro C, Munn A, Schwarm A, Hummel J, et al. Assessing the Jarman-Bell Principle: scaling of intake, digestibility, retention time and gut fill with body mass in mammalian herbivores. Comp Biochem Physiol A: Mol Integr Physiol. 2013;164:129–40.
31. Cattlemen's Beef Board and National Cattlemen's Beef Association. Explore beef-fact sheets: modern beef production; 2009. http://web.archive.org/web/20150511184326/, http://www.explorebeef.org/cmdocs/explorebeef/factsheet_modernbeefproduction.pdf. Accessed 21 Jan 2017.
32. Speer N. Industry at a glance: placement weight trends indicate feedyards are feeding longer. Beef Mag. 2015. http://www.beefmagazine.com/feedlots/industry-glance-placement-weight-trends-indicate-feedyards-are-feeding-longer. Accessed 21 Jan 2017.
33. Indiana Agriculture Report. Great Lakes Region: USDA NASS; 2014. https://www.nass.usda.gov/Statistics_by_State/Indiana/Publications/Ag_Report/2014/iar1411.pdf.
34. IPCC. Chapter 3: Uncertainties. In: Eggleston H, Buendia L, Miwa K, Ngara T, Tanabe K, editors. 2006 IPCC Guidelines National Greenhouse Gas Inventory vol. 1 General Guidelines Report. Kanagawa: Institute for Global Environmental Strategies; 2006. http://www.ipcc-nggip.iges.or.jp/public/2006gl/vol1.html.
35. Taylor JR. An introduction to error analysis: the study of uncertainties in physical measurements. 2nd ed. Mill Valley: University Science Books; 1997. https://archive.org/details/TaylorJ.R.IntroductionToErrorAnalysis2ed.
36. FAO. Food and Agriculture Organization of the United Nations Statistics Division (FAOSTAT). 2014. http://faostat.fao.org/. Accessed 1 Dec 2015.
37. USDA. Crop explorer for major crop regions. Foreign Agric Serv. 2011. http://www.pecad.fas.usda.gov/cropexplorer/. Accessed 1 Jan 2013.
38. FAO. Global livestock production and health atlas (GLiPHA). Food Agriculture Organisation United Nations Animal Production Health Division Global Livestock Production Health Atlas GLiPHA. 2014. http://kids.fao.org/glipha/. Accessed 1 Jan 2014.
39. USDA. Quickstats 2.0. National Agricultural Statistics Service. 2013. http://quickstats.nass.usda.gov/. Accessed 26 Nov 2013.
40. Krausmann F, Erb K-H, Gingrich S, Lauk C, Haberl H. Global patterns of socioeconomic biomass flows in the year 2000: a comprehensive assessment of supply, consumption and constraints. Ecol Econ. 2008;65:471–87.
41. West TO, Le Page Y, Huang M, Wolf J, Thomson AM. Downscaling global land cover projections from an integrated assessment model for use in regional analyses: results and evaluation for the US from 2005 to 2095. Environ Res Lett. 2014;9:064004.
42. Bouwman AF, Van der Hoek KW, Eickhout B, Soenario I. Exploring changes in world ruminant production systems. Agric Syst. 2005;84:121–53.

43. Capehart T. Feed grains data: yearbook tables. USDA Economic Research Service; 2016. http://www.ers.usda.gov/data-products/feed-grains-database/feed-grains-yearbook-tables.aspx.
44. Renewable Fuels Association Co-Products. Historic distillers grains production from U.S. ethanol biorefineries. http://ethanolrfa.org/resources/industry/co-products/#1456865649440-ae77f947-734a. Accessed 8 Nov 2016.
45. Three factors led to 1996 grain price shock. Economic Daily US Bureau Labor Statistics. https://www.bls.gov/opub/ted/1998/nov/wk1/art01.htm. Accessed 6 Mar 2017.
46. Neuman W. Corn prices raise worry over expensive food. New York Times; 2010. http://www.nytimes.com/2010/10/13/business/13corn.html. Accessed 12 Oct 2010.
47. Lehmkuhler J, Burris W. Distillers grain coproducts for beef cattle. University of Kentucky; 2011. http://www2.ca.uky.edu/agcomm/pubs/asc/asc186/asc186.pdf.
48. Hales KE, Cole NA, MacDonald JC. Effects of increasing concentrations of wet distillers grains with solubles in steam-flaked, corn-based diets on energy metabolism, carbon-nitrogen balance, and methane emissions of cattle. J Anim Sci. 2013;91:819–28.
49. Tian H, Lu C, Ciais P, Michalak AM, Canadell JG, Saikawa E, et al. The terrestrial biosphere as a net source of greenhouse gases to the atmosphere. Nature. 2016;531:225–8.
50. EPA. Inventory of U.S. greenhouse gas emissions and sinks: 1990–2013. Washington, DC: U.S. Environmental Protection Agency, Climate Change Division. http://www.epa.gov/climatechange/ghgemissions/usinventoryreport.html. Accessed Apr 2015.
51. Jeong S, Newman S, Zhang J, Andrews AE, Bianco L, Bagley J, et al. Estimating methane emissions in California's urban and rural regions using multitower observations. J Geophys Res Atmos. 2016;121:2016JD025404.
52. Owen JJ, Silver WL. Greenhouse gas emissions from dairy manure management: a review of field-based studies. Glob Change Biol. 2015;21:550–65.
53. Owen JJ, Silver WL. Greenhouse gas emissions from dairy manure management in a Mediterranean environment. Ecol Appl. 2017;27:545–59.
54. Dlugokencky EJ, Houweling S, Bruhwiler L, Masarie KA, Lang PM, Miller JB, et al. Atmospheric methane levels off: temporary pause or a new steady-state? Geophys Res Lett. 2003;30:1992.
55. Bergamaschi P, Houweling S, Segers A, Krol M, Frankenberg C, Scheepmaker RA, et al. Atmospheric CH4 in the first decade of the 21st century: inverse modeling analysis using SCIAMACHY satellite retrievals and NOAA surface measurements. J Geophys Res Atmos. 2013;118:7350–69.
56. Tixier-Boichard M, Leenstra F, Flock DK, Hocking PM, Weigend S. A century of poultry genetics. Worlds Poult Sci J. 2012;68:307–21.
57. Poultry Manure Management Planning. West Lafayette: Purdue University Cooperative Extension Service and Indiana Soil Conservation Service; 1999. https://www.extension.purdue.edu/extmedia/ID/ID-206.html. Accessed Sep 1999.
58. Beldman ACG, Daatselaar CHG, Prins AM. Developments in dairying worldwide, from a dairy farmer's perspective. In: Rozstalnyy A, Rozstalnyy A, Keane G, editors. Cattle husbandry in Eastern Europe and China: structure, development paths optimisation. Wageningen: Wageningen Academic Publishers; 2014. p. 61–9.
59. Gao Z, Lin Z, Yang Y, Ma W, Liao W, Li J, et al. Greenhouse gas emissions from the enteric fermentation and manure storage of dairy and beef cattle in China during 1961–2010. Environ Res. 2014;135:111–9.
60. China beef and cattle production market 2013|China. 2015. http://web.archive.org/web/20150318220516/, http://beforeitsnews.com/china/2013/06/china-beef-and-cattle-production-market-2013-2446620.html. Accessed 18 Mar 2015.
61. Larson C. China's lakes of pig manure spawn antibiotic resistance. Science. 2015;347:704.
62. McOrist S, Khampee K, Guo A. Modern pig farming in the People's Republic of China: growth and veterinary challenges. Revue Scientifique et Technique-OIE. 2011;30:961–8.
63. Mendoza Huaitalla R. Intensive pig production and manure management in Beijing, North China Plain; 2014. http://opus.uni-hohenheim.de/volltexte/2014/972/. Accessed 15 May 2015.
64. Feeding China's pigs: implications for the environment, China's smallholder farmers and food security|Institute for Agriculture and Trade Policy; 2015. http://web.archive.org/web/20150318220942/, http://www.iatp.org/documents/feeding-china%E2%80%99s-pigs-implications-for-the-environment-china%E2%80%99s-smallholder-farmers-and-food. Accessed 18 Mar 2015.
65. Hamilton D. Lagoons for livestock waste treatment. Stillwater: Oklahoma Cooperative Extension Service. http://pods.dasnr.okstate.edu/docushare/dsweb/Get/Document-7615/BAE-1736web2011.pdf.
66. What U.S. should know about South American beef production-progressive cattleman. http://www.progressivecattle.com/topics/management/6095-what-us-should-know-about-south-american-beef-production. Accessed 9 Oct 2016.
67. Gurian-Sherman D. CAFOs uncovered the untold costs of confined animal feeding operations. Union of Concerned Scientists; 2008. http://www.ucsusa.org/assets/documents/food_and_agriculture/cafos-uncovered.pdf.
68. Yang Q, Tian H, Li X, Ren W, Zhang B, Zhang X, et al. Spatiotemporal patterns of livestock manure nutrient production in the conterminous United States from 1930 to 2012. Sci Total Environ. 2016;541:1592–602.
69. National Conference of State Legislatures. Concentrated animal feeding operations. http://www.ncsl.org/research/agriculture-and-rural-development/concentrated-animal-feeding-operations.aspx. Accessed 27 Apr 2017.
70. United Nations Framework Convention on Climate Change. National Inventory Submissions; 2014. http://unfccc.int/national_reports/annex_i_ghg_inventories/national_inventories_submissions/items/8108.php. Accessed 27 May 2017.
71. EEA (European Environment Agency). Annual European Union greenhouse gas inventory 1990–2012 and inventory report 2014—European Environment Agency (EEA). http://www.eea.europa.eu/publications/european-union-greenhouse-gas-inventory-2014. Accessed 21 Apr 2015.
72. Madalena F, Peixoto M, Gibson J. Dairy cattle genetics and its applications in Brazil. LRRD Newsl. 2015. http://web.archive.org/web/20150423220801/, http://www.lrrd.org/lrrd24/6/made24097.htm.
73. de Leis CM, Cherubini E, Ruviaro CF, da Silva VP, do Nascimento Lampert V, Spies A, et al. Carbon footprint of milk production in Brazil: a comparative case study. Int J Life Cycle Assess. 2015;20:46–60.
74. United Nations Framework Convention on Climate Change. Submitted National Communications from Non-Annex I Parties. http://unfccc.int/national_reports/non-annex_i_natcom/items/10124.php. Accessed 27 May 2017.
75. Powell JM, MacLeod M, Vellinga TV, Opio C, Falcucci A, Tempio G, et al. Feed-milk-manure nitrogen relationships in global dairy production systems. Livest Sci. 2013;152:261–72.
76. Powell JM, Li Y, Wu Z, Broderick GA, Holmes BJ. Rapid assessment of feed and manure nutrient management on confinement dairy farms. Nutr Cycl Agroecosyst. 2008;82:107–15.
77. Duncan AJ, Teufel N, Mekonnen K, Singh VK, Bitew A, Gebremedhin B. Dairy intensification in developing countries: effects of market quality on farm-level feeding and breeding practices. Animal. 2013;7:2054–62.
78. Singhal K, Mohini M. Methane emission estimates from enteric fermentation in Indian livestock: dry matter intake approach. Curr Sci. 2005;88(1).
79. Patra AK. Estimation of methane and nitrous oxide emissions from Indian livestock. J Environ Monit. 2012;14:2673–84.
80. Brazil: Livestock and products annual|USDA Foreign Agricultural Service. https://www.fas.usda.gov/data/brazil-livestock-and-products-annual-2. Accessed 29 May 2017.
81. Ten Years Later: a look at 10 years' worth of beef-production systems in South America. http://www.beefmagazine.com/foreign-trade/0401-south-america-beef-production-systems. Accessed 29 May 2017.
82. Eastern Research Group, Inc., Tetra Tech, LOGICarbon. resource assessment for livestock and agro-industrial wastes—Brazil. Global Methane Initiative; 2010. https://www.globalmethane.org/documents/ag_brazil_res_assessment.pdf.
83. Ji ZY, Cao Z, Liao XD, Wu YB, Liang JB, Yu B. Methane production of growing and finishing pigs in southern China. Anim Feed Sci Technol. 2011;166–167:430–5.
84. Eastern Research Group, Inc., Tetra Tech, LOGICarbon. Resource assessment for livestock and agro-industrial wastes—India. Global Methane Initiative; 2010. https://www.globalmethane.org/documents/ag_brazil_res_assessment.pdf.

85. Huaitalla R, Gallmann E, Zheng K, Liu X, Hartung E. Pig husbandry and solid manures in a commercial pig farm in beijing, China. Int J Biol Biomol Agric Food Biotechnol Eng. 2010;4:217–26.
86. Wang F, Dou Z, Ma L, Ma W, Sims JT, Zhang F. Nitrogen mass flow in china's animal production system and environmental implications. J Environ Qual. 2010;39:1537.
87. Bai ZH, Ma L, Qin W, Chen Q, Oenema O, Zhang FS. Changes in pig production in China and their effects on nitrogen and phosphorus use and losses. Environ Sci Technol. 2014;48:12742–9.
88. EPA. Inventory of U.S. greenhouse gas emissions and sinks: 1990–2010. US EPA; 2012. https://www.epa.gov/ghgemissions/inventory-us-greenhouse-gas-emissions-and-sinks-1990-2010. Accessed 30 May 2017.
89. EPA. Inventory of U.S. greenhouse gas emissions and sinks: 1990–2014. US EPA; 2016. https://www.epa.gov/ghgemissions/inventory-us-greenhouse-gas-emissions-and-sinks-1990-2014.
90. Turner AJ, Jacob DJ, Benmergui J, Wofsy SC, Maasakkers JD, Butz A, et al. A large increase in U.S. methane emissions over the past decade inferred from satellite data and surface observations. Geophys Res Lett. 2016;43:2218–24.
91. EPA. Inventory of U.S. greenhouse gas emissions and sinks: 1990–2011. Washington, DC: U.S. Environmental Protection Agency, Climate Change Division; 2013. http://www.epa.gov/climatechange/ghgemissions/usinventoryreport.html.
92. EPA. Inventory of U.S. greenhouse gas emissions and sinks: 1990–2006. US EPA; 2008. https://www.epa.gov/ghgemissions/inventory-us-greenhouse-gas-emissions-and-sinks-1990-2006. Accessed 30 May 2017.

# Permissions

 Every chapter published in this book has been scrutinized by our experts. Their significance has been extensively debated. The topics covered herein carry significant findings which will fuel the growth of the discipline. They may even be implemented as practical applications or may be referred to as a beginning point for another development.

The contributors of this book come from diverse backgrounds, making this book a truly international effort. This book will bring forth new frontiers with its revolutionizing research information and detailed analysis of the nascent developments around the world.

We would like to thank all the contributing authors for lending their expertise to make the book truly unique. They have played a crucial role in the development of this book. Without their invaluable contributions this book wouldn't have been possible. They have made vital efforts to compile up to date information on the varied aspects of this subject to make this book a valuable addition to the collection of many professionals and students.

This book was conceptualized with the vision of imparting up-to-date information and advanced data in this field. To ensure the same, a matchless editorial board was set up. Every individual on the board went through rigorous rounds of assessment to prove their worth. After which they invested a large part of their time researching and compiling the most relevant data for our readers.

The editorial board has been involved in producing this book since its inception. They have spent rigorous hours researching and exploring the diverse topics which have resulted in the successful publishing of this book. They have passed on their knowledge of decades through this book. To expedite this challenging task, the publisher supported the team at every step. A small team of assistant editors was also appointed to further simplify the editing procedure and attain best results for the readers.

Apart from the editorial board, the designing team has also invested a significant amount of their time in understanding the subject and creating the most relevant covers. They scrutinized every image to scout for the most suitable representation of the subject and create an appropriate cover for the book.

The publishing team has been an ardent support to the editorial, designing and production team. Their endless efforts to recruit the best for this project, has resulted in the accomplishment of this book. They are a veteran in the field of academics and their pool of knowledge is as vast as their experience in printing. Their expertise and guidance has proved useful at every step. Their uncompromising quality standards have made this book an exceptional effort. Their encouragement from time to time has been an inspiration for everyone.

The publisher and the editorial board hope that this book will prove to be a valuable piece of knowledge for researchers, students, practitioners and scholars across the globe.

# List of Contributors

**Matthew B. Russell**
Department of Forest Resources, University of Minnesota, St. Paul, MN 55108, USA

**Grant M. Domke and Christopher W. Woodall**
USDA Forest Service, Northern Research Station, St. Paul, MN 55108,USA

**Anthony W. D'Amato**
Rubenstein School of Environment and Natural Resources, University of Vermont, Burlington, VT 05405

**Rajesh Bahadur Thapa, Takeshi Motohka, Manabu Watanabe and Masanobu Shimada**
Earth Observation Research Center, Japan Aerospace Exploration Agency (JAXA), 2-1-1 Sengen, Tsukuba, Ibaraki 305-8505, Japan

**Chihiro Kayo**
Department of Environment Conservation, Graduate School of Agriculture, Tokyo University of Agriculture and Technology, 3-5-8 Saiwai-cho, Fuchu, Tokyo 183-8509, Japan

**Yuko Tsunetsugu**
Department of Wood Engineering, Forestry and Forest Products Research Institute, 1 Matsunosato, Tsukuba, Ibaraki 305-8687, Japan

**Mario Tonosaki**
Shikoku Research Center, Forestry and Forest Products Research Institute, 2-915 Asakuranishi-cho, Kochi, Kochi 780-8077, Japan

**Vanessa S. Mascorro and Nicholas C. Coops**
Faculty of Forestry, Department of Forest Resources Management, University of British Columbia, 2424 Main Mall, Vancouver, BC V6T 1Z4, Canada

**Werner A. Kurz**
Canadian Forest Service, Natural Resources Canada, 506 West Burnside Road, Victoria, BC V8Z 1M5, Canada

**Marcela Olguín**
Proyecto Mexico-Noruega, Comision Nacional Forestal, Coyoacán, D.F. Col. del Carmen Coyoacán, CP 04100 Mexico, Mexico

**David C. Marvin and Gregory P. Asner**
Department of Global Ecology, Carnegie Institution for Science, 260 Panama St., Stanford, CA 94305, USA

**Jinxun Liu**
Western Geographic Science Center, USGS, Menlo Park, CA 94025, USA
San José State University Research Foundation, San José, CA 95112, USA

**Benjamin M. Sleeter, Tamara S. Wilson and Jason Sherba**
Western Geographic Science Center, USGS, Menlo Park, CA 94025, USA

**Zhiliang Zhu**
USGS, 12201 Sunrise Valley Drive, Reston, VA 20192, USA

**Linda S. Heath**
USDA Forest Service, Northern Research Station, Durham, NH 03824, USA
Present Address: USDA Forest Service, Washington Office R&D, Washington DC 20250, USA

**Zhengxi Tan**
Arctic Slope Regional Corporation, Contractor to USGS EROS, Sioux Falls, SD 57198, USA

**Decheng Zhou**
Jiangsu Key Laboratory of Agricultural Meteorology, Nanjing University Of Information Science And Technology, Nanjing 210044, China

**Camille Piponiot**
Université de la Guyane, UMR EcoFoG (AgroParisTech, CNRS, Inra, Université des Antilles, Cirad), Campus agronomique, 97310 Kourou, French Guiana
CNRS, UMR EcoFoG (AgroParisTech, Inra, Université de la Guyane, Université des Antilles, Cirad), Campus Agronomique, 97310 Kourou, French Guiana

**Antoine Cabon**
Centre Tecnològic Forestal de Catalunya, Crta. de Sant Llorenç de Morunys, Km.2, 25280 Solsona, Spain

**Laurent Descroix**
ONF-Guyane, Réserve de Montabo, 97307 Cayenne, French Guiana

**Aurélie Dourdain and Bruno Hérault**
Cirad, UMR EcoFoG (AgroParisTech, CNRS, Inra, Université de la Guyane, Université des Antilles), Campus agronomique, 97310 Kourou, French Guiana

**Lucas Mazzei**
EMBRAPA Amazônia Oriental, Trav. Dr. Enéas Pinheiro, Belém, Brazil

**Benjamin Ouliac**
Guyane Energie Climat, 16 rue Victor Schoelcher, 97300 Cayenne, French Guiana

**Ervan Rutishauser**
CarboForExpert, 1248 Hermance, Switzerland

**Plinio Sist**
Cirad UR Forêts et Sociétés, 34398 Montpellier, France

**Roberto Pilli, Giacomo Grassi and Raúl Abad Viñas**
European Commission, Joint Research Centre, Directorate D - Sustainable
Resources - Bio-Economy Unit, Via E. Fermi 2749, 21027 Ispra, VA, Italy

**Werner A. Kurz**
Natural Resources Canada, Canadian Forest Service, Victoria, BC V8Z 1M5, Canada

**Jose V. Moris**
Department of Agriculture, Forest and Food Sciences, University of Torino, Via Leonardo Da Vinci 44, 10095 Grugliasco, TO, Italy

**Evans Dawoe and Emmanuel Acheampong**
Faculty of Renewable Natural Resources (FRNR), Kwame Nkrumah University of Science and Technology (KNUST), Kumasi, Ghana

**Winston Asante**
SNV Netherlands Development Organization, Accra, Ghana

**Paul Bosu**
Forestry Research Institute of Ghana (FORIG), Council for Scientific and Industrial Research (CSIR), Kumasi, Ghana

**Jinfeng Chang**
Laboratoire des Sciences du Climat et de l'Environnement, UMR8212, CEACNRS—UVSQ, 91191 Gif-Sur-Yvette, France
Sorbonne Universités (UPMC), CNRS-IRD-MNHN, LOCEAN/IPSL, 4 Place Jussieu, 75005 Paris, France

**Philippe Ciais and Nicolas Viovy**
Laboratoire des Sciences du Climat et de l'Environnement, UMR8212, CEACNRS—UVSQ, 91191 Gif-Sur-Yvette, France

**Jean-François Soussana**
INRA, UAR0233 CODIR Collège de Direction. Centre-Siège de l'INRA, Paris, France

**Katja Klumpp**
Grassland Ecosystem Research Unit, French National Institute for Agricultural
Research (INRA), 63100, Clermont-Ferrand, France

**Benjamin Sultan**
Sorbonne Universités (UPMC), CNRS-IRD-MNHN, LOCEAN/IPSL, 4 Place Jussieu, 75005 Paris, France

**Carlos Alberto Silva**
Department of Natural Resources and Society, College of Natural Resources, University of Idaho, (UI), 875 Perimeter Drive, Moscow, ID 83843, USA. US Forest Service (USDA), Rocky Mountain Research Station, RMRS, 1221 South Main Street, Moscow, ID 83843, USA

**Andrew Thomas Hudak and Carine Klauberg**
US Forest Service (USDA), Rocky Mountain Research Station, RMRS, 1221 South Main Street, Moscow, ID 83843, USA

**Lee Alexandre Vierling**
Department of Natural Resources and Society, College of Natural Resources, University of Idaho, (UI), 875 Perimeter Drive, Moscow, ID 83843, USA

**Carlos Gonzalez-Benecke**
Department of Forest Engineering, Oregon State University, 269 Peavy Hall, Corvallis, OR 97331, USA

**Samuel de Padua Chaves Carvalho**
College of Forestry, Federal University of Mato Grosso, Av. Fernando Correa da Costa, 2367, Boa Esperança, Cuiabá, MT 78060-900, Brazil

**Luiz Carlos Estraviz Rodriguez**
Department of Forest Sciences, College of Agriculture Luiz de Queiroz (ESALQ), University of Sao Paulo (USP), Av. Pádua Dias, 11, Piracicaba, SP 13418-900, Brazil

**Adrián Cardil**
Tecnosylva, Parque Tecnológico de León, 24009 León, Spain

**B. P. Smiley**
Natural Resources Canada, Canadian Forest Service, 506 West Burnside Road, Victoria, BC V8Z 1M5, Canada

**J. A. Trofymow**
Natural Resources Canada, Canadian Forest Service, 506 West Burnside Road, Victoria, BC V8Z 1M5, Canada
Biology Department, University of Victoria, Victoria, BC V8W 3R4, Canada

**Julie Wolf**
USDA-ARS, Adaptive Cropping Systems Laboratory, 10300 Baltimore Ave., Building 001, Room. 342, BARC-WEST, Beltsville, MD 20705, USA

**Ghassem R. Asrar**
Joint Global Change Research Institute, 5825 University Research Court, Suite 3500, College Park, MD 20740, USA

**Tristram O. West**
Joint Global Change Research Institute, 5825 University Research Court, Suite 3500, College Park, MD 20740, USA
US Department of Energy, SC-23, 1000 Independence Ave., Washington, DC 20585, USA

# Index